Total Cold War

Total Cold War

Eisenhower's Secret Propaganda
Battle at Home and Abroad

Kenneth Osgood

UNIVERSITY PRESS OF KANSAS

Published by the University Press of Kansas (Lawrence, Kansas 66049),
which was organized by the Kansas Board of Regents and is operated and
funded by Emporia State University, Fort Hays State University, Kansas State
University, Pittsburg State University, the University of Kansas, and Wichita
State University

Library of Congress Cataloging-in-Publication Data

Osgood, Kenneth.
   Total Cold War : Eisenhower's secret propaganda battle at home and abroad /
Kenneth Osgood.
      p.   cm.
   Includes bibliographical references and index.
   ISBN 0-7006-1445-1 (cloth : alk. paper)
   1. United States—Foreign relations—Soviet Union.   2. Soviet Union—Foreign
relations—United States.   3. Eisenhower, Dwight D. (Dwight David), 1890–1969
—Political and social views.   4. Cold War.   5. United States—Foreign relations—
1953–1961.   6. Propaganda, American—History—20th century.   7. Psychological
warfare—United States—History—20th century.   I. Title.

   E183.8.S65O82 2006
   303.3'75097309045—dc22   2005027603

British Library Cataloguing-in-Publication Data is available.
Printed in the United States of America

10 9 8 7 6 5 4 3 2 1

For Mom and Dad

CONTENTS

ACKNOWLEDGMENTS

When I first began researching the history of U.S. propaganda and cultural diplomacy ten years ago, the pertinent historical literature was dauntingly thin. In trying to make sense of the thousands of declassified documents on the subject, I occasionally felt like I was wandering through the woods without a map or compass. Fortunately, as the years passed, more scholars turned their attention to this aspect of American foreign relations. I therefore owe a special debt to my colleagues who forged a path and inspired me, through their work, to move forward. I am especially obliged to Nancy Bernhard, Shawn J. Parry-Giles, Walter Hixson, Scott Lucas, Robert McMahon, Gregory Mitrovich, Frank Ninkovich, and Emily Rosenberg for charting the course with their pioneering research. I have sought in the pages that follow to build on what they have done, occasionally to challenge them, but mostly to provide others what they provided to me: a trail marker that will lead to more and exciting inquiries into the role of propaganda and cultural diplomacy in international relations.

Although space limitations—and undoubtedly my atrociously inadequate memory—prevent me from acknowledging everyone by name, I would like to thank the many friends and colleagues who assisted and encouraged me as I worked on this project over the years. My Ph.D. advisors at the University of California at Santa Barbara deserve special thanks for their friendship and guidance. By their instruction and example, they set a high standard of integrity, commitment, and dedication. Jane DeHart encouraged me to think broadly about American culture, directing me to important work beyond my field and enriching my thinking along the way. Tsuyoshi Hasegawa pushed me to be rigorous and comprehensive in my research, and he tolerated with good humor spontaneous "outbursts" in seminar. Fredrik Logevall was a true mentor. From the first research paper to the final manuscript, he replied thoughtfully to many long e-mails and responded cheerfully to many ill-timed phone calls. He guided me with sage advice, prodded me with critical commentaries, and, most importantly, inspired me with his boundless enthusiasm for the study of American diplomacy. Thanks, Doc.

A long list of people read and commented on the manuscript at various stages of completion. Nancy Jackson, my editor at Kansas, was a pleasure to work with and was very understanding of everything that slowed me down, from illness to hurricanes. Larisa Martin, Susan Schott, and Karen Hellekson also deserve special thanks for their help in the production process. Ken Moure, Jack Talbott, Ira Chernus, David Snyder, Gregory Mitrovich, Stephen Whitfield, Jason Parker, Robert McMahon, Jeremi Suri, Nate Citino, and H. W. Brands read proposals, drafts, and sections of the book. Allan Winkler, Peter Hahn, Michael Krenn, John Prados, and Jerald Combs contributed with their thoughtful comments on conference papers drawn from the manuscript. Chester Pach and Brian Etheridge were especially conscientious readers. Their critical commentaries pushed the manuscript in new and better directions, and I am grateful beyond words. Kathryn Statler and Andrew Johns made great research companions and on-call editors who made me reconsider everything from cliché titles to "dark and stormy night" introductions.

Several scholars trusted me with advance copies of works in progress. Their contributions, cited in the bibliography and in the notes, helped tremendously. Special thanks to Wilson Dizard and Nick Cull for sharing their work on the USIA. For research assistance, I am grateful to Martin Manning, who made available to me key documents from the USIA's Historical Collection. The Freedom of Information Act staff at the State Department was extraordinarily responsive in locating missing and classified records from the USIA. The late Abbott Washburn helped arrange interviews with several USIA veterans; I am grateful to him and to them for their insights and perspectives. Ellen Fisher and Bill Schriver provided me with images from the Advertising Council archives; Lori Allesee hunted down historic pictures of Project Hope; Melinda Schwenck graciously shared her collection of USIA films; and Bob Allen scanned his Atoms for Peace stamp. At the Eisenhower Library, Bonita Mulanax toiled through my many mandatory declassification review requests, expeditiously declassifying hundreds of pages of documents for this project. Archivist James Leyerzapf guided me to numerous collections I might not have seen. The Little Apple Brewing Company in faraway Manhattan, Kansas, provided a much-needed culinary escape during long research stints in Abilene—thanks for being there.

I was also fortunate to have received generous financial assistance from a number of sources. I am obliged to the Institute for Global Conflict and Cooperation for two years of research support, and to U.C. Santa Barbara's Interdisciplinary Humanities Center for a grant that broadened my disciplinary focus. At an important stage late in the project, a travel grant from the Eisenhower Foundation helped me address gaps in my research. A postdoctoral fellowship from the Mershon Center at Ohio State was invaluable. I cannot imagine a better place to study international relations. Thanks to Richard Herrmann, Peter Hahn, Mitch Lerner, Michael Hogan, Mark Jacobson, Maria Fanis, Chris Fettweis, Matt Masur, and the Mershon scholars who made the year at OSU so productive.

Florida Atlantic University provided a very supportive environment for completing this project. FAU became both workplace and home away from home, in large part because of the exceptional collegiality of the people that work there. I owe special thanks to Steve Engle, Sandy Norman, Jeff Galin, Prisca Augustyn, Jason Schwartz, Eric Hanne, Patty Kollander, and Bill Covino. I am also grateful to Steve Triana, who provided research assistance, and Zella Lynn, who saved me on many occasions—mostly from myself.

Others helped in various ways. My uncle, Carlyle Thayer, provided helpful advice, and my aunt Bibi encouraged me even though I wasn't writing something less serious. Nana Thayer deserves many thanks for sustaining me morally and materially; I only wish she could comprehend the final product of her support. My brother Matt was a great friend who also saved me hundreds of dollars on meals and hotel rooms. I am also indebted to the Murphys and the Kristensens for sharing their homes with me during long research stints at the National Archives. Andrew and Lisa Frank provided shelter and electricity through the worst hurricane season in recent memory. John and Rebecca Farrell kept hounding me to complete the book even as they kept me from working on it. Earlier in life, Deborah Nordlie and Michael "The DWAB" Ferris led me to love writing. At Notre Dame, Wilson Miscamble, Jim Langford, and Laura Crago developed my interest in history, and trained me to do as Charles Beard advised: think otherwise.

I dedicate this book to my parents, for without them, it would not have been possible. Mom and Dad edited sections of the manuscript and provided me a place to write during a critical time. More than that, they were true friends. They encouraged me every day, and they cheered me on when I needed it most. Mom sustained me with batches of meatballs and loaves of pumpkin bread that kept me eating well when I was all too tempted to let myself go. Dad was there at the drop of a hat to read a chapter or a paragraph or a sentence and show me ways to improve it. Both taught me to appreciate history, to cherish intellectual inquiry, and to value serious debate about meaningful issues. They taught me to ask questions and to challenge orthodoxies, especially those that are easiest to accept uncritically. For this and so much more, I am thankful.

Finally, I thank my wife Rachel. I feel like the luckiest historian in town, and not just because she edited my footnotes, corrected my bibliography, and humored me when I wanted to name our dog "Laika" in honor of the most famous canine victim of the battle for hearts and minds. Her unflagging love, support, and friendship made each day that I spent finishing this manuscript a happy one—especially the day I learned that by the time the book appears in print, Little Joe will be calling me daddy.

ABBREVIATIONS

AEC	Atomic Energy Commission
AFIE	Armed Forces Information and Education
AWF	Ann Whitman File
BBC	British Broadcasting Company
BCIU	Business Council on International Understanding
CBS	Columbia Broadcasting System
CEP	Citizen Education Project
CIA	Central Intelligence Agency
CIAA	Coordinator for Inter-American Affairs
DOD	Department of Defense
DOS	Department of State
EDC	European Defense Community
EL	Eisenhower Library
FCDA	Federal Civil Defense Administration
FEBC	Far Eastern Broadcast Company
FO	Foreign Office
FOA	Foreign Operations Administration
FOIA	Freedom of Information Act
FRUS	Foreign Relations of the United States
ICA	International Cooperation Administration
ICBM	Intercontinental Ballistic Missile
ICL	International Christian Leadership
IGY	International Geophysical Year
IIA	International Information Administration
IMG	Informational Media Guaranty
IOC	Office of Private Cooperation
IPS	International Press Service
IRBM	Intermediate Range Ballistic Missile
JCS	Joint Chiefs of Staff
NA	National Archives
NASA	National Aeronautics and Space Administration
NATO	North Atlantic Treaty Organization
NBC	National Broadcasting Company

NDEA   National Defense Education Act
NGO   nongovernmental organization
NIE   National Intelligence Estimate
NRC   National Records Center
NSA   National Security Archive
NSC   National Security Council
OCB   Operations Coordinating Board
OPC   Office of Policy Coordination
OSANSA   Office of the Special Assistant for National Security
             Affairs
OSS   Office of Strategic Services
OWI   Office of War Information
PAO   Public Affairs Officer
PCG   Planning Coordination Group
PCIAA   President's Committee on International Activities Abroad
POW   prisoner of war
PPS   Policy Planning Staff
PRO   Public Record Office
PSAC   President's Science Advisory Council
PSB   Psychological Strategy Board
PWD   Psychological Warfare Division
PWD/SHAEF   Psychological Warfare Division/ Supreme
             Headquarters, Allied Expeditionary Force
R&D   Research and Development
RAC   Rockefeller Archive Center
RFE   Radio Free Europe
RG   Record Group
RL   Radio Liberty
SAC   Strategic Air Command
SHAEF   Supreme Headquarters, Allied Expeditionary Force
TCP   Technological Capabilities Panel
UN   United Nations
USBE   United States Book Exchange
USIA   United States Information Agency
USIS   United States Information Service
USNC-IGY   U.S. National Committee for the International
             Geophysical Year
VOA   Voice of America
WHCF   White House Central Files
WHO   White House Office

Introduction

There is but one sure way to avoid total war—and that is to win the cold war.

—Dwight D. Eisenhower, State of the Union Address, February 2, 1953

Total war was the defining experience for the generation of Americans who led the United States during the early Cold War. The principle of total war—that wars were no longer fought just by armies in the field, but by the entire nation—erased distinctions between the front line and the home front and made the mobilization of the masses an indispensable feature of modern conflict. People who were thirty years old at the onset of the Cold War had lived through not one, but two such total wars—global conflicts that had claimed millions of lives and dramatically changed American society and culture. According to a series of Gallup polls taken in the first decade of the Cold War, most Americans—at times over 70 percent of them—expected to see yet another world war in their lifetimes. They expected this third world war to produce a new level of horror and destruction. At any given moment during the decade, roughly half of all Americans thought it likely that their communities would be attacked by nuclear weapons in the event of war.[1]

Yet in an apparent contradiction, most Americans did not identify the Cold War primarily as a military confrontation. At recurring intervals during the first Cold War decade, Gallup pollsters asked Americans to explain what the term "cold war" meant to them. Those who could identify the term—and a surprising number of them could not—defined it as a war of nerves, a war of words, a war of ideas, and a war of propaganda. In 1955, most Americans told the pollsters that the term "cold war" meant such things as "war through talking," "a subtle war," "a diplomatic war," "war without actual fighting," "political war," and "war propaganda." It was a "battle of wits" and a "battle of words among powers to gain prestige." Some respondents, whom Gallup labeled as incorrectly answering the question, identified cold war as being "just like a hot war," a "war where

1

no war is declared," and "real war all over the world." Most Americans, in other words, perceived the Cold War *as a war*, but as a different kind of war—one that was difficult to define, one that was fought not so much with guns and tanks and atom bombs, as with words and ideas and political maneuvers all over the world.[2]

This view of the conflict as a war has faded into the background of historical understanding. For some historians the most remarkable feature of the Cold War was the fact that it remained cold. As it intensified, the Cold War produced crisis after crisis, a spiraling arms race, dozens of covert and military interventions, and "hot" wars in Korea, Vietnam, and Afghanistan. Yet despite numerous close calls, the third world war anticipated by Americans never came. The fifty-year conflict that might have produced Armageddon instead delivered a peculiarly long peace marked by the absence of great power wars and a certain level of stability in the international system.[3]

To Americans in the 1950s, however, the Cold War was not a unique state of peaceful competition, but a war waged by other means. Americans might not have been able to define precisely what the Cold War was, but as it intensified, they increasingly perceived it as a total conflict. It was at once both like and unlike anything they had experienced—different in the absence of great power war, yet similar in its demands on the American people. The memory of the horrific costs of World War II and the unthinkable consequences of nuclear warfare meant that the Cold War, more than any other conflict in human history, was channeled into nonmilitary modes of combat, particularly ideological and symbolic ones. The importance of ideological and symbolic factors in this conflict, then, in turn, made the Cold War even more all-embracing. Virtually every aspect of the American way of life—from political organizations and philosophical ideals, to cultural products and scientific achievements, to economic practices and social relationships—was exposed to scrutiny in this total contest for the hearts and minds of the world's peoples.

This study began as a fairly limited inquiry into U.S. psychological warfare programs in the early Cold War. The massive declassified documentary record in the archives, however, indicated that the subject I was investigating was much larger than I expected. The untold story of America's cold war of words did not lie in the tale of psychological operations to foment unrest behind the Iron Curtain, as I first supposed. Rather, it was in the broader effort to win the hearts and minds of people on the other side of that curtain, in the areas of the world that were noncommunist, neutral, or tied to the United States through formal alliances—the area that Americans liked to call the "free world."

This was truly a global effort, one that extended well beyond the activities of the U.S. Information Agency, the official propaganda arm of the American government. The United States pursued a wide range of activities to shape, alter, and manipulate the perceptions and politics of allied and neutral nations alike. Here, too, the story extended further into the private sphere. Total war made distinctions between propaganda intended for "domestic" and "international" audiences meaningless. U.S. psychological operations were premised upon the total war notion that public opinion at home needed to be agitated and organized to sustain national morale for the long struggle that lay ahead. Many psychological warfare campaigns thus expressly targeted the American people and, at the same time, called on them to participate in the war of words being waged abroad.

The account that follows chronicles U.S. psychological warfare programs developed to win the hearts and minds of the free world. It traces some of the ways these programs intersected with the domestic sphere and simultaneously affected the broader pattern of U.S. foreign relations. Although aspects of this story have been explored in recent scholarship, its full dimensions have yet to be assessed. The present effort attempts to do just that while still acknowledging that more work needs to be done. To the extent that U.S. psychological programs had a global reach, the story in these pages is an international one. But the focus remains on assessing the meaning of these efforts for the United States and its history. I seek to complicate our understanding of the American experience during the Cold War on several levels—diplomatic, social, and cultural—by beginning with the state and looking outward at U.S. efforts to mobilize people at home and abroad for Cold War advantage.[4]

The Fourth Weapon

This study's investigation of America's global battle for hearts and minds develops five interrelated themes. First, it seeks to place the story of U.S. psychological warfare in its broader context: the changing nature of international relations as a result of the communications revolution and the age of mass politics and total war. It is often overlooked that the Cold War coincided with a moment in world history when media technologies and information resources were everywhere exploding. In societies around the world, more and more people were becoming politically active, attentive on some level to domestic politics and international relations. Not just democracies, but authoritarian regimes were concerned with public opinion, pursuing a form of mass politics that relied on formalized ideologies as well as coercion for the maintenance of political power. Mass media, mass politics, mass movements, mass ideologies, and mass societies had a totalizing effect on the Cold War. If politically activated segments of society could be captured by ideas, symbols, images, rhetoric, and propaganda, they could be harnessed and mobilized for foreign policy advantage. It was often here, in the struggle to win support from mass publics and ruling elites, that the Cold War played out.[5]

In this context, American policy makers increasingly perceived that communication techniques—propaganda, public relations, and media manipulation—would serve U.S. foreign policy interests. Officials reasoned that if the United States could get popular opinion on its side, it would exert pressure on foreign governments, which would in turn create a favorable atmosphere for U.S. policies. A report published by the House Foreign Relations Committee in 1964, entitled *Winning the Cold War: The U.S. Ideological Offensive*, captures this sentiment well and deserves lengthy quotation:

> For many years military and economic power, used separately, or in conjunction, have served as the pillars of diplomacy. They still serve that function but the recent increase in influence of the masses of the people over government, together with greater awareness on the part of the leaders of the aspirations of people . . . has created a new dimension of foreign policy operations. Certain foreign policy objectives can be pursued by dealing directly with the people of foreign countries, rather than with their governments. . . . Through the use of modern instruments and techniques of communications it is possible today to reach large or influential segments of national populations—to inform them, to influence their attitudes, and at times perhaps even to motivate them to a particular course of action. These groups, in turn, are capable of exerting noticeable, even decisive, pressures on their government.

As influencing international public opinion evolved into a major objective of international relations, propaganda emerged as a significant component of foreign relations. Indeed, some foreign policy experts referred to propaganda and any other action done to affect public opinion as the "fourth weapon" of American foreign policy, as important to international relations as the political, economic, and military components of policy.[6]

The second theme developed in these pages explores the ways in which the imperative of shaping, influencing, and at times manipulating popular sentiment infused a wide range of policies with psychological significance. The battle for hearts and minds was waged not just with words, but also with deeds—actions calculated to have an impact on public perceptions. Such "psychological considerations" extended beyond mere propaganda to affect a wide range of activities relating to U.S. foreign relations, including covert operations; cultural and educational exchanges; space exploration and scientific cooperation; book publication, translation, and distribution programs; nuclear energy and disarmament; diplomatic negotiations; and the daily operations of foreign affairs personnel abroad.

The third theme of this study looks past the decisions of policy makers in Washington to explore the implementation of psychological warfare directives by American officials in the field. Borrowing interdisciplinary insights from communications literature, this study analyzes the strategies, tactics, and themes developed by psychological strategists. It is concerned with what intelligence operatives would call the "sources and methods" of psychological operations. By exploring the concrete operations carried out to wage the battle for hearts and minds, this study tracks the correlation between policy and action, a method of inquiry sometimes overlooked by diplomatic historians concerned with the making, rather than the implementation, of policies.

This perspective has yielded some important insights into the nature of American propaganda strategy. Psychological warfare experts developed a "camouflaged" approach to propaganda that used the independent news media, nongovernmental organizations, and private individuals as surrogate communicators for conveying propaganda messages. The imperative of camouflaging U.S. propaganda efforts behind a private facade shaped U.S. persuasion campaigns on many levels. Psychological warfare planners particularly worked to enlist the services of ordinary Americans, prominent citizens, civic organizations, women's groups, labor organizations, and virtually every arm of government in their propaganda campaigns. In the process, they blurred any lingering distinctions between "domestic" and "international" propaganda by both "targeting" the American people and enlisting them as active participants in the war of persuasion being waged abroad.

This "state-private" and "domestic-international" nexus of Cold War propaganda comprises the fourth theme of the book. By exploring how campaigns of persuasion ostensibly intended for international audiences also influenced and reflected ideas and values at home, it reveals a close relationship between propaganda intended for international and domestic audiences. It also broadens our understanding of the social and cultural aspects of the Cold War by highlighting a largely unexplored aspect of American life in the 1950s: the participation of private groups and individuals in Cold War propaganda campaigns. This study serves as a reminder that it is important to look at the role of the state in contributing to the cultural context of the Cold War. This is one area, as Nancy Bernhard recently noted, "where diplomatic history not only has much to gain from opening itself to cultural history but also has much to offer."[7]

The fifth and final theme of the book engages the historical literature on President Dwight D. Eisenhower. Eisenhower, who believed that psychological warfare was a potent weapon in the American Cold War arsenal, ensured that psychological considerations exerted a profound influence on the overall direction of U.S. foreign policy. Although Eisenhower's affinity for psychological warfare has not gone unnoticed by historians, the subject has received little in-depth investigation. There exists no systematic study of the impact of political warfare on Eisenhower's foreign policy. Even those historians who have studied the subject downplay its importance, asserting that Eisenhower subordinated psychological warfare to other national security considerations. Contrary to this perception, Eisenhower believed deeply in the importance of psychological warfare. Time and again he insisted that the risk of an "out and out shooting war is far less than the danger we face on the political warfare front." The president emphasized the need for a powerful and carefully coordinated psychological warfare effort on a national scale, and he encouraged all of his advisors to consider psychological factors in the policy-making process.[8]

Eisenhower's leadership in matters relating to propaganda and psychological warfare sustains the revisionist view that he presided actively over his administration. Indeed, the president was surprisingly involved in shaping U.S. propaganda strategy. Eisenhower appears in these pages not as a disinterested figurehead, as he was once portrayed, but as an activist president who left a mark on national security affairs. Largely as a result of Eisenhower's personal leadership, psychological warfare assumed a place of prominence in the making of U.S. foreign policy in the 1950s. Yet Eisenhower's commitment to psychological warfare raises questions about other aspects of the revisionist interpretation of his presidency. Historians such as Robert Divine, Stephen Ambrose, and Richard Immerman have suggested that the overall objective of his foreign policy was pursuing détente with the Soviet Union. Eisenhower, they say, wanted to end the Cold War through a negotiated settlement. But revisionists have been able to make this argument only by ignoring or sidelining the massive documentation on Eisenhower's use of psychological warfare. Close examination of this evidence indicates that shrewd Cold War calculations took precedence over an altruistic desire for world peace. Eisenhower attached far greater value to waging and to winning the Cold War than to ending it through negotiations.[9]

Propaganda and Its Euphemisms

To talk about the "battle for hearts and minds" is to use one of the dozens of euphemisms for propaganda—a "much maligned and often misunderstood word," as the infamous propaganda mastermind Joseph Goebbels noted. Propaganda is notoriously difficult to define, but at its core, the term refers to any technique or action that attempts to influence the emotions, attitudes, or behavior of a group, usually to serve the interests of the sponsor. Propaganda is often concerned with shaping the opinions and attitudes of the masses. The purpose of propaganda is to persuade—to change or to reinforce existing attitudes and opinions. Yet propaganda is also a manipulative activity. It generally disguises the secret intentions and goals of the sponsor; it seeks to inculcate ideas rather than to explain them; and it aspires to modify or control opinions and actions to benefit the sponsor rather than the recipient. As such, propaganda also functions as an instrument of coercion: it seeks to compel the target audience to submit to the will of the propagandist.[10]

For this reason, propaganda has a negative connotation as a treacherous and deceitful practice. This negative connotation has led to numerous misconceptions about the nature of propaganda. Although manipulative, propaganda is not necessarily untruthful, as is commonly believed. In fact, many specialists believe that the most effective propaganda employs different layers of truth—from half-truths, to the truth out of context, to the just plain truth. Propagandists have on many occasions used lies, misrepresentations, or deceptions, but propaganda that is based on fact and that rings true to the intended audience is more likely to be persuasive than baldfaced lies. Another common misconception identifies propaganda narrowly by its most obvious manifestations: radio broadcasts, posters, leaflets, and films. Yet propaganda experts use a much wider range of symbols, ideas, and activities to influence the perceptions of others, including such disparate modes of communication and human interaction as cultural attractions, books, slogans, monuments, museums, and staged media events.

In the first half of the twentieth century, the terms "psychological warfare" and "political warfare" entered the American lexicon as mysterious and supposedly powerful instruments of national policy. Interest in these concepts developed during World War I and World War II, when psychological warfare was viewed as an accessory to military operations that could save lives and expedite victory. During the early Cold War, psychological warfare specialists defined their craft broadly to include any nonmilitary action taken to influence public opinion or to advance foreign policy interests. Psychological warfare became an all-inclusive formula for describing disparate modes of intervention into the internal affairs of both hostile and friendly nations. It was often equated with all unorthodox and unofficial measures used in the Cold War effort, becoming, in a sense, a synonym for "cold war."[11]

This expansive concept of psychological warfare requires us to consider the psychological dimension of American diplomacy on two levels. At the most obvious level, propaganda as it is conventionally understood (the use of communication techniques to influence beliefs and actions) was employed as a distinct instrument of foreign policy. Through the USIA, CIA, and other mechanisms, the United States waged a war of words to influence friends, woo neutrals, and alienate enemies. On another level, the awareness that international public opinion had become a major factor in the conduct of diplomacy meant that psychological warfare considerations intruded on the policy-making process itself. Policy makers at the time called this "psychological strategy": the shaping of policies to influence the thoughts, beliefs, perceptions, and actions of public opinion at home and abroad.

Government officials widely used the terms "psychological warfare," "political warfare," "propaganda," and "psychological strategy" more or less interchangeably in their classified communications. In public, they preferred the euphemism "information." The idea of propaganda as information conformed to the view psychological warfare planners had of themselves: they believed that they were not propagandizing but "explaining the facts," "educating," and "informing people of the realities they faced." Undoubtedly they were protesting too much, for propaganda differs from "education" and "information" in usage and intent. As David Welch points out, "propaganda is distinct from information—which seeks to transmit facts objectively—and from education, which hopes to open its students' minds. The aim of propaganda is the opposite: to persuade its subject or public of one point of view; and to close off other options."[12]

In the narrative that follows, I use many of these terms interchangeably. This is done in part for stylistic reasons and in part because it reflects common usage during the 1950s. Somewhat hesitantly, I use the term "information" to refer to propaganda efforts because the term was so commonly used at the time and because it accurately reflects the mind-set of psychological warfare planners. Not without reservation, I also use the term "free world" to refer generally to all noncommunist countries. Naturally, such usage includes many countries that were not particularly free, most obviously the oppressive dictatorships allied with the United States. The term "free world" is used here only as a convenient shorthand. It is worth remembering, however, that during the Cold War the very idea of the "free world" was used as a rhetorical device to draw a moral dividing line between communist and noncommunist countries and to isolate the Soviet Union and its allies morally and psychologically.[13]

Methodology and Organization

This book focuses on the major propaganda themes and campaigns developed to influence public opinion in the "free world." It does not address in detail psychological warfare against the Soviet Union, which has been analyzed elsewhere, and it does not examine the daily propaganda effort to spin international developments as they occurred.[14] Because U.S. psychological programs were planned at the highest levels of the American government, my study has drawn predominantly from high-level planning documents, many of which were only recently declassified, often at my request. To connect the planning in Washington with U.S. field operations, I also researched U.S. propaganda materials prepared for dozens of countries around the world. Such an undertaking has required distilling a massive body of documentation. To make the project manageable, it was necessary to develop a research strategy privileging some countries over others. I examined records from roughly three dozen countries, focusing particularly on the following: Germany, France, Italy, Iceland, Japan, Thailand, Indochina (Vietnam), Philippines, Indonesia, Mexico, Guatemala, Brazil, India, Iran, and Egypt. These were selected for their geographical diversity and/or importance to U.S. propagandists, as determined by resources allocated, strategic value, or their perceived vulnerability to communist influence. I also surveyed in a less systematic way several countries in Africa that emerged as targets of American propaganda only at the end of the decade.

Lamentably, this study does not attempt to assess the effectiveness of U.S. psychological warfare efforts. Even with today's highly developed methods of public opinion assessment, it is exceedingly difficult to ascertain with any certainty the effectiveness of advertising, public relations, and propaganda campaigns. It is even more difficult when looking back in time. There was no 1950s equivalent of the recent Pew Global Attitudes Project by which to gauge international public opinion trends, and USIA polling abroad in those days was episodic, unsystematic, and geographically limited. Moreover, my intention was to paint a picture of an effort that was truly global in scale. An assessment of effectiveness would require a rigorous examination of media and opinion at national or local levels—a daunting task for a global project. It would also demand paying attention not just to propaganda content, but also to the pace of the news cycle and the flow of events outside the control of the spin masters. To have speculated on effectiveness without a rigorous methodology and evidentiary base would have been misleading. It is my hope that specialized studies on the local impact of U.S. psychological operations will shed light on the important question of effectiveness.[15]

This book relates the history of U.S. psychological operations through ten topical chapters, divided into two sections. Part I of the book explores the theory and practice of propaganda and psychological warfare. Chapter 1 traces the development of propaganda expertise from the beginning of the twentieth century to the end of the Truman administration. It explores the ways in which communication techniques, developed by public relations professionals and propaganda specialists during this age of mass politics and total war, were refined and adapted to the unique circumstances of the Cold War. Chapter 2 argues that psychological factors achieved special significance during the years that coincided with Eisenhower's presidency, 1953 to 1961, in part because of the role played by Eisenhower himself, and in part because of broader changes in the nature of the Cold War. Chapter 3 examines the evolution of propaganda strategy during the Eisenhower administration, analyzing as well Eisenhower's attempt to institutionalize "psychological strategy" in the national security policy-making machinery. Chapter 4 assesses the global reach of U.S. psychological operations. It surveys regional and national trends in political warfare programs and argues that operations to manipulate foreign perceptions and politics became routine features of U.S. diplomatic activity worldwide.

Part II of the book recounts the major themes, campaigns, and operations of America's battle for hearts and minds. Chapter 5 analyzes the Atoms for Peace campaign. The largest and most concerted propaganda campaign of the period, Atoms for Peace endeavored to ease public fears of nuclear annihilation by advertising the peaceful applications of atomic energy. It also was a political warfare move designed to contrast Soviet intransigence on disarmament with American flexibility and ingenuity. It thus highlights the close relationship between propaganda and diplomacy during this period, the subject of Chapter 6. It explores the symbolic and psychological dimensions of disarmament negotiations by reexamining Eisenhower's 1955 Open Skies proposal and the test ban negotiations at the end of the decade.

The next three chapters analyze the cultural dimensions of American propaganda programs. Chapter 7 discusses the participation of private groups and individuals in Cold War propaganda campaigns by looking at U.S. cultural exchange and trade fair programs and by examining the People-to-People program. Chapter 8 relates how the presentation of everyday life in America became a major theme of the USIA's global propaganda output. Chapter 9 tells the story of a secret program to wage "ideological warfare" through books and educational activities. Together, these chapters indicate the far reach of propaganda concerns. American cultural achievements, intellectual products, business practices, philosophical ideals, and social relationships were considered vital components of the Cold War struggle.

The final chapter reexamines the early years of the space race in light of U.S. psychological strategy and the broader ideological war between the superpowers. It focuses especially on the greatest propaganda challenge of the early Cold War: the Soviet launch of Sputnik in October 1957. Eisenhower interpreted this Soviet venture into outer space as a direct propaganda challenge with implications far beyond the military and technological spheres. In addition to ordering a large-scale effort to defend America's reputation as the world leader in science and technology, he called upon the nation to wage "total cold war" by harnessing all aspects of American life to meet the communist ideological challenge.

Eisenhower's declaration of total cold war, delivered in his 1958 State of the Union address, served as the inspiration for this book. It illustrated how deeply and how uniquely the total war mind-set had affected the American approach to the Cold War. It pointed the way to the overarching theme of this book: Far from being a peripheral aspect of the U.S.-Soviet struggle, the competition for hearts and minds—the cold war of words and of deeds—was one of its principal battlegrounds.

Part I
Theory and Practice

Chapter 1

Regimenting the Public Mind

The Communications Revolution and the Age of Total War

We are witnessing the growth of a world public, and this public has arisen in part because propaganda has at once agitated and organized it. . . . There is no doubt that democratic governments must assume the task, regardless of all complicating difficulties, of mobilizing minds as well as men and money in war.

— Harold Lasswell, *Propaganda Technique in the World War* (1927)

In the aftermath of World War I, famed political scientist Harold Lasswell made a prophetic prediction. Observing how the masses had been mobilized by their national governments for the twentieth century's first total war, he announced that propaganda would become an increasingly prominent feature of modern life. "Propaganda has become a profession," he declared. "Propaganda . . . is developing its practitioners, its professors, its teachers and its theories," he continued, adding his prediction: "It is expected that governments will rely increasingly upon the professional propagandists for advice and aid."[1]

Lasswell correctly perceived that he was living at the dawn of a new age of propaganda, an age of public relations experts, psychological warfare specialists, image consultants, and spin doctors. The rise of propaganda witnessed by Lasswell was fueled by an ongoing communications revolution. Beginning in the late nineteenth century, one breakthrough in communications technology followed another, yielding, at nearly regular intervals, telegraphs, telephones, mass-circulation periodicals, radio, film, television, electronic mail, the Internet, and other associated technologies. This explosion of communications media touched nearly every aspect of human development. It shaped the way ordinary people perceived and reacted to the world around them. It changed the relationship of individuals to the state. And it dramatically affected international relations in ways still barely understood by those who implement and study national foreign policies. The communications revolution transformed international politics, pulling diplomacy away from the private and secret practice it had been in the nineteenth century and exposing it to unprecedented public scrutiny in the twentieth.

In this context, propaganda and public relations emerged as integral components of statecraft, altering the ways foreign policies were formulated, implemented, and presented. Foreign policy experts increasingly acknowledged that diplomacy needed to be shaped and presented to exert an impact on key segments of public opinion. There had long been a connection between foreign and domestic affairs, of course, but foreign policy professionals in the twentieth century developed an acute awareness that public opinion—both at home and abroad—should become a principal target of diplomacy and a resource for achieving diplomatic ends. Such attention to public opinion was understood in those days as the "psychological" dimension of international affairs—an understanding that has since given way to concepts such as "cultural diplomacy" and "public diplomacy." Psychology was applied to international affairs as a way for foreign policy elites to harness the power of the media in order to mobilize mass support for desired policies. The unprecedented demands of "total war" compelled the change. Propaganda specialists, professing to understand the psychology of the masses, advocated the regimentation of the public mind through propaganda on a massive scale. On the battlefield and in allied and neutral territories, they called for waging "psychological warfare" to break the will of the enemy and to earn the support of friends and neutrals.

The Cold War arrived at the midpoint of the twentieth century's communications revolution. The earliest and most dangerous years of the Cold War happened just as radio, press, and film media were reaching maturity, coinciding with the advent of television and the expansion of the public relations and advertising professions. These developments provided many of the media technologies and communications techniques that made the Cold War so ideological, so all-encompassing, so total. In America, virtually all social, cultural, and political developments were filtered through the language and images of the anticommunist crusade, a process aided and abetted by the media explosions that had transpired over the preceding decades. As nuclear weapons made all-out war between the superpowers unthinkable, the concept of psychological warfare, forged in total war, was transferred to the Cold War. It appeared to be a way to "win" the conflict without firing a shot. Americans perceived the Cold War as they had the two world wars that preceded it: as a total contest for national survival. But in the Cold War, the battle for hearts and minds loomed larger. It was in the realms of politics, ideas, and symbols that progress toward victory would be measured and geopolitical goals would be obtained. The imperative of waging total Cold War and the practice of psychological warfare would become inextricably linked, blurring the lines between statecraft and propaganda in a global battle for hearts and minds.

Mass Society and the Communications Revolution

The practice, if not the concept, of propaganda is as old as human history. Rulers, leaders, and revolutionaries have always attempted to influence the ways their followers perceived the world. From ancient times to the present day, religious and secular leaders have used songs, legends, poems, monuments, illustrations, rituals, speeches, and publications to legitimize their authority, to mobilize their supporters, and to neutralize their opponents. Similarly, war propaganda is as ancient as war itself. Primitive peoples used pictures and symbols to impress others with their hunting and fighting capabilities. The Assyrian, Hellenistic, and Roman empires used a wide range of communication techniques and terrorizing tactics to mobilize their armed forces and demoralize their enemies. As early as the fifth century B.C., the Chinese military philosopher Sun Tzu advocated various strategies to maintain morale and to destroy the enemy's will to fight. In the nineteenth century, famed German military strategist Carl von Clausewitz identified psychological forces—especially morale—as critical elements of modern war.[2]

Nevertheless, the twentieth century truly inaugurated an "age of propaganda." The steady expansion of communications media, coupled with an increasingly scientific approach to psychological manipulation, made propaganda more pervasive and more sophisticated than before. Industrialization, mechanization, and urbanization brought sweeping social, cultural, and economic changes to the western world. At the same time, a dramatic revolution in communications was under way. Before the advent of the telegraph in the 1840s, information traveled only as fast as the messenger who carried it, but the spread of telegraph, telephone, and then radio networks meant that communication over long distances was no longer tied to the available means of transportation. The increased speed of communication, combined with rising literacy rates and the spread of mass-circulation periodicals, created increasingly well-informed and active publics that were "demanding an unprecedented accountability from leaders in business and government." Observers from all walks of life were acutely aware that public opinion was becoming a force to be reckoned with. The perceived power of public opinion was tied to the ongoing media revolution, which made the public at once more informed and more visible.[3]

Many elite observers developed an exaggerated perception of the power of public opinion, issuing alarmist warnings about the perilous implications of mass politics. Several social theorists cautioned that the new century had produced a new and frightening form of "mass society." Urbanization and industrialization had eroded traditional bonds of locality and kinship, they theorized, producing a vast workforce of atomized and isolated individuals comprising an ignorant, irrational public that was acquiring unprecedented power to shape the world around them. "The voice of the masses has become preponderant," the French theorist Gustave LeBon announced in 1895: "The destinies of nations are elaborated at present in the heart of the masses, and no longer in the councils of princes." In his internationally acclaimed treatise *The Crowd*, LeBon cautioned that the old order was degenerating into an "era of crowds," an age where the preponderant power of ignorant masses was supplanting informed aristocratic rule. Impervious to reason and moved only by images, illusions, and emotional impulses, this crowd threatened to undo the social fabric and destroy civilization. If a reasonable form of social order was to be preserved, LeBon cautioned, elites had better understand the irrational and impulsive instincts of the masses.[4]

LeBon's theory made an indelible impression on a generation of social scientists who "saw the study of society as a tool by which a technocratic elite could help serve the interests of vested power." In the ensuing decades, an international cohort of intellectuals echoed LeBon's concerns about the new mass society. In two highly influential studies written in the 1920s, *Public Opinion* and *The Phantom Public*, the American journalist Walter Lippmann anguished over the potential power of "the mass of absolutely illiterate, of feeble-minded, grossly neurotic, undernourished and frustrated individuals." He captured well the sentiments of an emerging class of propaganda specialists when he suggested that elites would have to master words, symbols, images, and emotive modes of communication in order to protect the social order from a democratic public run amok. In reaching this conclusion, Lippmann and others adapted the insights of the emerging field of psychology. Sigmund Freud's conclusion that humans were motivated by subconscious fantasies and fears, rather than by logic and reason, raised the possibility of harnessing new technologies of mass communication to control the "mass mind" of the crowd by using words, ideas, and symbols to manipulate the invisible forces that governed hu-

man behavior. Elite experts, who used new instruments of mass communications and social science research, could tame what these intellectuals
openly derided as the "herd."[5]

Underlying all these analyses was the assumption that leaders—whether in business, government, or other enterprises—needed to adapt to the
challenges posed by mass society and the communications revolution by
harnessing the power of propaganda to manage an increasingly complex
world. Over time, the communications revolution would dramatically
change the way business, politics, warfare, and eventually diplomacy were
conducted as elites in these fields came to acknowledge that they would
have to accommodate their practices to take into account the heightened
public interest in and awareness of their activities.

The Rise of Public Relations

Business was first to adapt to the communications revolution. Industrial
firms gradually turned to public relations experts to manage their images.
William Henry Vanderbilt's historic phrase "the public be damned" had
aptly characterized the attitude of business in the Gilded Age, as most
industries of the day operated secretly, guided by the belief that the less
the public knew of their operations, the better. But it was hard to hold
such attitudes under the watchful gaze of a muckraking press and an increasingly articulate public. Under attack from organized labor, progressive reformers, and agrarian populists, and finding their activities increasingly constrained by government regulations, many businesses sought to
mollify or seduce the public through public relations. Beginning in the
1880s, businesses such as AT&T, Mutual Life Insurance Company, and
Westinghouse Corporation established in-house publicity departments to
write publications to mold public perceptions of their activities. In 1900,
the first publicity agency was established in Boston, and a few years later,
renowned public relations pioneer Ivy Ledbetter Lee opened an office in
New York City.[6]

Especially instrumental in the development of public relations and
propaganda strategy was Edward Bernays, the father of spin if ever there
was one. By his example and from his many writings—including the still-
consulted *Crystallizing Public Opinion* (1923) and *Propaganda* (1928)—Bernays instructed several generations of image makers on the new techniques
of persuasion. Deeply influenced by the mass society theorists and by his
uncle, Sigmund Freud, Bernays believed that "the conscious and intelligent

manipulation of the masses" had become an indispensable feature of democratic society. Bernays did more than concoct abstract theories about psychological manipulation; he developed programs for action and demonstrated their effectiveness. He was, as PR historian Stuart Ewen observes, "the most important theorist of American public relations," but he was also the technocratic guide for "practitioners in the trenches."[7]

For the historian of propaganda, two aspects of Bernays's approach to public relations stand out. First was his emphasis on the indirect approach to perception management. Rather than the hard-sell tactics of most advertising campaigns, Bernays's public relations campaigns sought to influence audiences through a secretive, covert approach that hid the involvement of the public relations counselor from the target audiences. His method was to skillfully manipulate symbols and trends through the regular media and established modes of communication so that most people were unaware it was happening. Bernays likened it to the game of billiards "where you bounce the ball off cushions, as opposed to pool, where you aim directly for the pockets." Second, Bernays believed that there was a direct link between action and influence. The PR person, he explained, was "responsible for every impingement of his principal's or client's action on every phase of public opinion." This included advising clients on how to act and present their policies, but it also went further. Bernays developed a formula for generating news through carefully orchestrated action. He did not view the news as something that just happened. It was something that could be made to happen. Bernays clandestinely staged events and created "circumstances" that the press considered newsworthy. As Larry Tye put it: "Bernays generated events, the events generated news, and the news generated a demand for whatever he happened to be selling."

Bernays's formula for generating news was simple, but his campaigns were sophisticated and complex, involving an intricate web of surrogate spokespersons, front groups, letter-writing campaigns, publicity stunts, and unlikely alliances between organizations and individuals sharing common interests. Bernays particularly mastered the use of front groups and third-party spokespersons to mask the identities of his clients. Rather than assaulting his audiences by direct attack, as advertisers did, Bernays got surrogate voices—often those of recognized "experts" and societal "leaders"—to carry his messages. This obscured the PR person's manipulating hand, and to Bernays, it also played to the herd instinct and the public's inclination to follow established leaders.[8]

Bernays thus established the basic paradigm for major public relations campaigns. These campaigns usually began with overt acts to attract the attention of the press. These staged happenings ranged from small-scale events, such as a luncheon with a featured speaker, to elaborate spectacles and full-fledged media events. For example, when Bernays was working for a tobacco company trying to induce women to take up smoking, he orchestrated a parade of prominent women who marched down Fifth Avenue in New York smoking "torches of freedom." Such initial media stunts were usually followed by "segmenting"—identifying particular target audiences for additional appeals. Then Bernays mobilized his front groups, the imaginary and allegedly independent organizations that spread his messages through many channels of communication. Most imaginatively, for his tobacco company clients, Bernays created the Tobacco Society for Voice Culture, which promoted the incredulous message that cigarette smoking was good for the voice. Fronts like these did such things as mail promotional literature, issue supportive statements, hold press conferences, create other associations to promote the chosen cause, and concoct media events of their own. Through it all, Bernays operated in the shadows. Although he was not entirely invisible, only insiders and experts perceived his hidden hand. He never disclosed the identities of his clients, and he allowed his front groups to do the job of distributing the information and ideas he wanted the public to absorb. All of this reflected his general belief that "emphasis by repetition gains acceptance for an idea, particularly if the repetition comes from different sources."[9]

Bernays articulated and put into practice a revolutionary approach to opinion management that altered public moods, perceptions, and desires in subtle, often imperceptible, ways. Synchronized messages and symbols would be distributed through diffuse modes of communication by willing and unwilling collaborators. Actions, policies, and programs would be implemented for their psychological impact. Staged spectacles would become media events; neutral-seeming front groups would pass on messages; and, ideally, the public would follow the suggestions of the psychological engineers, blissfully unaware of the calculations of the new spin masters. Most importantly, the basic premise underlying this approach to public relations—that the task of the PR counselor was to advise clients on how best to win public opinion through their actions rather than through words alone—would later influence the thinking of the Cold War's psychological warfare experts who insisted on having a role that went beyond the dissemination of propaganda to include advising on U.S. foreign policies as a whole. It would take some time, however, for this lesson to sink in.

Total War and Mass Mobilization

Although business increasingly employed public relations tactics, the American government turned to PR more gradually. Foreign policy particularly remained an elite affair, aloof from the daily concerns of most Americans. To be sure, the Spanish-American War had demonstrated the power of the press to fuel wartime passions, and both imperialists and antiimperialists turned to the media to press their cases. But there was no systematic or concerted effort by the government to harness the power of public opinion to achieve foreign policy objectives until World War I created an immediate need for mass mobilization.[10]

Long before the United States joined the conflict in 1917, it was apparent that industrialization, the communications revolution, and the rise of mass societies had made this war unlike any other in human history. It was widely perceived as a "total war": an all-encompassing contest for national survival requiring the total mobilization of all the nation's resources—military, political, economic, and psychological. To a greater extent than ever before, the new warfare affected the lives of ordinary citizens who served in the mass armies, worked in the industrial plants, experienced dramatic social changes induced by wartime conditions, and occasionally themselves became targets of military reprisals. People at home also experienced warfare in a new way. The burgeoning mass media created a sense of intimate involvement with the conflict by delivering news of far-off developments as quickly as they happened. Government officials in all the warring states were guided by the total war notion that wars were no longer won merely by armies in the field, but by the morale of the entire nation. Accordingly, it was axiomatic that total warfare "necessitated the mobilization of the civilian mind" and that "no government could have a united nation behind it unless it controlled the minds of its people." Keenly aware of the power of the mass media to affect national morale, all the warring governments turned to propaganda to mobilize civilians, boost the confidence of their soldiers and allies, demoralize enemies, and woo neutrals. For democracies, the concept of total war provided a legitimating rationale for the systematic use of propaganda as an instrument of statecraft, an implement of combat, and a means of mobilizing public opinion.[11]

The convergence of total war and the communications revolution transformed international relations. Casting aside the old notion that governments do not meddle in the internal affairs of others, the major belligerents employed all kinds of propaganda tactics to influence the attitudes of civilian populations in neutral and enemy nations. In the early years of the conflict, the United States was the prime target of a British-German propaganda war as each side tried to prevent American arms, dollars, and assistance from going to the other. Indeed, one of the Royal Navy's first acts in the war was to cut off German cable access to the United States to ensure that Britain, not Germany, remained the principal supplier of war news to America. British domination of communications networks and war information gave London a tremendous advantage in the fight for American opinion. It also meant that once the Americans joined the war, the U.S. would have to contest British communications domination to realize American war aims.[12]

President Woodrow Wilson understood that the United States had to break foreign control of information if he was to fulfill his vision of a new world order of democratic government, free trade, open diplomacy, and collective security. He also appreciated the need to unite a nation divided by race, class, and ethnicity behind the war effort, particularly in the absence of a clear and present danger to American national security. Making communications an "urgent national priority," Wilson worked with private industry to improve U.S. capabilities in wireless telegraphy, news services, motion pictures, air travel, and cable communications. More immediately, a week after declaring war, Wilson created the first official propaganda agency of the American government: the Committee on Public Information (CPI). Headed by the progressive journalist George Creel, the CPI was responsible for censorship, propaganda, and general information about the war effort. It became a sprawling organization, comprising more than twenty divisions and bureaus, with offices across the United States and abroad. The committee employed hundreds of professionals, most with experience in advertising, public relations, and journalism, many of them sharing Creel's progressive vision of carrying "the gospel of Americanism to every corner of the globe."[13]

Creel devoted most of his resources to domestic propaganda. He attacked the "shapelessness of public opinion" by presenting the conflict as a war of self-defense against a "rising tide of militarism." The CPI encouraged Americans to ration food and other resources, to purchase war bonds, to contribute to war industries and relief work, and to support recruitment and conscription. Abroad, the committee's propaganda effort was extraordinary, featuring operations around the globe in more than thirty countries. To challenge international domination of news, Creel created a government-owned news agency, Compub, that allowed the CPI to disseminate news about the U.S. war effort, official statements, and various "human interest" features for publication in foreign media outlets. The CPI also used motion pictures, photographs, cartoons, posters, and signboards to promote its messages. The committee established reading rooms abroad, brought foreign journalists to the United States, crafted special appeals for teachers and labor groups, and sponsored lectures and seminars. In some countries, the CPI borrowed the missionary technique of "using language to sell an ideological message" and sponsored free classes in English. The Creel Committee also enlisted private support from sympathetic individuals and groups in crafting and distributing American propaganda, thus developing an enduring strategy of "private cooperation" in the government's persuasive campaigns.[14]

The United States may have been last to enter the great propaganda war, but it had nonetheless adapted quickly. Professionals in advertising, public relations, and journalism supplied the Creel Committee with a corps of able-minded experts that were highly skilled in the image-making techniques of the time. More importantly, perhaps, the United States had in President Wilson "the great generalissimo on the propaganda front."[15] In American democracy, the president's ability to command the attention of the media makes him the nation's chief propagandist: the words and images emanating from the White House provide a centralized message reaching audiences at home and abroad. Over the course of the twentieth century, this power to attract media attention and to frame public debate significantly contributed to the growth of the "imperial presidency," vastly expanding presidential power, particularly in foreign affairs. Wilson appeared to foresee this and furnished the ideas, symbols, and words that gave American propaganda much of its force. Most of his speeches were translated and transmitted around the globe within twenty-four hours, making Wilson the first leader to address the world and establishing Wilson as *the* spokesperson for the Allies. Wilson's wartime ideas and rhetoric, as Frank Ninkovich notes, rejuvenated the Allies "by turning the war into an ideological conflict whose stirring message the nationalistic Germans could not hope to match."[16]

Wilson's idealism and faith in democracy led him to perceive sooner than most statesmen that world opinion—"the thoughts of the mass of men"—was a powerful force that was changing international diplomacy. "If my convictions have any validity," he said in 1915, "opinion ultimately governs the world." He perceived that the new mass media had tremendous power for advancing his foreign policy agenda: public opinion could be harnessed as an instrument of diplomacy in both war and peace. Knowing that his concept of a fair, nonputative postwar settlement and his dream of a new world order based on collective security and open diplomacy would run afoul of the nationalist ambitions of the Allies, Wilson endeavored to mobilize world public opinion to pressure governments to implement his vision of the postwar world. He blended traditional diplomacy—negotiations between governments—with a new form of what is now called "public diplomacy": appealing to international public opinion to achieve foreign policy objectives. Wilson's public diplomacy, together with the CPI's global propaganda campaigns, marked a significant departure in the conduct of international relations. Here was a government advancing its foreign policy agenda by blanketing idealistic appeals over heads of state directly to mass audiences around the globe. Whereas other governments had employed propaganda principally to further their war efforts, Wilson and the CPI were assertively using propaganda to advance the peacetime foreign policy goals of the United States. That Wilson failed to see all of his vision realized should not obscure the basic fact that he anticipated instinctively, if not expressly, the changing nature of international relations resulting from the communications revolution and the age of total war. Unfortunately for Wilson, in this understanding he was nearly alone. Throughout the war, Congressional leaders had been anxious about granting propaganda power to the executive branch. When the war ended, Congress immediately—almost violently—shut down the CPI.[17]

Overcoming Psychic Resistance to War

The experience with propaganda in wartime awakened Americans to the power of persuasive approaches, and it caught many of them by surprise. As wartime jingoism faded into postwar disillusionment, many Americans came to believe they had been tricked by propaganda into fighting an unnecessary war. They became "extremely anxious—even paranoid—about the influence of propaganda in modern social and political life." Postwar revelations about fabricated atrocity stories—such as the infamous story circulated by British agents that the Germans were turning human corpses into soap—permanently transformed "propaganda" into a dirty word, synonymous with "lies."[18]

Yet many social scientists, public relations practitioners, and propaganda theorists took from the Great War a different lesson. The ease with which civilian populations had been made to despise the enemy had proven the effectiveness of propaganda as a tool for manipulating popular opinion. This raised the possibility of using propaganda to affect all kinds of psychological changes. The apparent success of propaganda during the war, as Edward Bernays put it, "opened the eyes of the intelligent few in all departments of life to the possibilities of regimenting the public mind." The war had a profound effect on the public relations and advertising professions by demonstrating the power of publicity techniques. In the interwar years, both professions ballooned into independent industries, employing thousands who increasingly influenced the conduct of American business. A veritable class of propaganda specialists emerged after the war, comprising an amalgamation of intellectuals, social scientists, journalists, and public relations practitioners. Many, such as Lippmann and Bernays, had themselves worked for the Committee on Public Information. That experience taught them how easily public opinion could be molded. Total war inspired their influential studies on public opinion management. Several hundred studies on the science of persuasion were written during the interwar years. A bibliography of "propaganda and promotional activities" compiled in 1935 included some 4,500 entries, almost all of them written after the outbreak of World War I. Total war had fueled a propaganda revolution.[19]

Of these many studies, Harold Lasswell's *Propaganda Technique in the World War* was particularly influential. This book, a comprehensive analysis of wartime propaganda, became a standard text for propagandists, communication specialists, and political scientists—a veritable blueprint for improving persuasive techniques. Lasswell's particular concern was the control of public opinion during wartime. Like many of the propaganda specialists of the day, he viewed propaganda quite literally as a tool, an instrument for achieving desired goals. Propaganda "is no more moral or immoral than a pump handle," he declared famously. Lasswell thus approached his subject from a pragmatic perspective. He wanted to identify the most successful techniques in order to establish rules, principles, and themes for effective propaganda during war.[20] He identified four broad strategic aims of war propaganda:

1. To mobilize hatred against the enemy;
2. To preserve the friendship of allies;

3. To preserve the friendship and, if possible, to procure the co-operation of neutrals;
4. To demoralize the enemy.

In Lasswell's formulation, appeals to "friendly" audiences were paramount. Psychological warfare against the enemy was an important accessory to military conflict, but maintaining the morale of one's own soldiers, citizens, and allies was indispensable. So too was it vital to keep neutrals from siding with the enemy. At home, public opinion was a problem requiring skillful management. Abroad, public opinion was a battlefield, a contested terrain that needed to be won.[21]

Lasswell cataloged the rhetorical stances, words, and appeals that would most resonate with public opinion in wartime, identifying the essential "psychological appeals" for future propagandists. To Lasswell, the public's desire for peace represented a particularly difficult psychological obstacle that needed to be overcome. In an age where "peace has come to be regarded as the normal state of society," the circulation of ideas by propaganda was necessary to transform the pacific public's inclinations into a thirst for vengeance. "So great are the psychological resistances to war in modern nations that every war must appear to be a war of defence against a menacing, murderous aggressor," he wrote. Skillful propagandists should therefore identify the nation's cause as the cause of peace while casting the onus for war on the wicked ambitions of an evil foe. Hatred and vengeance were not sufficient to maintain popular support for war, Lasswell emphasized. Populations also needed to have a sense of higher purpose. They "must be furnished with war aims of a highly rationalized and idealistic type." War should be fought for civilization, for international law, for freedom, and other noble causes. Equally important to the maintenance of morale, propagandists needed to hammer home the certainty of victory. "The enemy may be dangerous, obstructive, and satanic," he observed, "but if [the enemy] is sure to win, the [morale] of many elements in the nation will begin to waver and crumble." Public fear needed to be managed to prevent defeatism and a collapse of morale: all wartime propaganda "must insist upon the ultimate success of the cause." Victory should be inevitable, but people should sacrifice and persevere to achieve it.[22]

Propaganda Technique crystallized the argument that propaganda was essential to modern war. According to Lasswell, the sociopolitical conditions arising from the democratization of world politics, the rise of mass societies, and the demands of total war meant that the management of opinion had become an indispensable element of international conflict. In total war, where civilian contributions were as vital to victory as military maneuvers, public attitudes needed to be organized and agitated by propaganda to overcome the "psychic resistances" to war. Just as soldiers needed to be drilled into becoming obedient fighting machines, so too was it necessary to "standardize" civilians. "Civilian unity is not achieved by the regimentation of muscles," he wrote. "It is achieved by a repetition of ideas rather than movements."[23]

Propaganda and World War II

By the 1930s propaganda had clearly become "an established fact of everyday life." Ideologically driven authoritarian states in Russia, Germany, Italy, and Japan rode to power on propaganda and agitation that exploited social and economic unrest. Combining the propaganda methods pioneered during World War I with the new technologies and techniques of psychological manipulation developed in the 1920s, communist, Nazi, and fascist governments conspicuously used propaganda to legitimize rule at home and to facilitate expansionist ventures abroad. International broadcasting developed into a major foreign policy instrument as the Soviet Union, Germany, Japan, and Britain waged a shortwave radio battle on a global scale.[24]

The United States reacted nervously and cautiously to the intensifying psychological warfare battles being waged abroad in the 1930s. Many Americans viewed Nazi propaganda as a sort of "superweapon" spread by "veritable armies of Nazi provocateurs." Some Americans became nearly hysterical about the threat of Nazi propaganda—"the voice of destruction and the instrument of conquest," as one writer described it. Congress created several committees to investigate communist and fascist propaganda activities in the United States, and in 1938 it passed a law requiring foreign propagandists to register with the government. Meanwhile, numerous public advocacy groups pressed the government to mobilize the American public psychologically and to mount a "total" propaganda attack on the Axis.[25] Nevertheless, no acceptable rationale had been developed for the peacetime use of propaganda as a foreign policy tool. Isolationist sentiment combined with lingering hostility to Creel's activities to delay the establishment of a new propaganda agency until war had become all but inevitable. President Franklin Roosevelt was a masterful communicator who expertly used the media for his own ends, but he was also deeply suspicious of propaganda. He hesitated to commit his political prestige to the creation of a new information agency. As the storm of yet another world war gathered on the horizon, he would relent, ultimately creating a vast bureaucratic apparatus that would spread American propaganda quite literally around the world.[26]

When Roosevelt acted to create a new propaganda agency, he would direct its efforts neither at Europe nor at Asia, where fighting had already begun, but at Latin America, where Nazi provocateurs were agitating against the United States. In 1938, FDR established the euphemistically titled Office of the Coordinator of Inter-American Affairs (CIAA), headed by Nelson Rockefeller, to "combat the Nazi lie." The CIAA employed many techniques used by the CPI, but it also inaugurated a new tradition in U.S. foreign policy: government sponsorship of educational and cultural exchanges. It sponsored tours by ballet, theater, and musical groups, archaeological expeditions, art exhibits, publishing enterprises, and academic conferences. Publicly, the CIAA's cultural programs were defended as promoting "international understanding." Behind closed doors, however, the CIAA frankly emphasized propaganda motives. It used its cultural initiatives to quell anti-American sentiment, to promote American business, and to encourage closer relations between Latin American countries and the United States. Rockefeller, an ambitious political operator and bureaucratic empire builder, transformed the CIAA from a relatively small outfit into a sprawling propaganda agency with nearly 1,500 employees. By 1943, its budget had soared from an initial $3.5 million to $38 million.[27]

In the early part of 1941, as war appeared imminent, Roosevelt created several additional agencies to disseminate propaganda at home and abroad. In 1942, these various information programs were combined into the Office of War Information (OWI) under the direction of the well-known journalist and broadcaster Elmer Davis. Roosevelt also established the Office of Strategic Services (OSS), the forerunner to the Central Intelligence Agency (CIA). A loose division of labor separated the two organizations. The OWI directed all government-sponsored propaganda at home and was the central clearinghouse for all "white," or officially acknowledged, propaganda abroad, including the Voice of America. Leaflets, posters, and radio broadcasts were the staples of OWI propaganda, as was factually based yet appropriately slanted news. OWI personnel were convinced that simple news "would in the long run be more corrosive to Nazi morale than any amount of agitational propaganda." The OSS, on the other hand, was an intelligence service involved in guerilla warfare, covert operations, and other unorthodox activities. It distributed "black" propaganda—secret, subversive propaganda that deliberately lied and slandered, purporting to come from enemy sources rather than the American government. The OSS created and disseminated covert propaganda behind enemy lines, and it "organized, trained, and sustained the morale of local resistance groups" in occupied territories. Its Morale Operations Division spread fabricated rumors, printed phony newspapers, and engaged in all manner of clever tactics designed to harass the enemy, foment divisions within the enemy leadership, and undermine the morale of enemy soldiers and civilians.[28]

The military services developed the concept of "psychological warfare" during the war. Defining it chiefly as an accessory to military operations, psychological warfare was used as "strategic propaganda" to demoralize populations in enemy and enemy-occupied territories; as "combat propaganda" to induce surrenders, affect morale, and facilitate cover and deception operations; and as "consolidation propaganda" to ensure cooperation from civilians in occupied areas. When the United States entered the war, the military services were remarkably ill-prepared for psychological operations—in 1941 there was only one officer with propaganda experience in the War Department—but by the end of the war, psychological warfare divisions or branches were established in all the major theaters of operation. In the fall of 1942, General Eisenhower created a Psychological Warfare Branch (PWB) to participate in the Allied invasion of North Africa. The PWB, a joint Anglo-American organization run by General Robert McClure and C. D. Jackson, began as a small outfit consisting of some forty-seven journalists, advertising professionals, OWI and OSS recruits, and British officers. Apparent successes in inducing enemy surrenders helped to win over formerly suspicious commanders, and by the end of the North African campaign, the PWB had grown to nearly a thousand men. Psychological warfare planning was increasingly integrated with military operations planning. It produced some remarkable successes. Most notably, at the battle for the island of Pantellaria, an alternate barrage of bombs and leaflets led the Italian garrison to surrender, sparing the Allies a costly ground assault. In early 1944, the combined Allied forces created a sprawling organization, the Psychological Warfare Division of the Supreme Headquarters of the Allied Expeditionary Forces (PWD/SHAEF), which prepared some 50 million leaflets for the invasion of Normandy. The United States moved much more slowly to develop a psywar capability for the Pacific theater, preferring to leave such efforts to the Australians and assuming that the "fanatical" Japanese could not be induced to surrender. In June 1944, however, General Douglas MacArthur created a psychological warfare branch that became more effective as the campaign progressed. Together, the Americans and the Australians dropped some 400 million leaflets in the Southwest Pacific and secured the surrender of nearly 20,000 Japanese soldiers.[29]

On the home front, the propaganda of total war was not a monolithic product of the state. It was forged by a network of government and private entities working in partnership. Popular entertainment on the airwaves and on the silver screen vilified the enemy, defended the cause, and promoted patriotic sentiment. Only the finest of lines separated wartime propaganda from advertising, as advertisers patriotically sold their products by selling the war. A consortium of advertising and public relations firms created the War Advertising Council, which donated creative and material resources to promoting war-related causes. Working with the OWI, the War Advertising Council urged Americans to collect scrap and rubber for war-related production; to save bacon grease in coffee cans; to look out for enemy aircraft, warships, and spies; to plant victory gardens; to reduce their gasoline consumption; to work overtime; and to buy war bonds. The war bond program comprised the centerpiece of domestic mobilization efforts. Advertisements for war bonds hyped American patriotism, the capriciousness of the enemy, and the totality of the Axis threat. Roosevelt's Treasury Secretary, Henry Morgenthau, developed the bond drive program as a way to "make the country war-minded." His plan was "to use *bonds* to sell the *war*, rather than *vice versa*." In a radio broadcast to the American people, Morgenthau explained that buying war bonds afforded "every one of you to a chance to have a financial stake in American democracy—an opportunity to contribute toward the defense of that democracy." As his words indicated, such activities as bond purchases, rationing, and scrap drives were psychological as much as practical: the simple fact of doing something for the war effort contributed to wartime morale by giving ordinary Americans a sense of personal participation in the struggle—a feeling that they had sacrificed, an awareness that they had something material at stake in the national cause.[30]

World War II was an ideological conflict with staggering implications for the development of propaganda techniques. By the end of the war, the Voice of America was broadcasting around the world on a twenty-four-hour-a-day basis in some forty languages. Over 30,000 people were directly involved in psychological operations either at home or abroad, and Congress appropriated funds for propaganda activities at the rate of $150 million a year. The numerous civilian, military, and intelligence agencies created for wartime propaganda work had trained thousands of Americans in psychological warfare techniques, producing a cadre of experts skilled in the techniques of media manipulation, propaganda, and clandestine psychological operations.[31]

Moreover, a generation of American military and political leaders—the generation that would lead the country through the Cold War—had grown up in this new age of propaganda. They believed that the power of their enemies—Nazi, communist, and otherwise—derived from fanatical ideologies inflicted on malleable publics by propaganda. Many American military and political leaders had learned that psychological warfare to demoralize enemies, woo neutrals, and maintain alliances was an indispensable element of modern war. They also believed that total war demanded unrelenting attention to public opinion. The American people needed to be organized and agitated to support costly ventures abroad. Their morale needed as much attention as that of American soldiers, and policy makers and military leaders needed to use propaganda and public relations to maintain that morale. Propaganda advocates championed a "strategy of truth," but they also conceded that lies, media manipulation, and the withholding of information were necessary for national security—not just to keep critical information from enemies, but also to preserve American morale and to maintain Allied unity. As Americans adapted their experiences with total war to the unique challenges of the Cold War, they modified the concept of psychological warfare to fit this new war of nerves, a contest that was above all psychological, ideological, and political.

Psychological Strategy and the Early Cold War

After the surrender of Japan, the Truman administration felt obliged to dismantle much of the vast propaganda apparatus constructed during the war. The OWI had become the target of acrimonious partisan attacks from conservatives in Congress who saw it as a promoter of the hated New Deal. President Truman swiftly liquidated the organization. What was left he transferred to the Department of State, announcing that "the nature of present-day foreign relations makes it essential for the United States to maintain information activities abroad as an integral part of the conduct of our foreign affairs." He hoped that locating the information program in an established department would spare it from being completely dismembered by Congress, thus acknowledging the continuing importance of the information program to U.S. foreign relations. In this he was supported by a small but influential segment of the foreign policy elite. The wartime experiences with propaganda and the perceived impact of the communications revolution on international relations persuaded these elites that the United States needed to continue to harness the power of international public opinion to serve American foreign policy interests. This signified a remarkable change in official attitudes toward propaganda as an instrument of foreign policy, because before the 1940s no one seriously considered an organized, government-sponsored effort to influence foreign peoples except during a national emergency.[32]

Several factors converged to revitalize American interest in propaganda during the Cold War. The desire of both the United States and the Soviet Union to avoid all-out war channeled their rivalry into nonmilitary spheres of competition that increasingly revolved around the symbolic dimensions of national power and progress. The fragile political and economic state of the postwar world provided fertile ground for meddling abroad, and the superpowers found that they could influence international politics through economic warfare, covert intervention, and propaganda at less cost and less risk than through direct military action. The Cold War reversed Clausewitz's famous dictum that war was an extension of politics: diplomacy became an extension of warfare, and propaganda became a critical weapon in this new type of international combat.

Mass communications, mass societies, and the democratization of politics further accentuated the ideational aspects of the Cold War competition. Communications networks continued to develop and expand during the early Cold War years. Both superpowers attached a high priority to the use of media for persuasion: to mobilize their domestic populations and to win international support. The political struggle for the control of core areas of industrial production and resources often took place in the realm of ideas. Ideas served as levers for moving individuals along ideological—and therefore political—lines. Both superpowers also mobilized private organizations, media outlets, and leadership groups in an effort to affect the political structure of contested areas of the world. In the United States, propaganda by government officials, private groups, and the American media helped to forge the Cold War consensus view that the United States had to lead worldwide opposition to communism.[33]

In such a context, Americans adapted their understanding of psychological warfare to fit the unique challenges of the Cold War. During the two world wars, psychological warfare was largely seen as an accessory to

Free World Speaks, a propaganda comic strip, was distributed in the late 1940s and early 1950s by U.S. information agencies. It contrasted communist repression with universally held democratic values. Source: National Archives.

military operations, but with the onset of the Cold War, psychological warfare enthusiasts defined the concept broadly to include any action taken to influence public opinion or to advance foreign policy interests by nonmilitary means. Psychological warfare was transformed into a catchall formula.

It went beyond mere propaganda to embrace a wide range of clandestine activities, including covert operations, trade and economic aid, and educational exchange programs. Psychological warfare became, in essence, a synonym for "cold war." It reflected the belief of many politicians and foreign policy analysts that the Cold War was a political, ideological, psychological, and cultural contest as well as a military and economic one.[34]

As Cold War tensions intensified, a widespread and persistent belief developed that the United States was losing the "war of ideas" to the Soviet Union's supposedly superior propaganda apparatus. Just as American observers had accorded tremendous power to Nazi psychological warfare, so too did they perceive communist propaganda as a finely honed weapon for subversion, revolution, and conquest. Alarmist warnings about the communists' skill as propagandists circulated in official circles and in the media. In popular culture, communist ideology was perceived as an invasive disease eroding the foundations of American society.[35]

The evolving national security strategy of the United States reflected deeply held concerns about the impact of communist ideology and propaganda. The Truman administration's national security policy, as Melvyn Leffler has argued, was "defined *not* in terms of numbers of troops, tanks, planes, and bombs, but in terms of control of industrial infrastructure, raw materials, and skilled labor." The challenge for American officials was to keep the industrial core and centers of raw materials out of the Kremlin's orbit and firmly under western control. The primary threat was not that the Soviet Union would seize territory through direct military intervention, but that it would capitalize on economic and social unrest, expanding its power through subversion and manipulation. For all the talk of the danger posed by the massive Red Army along the borders of Western Europe, American officials generally believed that the greatest threats lay in the political, economic, and psychological arenas. American intelligence consistently predicted that the Soviet Union was "unlikely to resort to open military aggression" while at the same time cautioning about the dangers posed by Soviet "political, economic, and psychological warfare." U.S. strategy therefore sought to turn American economic potential into preponderant power while escalating its psychological warfare and covert operations campaigns in order to destabilize the Soviet bloc and consolidate a liberal-democratic world order in areas not under Kremlin control.[36]

This strategy, identified at the time as a policy of "containment," crystallized during 1947–1948 with the declarations of the Truman Doctrine and the Marshall Plan. The USSR's creation of the Cominform, its in-

The outbreak of the Korean War led the United States to significantly expand
its propaganda operations overseas. *The Korea Story*, a comic book prepared by
the International Information Administration, presented the conflict as a series of
atrocities perpetrated by the communists on the innocent people of Korea.

ternational propaganda organization, and the consolidation of communist governments in Eastern Europe led American officials to conclude
that they would have to employ communist methods in order to thwart
what they saw as the Soviet drive for world domination. Within the
American government, plans emerged to revitalize psychological operations. The first meetings of the National Security Council in the fall of
1947 addressed psychological warfare. In December, the council approved
NSC-4, the first direct action to improve postwar information services.
NSC-4 warned that the USSR was "employing coordinated psychologi-

cal, political, and economic measures designed . . . to weaken and divide world opinion to a point where effective opposition to Soviet designs is no longer attainable by political, economic, or military means." It called for "the immediate strengthening" of programs designed to influence foreign opinion, and it placed the State Department in charge of coordinating the various information programs scattered throughout the government. The National Security Council also approved an annex to NSC-4 directing the Central Intelligence Agency to plan and implement covert psychological operations.[37]

Thus, as during World War II, the American Cold War propaganda apparatus developed on two separate tracts: one overt, the other covert. The overt information program received legislative sanction with the passage of the Smith-Mundt Act in January 1948. The legislation provided the Truman administration with the legal and financial foundation for a global propaganda program that used a number of modes of modern communications, including print, radio, film, exchange programs, and exhibitions. The Department of State retained control of overt information and exchange programs, and funding for information activities promptly doubled to $31.2 million. After the passage of the Smith-Mundt Act, the overt information program developed an increasingly hard-hitting style. Voice of America broadcasts became especially strident. Information officers in the State Department were instructed to "openly and explicitly" respond to communist "lies" and to "point the accusing finger at the Russians." The early postwar emphasis on objective-sounding news and information shifted to hard-hitting propaganda in its most obvious form: cartoons depicting bloodthirsty communists, vituperative anticommunist polemics, and sensational commentary.[38]

Although the Smith-Mundt Act authorized an overt and official information program akin to the OWI, the Truman administration secretly developed a covert and subversive psychological warfare program modeled on the OSS and assigned to the newly created Central Intelligence Agency. Clandestine psychological operations had many advantages, not least of which was freedom from Congressional scrutiny. Covert political action programs also allowed the United States to manipulate developments abroad without provoking military reprisals, arousing nationalist sentiment, or stimulating an anti-American backlash.

The earliest covert programs focused on Western Europe, where American officials feared that communists would capitalize on the postwar economic and political turmoil to seize power legally or illegally. To forestall such a development, the CIA orchestrated "political action" campaigns to ensure liberal-democratic victories in France and Italy during the 1948 elections. In France, the CIA promoted a schism in the labor movement to undermine the influence of communist labor leaders, and in Italy the Truman administration spent several million dollars on bribes, local election campaigns, and anticommunist propaganda. Vast sums of Marshall Plan aid were also diverted to covert action programs and propaganda in Western Europe. The Economic Cooperation Administration, which administered the Marshall Plan, "swamped European media" with pro-American news, and it participated in covert operations to influence the political orientation of European labor unions.[39]

The apparent success of the operations in France and Italy encouraged additional ventures. Active intervention in the internal affairs of sovereign states became a core feature of U.S. foreign policy during the Cold War. George F. Kennan, known for his role in developing the policy of containment, was instrumental in further developing U.S. covert psychological warfare capabilities. "We must accept propaganda as a major weapon of policy, tactical as well as strategic, and begin to conduct it on modern and realist lines," he wrote. "No important step should be decided without a simultaneous determination of the nature of its propagandistic development." In early 1948, Kennan, who was then serving as the head of the State Department's Policy Planning Staff, developed a plan for "organized political warfare" against communism. "We cannot afford to leave un-mobilized our resources for covert political warfare," Kennan declared. "What is proposed here is an operation in the traditional American form: organized public support of resistance to tyranny in foreign countries. Throughout our history, private American citizens have banded together to champion the cause of freedom for people suffering under oppression. . . . Our proposal is that this tradition be revived specifically to further American national interests in the present crisis." He urged the NSC to approve a comprehensive assault on communist power behind the Iron Curtain. According to his plan, private liberation committees, composed of political refugees from the Soviet bloc, would be organized by U.S. intelligence to carry out propaganda and agitation against communist regimes and to provide a source of inspiration for popular resistance to communist rule. He also proposed active intervention in "threatened countries of the free world," by extending covert U.S. assistance to indigenous anticommunist elements abroad and by engaging in "preventive direct action in free countries."[40]

Kennan's plan led to the development of NSC 10/2, the original charter for American covert operations that fundamentally altered postwar thinking about psychological warfare. It authorized a virtually limitless array of covert operations, including "propaganda, economic warfare; preventive direct action, including sabotage, anti-sabotage, demolition and evacuation measures; subversion against hostile states, including assistance to underground resistance movements, guerrillas and refugee liberation groups; and support of indigenous anti-communist elements in threatened countries of the free world." According to NSC 10/2, covert actions were to be constrained only by the limits of plausible deniability, the capacity to conceal credibly U.S. sponsorship. President Truman approved NSC 10/2 in June 1948 and authorized the creation of the Office of Policy Coordination (OPC), a euphemistically titled organization attached to the CIA and authorized to engage in all sorts of clandestine operations. Over the next three years, the OPC grew exponentially. In 1949, it employed only 302 persons, but by 1952 the number had grown to 2,812, with an additional 3,142 operatives under contract. The OPC's budget likewise exploded from $4.7 million in 1949 to $82 million by 1952.[41]

In the context of the Cold War, psychological warfare was closely associated with the policy of rollback, or the employment of nonmilitary means to force the retraction of Soviet power and the "liberation" of Eastern Europe. The National Security Council codified its commitment to liberation by approving the landmark document NSC 20/4 in November 1948. Authorizing a wide range of actions to destabilize the Soviet bloc, it declared the aim of U.S. policy to be the "gradual retraction of undue Russian power and influence from the present perimeter areas around the traditional Russian boundaries and the emergence of the satellite countries as entities independent of the USSR." Although publicly the United States was committed to a policy of containment—portrayed as a defensive reaction to Soviet expansion—a parallel policy of liberation was being pursued covertly through psychological warfare. As one psychological strategist wrote, "Though it seems to be widely believed . . . that our policy is simply 'containment,' it may be doubted whether it is or has been, except for public consumption." Thus, contrary to conventional wisdom, the Truman administration secretly inaugurated a robust policy of rollback long before it became an issue in the 1952 presidential campaign. As recent scholarship suggests, the administration's efforts to liberate areas under Moscow's control indicate that American foreign policy in the early Cold War was not as defensive and fundamentally nonaggressive as the term "containment" implies or as earlier historiography suggested.[42]

Under the authorization of NSC 20/4, the Office of Policy Coordination pursued an aggressive program of clandestine warfare against the Soviet bloc. Frank G. Wisner, the office's director and a veteran of the OSS, recruited many of his former colleagues into the organization, applying their wartime expertise in organizing underground resistance groups, sabotage, and partisan warfare to the Cold War. The OPC embarked on an unsuccessful effort to detach Albania from the Kremlin's grip, launched leaflet-dropping operations via enormous unmanned hot air balloons, and encouraged defections from communist regimes in Eastern Europe. The OPC also organized guerrilla units, sabotage forces, and other subversive operations to support resistance movements behind the Iron Curtain. Émigrés from Russia, Ukraine, Romania, Hungary, Albania, Czechoslovakia, Poland, and elsewhere were dropped by unmarked U.S. war planes at scattered locations across Eastern Europe to engage in various kinds of paramilitary operations. These operations were provocative but ineffective. They nearly always ended in disaster. In case after case, the American-sponsored émigrés were discovered by the authorities, arrested, and sentenced to decades in Siberian labor camps; in many cases, they were summarily executed.[43]

A more influential form of anti-Soviet propaganda came in the form of Radio Free Europe and Radio Liberty (RFE/RL), which broadcast "a muscular brand of political warfare" to Eastern Europe and Russia, respectively. Covertly established by the Central Intelligence Agency, the radios were staffed by émigrés and exiled political leaders from the Soviet bloc. The CIA maintained control over their broadcasts through the National Committee for a Free Europe, an ostensibly private organization established by the agency to support émigré anticommunist activities and run by the longtime propaganda advocate and World War II psychological operations specialist C. D. Jackson. The goal of RFE/RL, as one of its policy manuals stated, was "to contribute to the liberation of the nations imprisoned by the Iron Curtain by sustaining their morale and stimulating in them a spirit of noncooperation with the Soviet-dominated regimes."[44]

The activities of the Free Europe Committee extended beyond mere radio propaganda in Eastern Europe. The RFE/RL operation also provided cover for a sophisticated propaganda campaign designed to drum up support for the Cold War at home. The CIA, legally barred from operating in the United States, orchestrated a massive effort to stimulate Cold War morale known as the Crusade for Freedom. The Crusade for Freedom purported to be a fund-raising drive to raise money for the RFE/RL broadcasts—to provide cover for the agency's radio stations by creating the appearance that they were funded by private contributions. But the crusade was also a domestic propaganda campaign. It sought to whip up anticommunist fervor and stir American patriotism by portraying the Cold War as a fight for the freedom of the "enslaved peoples" of Eastern Europe. The Advertising Council financed a multimillion-dollar advertising campaign asking Americans to "help truth fight Communism" by donating "freedom dollars" to the crusade. Appeals from celebrities were heard on the radio; posters were plastered on buses and subways; and parades were organized stressing the theme of freedom for the satellite peoples. Christopher Simpson has suggested that the CIA's contributions to the Crusade for Freedom "made the CIA the largest single political advertiser on the American scene during the early 1950s." The crusade helped transform a distant geopolitical battle into a high-minded struggle for freedom, and it helped entrench the goal of "liberation" firmly within American domestic political culture, thus making it exceedingly difficult for any American political leader to pursue a settlement with the USSR that acknowledged a Soviet sphere of influence in Eastern Europe.[45]

Momentum for a broad psychological warfare effort picked up in early 1950, after the victory of the communists in the Chinese revolution and the Soviet detonation of an atomic bomb. Reacting to these developments, the Truman administration commenced a reappraisal of its national security policies that culminated in the historic policy paper NSC-68, calling for intensified programs to build up the political, economic, and military strength of the "free world."[46] Paul Nitze, who succeeded Kennan as director of the Policy Planning Staff and the principal author of the document, urged an exponential increase in defense spending in order to maintain American military superiority. Nitze prescribed such a buildup not because he feared an imminent military clash with the USSR—he admitted that the likelihood "of a total war started deliberately by the Soviets is a tertiary risk"—but because of the psychological implications of preponderant power, which would affect the ability of the United States to circumscribe Soviet behavior through diplomatic maneuver, as well as the U.S. penchant for risk-taking through covert action and psychological warfare. The military buildup, as a government task force declared, was "a shield behind

A float advertises the Crusade for Freedom in a New York City parade. The "freedom bell" in the center of the float was used to kick off the Crusade for Freedom in Berlin. It subsequently traveled from city to city to symbolize America's commitment to fight for the freedom of Eastern Europe. A placard on the float urges Americans to "Help Lift the Iron Curtain *Everywhere!*"

which we must deploy all of our nonmilitary resources in the campaign to roll back the power of the USSR and to frustrate the Kremlin design."

In line with such a strategy, NSC-68 called for a massive expansion of American programs for overt and covert psychological warfare against the Kremlin and within the free world. It urged "the development of programs designed to build and maintain confidence among other peoples in our strength and resolution, and to wage overt psychological warfare calculated to encourage mass defections from Soviet allegiance and to frustrate the Kremlin's design in other ways." It also called for the "intensification . . . in the fields of economic warfare and political and psychological warfare with a view to fomenting and supporting unrest and revolt in selected strategic satellite countries." NSC-68 cast the Cold War as a "total" struggle demanding sacrifices and contributions from all Americans and further advocated intense campaigns of public persuasion at home "to strengthen the moral fiber of the people."[47]

Plans for a "total information effort abroad" and a "psychological 'scare campaign'" at home were being developed simultaneously by the State Department. On April 20, 1950, four days after Nitze submitted his report to the NSC, Truman initiated a public relations blitz to pitch an expanded informational program to Congress and the American people. In a speech to the American Society of Newspaper Editors, the president called for a "great campaign of truth . . . a sustained, intensified program to promote the cause of freedom against the propaganda of slavery."[48] Truman's speech, the outbreak of the Korean War days later, and the subsequent approval of NSC-68 prepared the way for a global propaganda crusade waged by overt and covert means. The State Department's budget for information activities jumped from around $20 million in 1948 to $115 million in 1952. By 1951, American propaganda programs were targeting 93 countries, producing some 60 million booklets and leaflets, and broadcasting VOA programming in 45 languages. Meanwhile, the covert psychological operations proposed in the wake of NSC-68 were of such a scale that they threatened to overwhelm the administrative capability of the CIA. All told, the focus of this campaign was not so much the truth as much as it was strident propaganda vilifying the Soviet Union in the strongest terms. In the Campaign of Truth, one VOA official recalled, "anything more subtle than a bludgeon was considered 'soft on communism.'"[49]

A sprawling apparatus for influencing foreign attitudes was in place by the late 1940s. Important information programs were being conducted by the State Department, the CIA, the military services, the Economic Cooperation Administration, and the Technical Cooperation Administration. It was organizational chaos, with overlapping and contradictory jurisdictions, duplication of effort, lack of coordination, and incessant interdepartmental squabbling. It would not be too much of an exaggeration to state that these organizations spent as much time battling each other as they did fighting communism. In an attempt to overcome the bureaucratic infighting, Truman created a Psychological Strategy Board (PSB) in April 1951 to produce unified planning for American psychological operations. This autonomous, interdepartmental body was composed of representatives from the Departments of State and Defense, the CIA, and the Joint Chiefs of Staff. The PSB intended to act as a coordinating body for all nonmilitary Cold War activities, including covert operations. It was an ambitious project. The PSB took a global perspective in planning U.S.

psychological strategy, emphasizing in particular "offensive action against Eastern Europe" and the "rollback of Soviet power." Working through the PSB and CIA, the Truman administration orchestrated a vast and provocative anticommunist psychological warfare program to break up the Soviet bloc by encouraging revolution behind the Iron Curtain. It also developed "psychological strategy" plans for dozens of countries in Western Europe, Asia, and the Middle East that detailed military, economic, political, and propaganda measures designed for their impact on international opinion. It also played a role in orchestrating psychological warfare for the Korean War and in devising propaganda campaigns for exploitation at the United Nations.[50]

The PSB was staffed with energetic psychological warriors who spoke hyperbolically and incredulously of "winning" the Cold War through psychological warfare. The board operated on the assumption that policy and propaganda were inextricably intertwined, because any action by the American government possessed "psychological" significance. Gordon Gray, the board's first director, had a sweeping vision of psychological strategy. He described it as "a cover name to describe those activities of the US in peace and in war through which all elements of national power are systematically brought to bear on other nations for the attainment of US foreign policy objectives. . . . The purpose of creating the PSB was not to act as planner or coordinator with respect to any one major effect, but to act as planner and coordinator with respect to all." Gray's assistant, Robert Cutler (who would later become Eisenhower's national security advisor), likewise perceived the PSB mandate broadly. The PSB was "a nerve-center for initiating, receiving, coordinating, and evaluating psychological impulses. This concept is more that of a command post than an information center." The logic that the PSB was a command center for devising policies for their psychological impact stemmed from public relations theory and from wartime experience. The PSB took from World War II the lesson that psychological warfare and covert action were most effective when they were coordinated with each other and integrated into an overarching strategy. The best propaganda, as Edward Bernays had shown, was propaganda of the deed, and the best way to manipulate the media was through orchestrated action, not through rhetorical ploys that merely put the best spin on policy. The "real purpose" of psychological warfare, a consultant to the PSB asserted, "was to make events happen."[51]

The conviction of PSB members that all decisions and actions of the government possessed psychological significance set off alarm bells in all of the U.S. bureaucracies involved in formulating U.S. foreign policy. Officials involved in psychological warfare were not, generally speaking, foreign policy professionals; rather, they had different backgrounds—in advertising, public relations, and covert operations. They represented a kind of expertise that career diplomats and foreign policy traditionalists neither understood nor appreciated. Efforts by psychological warriors to inject propaganda concerns into policy making encountered intractable resistance. State Department officials had different ideas about the "real" purpose of the PSB: it should monitor propaganda content, not devise or even influence policy. Paul Nitze's comment to Gordon Gray speaks volumes about his department's attitude toward the board's "psychological strategy" plans: "Look, you just forget about policy, that's not your business; we'll make the policy and then you can put it on your damn radios." At every turn the PSB was stymied by State Department opposition. Ultimately, as Walter Hixson noted, "the PSB produced reams of studies, but failed to marshal the national security bureaucracy behind a coordinated effort." Operating in a bureaucratic limbo, without the clear support of President Truman, the PSB died from bureaucratic strangulation by the time of the 1952 election.[52]

The PSB's concept of psychological strategy represented a significant evolution in U.S. attitudes toward propaganda since World War I. In an age of virtually instantaneous communications that exposed government actions to endless scrutiny, policy could not be separated from presentation. Winning hearts and minds would depend as much on what the United States did as on what it said. Such thinking lingered behind key policies of the early Cold War like the Marshall Plan, but the notion of a national security bureaucracy organized for political warfare was not widely accepted within the Truman administration. The belief that the Cold War was a war of ideas was widely shared, but there was little coordination and no central leadership. Truman himself evinced little interest in or understanding of psychological warfare, preferring to leave such matters to the bureaucracy. The next American president, Dwight D. Eisenhower, would make psychological strategy a key element of his election campaign and a priority of his administration. Under his leadership, many PSB plans would be salvaged, updated, and implemented.

Chapter 2

A New Type of Cold War
Eisenhower and the Challenge of Coexistence

The world struggle is shifting more than ever from the arena of power to the
arena of ideas and international persuasion.

— Nelson Rockefeller to Eisenhower, December 2, 1955

In October 1952, Republican presidential candidate General Dwight D.
Eisenhower delivered a campaign speech in San Francisco on the subject
of American foreign policy. It was in many ways a typical Cold War moment—one presidential candidate lambasting another for being soft on
defense. But it was also an unusual speech, for the subject of this particular address was psychological warfare. "Don't be afraid of that term just
because it's a five-dollar, five syllable word," Eisenhower advised his audience. "Psychological warfare is the struggle for the minds and wills of
men." Blasting the Truman administration for neglecting this important
dimension of the Cold War struggle, Eisenhower went on to promise that,
if elected, he would make psychological warfare a central focus of U.S.
national security strategy.[1]

Eisenhower's speech reflected the interest in propaganda, brainwashing, and other forms of mind control that had developed during the preceding decades. But it was still a remarkable moment: here was a presidential candidate, making *psychological warfare* an issue in a political campaign.
The speech spoke to widespread public fears that the United States was
losing the war of words against the communists, fears exacerbated by apparent communist propaganda successes in Korea. At the time, the Chinese and Soviets were conducting an apparently well-coordinated campaign to convince international observers that U.S. forces were conducting
bacteriological warfare, and tales of the brainwashing of American POWs
were circulating through the media. The Truman administration appeared
to have blundered into the war by conspicuously omitting Korea from the
list of countries protected by America's "defensive perimeter." Such short-
sightedness, Eisenhower declared, was "psychological strategy in reverse."[2]

Such heated rhetoric aside, Eisenhower's speech was more than political posturing: it reflected his deeply held conviction that psychological forces were critical elements of American leadership in the world. As the general who led the Allies to victory against the Nazi war machine, Eisenhower had come to believe that psychological warfare was a vital component of modern total war. He had concluded as well that the battle for hearts and minds was one of the most critical dimensions of the Cold War struggle. In the October speech, as in his private writings and diaries, Eisenhower articulated a notably coherent and sophisticated understanding of psychological warfare. To Eisenhower, psychological warfare extended beyond the official propaganda agencies of the American government to embrace any word or deed that affected the hearts and minds of the world's peoples. "As a nation," he admonished in his October speech, "everything we say, everything we do, and everything we fail to say or do, will have its impact in other lands." To win the struggle for hearts and minds, he emphasized, the United States needed to marshal the energies of a wide range of activities relating to American foreign relations, including diplomacy, economic assistance, trade, person-to-person contacts, and ideas. Eisenhower promised a coherent national security strategy that accorded paramount significance to psychological considerations. His personal commitment to psychological warfare ensured that psychological strategy—the shaping of policies and programs for their impact on public attitudes at home and abroad—would exert a greater influence on his foreign policies than on those of any other presidential administration.[3]

International developments coinciding with Eisenhower's presidency further accentuated the ideological, political, and symbolic dimensions of the Cold War. By mid-1953, both superpowers had developed workable hydrogen bombs, weapons infinitely more destructive than atomic bombs. Thermonuclear weapons made a third world war all but unthinkable while at the same time elevating symbolic displays of national power. The accelerating pace of decolonization in the third world raised the importance of the nonmilitary dimensions of the Cold War, as both superpowers competed for the political allegiance and vital economic resources of countries in Asia, Africa, and Latin America. Changes in Soviet foreign policy after the death of Joseph Stalin in March 1953 further channeled the Cold War competition into political, cultural, and ideological arenas. The post-Stalin leadership took steps to repudiate many of the most egregious elements of Stalin's rule and foreign policy, announcing a new emphasis on "peaceful coexistence" and moving to expand Soviet contacts with the outside world.

As the fear generated by Stalin's menacing tactics dissipated, and as Stalin's successors cultivated a reformist image, the excesses of American anticommunism appeared at times to be the greater threats to international peace and stability. This challenged Americans to demonstrate the necessity of their anticommunist foreign policy, to prove their peaceful intentions, and to persuade others of the ideological and cultural superiority of the "American way of life." Eisenhower and many of his advisors viewed the challenge of peaceful coexistence as representing a "new type of Cold War," a war of persuasion that would be won or lost on the plain of public opinion. The world struggle appeared to be shifting from the arena of military power to the arena of ideas. Psychological factors seemed more important than ever to winning this total contest for hearts and minds.

Eisenhower and Psychological Warfare

Eisenhower developed his belief in the power of propaganda during World War II. In his own words, he had "learned the importance of truth as a weapon in the midst of battle." Within the U.S. Army, Eisenhower was one of the earliest and most consistent supporters of psychological warfare. Most American military officers derided psychological warriors as "feather merchants," viewing leaflet bombings and radio broadcasts as wasteful drains on manpower and resources. To battle-hardened soldiers, guns and planes and tanks held the keys to victory, not psychological hocus pocus. General Eisenhower was different. Unlike other officers who dismissed psychological warfare as "just another 'crackpot' conception," he believed that it had "tremendous importance" as an instrument of total war. According to C. D. Jackson, who planned psychological warfare for the North African and European campaigns, Eisenhower's support for psychological warfare was crucial to its success during the war: "Too much credit cannot be given to Eisenhower and his staff for having overcome the original hurdle of soldierly distrust, and for having accepted [psychological warfare] as something more than a newfangled nuisance."[4]

Psychological operations figured prominently in the North African campaign in 1942 and again in the 1944 invasion of Normandy. Eisenhower personally supervised the psychological components of these campaigns. He integrated psychological warfare planning into his overall war plans, coordinating propaganda messages with military operations, cover and deception requirements, and political circumstances. All told, some 8 billion leaflets were dropped in the Mediterranean and European theaters

Eisenhower visits the new Voice of America headquarters in New York City.
Source: Eisenhower Library.

under Eisenhower's command—"enough to have given every man, woman
and child on earth four leaflets." By war's end, Eisenhower was convinced
that psychological warfare had contributed measurably to Allied victory.
He wrote in 1945:

> In this war, which was Total in every sense of the word, we have seen
> many great changes in military science. It seems to me that not the
> least of these was the development of psychological warfare as a spe-
> cial and effective weapon. . . . I am convinced that the expenditure
> of men and money in wielding the spoken and written word was an
> important contributing factor in undermining the enemy's will to
> resist and supporting the fighting morale of our potential Allies in
> the Occupied Countries. . . . Without doubt, psychological warfare
> has proved its right to a place of dignity in our military arsenal.

The critical importance of psychological factors to the war effort was a
theme Eisenhower often returned to. "Anybody who was a part of World
War II does not need to have re-emphasized to him the importance of the
psychological factor," he told the National War College in 1962.[5]

Eisenhower's belief in the power of propaganda stemmed from his deep understanding of the importance of psychological forces to modern war. He was constantly concerned with morale—of his soldiers, and of civilians back home. "Americans either will not or cannot fight at maximum efficiency unless they understand the why and wherefore of their orders," he wrote, instructing his subordinate commanders to indoctrinate their troops on the cause of democracy for which they were fighting. He repeatedly stated that "an Army fights just as hard as the pressure of public opinion behind it" and that "wars are conducted by public opinion." To help maintain morale at home, Eisenhower worked earnestly at public relations, effectively manipulating the press to cover the war effort positively. Harry Butcher, one of Eisenhower's aides, recorded in his diary that Ike was "the keenest with the press I've ever seen, and I have met a lot of them, many of whom are phonies." Eisenhower addressed war correspondents as if they were his men—"quasi members of my staff"—regularly telling them that their work was a crucial element of the struggle. He encouraged them to write stories and to take positions that would help the cause, and he used the threat of censorship to make sure they followed the line.[6]

As the supreme commander of the Allied forces, Eisenhower was particularly attuned to the need for maintaining unity within the alliance. He consciously used the press to promote a positive Allied relationship. According to Butcher, Eisenhower worked tirelessly at this: "He carries the idea [of Allied unity] into all of his press conferences, communiqués, and into every contact with the people." Highly sensitive to Nazi propaganda that sought to sow divisions between the Americans and the British, he instructed his commanders to prevent incidents which might provide grist for the enemy propaganda mill. He flatly prohibited war correspondents from writing stories that jeopardized Allied unity. All told, Eisenhower's wartime actions revealed an understanding of psychological warfare that went beyond mere combat propaganda directed at enemy troops: it involved a systematic effort to encourage Allied solidarity, to boost fighting morale, and to maintain the resolve of the home front. It was a far-reaching conception of psychological strategy and strategic communications that he would carry with him to the White House.[7]

After the war ended, while serving as Army chief of staff, Eisenhower worked to preserve and enhance military expertise in psychological operations. In 1947, he sent a memorandum to Secretary of Defense Forrestal urging him to "give continued impetus to the organization and realistic functioning of this important activity." He pushed the Joint Chiefs of

Staff to carefully study the world war experience in psychological warfare and urged the military establishment to "keep alive the arts of psychological warfare." As Cold War tensions escalated, General Eisenhower also became convinced of the importance of peacetime propaganda. A self-described "ardent supporter" of the Smith-Mundt Act, Eisenhower twice testified on behalf of the legislation to establish a U.S. information service abroad. His testimony presaged the multifaceted approach to psychological strategy he would pursue as president. "There is no single field of this thing that should be neglected," he stressed. "We should disseminate the truth about the United States to the peoples of the world by every means at our disposal." He argued then, as he would many times thereafter, that the battle for hearts and minds would be a long-term enterprise, involving a wide range of activities to enhance U.S. influence and spread American ideals of democracy and free enterprise. "It is not merely the beaming out of facts. I would encourage the exchange of students, of scientists, of doctors, of instructors, of even theologians; anything you could think of that would tend to carry back into these various countries an understanding of what we are doing and just how we live . . . I believe it should go in the fields of art, science, and everything." It should be a total effort, he argued.[8]

During Eisenhower's brief retirement from active military service, when he served as president of Columbia University, he continued to support psychological operations against the Soviet bloc. He maintained contact with C. D. Jackson, Walter Bedell Smith, Allen Dulles, Frank Wisner, and others active in developing covert psychological warfare initiatives for the Truman administration. In October 1950 he helped launch the Crusade for Freedom, the fund-raising drive to benefit Radio Free Europe and the Free Europe Committee. In December, President Truman asked Eisenhower to serve as commander of NATO. Truman's request gave Eisenhower the president's attention. Eisenhower capitalized on the moment to pen a letter to Truman urging him to devote more resources and energy to the propaganda war in Europe: "Because the opponent is militant in the political world as well as in the field of force, he develops a many-sided and complex system of attack, using force, deceit, propaganda, and subversion. It is against this kind of relentless world-wide attack that the free world must defend itself." Indicating his preference for covert forms of psychological warfare, Eisenhower advised that "both subversion and propaganda weapons should be developed, for use throughout the world, to a high degree of efficiency. I think that 200 million yearly could profitably be spent on propaganda and OSS." Scribbling in the margins, he signaled that the

U.S. government should hide its involvement in overseas propaganda, noting that there should be "no govt connection for most of this!"[9]

Eisenhower's appointment to NATO added his considerable personal prestige to the task of European rearmament. It also made him a target of vicious Soviet propaganda attacks, an experience he took personally. "Soviet propaganda is relentless," he complained in his memoirs, recalling his tour of NATO capitals in January 1951. "The assaults on me were bitter," Eisenhower wrote. "They charged that I was a cruel, military type, coming to Europe to start an aggressive war no matter what the cost in European lives." Such experiences reinforced his growing conviction that the United States needed to find a way to effectively counter communist propaganda. In his diary, Eisenhower expressed his belief that all NATO countries should develop PR campaigns to sell the alliance to public opinion. He also advocated an effort to alert the European working class to the emptiness of communist promises. "It could be fatal if the workmen of Italy, France, and Germany came to believe that workmen in Poland, Czechoslovakia, Hungary, and so on were as well or even better off in their daily lives as they, themselves," he wrote to Averell Harriman, adding, "there should be watchful regard for every kind of advance that will convince populations of this truth."[10]

Acutely sensitive to the divisions that separated European allies from each other, he agonized that the "ancient fears and prejudices" of the Europeans made them "easy prey" of Soviet propaganda. Communist success in Italian elections, for example, testified to the need for better information programs: "People are misled and misinformed. This fact is obvious after the most cursory examination of the problem, and should indicate to all concerned the very urgent necessity for skillful propaganda on the part of the Allies, of the highest quality and of adequate volume." Writing to Walter Bedell Smith, then director of Central Intelligence, Eisenhower stated his belief that "the CIA should have its men here and there working quietly in foreign countries fostering political movements and propaganda." (Smith, who was better informed about CIA activities in Western Europe, responded, "What the hell do you think I've been doing for the last year?")[11]

By the time Eisenhower declared his candidacy in the 1952 presidential election, he had come to believe strongly in the importance of political warfare to the American Cold War effort. Eisenhower made the promotion of psychological warfare a prominent feature of his campaign. In a series of speeches from August to October, Eisenhower repeatedly called for the "skillful and constant use of the power of truth" to combat the "hideous disease" of communism. Emphasizing that "the struggle between communism and freedom is a struggle of ideas," he argued that the United States "must fully develop . . . every psychological weapon that is available to us."[12]

Eisenhower's speech in San Francisco on October 8 was the most comprehensive of these campaign pronouncements. In the speech, he articulated the far-reaching conception of psychological warfare that he later integrated into his administration. He lectured his audience on the real meaning of psychological warfare: "Many people think 'psychological warfare' means just the use of propaganda, like the controversial 'Voice of America.' Certainly the use of propaganda, of the written and spoken word, of every means known to transmit ideas, is an essential part of winning other people to your side. But propaganda is not the most important part in this struggle." Real psychological warfare, he explained, extended beyond government propaganda agencies to include such factors as "diplomacy, the spreading of ideas through every medium of communication, mutual economic assistance, trade and barter, friendly contacts through travel and correspondence and sports." What was needed was a "psychological effort put forth on a national scale"—a comprehensive national security strategy that integrated psychological considerations with other elements of U.S. foreign policy. "We are not going to win the struggle for men's minds merely by tripling Congressional appropriations for a super-loud Voice of America," he explained. "Rather it will be the planned and effective use of every means of appeal to men and women everywhere." This required sound government policies designed to attract the allegiance of the world's peoples. Eisenhower argued that "every significant act of government" should be so planned and coordinated to produce "maximum effect" in the Cold War arena, and he pledged to make this form of psychological strategy the focus of his administration.[13]

The Great Equation

Eisenhower's campaign statements were more than mere rhetoric; they reflected his sincere commitment to psychological warfare. Presidential subordinates believed that Eisenhower possessed "a natural understanding of overseas opinion and world opinion, and the struggle that was going on . . . for people's minds." Eisenhower lamented the "propaganda disadvantage" of the United States and complained about the tendency of western powers to "sit quietly" while the enemy enjoys a "propaganda feast . . . at our expense." According to Eisenhower, "The Russians spent about $2 billion a year on their propaganda and . . . it was ridiculous for us to spend only a small amount." He was shocked and disgusted when told of twenty-one Korean War POWs who voluntarily chose communism. He complained in his diary about the "need to educate others in the beauty of liberty" and about "tyrants who thrive in the ignorance of others and employ curtains of iron or of oratory to deepen and prolong that ignorance."[14]

Eisenhower's diaries, private letters, and official papers reveal his abiding faith in psychological warfare. In part this faith grew from his personal experience with Nazi propaganda in World War II, but it also stemmed from his personal philosophy of leadership and belief in the powers of persuasion. He came to the White House with a formula for world leadership. He called it the Great Equation: "Spiritual force, multiplied by economic force, multiplied by military force, is roughly equal to security." The spiritual factor, never well defined and usually forgotten in the literature on the Eisenhower presidency, suggested that the United States could accomplish its foreign policy goals through persuasion and moral leadership. It was bound up with Eisenhower's uniquely expansive concept of psychological warfare as the mobilization of public opinion at home and abroad by virtually any means. A "basic truth," he wrote, is that "humans are spiritual beings; they respond to sentiment and emotions as well as to statistics and logic." As such, "the minds of all men are susceptible to outside influences." Eisenhower looked upon these psychological forces much as a soldier would look upon the morale of his troops. In his diary, he noted that civilian leaders look upon public opinion as something to be followed, while military leaders know that opinions can be changed: "Civilian leaders talk about the state of morale in a given country as if it were some kind of uncontrollable event or phenomenon, like a thunderstorm or a cold winter. The soldier leader looks on morale as one of the great factors (or greatest) in all his problems, but also as one about which he can and must do something."[15]

For him it was a question of leadership, of education, and of conveying a "basic understanding of the facts." Those that subscribed to communist ideology were "misguided," in Eisenhower's view, for not recognizing that communism promised only "slavery" and domination by the Kremlin. Speaking of the developing world, Eisenhower wrote in his diary, "We must show the wickedness of purpose in the communist promises and convince dependent peoples that their only hope of maintaining independence, once attained, is through cooperation with the free world." Eisenhower had long emphasized, in private and in public, that American power relied not only on economic and military factors, but also on the nation's moral strength. Eisenhower never defined precisely what he meant by this last factor, but it connoted a belief in the superiority of American values and institutions, a belief that the greater appeal of freedom would win in the long run. "The greatest weapon that freedom has against the Communist dictatorship," Eisenhower wrote, "is its ultimate appeal to the soul and spirit of man. . . . The appeal of our own system will, in the long run, triumph over the desperate doctrines of Communism."[16]

Upon winning the presidential election, Eisenhower moved swiftly to implement his vision of psychological strategy. He appointed C. D. Jackson to a post in the White House, designating him as his special assistant for psychological warfare and placing him in control of the Psychological Strategy Board. Eisenhower staffed his administration with other experienced psychological warriors, including Allen Dulles, Gordon Gray, Robert Cutler, Abbott Washburn, Lloyd Berkener, Nelson Rockefeller, and Walter Bedell Smith. He also appointed a high-level committee to make recommendations on how to strengthen and centralize U.S. psychological warfare activities. Within days of his inauguration, he established the President's Committee on International Information Activities, directing it to investigate all aspects of the American Cold War effort and mandating an official report by the summer. Before the committee could release its findings, however, events intervened that made their work even more important. The death of Joseph Stalin in March 1953 seemed to offer Eisenhower his first opportunity for all-out psychological warfare.

The Death of Stalin

When Eisenhower moved into the White House in January 1953, every indication suggested that Cold War tensions were going to get worse. An atmosphere of hysteria and suspicion gripped life in the world's two remaining great powers. Senator Joseph McCarthy and his supporters searched for communists in the State Department, Hollywood, and beyond, while an aging Joseph Stalin initiated yet another paranoid campaign of "vigilance" against internal enemies, an anti-Semitic purge resting on allegations of a "doctor's plot" against leading Communist Party officials. The United States had recently tested its first thermonuclear weapon, and the new secretary of state, John Foster Dulles, was promising a new "policy of boldness" and a deterrence strategy resting on "massive retaliation." Soviet officials interpreted Eisenhower's election as measurably increasing the danger of war, their fears exacerbated by Eisenhower's bellicose pledge to "win the Cold War" in his first State of the Union address. The Cold War had spread to Asia, where the war in Korea had coagulated into an agonizing stalemate and communist forces were gaining ground in Indochina. U.S.-Soviet relations, already at an all-time low, were further poisoned by the Soviet government's "hate-America" propaganda campaign, a propaganda blitz without "equal in viciousness, shamelessness, mendacity, and intensity." To some observers, the campaign indicated that the Soviet government was preparing its citizens for war. So chilly were U.S.-Soviet relations in the winter of 1952–1953 that the United States did not even have an ambassador in Moscow. Ambassador George Kennan had been declared persona non grata by the Soviet government and the appointment of his successor, Charles Bohlen, was delayed for weeks by McCarthyite investigations into his supposed "disloyalty."[17]

Yet before the first thaw of spring, as British Prime Minister Winston Churchill noted, a "new breeze [was] blowing on the tormented world." On March 6, 1953, Soviet newspapers announced the death of Joseph Stalin. Within days, Stalin's heir apparent, Georgii Malenkov, made an unexpected appeal for peace. Speaking before the Supreme Soviet of the USSR on March 16, he announced his government's willingness to reach an accord on outstanding East-West issues. Over the course of the next five months, the Soviet Union relinquished its territorial claims on Turkey, reestablished diplomatic relations with Israel, helped bring the Korean War to a close, worked to improve relations with Yugoslavia and Greece, and, in general, emphasized "peaceful coexistence" with the West.[18]

These and other changes wrought by Stalin's death raised the stakes in the psychological, political, and cultural fields of the Cold War. The softening of the Soviet image engendered by its more constructive approach to diplomacy enhanced the prestige of the Soviet Union, while the apparent economic and scientific progress enjoyed by the USSR during the post-Stalin years made the Soviet model more attractive to leaders in the newly emerging states in the developing world. Stalin's death, by freeing communism from the image of Stalinist repression, increased the appeal of Marxist thought outside the communist bloc. Not only in places experiencing the foment of postcolonial revolution, but also in Western Europe, Japan, and the United States, communist ideas won increased acceptance. In the cultural realm, too, the Soviets appeared to be gaining ground in the eyes of the world: cultural tours suggested that life under communism produced wonderful art, music, ballet, and theater. The "thaw" and liberalizing trends within the Soviet bloc dulled the image of the Iron Curtain. To American observers, all these developments were a fertile field for Soviet propagandists and portended a new phase in the Cold War, potentially more dangerous than the first—when Stalin could be so easily demonized as the evil figure in the Kremlin "pulling the strings behind every evil happening in the world."[19]

Scholarly analysis of the period after the death of Stalin in March 1953 has been overwhelmingly preoccupied with questions of missed opportunities. Did the Soviet "peace offensive" signal a genuine interest in reaching an accommodation with the West? Did the United States miss an opportunity for a resolution of the German question? For arms control? For a general relaxation of tensions, détente, or even an end to the Cold War itself? These are important and provocative questions, but they are not ones that President Eisenhower or his advisers considered carefully at the time. No senior officials within the administration perceived the Stalin succession as an opportunity for a relaxation of tensions or for resolving, through negotiation, outstanding Cold War differences. Instead, Eisenhower and his advisors perceived Soviet peace protestations as menacing political warfare tactics that threatened to weaken western resolve at a critical moment. They feared that Soviet conciliatory measures would take the wind out of the sails of the American Cold War effort. Should the U.S. and its allies let down their guard for short-run concessions, the West would pay the price over the long haul as American allies and domestic audiences, eager for a relaxation of tensions, pressed hard for a reduction in defense expenditures while the Soviets continued to expand their military forces. Unsettled by the changes in Soviet diplomacy, Eisenhower launched a U.S. peace counteroffensive to expose the hollowness of Soviet overtures, bolster free world morale, and rejuvenate American leadership.[20]

A Dramatic Psychological Move

Stalin's death appeared to be the moment U.S. psychological warriors had been looking for. If ever there were an opportunity for rollback, this was it. The Truman administration had tried all manner of covert psychological warfare activities to incite an uprising behind the Iron Curtain, but Stalin's grip was too strong. Now Moscow was vulnerable. Undoubtedly a power struggle was taking place behind the Kremlin's walls, and the captive nations of

Eastern Europe might sense weakness and revolt. C. D. Jackson was ec-
static. Stalin's death presented the United States with "the greatest opportunity . . . in many years to seize the initiative," he wrote to NSC executive
secretary Robert Cutler. "Shouldn't we do everything possible to overload
the enemy at the precise moment when he is least capable of bearing even
his normal load," he asked, predicting that well-planned political warfare
might lead to victory in the Cold War and the "disintegration of the Soviet
empire."[21]

Jackson immediately assembled his colleagues on the Psychological
Strategy Board to draft a plan of attack. A few months earlier, the Truman administration's PSB had prepared a psychological strategy plan to
prepare for Stalin's death, but Jackson found the plan worthless. It was
not nearly bold enough, he concluded. A big idea was needed. But what?
Should the United States come out and encourage a revolt? Soviet expert
Charles Bohlen, temporarily serving as State Department counselor, advised against such an approach. A "direct frontal political or psychological
assault" on the Kremlin would merely help the new leadership consolidate
its position. Instead, Bohlen proposed, the United States should offer the
new Soviet leadership an opportunity to discuss outstanding issues in a
conference of foreign ministers. The United States would use this peace
proposal for political warfare, advancing new policy initiatives to encourage
internal disputes and exacerbate personal rivalries within the USSR during
the succession. When questioned by Jackson if the United States "could be
sure of gaining the advantage from a meeting," Bohlen pointed out that the
U.S. "had been more successful on the propaganda side than the Russians
in international conferences [because] we had learned the technique of
finding the limits of the Soviet instructions and then boxing them in." He
added that the new Soviet leaders "would find a meeting very difficult right
now and we could play on that [and we] could also play on their refusal to
meet." There would be plenty of opportunities for propaganda.[22]

Bohlen's idea for using a "peace plan" as political warfare had been
circulating in official circles, and Jackson had discussed a similar concept
at a psychological warfare brainstorming meeting held at Princeton at the
end of 1952. It was the bold initiative Jackson wanted. President Eisenhower would deliver a high-profile presidential speech, extending the possibility of peace and proposing a foreign minister's conference as a first
step toward a broader relaxation of tensions. According to Jackson, such a
speech would be a "dramatic psychological move" that would "capitalize
on the dismay, confusion, fear and selfish hopes" undoubtedly plaguing

the Kremlin leadership. Jackson and the PSB went to work devising a new psychological strategy plan revolving around the presidential speech and the proposed foreign minister's conference. A new PSB draft plan prepared on March 9 saw the speech and proposal as a means of widening divisions within the new Soviet leadership. It would "confront the communist rulers with difficult major choices in a way that tends to divide its power components and unite humanity, especially the free world, with us." If one of the contenders for power could be pulled from Moscow to attend the meeting in the middle of the succession, for example, it might encourage rivals to take advantage of his absence to undermine his position. It would also serve to arouse suspicion, jealousy, and maneuvering among the non-invitees. Moreover, the PSB reasoned, by confronting the successors with a chance to negotiate with the West, the proposed meeting could strain Moscow's relations with China and the satellite governments. This tactic might also encourage a power struggle between the key Communist Party leaders and other major elements of power within the USSR, such as the military or internal security forces. These concerns revealed the interest of the psychological warriors in using high-level diplomacy as an offensive weapon to capitalize on Soviet vulnerabilities and generate trouble for the new leadership.[23]

The PSB program also ordered a vigorous program of covert propaganda action to support U.S. objectives. The PSB plan gave the CIA carte blanche to pursue subversive propaganda activities limited only by a "test of plausibility" and explained that there should be a "broad scope and wide latitude of action in this field." These operations "should be directed at the goal of applying pressure at many points by harassment and the sowing of doubt, confusion and suspicion." The plan was also designed to encourage defections of Soviet and satellite officials, encourage confusion and uncertainty in the ranks of the communist parties, and, if possible, reach ruling groups of the USSR as well. Subversive rumors, contrived events, and half-truths would be planted by the CIA, picked up by the media, and reported as fact. Reflecting C. D. Jackson's hyperbolic optimism, the PSB predicted that U.S. measures would force a "climax in which the communist system would break into open internal conflict."[24]

No one in the Eisenhower administration questioned the need for the United States to seize the initiative and exploit Stalin's death for Cold War gain, least of all the president. At an NSC meeting to consider the administration's response to Stalin's death, Eisenhower expressed his desire to exploit the situation for psychological warfare purposes. Stressing "that the psychological effects not be overlooked," he mused that "the moment was propitious for introducing the right word directly into the Soviet Union. The Russians would be so interested in the reaction of the rest of the world that it would be possible on this occasion to penetrate the Iron Curtain." This was indeed a fortuitous moment: an unfettered opportunity to appeal directly to the people of Russia and Eastern Europe.[25]

Yet despite agreement on the need to exploit the situation in Moscow, there was wide disagreement on what form this exploitation should take. John Foster Dulles and the State Department balked at the idea of a four-power meeting. Dulles agreed that Stalin's death afforded the U.S. with a unique opportunity, but stressed that a foreign minister's meeting would have "quite disastrous" effects on the allies who "would believe our leadership erratic, venturous, and arbitrary." Dulles emphasized that negotiations with the new Soviet leadership would defeat plans for the European Defense Community (EDC), which permitted German rearmament under a European army of mixed nationalities. Dulles was firmly committed to West German participation in a military pact and was convinced that the EDC provided the only viable means to accomplish this goal. But the EDC had been plagued by ratification difficulties in European parliaments, and Dulles believed Jackson's plan for a four-power meeting "was tantamount to inviting the fall of French, German, and Italian governments" that had staked their political futures on the EDC. The conference could not produce the psychological advantages Jackson predicted because the "soviets would simply dig up all their old plans for Foreign Ministers meeting [and] resort to all their devices for delay and obstruction." Dulles established a pattern that continued throughout the year. He vigorously opposed negotiation with the Soviets on the grounds that it would ruin any hope of EDC ratification and it would impede rearmament efforts at home and abroad by raising false hopes.[26]

Eisenhower shared Dulles's concerns. He also doubted whether an appeal for a foreign minister's conference would have the electrifying effect on public opinion that Jackson predicted. Tired old issues like German unification held little propaganda appeal in the world, he believed. The United States should concentrate on a more imaginative approach that would excite the people of the world. Speech writer Emmet Hughes captured this sentiment well when he recorded Eisenhower as saying, "Look, I am tired—and I think everyone is tired—of just plain indictments of the Soviet regime. I think it would be wrong—in fact, asinine—for me to get up before the world now to make another one of those indictments. Instead, just one thing matters: what have *we* got to offer the world?" As Eisenhower explained in a meeting of the National Security Council after Stalin's death, "We . . . need something dramatic to rally the peoples of the world around some idea, some hope, of a better future." According to Eisenhower, such sentiment could be found in the American determination to raise the general standard of living throughout the world: "the common man's yearning for food, shelter and a decent standard of living . . . was a universal desire and we should respond to it." Eisenhower reasoned such an approach would appeal especially to Soviet citizens, who had been promised in vain that each successive Five-Year Plan would meet their personal needs and aspirations. It would focus attention on the fact that the Soviet government's expenditures on armaments were costing ordinary people their livelihoods and wasting resources that should be used to raise living standards and provide for consumer goods. It also might remind the East Europeans of the Marshall Plan aid Stalin had denied them. Furthermore, such an approach would reinforce in the minds of allies and neutrals the prudence and goodwill of American leadership. Accordingly, at an NSC meeting on March 11, Eisenhower instructed Jackson to develop this new "psychological plan."[27]

Eisenhower had scarcely set the wheels in motion when the new Soviet premier, Georgii Malenkov, delivered a major peace address of his own. Speaking before the Supreme Soviet of the USSR on March 16, Malenkov declared the peaceful intentions of the Soviet people. In a startling reversal of Stalin's "no concessions" maxim, Malenkov declared that no problems divided East and West that could not be solved by negotiation: "At the present time there is no disputed or unresolved question that cannot be resolved by peaceful means, on the basis of mutual agreement . . . States interested in preserving peace may be assured, both now and in the future, of the firm peaceful policy of the Soviet Union." Malenkov's statement captured the attention of the world press, which focused its attention on the apparent change of attitude in the Kremlin and the promise of peace it suggested.[28]

Malenkov had stolen Eisenhower's thunder. A barrage of measures to demonstrate Soviet goodwill followed. The State Department's Policy Planning Staff noted two weeks after the speech that "there have been more Soviet gestures toward the West than at any other similar period." These measures included an agreement to exchange sick and wounded

POWs from the Korean conflict, a proposal for the resumption of Korean armistice talks, a proposal for British-Soviet talks in Berlin to reduce air incidents in Germany, Soviet admission in propaganda that the United States had a hand in the defeat of Germany in 1945, permission for a group of American correspondents to enter Russia, and the hint of a possible meeting between Eisenhower and Malenkov to discuss disarmament. Other developments suggested that the new Soviet leadership was attempting to break from Stalin's legacy. The gradual disappearance of ubiquitous references to Stalin in the Soviet press provided the most obvious sign. Furthermore, Malenkov declared the intention of the Soviet regime to devote more of its resources to producing consumer goods to improve the standard of living for the populace, a possible diversion of resources from heavy to light industry signaling a reduction in arms expenditures. The most surprising departure, however, came in April when *Pravda* renounced the so-called Doctor's Plot. As one State Department official observed, "this startling event, perhaps more than any other, provides [the] most concrete evidence thus far of [the] present regime's break with Stalinism."[29]

These initiatives potentially signaled a sharp break in Soviet policy from its Stalinist roots, but American analysts responded to these developments with alarm. The NSC recognized that Soviet actions since March marked a "clear departure" from the tactics of the Stalin regime but saw this as a menacing development. As a State Department memorandum observed, "it is necessary to assume that the peace offensive is a treacherous stratagem of as yet indiscernible proportions." According to most analyses, the Soviet peace campaign threatened to undermine free world morale by dissipating the element of fear that glued U.S. alliances together and maintained domestic support for rearmament. By offering meaningless gestures of goodwill, the Soviets threatened to weaken the resolve of the American public and of western allies. As the State Department noted, "The purpose for such campaign is painfully obvious: by this method the Communists hope to crack the wall of resistance which the West has been constructing, and to bring about an eventual slowing-down of the armaments program of the Free World." Because "the world-wide hope for peace is certain to far overshadow the fears of long-range communist designs," public opinion in the West would not support the armaments programs necessary to keep pressure on the Kremlin leadership. The momentum of international opinion was shifting from rearmament and confrontation to disarmament and conciliation. To American officials, the Soviet "peace offensive" represented the most worrisome kind of psychological warfare.[30]

The perceived psychological threat to the morale and unity of the free world precipitated a general reappraisal of the nature of the U.S. response to the Stalin succession. Initially preoccupied with how best to use diplomacy and psychological warfare to stir up trouble behind the Iron Curtain, administration officials turned their attention to forestalling wishful thinking on the part of allies and domestic audiences. In light of Soviet peace maneuvers, the presidential speech assumed renewed urgency as a means of inspiring caution in free world attitudes toward the Soviet Union. A revised version of the PSB plan explained that in the presidential speech and subsequent propaganda the United States should "combat any wishful thinking in the free world [and] expose vigorously the motives and pitfall of any false 'peace' campaigns." Toward this end, the speech should set forth the issues preventing a genuine relaxation of tensions. It should place the onus squarely upon the Soviet government for failure to resolve these issues. Rather than signal an interest in meeting the new leadership at the bargaining table, the speech should "make the Kremlin assume maximum liability if it does not accept the proposals in the Presidential speech." It would be an indictment of the Soviet regime, but it would be framed as an opportunity, a chance for peace.[31]

A Chance for Peace

After nearly a month of revision and internal debate, Eisenhower delivered the speech before the American Society of Newspaper Editors on April 16, 1953. Dubbed "A Chance for Peace," the speech addressed the American commitment to peace, dedication to disarmament, and desire to raise living standards across the world. With timeless rhetorical flourish, Eisenhower reminded his audiences of the costly consequences of a strategic arms race: "Every gun that is made, every warship launched, every rocket fired signifies, in the final sense, a theft from those who hunger and are not fed, those who are cold and are not clothed. . . . Under the cloud of threatening war, it is humanity hanging from a cross of iron." These words, among Eisenhower's most widely quoted, harkened back to wartime propaganda themes. American propaganda against Axis soldiers had publicized deprivations of the enemy, suggesting that a better standard of living would be found with the cessation of hostilities. The Chance for Peace address sought to make the same connections to those living under communist rule, albeit in an oblique fashion. By highlighting the costs of the strategic arms race in human terms—in schools, in health care, in food and luxury goods—it sought to convey to East European listeners the costs they were paying to maintain the Soviet war economy.[32]

Many observers at the time and since interpreted Eisenhower's speech as representing a genuine opportunity for peace. It appeared to signal his willingness to end the arms race and work for an early end to the Cold War, a view still held by some historians. The overall tone and message of the speech, however, elevated the cause of waging Cold War over that of making peace. As communication specialist Robert L. Ivie notes, the speech rhetorically constructed the Cold War "as a profound struggle between absolute good and absolute evil, with the enemy's unconditional surrender as the only acceptable result." Evoking the spirit of the Truman Doctrine, Eisenhower explained the Cold War as a moral contest between two ways of life, one good and one evil, each represented by different roads. The road followed by the U.S. was one of cooperation, fellowship, and justice, but the Soviet Union followed a road of "force: huge armies, subversion, rule of neighbor nations." Whereas the United States believed that "any nation's right to a form of government and economic system of its own choosing is inalienable," the Soviet Union sought security "by denying it to all others." Having assumed the moral high ground, Eisenhower warned his listeners not to accept Soviet peace overtures too eagerly. He explained that the new leaders must prove their goodwill through deeds: "We welcome every honest act of peace. We care nothing for mere rhetoric." Several times Eisenhower cautioned of the need for free world unity, urging his audiences to wait for proof of sincerity from the Soviets. The free world, he explained, neither should let its desire for peace nor its fear of war impede its resistance to the Soviet threat. It "must, at any cost, remain armed, strong, and ready for the risk of war." Eisenhower assigned the moral responsibility for prolonging the Cold War to the Soviet Union. He noted that if the peace fails, at least the world will know clearly "who has condemned humankind to this fate."[33]

The speech sought to identify world hopes for peace with U.S. aims in the Cold War, to define a hope for the future that could sustain free world morale for the long haul. Eisenhower emphasized that peace required much more than a reduction of tensions. He cautioned against a status quo settlement merely to preserve the peace. At a minimum, Eisenhower suggested, the Soviet leadership needed to prove its good faith through an "honorable" armistice in Korea, an end to hostilities in Indochina and Malaya, and a peace treaty with Austria. These "deeds" of good faith needed to include a united Germany, one free to participate in NATO and the

EDC. Significantly, Eisenhower also called for the full independence of East European nations no fewer than three times. C. D. Jackson's deputy Abbott Washburn explained that Eisenhower's speech sought to "build a climate of world opinion that will place a moral obligation on the Soviet Union to accept these points and act on them." Although eloquent, it was hardly the "serious bid for peace" depicted in much of the literature on the Eisenhower presidency; it was a skillfully executed exercise in political warfare designed to wrestle the peace initiative away from the Kremlin. While requesting proof of Soviet sincerity, Eisenhower offered neither concessions from the West nor proposals for negotiation. Rather than promote peace *from* the Cold War, Eisenhower promoted a vision of peace that could only be achieved *through* it. By delineating the issues that divided East from West, he was reminding domestic and international audiences of the purposes for which the Cold War was being fought. He explained in the speech that the world faced a choice between only two bleak possibilities—a new world war or a prolonged Cold War—thus rhetorically guiding his audiences to the conclusion that sustained effort in the Cold War was the only real option before them. Far from offering a chance for peace, Eisenhower was urging the world's peoples to keep their chins up, unite behind American leadership, and work for victory. It was the genius of Eisenhower's rhetoric that framed this Cold War crusade as a quest for peace.[34]

To ensure that audiences at home and abroad interpreted the address as a serious peace proposal, C. D. Jackson and the Psychological Strategy Board orchestrated a massive follow-up campaign. Capitalizing on a network linking government information officers and journalists, the PSB exploitation plan used the speech to encourage and shape international press coverage of the address. The plan ensured, as the State Department subsequently noted, that the address elicited "more public interest and excited more favorable comment throughout the world than any official statement of high policy" since the Marshall Plan. Under PSB supervision, the government information apparatus broadcast the speech to radio stations around the world, distributed over 3 million leaflets containing the speech text, and produced motion picture films and documentaries containing references to the address for widespread distribution on newsreels and in theaters. Diplomatic posts overseas, meanwhile, presented copies of the speech to the Foreign and Prime Ministers of each nation, calling "attention personally to specific points in the address which would most interest" them. Information officers sent copies of the speech to newspapers and magazines at home and abroad, and utilized their contacts to persuade

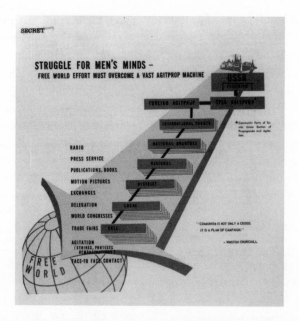

This illustration from a classified study on the "Psychological Aspects of U.S. Strategy" emphasizes the threat to the free world posed by the Kremlin's psychological warfare apparatus. Source: Eisenhower Library.

editorialists to discuss the speech approvingly. Private organizations, such as the Common Council for American Unity and the International Confederation of Free Trade Unions, contacted foreign-owned newspapers to encourage them to present summaries of the speech in approving editorials. Meanwhile, businessmen, labor leaders, and women's groups were all requested to spread awareness of the speech within their circles. These measures, the State Department explained, sought "to throw back the so-called Soviet 'peace-offensive' and [turn] it into a 'peace defensive.'"[35]

The Chance for Peace campaign was a dress rehearsal for the administration's notion of psychological strategy. In some ways, it paralleled the public relations strategies developed by Edward Bernays: it began with an overt act designed to capture media attention and was followed by a series of lesser acts to keep the message before the international media. Every phrase of the speech was laden with psychological goals; every word had been scrutinized for its psychological suggestions and implications. On one level, the administration's use of the presidency as a bully pulpit for international communication had been successful. Eisenhower's message was widely disseminated and became the media event that the psychological warriors had hoped it would become. The rhetoric and language were so artfully crafted that, despite a wide body of declassified evidence to the contrary, observers still hear in Eisenhower's words a sincere attempt at international reconciliation. His words had convinced many that the United States was offering the Soviet Union a genuine and realistic Chance for Peace.

How successfully the speech worked as a morale-building exercise is less certain. Allied unity—a paramount objective of the speech—became worse. The western alliance was deeply divided over how to deal with Moscow's conciliatory stance. From the moment Stalin died, British Prime Minister Winston Churchill had been pressuring Eisenhower to meet with the new Soviet leadership. Unlike Eisenhower, Churchill was not convinced of Soviet insincerity and believed the West should test Soviet intentions. On May 11, Churchill publicly called for a high-level Summit meeting, a clear challenge to the more cautious American approach. From Eisenhower's perspective, Churchill seemed to be playing into the hands of the Kremlin. He grumbled to the NSC that the United States was having "considerable difficulties" with its European allies, complaining that relations with the United Kingdom "had become worse in the last few weeks than at any time since the end of the war." The State Department's Policy Planning Staff observed, "Though Stalin's death enhanced the possibility of exploitable conflicts developing within the communist camp, subsequent months have chiefly brought evidence of widening cleavages in the free world instead, some of them evidently due to differing views of the situation created by Stalin's death." American officials concluded that their worst fears were being realized: Soviet divisive psychological warfare was working all too well.[36]

A New Type of Cold War

The Soviet leadership, although divided by a behind-the-scenes struggle for power, pressed forward with its "peace" blitz. The armistice talks on Korea, which had been stalled for months, resumed on April 26. By July, a truce effectively brought the war in Korea to an end. A month later, Malenkov, while announcing his country's thermonuclear capability, made an important doctrinal revision in favor of coexistence, arguing that nuclear weapons forced both the socialist and capitalist camps to seek world peace. In July 1954, French and Vietminh forces in Vietnam signed a cease-fire, temporarily ending the war in Southeast Asia. Later that fall the Soviet Union and China issued a joint declaration stating their willingness to normalize relations with Japan, a move that raised the possibility of a Soviet-Japanese peace treaty. Malenkov's ouster and the rise of Nikita Khrushchev in February 1955 produced a flurry of diplomatic activity that suggested that under Khrushchev's leadership the USSR would continue the soft line that had dominated Soviet foreign policy since Stalin's death. Khrushchev pursued rapprochement with Yugoslavia, signed a peace treaty with Austria, and adopted a more flexible stance on disarmament. These moves were accompanied by liberalizing trends within Soviet borders, an expansion of East-West contacts and trade, and a marked increase in Soviet cultural exchange activities. To be sure, the Soviet Union still retained firm control over Eastern Europe, Germany remained divided, and the arms race continued. But all the same, by the end of 1955, Stalin's successors had met most of the tests of good faith requested by Eisenhower in his April speech. The regime's break with Stalin was unmistakable, particularly after the 20th Party Congress in February 1956, when Khrushchev delivered his "secret speech" denouncing the worst evils of Stalin's oppression.[37]

One might expect to find U.S. officials breathing sighs of relief. After all, Stalin was dead and his successors had embarked on a more cooperative, less confrontational foreign policy. Whatever the limitations of "peaceful coexistence," Stalin's death had produced a more flexible regime, one less menacing, less hostile, and more open to the outside world. The new Soviet leadership demonstrated—in Korea, Austria, Indochina, and Japan—that it was capable of constructive, if imperfect, negotiations. Were these developments hopeful signs that perhaps a thaw in the Cold War was possible? Could the West do "serious business" with the Kremlin and negotiate an early détente, or perhaps even an early end to the Cold War? Not according to officials in the Eisenhower administration. As Raymond Garthoff notes, "Soviet flexibility and moves toward more cooperative and less threatening policies were regarded not only with suspicion, but as an obstacle to American interests rather than steps toward accommodation."[38]

Far from seeing the policy of "peaceful coexistence" as cause for optimism, Eisenhower and his advisors viewed it as a greater threat to the free world than Stalinism. To American officials, peaceful coexistence represented a menacing political warfare strategy of the most treacherous kind: it raised doubts about the entire Cold War enterprise, and it bred "false hopes" that a negotiated settlement might be possible. How could the U.S. continue to pursue policies of strength, through rearmament and alliance building, while fear of the Soviet Union was subsiding? Fear had been the psychological glue that held free world coalitions together, but now that fear, along with Allied unity and resolve, was dissipating. A new psychological message was needed, a new moral mission, to cohere the free world into a united front. Eisenhower explained these sentiments to Dulles, saying, "During the Stalin regime, the Soviets seemed to prefer the use of force—or the threat of force—to gain their ends. . . . So long as they used force and the threat of force, we had the world's natural reaction of fear to aid us in building consolidations of power and strength in order to resist Soviet advances." Because Soviet peace gestures muddled free world perceptions of a Soviet threat, as Nelson Rockefeller explained in a letter to Eisenhower, "the U.S. must find some other motivation than fear with which to inspire the efforts of free men for the long pull."[39]

This perspective was likely influenced by pessimistic reporting from the intelligence community. The CIA predicted that the free world's international position would deteriorate as a result of the new Soviet tactics. A National Intelligence Estimate produced in October 1953 cautioned that "the USSR will continue its cold war against the Free World, largely through a vigorous political warfare campaign." The report emphasized that Soviet conciliatory tactics represented a "new challenge." The reduction in threat perceptions engendered by "peaceful coexistence" threatened "to relax the vigilance of some Western states, to encourage dissension between the US and its allies, and to delay the progress of Western rearmament." With Stalin running the Soviet Union, the United States faced no such problem. Stalin's domestic repression and confrontational tactics in Berlin, Korea, and Eastern Europe reinforced the hostility of the world to Soviet communism. But the actions of Stalin's successors to repudiate the worst excesses of his policies enhanced their credibility as promoters of peace and, to many observers, made rigid U.S. anticommunism appear the greater threat to world peace and stability.[40]

Moreover, the growth of Soviet military power and the accelerating pace of decolonization added new psychological variables to the equation. The intelligence community warned that key allies, faced with the prospect of becoming embroiled in a Soviet-American thermonuclear holocaust, might adopt a more neutral position in the Cold War and thus would be less likely to oppose communist expansion in the periphery. These fears were exacerbated by doubts about the "the stability, moderation, and maturity of U.S. policy" among allies who feared that "the U.S. may go it alone and involve the world in global war." Such a threat to free world morale offered "a fertile field for Soviet divisive tactics, and the new Soviet regime may be more successful than Stalin in exploiting them." In addition, policy makers worried that the Soviets would profit from the dismantling of European empires to alter the international status quo in their favor. A new course in Soviet foreign policy actively courted the newly independent states in the developing world, utilizing a combination of flexible diplomacy and liberal trade and aid offers. With Stalin gone and his successors apparently demonstrating their peaceful intentions, administration officials feared that previously "peripheral" states would voluntarily gravitate into the Soviet orbit as a result of sympathy to communist ideology, lingering hostility toward European imperialists, material necessity, or admiration for Soviet industrial and technological feats.

As a National Intelligence Estimate concluded at the end of the decade, peaceful coexistence was "a strategy to defeat the West without war," involving a political struggle to capture the support of peoples across the world. By manipulating issues of peace, disarmament, anticolonialism, and economic development, by dramatizing the growth of Soviet power, and by capturing the imagination of the world's peoples with their technical prowess, the Soviets threatened to attract the allegiance of the underdeveloped and uncommitted states against the West while sending U.S. allies into neutralism. Although the United States enjoyed a clear preponderance of power vis-à-vis the Soviet Union throughout the 1950s, intelligence analysts saw these developments as "an even more serious threat to the Free World than . . . Stalin's aggressive postwar policies."[41]

By the middle of the decade Eisenhower and his advisors concluded that the very nature of the Soviet threat had changed. Drifting from the previous administration's emphasis on the military and geostrategic aspects of the Soviet threat, Eisenhower instead emphasized the political, ideological, economic, and psychological challenges posed by Soviet foreign policy changes. From Eisenhower's perspective, "peaceful coexistence" portended a serious crisis of morale among free world nations, which was signified by doubts about U.S. peaceful intentions, pressure for negotiation, unwillingness to support defense expenditures, and lack of faith in the administration's prosecution of the Cold War. Eisenhower expressed frustration that "the seductive quality of Soviet promises and pronouncements . . . make us appear before the world as something less than persuasive in proclaiming our peaceful purposes." "The Communist threat has not diminished," he wrote to a friend in 1957. "In fact since it has abandoned the single line of using force and threat of force to gain its ends, and has instead broadened its front to include economic, political and cultural attacks, I think that the cause of the Western world has suffered. . . . So—while we can boast of the finest civilization the world has known, the Soviets are convincing many peoples that they are the true exponents of progress, peace and of freedom! What a tragedy!" According to Eisenhower, the new Soviet tactics signi- fied a "new type of cold war," a war of persuasion where the battlegrounds were shifting to the economic and propaganda fields.[42]

Psychological Warfare and National Security Policy

It was thus in the psychological arena that Eisenhower sensed the biggest problems facing the United States, a view reflected in the national security policies adopted by his administration. Eisenhower's basic national secu- rity strategy, known as "the New Look," was shaped at all levels by the personal convictions of the president. Of central importance were three propositions: the Cold War was going to last a very long time; the United States must remain strong both economically and militarily; and because the Soviets were unlikely to initiate a suicidal world war, the brunt of the Cold War effort should be directed at besting the Soviets in the political and psychological fields.

Eisenhower, like George Kennan and others, anticipated that the So- viet regime would eventually crumble beneath the weight of its own op- pressive and backward system. But in contrast to C. D. Jackson and other apostles of "liberation," Eisenhower did not expect to see the Iron Curtain rolled back any time soon. For the moment, at least, the communists re- tained firm control over Russia and the peoples of Eastern Europe. The key for Eisenhower was therefore to prepare for what he called "the long haul." The United States should retain its overall military superiority to put maximum pressure on the Kremlin and to deter general war, but at the same time the United States should not spend so recklessly on de- fense that it undermined the basic soundness of the American economy. Eisenhower understood that the overall economic health of the nation was the most important source of U.S. power, and he also believed that American prosperity—the ability of the United States to provide for its people—comprised a core ingredient of American appeal in the world. In pursuit of balanced budgets and economic solvency, Eisenhower sought to cut military expenditures by trimming expensive conventional forces and relying instead on nuclear weapons to deter a Soviet attack. Eisenhower's

"Peace Without Fear," a USIA pamphlet, advertises America's peaceful intentions.
Source: Eisenhower Library.

New Look thus called for a vast build-up of nuclear arms to threaten the
Soviets with "massive retaliation" should they embark on any form of overt
aggression.[43]

When Eisenhower was in office and for years afterward, the New Look
was misunderstood as involving little more than "massive retaliation" and
balanced budgets, but these elements, important as they were, deflected
attention from the important nonmilitary components of the New Look.
Although nuclear weapons deterred overt military aggression, the United
States would wage the Cold War assertively through nonmilitary means in
the political and psychological arenas, fighting the spread of communism
everywhere outside the Soviet orbit. It would do so by building a system of
formal and informal alliances to bind core regions of the world to American leadership politically, economically, and militarily; and by fighting local communist movements through robust covert action and psychological
warfare programs.[44]

Eisenhower frequently expressed his belief that the Soviet leadership
would not take risks involving general war with the United States. Instead,
he believed that they would pursue their expansionist agenda through

political, psychological, and economic "subversion." When John Foster Dulles first publicized the theory of "massive retaliation" in *Life* magazine, Eisenhower responded to Dulles's formulation of the New Look as being too narrowly focused on military aggression. Highlighting the importance of the psychological and political challenges posed by the USSR, he wrote: "There is only one point that bothered me. . . . It is this: What should we do if Soviet *political* aggression, as in Czechoslovakia, successively chips away exposed positions of the free world? . . . Such an eventuality would be just as bad as if the area had been captured by force. To my mind, this is the case where the theory of 'retaliation' falls down." Eisenhower reminded Dulles later, "exclusive reliance upon mere power of retaliation is not a complete answer to the broad Soviet threat." A comprehensive strategy for besting the Soviet Union in the political and psychological competition was vital.[45]

Eisenhower's views were reflected in National Security Council document 162/2, which guided the implementation of the New Look. Discounting the danger of overt military aggression, NSC 162/2 defined the Soviet threat overwhelmingly in political and psychological terms. Even though "the capability of the USSR to attack the United States . . . has been continuously growing," the report noted, "the USSR does not seem likely deliberately to launch a general war against the United States." Instead, NSC 162/2 advised,

> The USSR will continue to rely heavily on tactics of division and subversion to weaken the free world alliances and will to resist Soviet power. Using both the fear of atomic warfare and the hope of peace, such political warfare will seek to exploit differences among members of the free world, neutralist attitudes, and anti-colonial and nationalist sentiments in underdeveloped areas. For these purposes, communist parties and other cooperating elements will be used to manipulate opinion and control governments wherever possible.

Cataloging a list of free world vulnerabilities to such political warfare, the report warned, "this aspect of the Soviet threat is likely to continue indefinitely and to grow in intensity." Of particular concern, NSC 162/2 cautioned, was the Soviet peace offensive, which "cost the Soviets very little in actual concessions" and was "designed to divide the West by raising false hopes and seeking to make the United States appear unyielding." Equally alarming were the attitudes of U.S. allies who perceived "the actual danger of Soviet aggression as less imminent than the United States." They feared

American policies would "involve Europe in general war" or "indefinitely prolong Cold War tensions." Accordingly, NSC 162/2 advised that allies, uncommitted nations, and the American people "must be informed of the nature of the Soviet-Communist threat."[46]

A major reappraisal of basic national security policy, formalized in the policy paper NSC 5501 and adopted in January 1955, repeated the warning that Soviet peace tactics, the approach of nuclear balance, and Soviet initiatives in the developing world together posed the foremost challenges to the United States. The report apocalyptically warned that peaceful coexistence "will probably present the free world with its most serious challenge and greatest danger in the next few years." Because greater receptivity to Soviet peace overtures and growing fears of atomic war encouraged third world neutralism and exacerbated strains within the alliance, the United States "should place more stress than heretofore on building the strength and cohesion of the free world" by using a "flexible combination of military, political, economic, propaganda, and covert actions." Moral and economic strength, the report suggested, rather than military force alone, were the keys to victory in this new kind of Cold War. As NSC 5501 stated, "The ability of the free world, over the long pull, to meet the challenge and competition of the Communist world will depend in large measure on the capacity to demonstrate progress toward meeting the basic needs and aspirations of its people." This meant encouraging modernization across the globe, fostering international trade, moderating disputes within the free world, and maintaining a sound economy. It also required strengthening U.S. information, cultural, education, and exchange programs to inspire confidence in American peaceful intentions and to shore up faith in American leadership.[47]

Eisenhower hoped that a vigorous propaganda operation could help stem the tide of Soviet influence and, in line with NSC 5501, propaganda programs emphasizing peace and free world unity were intensified. Classified reviews of U.S. propaganda activities explained that the "most significant development" for U.S. policy "was the unfolding of the post-Stalinist strategy of the Soviet Union in a series of tactical moves designed to impress peoples abroad with the peaceful intent of the new regime, and to increase the respectability of the Communist system in the eyes of Free World countries." Such tactics "were highly effective in strengthening tendencies toward relaxation in Free World countries, and in reinforcing neutralist opinion." The 1955 review of U.S. propaganda operations noted, "The favorable impression created by the new Soviet tactics has increased the difficulties . . . in persuading other peoples to accept American policies of building up Free World strength to counter the threat posed by the massive military power of the Soviet bloc. As the Soviet posture appeared to grow less threatening, and the danger of war seemed less immediate, it became increasingly difficult to persuade Europeans that the necessary sacrifices demanded by our military counter-measures were a matter of immediate urgency."[48]

To counter the appeal of coexistence, the Eisenhower administration developed a global emphasis in American propaganda to convince audiences abroad that the United States "stands and works for peace, and for a peace which is more meaningful than simple co-existence of two blocs of nations." Indeed, convincing the world of American peaceful intentions became the highest priority of American psychological strategy throughout the decade. All of the major propaganda campaigns discussed in the remaining chapters relate in some way to the broader objective of proving to the world American peaceful purposes and the fitness of the United States for world leadership.[49]

Stalin's death thus represented far more than the first great challenge of the Eisenhower administration. Coinciding as it did with the thermonuclear revolution and the emergence of the developing world as a significant force in international affairs, the year Stalin died marked a watershed in the Cold War. The future of the competition increasingly appeared to be about demonstrating the superiority of competing ways of life—political systems, economic organizations, ideological foundations, cultural and artistic accomplishments, scientific and technological progress, and relations between races, classes, and genders. In this context, psychological warfare achieved increasing prominence as a critical component of the Eisenhower administration's evolving Cold War strategy. Psychological warfare became not only an instrument of this conflict, but an influence (if not always a decisive one) on the administration's entire approach to foreign relations. It shaped diplomatic, economic, and military policies as well as scientific exploration, cultural interchange, tourism, production of ideas, and, indeed, everyday life. The global competition for hearts and minds was increasingly recognized as the principal battleground of the Cold War.

Chapter 3

Camouflaged Propaganda
Psychological Warfare's New Look

True psychological warfare, properly defined, is so bound up with the conduct and
demeanor of the whole American Government, that you cannot establish a
separate department of psychological warriors.

—*Washington Post*, May 24, 1953

As the crisis over Stalin's succession unfolded in the early months of
Eisenhower's first year in the White House, no fewer than five government committees were investigating U.S. information programs abroad.
All of them called for sweeping changes. Serious questions about the effectiveness of U.S. psychological programs had been raised by the apparent
strength of communist parties and peace campaigns in Europe, the bacteriological warfare charges and brainwashing scares of the Korean War,
and the fratricidal infighting that had paralyzed the Psychological Strategy
Board in the waning days of the Truman presidency. The International
Information Administration (IIA), run by the Department of State, appeared incapable of stemming the tide of neutralist, procommunist, and
anti-American sentiment abroad. In the summer and fall of 1952, inquiries
into the propaganda problem were initiated by committees in the House
and Senate and by the U.S. Advisory Commission on Information, a quasi-
governmental body that prepared semiannual reports on U.S. information
programs. After his inauguration, Eisenhower promptly established two
additional committees to consider ways of improving overseas propaganda
work: the President's Committee on Governmental Organization, chaired
by the wartime czar of U.S. cultural diplomacy in Latin America, Nelson
Rockefeller; and the President's Committee on International Information
Activities, directed by William H. Jackson, an investment banker with intelligence experience.[1]

Although the recommendations of all these committees varied, they
agreed that two years of hard-hitting anticommunist propaganda under
the Campaign of Truth had been ineffective, if not counterproductive. Far

from convincing foreigners of the Red Menace, the IIA's crude anticommunist exhortation alienated international audiences who were saturated by propaganda from abroad and suspicious of proclamations by foreign governments. U.S. information programs "were too strident, angry, and antagonistic in tone and were placing too much emphasis on American power," the Advisory Commission complained. A more sophisticated approach to psychological warfare was needed—one that was less propagandistic, less emotional, less obvious, and less reminiscent of Joseph Goebbels.[2]

Eisenhower could not have agreed more. Even before he moved into the White House, as C. D. Jackson noted, Eisenhower had come to the "*considered* conclusion" that the existing psychological warfare apparatus "was poor, disorganized, diffuse, ineffective, and not what he wanted." He directed the Jackson and Rockefeller committees to develop his ideas of psychological strategy, to produce a national security machinery that was able and willing to wage a coordinated political warfare assault against communism worldwide. Under Eisenhower's direction, the United States ushered in a new and more sophisticated approach to psychological warfare aimed at international and domestic audiences. Propaganda strategies and techniques were refined and improved, and new bureaucratic entities were developed to centralize psychological warfare planning in the White House.[3]

To a remarkable extent, Eisenhower involved himself personally in the adoption of psychological warfare strategies intended to make U.S. propaganda more persuasive and credible. Eisenhower believed that for propaganda to be effective, "the hand of government must be carefully concealed, and, in some cases I should say, wholly eliminated." He perceived that audiences would be more receptive to the American message if they were kept from identifying it as propaganda. Avowedly propagandistic materials from the United States might convince few, but the same viewpoints presented by seemingly independent voices would be more persuasive.[4]

Eisenhower's "New Look" for the U.S. psychological warfare program therefore rested on a distinction between overt and covert propaganda strategies. Overt propaganda, which openly acknowledged U.S. sponsorship, included VOA broadcasts, films, and press items. Eisenhower insisted that all such materials officially "attributed" to the United States abandon the brazenly anticommunist tone that marked the Campaign of Truth. He wanted official voices of America to sound more like the seemingly neutral British Broadcasting Corporation (BBC) than the propagandistic *Pravda*.

Although official voices were restricted to positive and factual messages, covert channels operated with no such restraints. Covert, or "unattributed," propaganda was used to condemn communism or to advance viewpoints that served American foreign policy interests without revealing U.S. government involvement. To a great extent, U.S. propagandists used the independent news media, nongovernmental organizations, and private individuals as surrogate communicators to convey propaganda messages. Recently declassified documents make clear that the overwhelming majority of the Eisenhower administration's propaganda activities were not attributed to the United States. According to communication specialist Shawn J. Parry-Giles, this "camouflaged" propaganda was a "more subtle propaganda with potentially greater power than that of totalitarian regimes, where governmental controls are more common and overt."[5]

The Eisenhower administration's approach to psychological warfare operated on another level as well. Eisenhower always considered propaganda, narrowly defined, a relatively minor component of the larger battle for hearts and minds. Words must be matched by deeds, he emphasized. In Eisenhower's view, the ideological competition for the allegiance of the world's peoples suffused all actions with psychological significance. Believing that psychological warfare was inseparable from other elements of national security strategy, Eisenhower worked to integrate "psychological considerations" into the very process of making and implementing foreign policy. From the highest levels of the national security establishment to the remotest diplomatic outposts abroad, political warfare became *the* organizing concept for American foreign policy during the Eisenhower presidency.[6]

The Jackson Committee

The high priority attached to the work of the Jackson Committee signaled that psychological warfare would assume a place of prominence in Eisenhower's administration. Eisenhower established the committee six days after his inauguration, but in fact it had begun working informally several weeks earlier. C. D. Jackson had proposed the establishment of such a body to the president-elect in November 1952, and the committee began its work soon thereafter. Although the committee received its informal name from its chairman, William H. Jackson, C. D. Jackson was the group's brainchild. Now serving as Eisenhower's psychological warfare advisor and head of the PSB, he imprinted his boundless energy and vision on the committee's final report. Other notable committee members included Robert Cutler (Eisenhower's national security advisor), Gordon Gray (former PSB director), Roger M. Keys (Deputy Secretary of Defense), and a few prominent businessmen with public relations experience and ties to the CIA. The committee's staff included members from the Departments of State and Defense, the Mutual Security Agency, and a disproportionately large number of CIA representatives. Abbott Washburn—C. D. Jackson's deputy, OSS veteran, and future deputy director of the USIA—served as the committee's executive secretary. Ostensibly commissioned to study information programs, the committee "was actually told in effect to study all aspects of our conduct of the cold war." It interviewed over 250 witnesses and had access to a wide range of highly classified intelligence estimates and national security documents. Its recommendations played a pivotal role in providing a new look for the information program and shaping U.S. propaganda strategy for the rest of the decade and beyond.[7]

When the committee submitted its final report to the president in June 1953, it warned emphatically that "the greatest danger of Soviet expansion lies in political warfare and local communist armed action." Emphasizing the nonmilitary character of the Soviet threat, the report conceded that the Soviet leadership wanted to avoid a military conflict with the West. Moscow would instead pursue its goal of "world domination" by "political warfare methods." "We expect an intensification of Soviet political warfare during the period immediately ahead," the report predicted. "We believe . . . that the Kremlin will avoid initiatives involving serious risk of general war, especially since it may hope to make additional gains by political warfare methods without such risk." Capitalizing on the sensitivity of the United States and its democratic allies to domestic opinion pressures, the Kremlin would use its political warfare to isolate the United States, promote dissension within U.S. alliances, manipulate neutralist sentiment, lure uncommitted countries into the Soviet orbit, and undermine domestic support for rearmament and other Cold War policies.[8]

The presumed vulnerability of the free world to Soviet political warfare meant that the bulk of the Jackson Committee report dealt with public opinion in allied and neutral nations. "In general," the report noted, "the free nations, because they are free, are necessarily more open to communist penetration and subversion than the Soviet system is to Western political warfare." Remarkably, the report conveyed an almost ambivalent attitude toward psychological operations targeting the Soviet Union and its East European satellites. The committee concluded that rollback operations

against the Soviet bloc "must be considered unsuccessful to date." Notably pessimistic about these rollback initiatives, the report left open the
option of abandoning certain anti-Soviet psychological warfare activities
altogether—a startling suggestion considering that committee members
were serving an administration that had condemned containment as "fu-
tile" and had advocated "liberation" as a more proactive alternative. For
the Jackson Committee, as for Eisenhower, the free world appeared to be
the most important theater for ideological warfare.[9]

Psychological Strategy

Propaganda experts in the first half of the twentieth century tended to
wildly exaggerate the power of propaganda, and the Jackson Committee's
grave warnings about the dangers of Soviet political warfare echoed such
sentiments. Yet in a curious sort of way, the committee's belief in the *power*
of propaganda was conditioned by an appreciation of the *limits* of propaganda. The committee felt that because all U.S. actions had psychological repercussions, slick wordplay and clever propaganda gimmicks alone
would not win the global battle for hearts and minds. Mere words could
only accomplish so much; they needed to be harmonized with deeds. Statements of principle needed to be backed up with hard evidence. Policies
and actions needed to reflect "psychological" considerations. According
to the committee's final report, "Propaganda cannot be expected to be the
determining factor in deciding major issues. The United States is judged
less by what it says through official information outlets than by the actions
and attitudes of the Government in international affairs and the actions
and attitudes of its citizens and officials, abroad and at home." In short, the
committee advised, "The cold war cannot be won by words alone. What
we do will continue to be vastly more important than what we say."[10]

The most significant conclusion of the Jackson Committee was therefore its finding that psychological warfare could not be separated from
other aspects of U.S. foreign policy. Psychological considerations should
intrude on the very policy-making process itself. This conclusion Eisenhower readily endorsed. In the NSC meeting to discuss the report, Eisenhower stressed that "we make sure that the psychological factor in important Government actions was not overlooked." He instructed the NSC
that he "wanted to be assured that someone was going to keep track of
the psychological side as of major importance." Abbott Washburn later
recalled that Eisenhower sometimes called this the "p-factor"—meaning

psychological, propaganda, or persuasion. According to Washburn, Eisenhower "wanted it cranked in at all levels of policy consideration—from the National Security Council on down." Eisenhower agreed with the Jackson Committee, as he explained in a letter to Nelson Rockefeller, that "every economic, security, and political policy of the government manifestly is one of the weapons (or should be) in psychological warfare."[11]

To Eisenhower and his political warfare advisors, "psychological strategy" signified a close relationship between international public opinion, persuasion, and national security policy. As William H. Jackson explained it, policies needed to be made, implemented, and presented with international public opinion in mind: "A profound concern for public opinion in the countries affected must be a consideration and an ingredient in the whole process of government, in the planning stage, in the formulation and determination of policy, in the coordination and timing of operations and finally in the last phase of enunciation, explanation and interpretation by government officials, and expression in information programs." Although this notion of psychological strategy held that psychological considerations should be factored in at all stages of the policy-making process, it did not suggest that the United States should consider international public opinion in the same way that a politician, sensitive to domestic opinion, might conduct a poll to decide his stance on a particular issue. Nor did it necessarily mean choosing policy options for their popularity abroad, or determining policy to fulfill a desire "to be liked." Rather, the focus became devising policies and programs to produce an effect on public opinion. It all boiled down to a belief that mere words were not enough: the whole posture of government must reflect "psychological" considerations.[12]

Calculations about international public opinion naturally ventured into the amorphous realm of the psyche. The process involved anticipating and identifying how American actions—and the presentation of those actions—would be perceived abroad, and how in turn those perceptions would affect the political decisions of governments. Psychological strategy also involved calculations pertaining to domestic political conditions abroad, anticipating how American actions and statements would affect the political fortunes of foreign leaders and parties. "Damage control" thinking also colored the administration's concept of psychological strategy. Officials evinced an acute sensitivity to communist propaganda, tracking propaganda lines closely. The appearance of enemy themes in noncommunist media was a constant worry, and psychological strategists shaped their recommendations to minimize the potential fallout from U.S. actions.

Eisenhower's Psychological Warfare Advisors

This broad concept of psychological strategy assimilated the underlying philosophy that had guided the Psychological Strategy Board. But whereas bureaucratic warfare had marginalized the PSB's psychological strategy planning, Eisenhower placed such planning at the heart of American national security policy. Eisenhower's appointment of an advisor to "keep track" of the psychological dimension of U.S. foreign policy represented one way of integrating political warfare planning into his administration. He had barely been in office a day when he appointed C. D. Jackson to this position, formally the Special Assistant for Cold War Planning but casually described in Washington as Eisenhower's psychological warfare advisor.

A paradigmatic cold warrior, Jackson's faith in psychological warfare had few limits. He spoke often and enthusiastically about using psychological warfare to win World War III "without having to fight it." Walter Bedell Smith described him as "the most successful psychological warrior he had every known." Others saw him as a "renaissance man," a "great idea man," and the "archpriest of psychological warfare." He had been an advocate of psychological warfare since before World War II, when he founded the Council for Democracy to press Franklin Roosevelt to launch an all-out counterpropaganda offensive against the Nazis. After his service as Eisenhower's psychological warfare chief in North Africa and Normandy, Jackson became involved in the National Committee for a Free Europe, where he devised various liberation schemes and expanded his already-formidable network of contacts in the intelligence community. Over these years, Jackson also worked his way up the *Time-Life* corporate ladder, developing a close friendship with media mogul Henry Luce and overseeing the publication of *Life* and *Fortune* magazines. Jackson's senior position in Luce's media empire provided the administration with another asset on the public relations front, as Eisenhower could count on *Time-Life* publications to sell his Cold War policies to the American public and to the wider readership abroad. Jackson wrote dozens of letters to Luce, funneling behind-the-scenes details about the inner workings of the administration and peddling story ideas that furthered the interests of the administration's Cold War priorities. He also maintained a broader network of contacts with journalists, advertisers, business leaders, and nongovernmental organizations, and he called on them repeatedly to contribute to his psychological warfare initiatives. Jackson remained in the White House for just over a year before returning to *Time-Life*—Henry Luce insisted that he

Nelson Rockefeller, Eisenhower's advisor for psychological strategy from 1955 to 1956, meets with the president. Source: Rockefeller Archive Center, Sleepy Hollow, New York.

return—but he continued to advise the president throughout Eisenhower's two terms, frequently firing off memos, cranking out ideas, and making suggestions for new psychological warfare campaigns.[13]

After Jackson's departure, Eisenhower wrote him several times expressing regret at his absence, revealing in his personal letters a genuine feeling of warmth and appreciation for his psywar chief. "For a long time I have been acutely conscious of the gap that has existed here since you left," Eisenhower wrote in May 1954. "Situations seem to arise constantly in which I need the counsel of someone whose job it is—and who has the ability—to weigh the probable effects upon our people and the world of incidents, ideas and projects." "Your enthusiasm is always contagious," he wrote to Jackson on another occasion, expressing admiration for his "strictly un-governmental and unjaundiced approach." Later the president wrote Jackson a personal note: "I miss you—your imagination, your energy, your refreshing point of view. . . . I often wish that I had someone around, like yourself, who is willing to tackle a large problem, eager to think it through, and to come up with a concrete suggestion for a coordinated plan of attack."[14]

Nelson Rockefeller succeeded Jackson and assertively tried to expand the powers of his position. Whereas C. D. Jackson had worked alone, functioning as his own "idea man," Rockefeller relied on a large staff paid for with his vast personal fortune. He seemed not to have an idea of his own; most initiatives originated with his staff or elsewhere. Rockefeller instead focused on bureaucratic solutions to psychological strategy problems. He embarked on a number of schemes to extend his influence in the policy-making arena, creating a Policy Coordination Group and then an Office of Non-Military Warfare to use as vehicles for his considerable ambitions. Such efforts, along with his formidable ego, made him Public Enemy Number One of the foreign policy bureaucracy. Largely because Rockefeller personally antagonized virtually everyone in the administration, by the time he left the White House, the position became more focused on coordinating and implementing existing operations and less on initiating and planning new ones. The men who succeeded Rockefeller and Jackson—William H. Jackson, Fred Dearborn, and Karl Harr—were later designated special assistants for "security operations coordination" rather than "cold war planning." These men were the Cold War's inheritors of Edward Bernays's legacy. They quietly orchestrated events, created news, and manipulated the international media in the service of American foreign policy. They were also the forerunners of the communication advisors who became permanent features of the presidential staff in subsequent decades. Although later communication advisors would focus on domestic affairs, the psychological warfare experts advising Eisenhower kept their eyes on the Cold War. Theirs was the business of winning hearts and minds at home and abroad.[15]

Their efforts to integrate psychological factors into the formulation of policy, however, put them on a collision course with the State Department. John Foster Dulles erected bureaucratic obstacles to every major initiative they proposed. His jealous guardianship of State Department control over foreign affairs was legendary among administration officials, and he opposed quite possibly every single initiative that originated outside his department. In principle, Dulles supported U.S. propaganda efforts, but his view of what constituted propaganda or psychological warfare was considerably more narrow than that of the president. Dulles tended to see propaganda in terms of its most obvious manifestations: leaflets, radio broadcasts, posters. He identified psychological warfare principally with the concept of "liberation," the campaign to foment unrest behind the Iron Curtain. Eisenhower took a much broader view. He repeatedly instructed Dulles, like a schoolmaster, on the real meaning of psychological warfare. He cautioned Dulles not to treat "'psychological warfare' in too narrow a fashion," instructing him that it was "a broad program" embracing more than the liberation of the satellites. "Psychological warfare can be anything," he lectured Dulles, "from the singing of a beautiful hymn up to the most extraordinary kind of physical sabotage." Eisenhower noted that "programs for informing the American public, as well as other populations, are indispensable if we are to do anything except to drift aimlessly, probably to our own eventual destruction." During the crisis over Stalin's death, and on other notable occasions discussed later, Eisenhower overruled the secretary of state to pursue propaganda initiatives suggested by his psychological warfare advisors. In the formulation of foreign policy, Eisenhower rarely challenged Dulles directly, but he worked around the secretary's stonewalling by keeping his psywarriors in the wings until Dulles softened his opposition. A pattern repeated itself over and over. Dulles initially rejected any idea that did not originate in his department, but reversed course when he saw Eisenhower's sincere interest in the idea, at which point Dulles acted as if he had been interested all along.[16]

If Dulles rarely saw eye to eye with Eisenhower on psychological warfare matters, neither did the Department of State. The Advisory Commission on Information noted that within the department the information services suffered from "inflexibility," "internal resistance and misunderstanding," and "lack of enthusiasm and imagination." Perhaps because of this attitude, Eisenhower believed that "developing public opinion of a positive kind both at home and throughout the world" could not be done wholly within the State Department: "In fact," he wrote to Dulles, "this is so clearly recognized throughout the government that we gave the function to the [Operations Coordinating Board]."[17]

The Operations Coordinating Board

The Eisenhower administration created the Operations Coordinating Board (OCB) in September 1953 to replace the Psychological Strategy Board as the center of psychological warfare planning in the government. The Jackson Committee felt that the PSB had been ineffective and that a new structure was needed. It recommended abolishing the PSB, arguing that it was "founded upon the misconception that 'psychological activities' somehow exist apart from official policies and actions." This reasoning was disingenuous. PSB members agreed with the Jackson Committee that the psychological aspects of policy were inherent in all diplomatic, economic, and military actions of the government. They had argued this point all along. The PSB's problem was not philosophical, but bureaucratic: it treaded on the State Department's turf. Precisely because the PSB considered political, military, economic, and psychological measures in formulating its plans, it was seen as a contender for power and a rival by the State Department. Placed outside the normal policy-making channels, and without sufficient statutory authority, the PSB provided easy prey for jealous State Department officials eager to maintain control of foreign policy decision making.[18]

During the Truman administration, State Department opposition had effectively neutralized the PSB. Eisenhower and the Jackson Committee, however, believed that the board's psychological planning function was important and needed to be revived. So the PSB was dissolved and reinvented as the OCB. The new board resembled its predecessor in several respects. Its membership remained mostly the same. OCB members initially included the undersecretary of state as chair, the psychological warfare advisor, the undersecretary of defense, and the directors of the Central Intelligence Agency and Foreign Operations Administration. Its mission also resembled that of the PSB. It was created to "get the psychological factor injected into all operations" of the U.S. government so that they contributed to a "climate of opinion" favorable to American foreign policy. The OCB continued to develop and implement psychological initiatives, many of them stemming from plans originally developed by its predecessor.[19]

Eisenhower and the Jackson Committee sought to insulate the OCB from the bureaucratic conflicts that had paralyzed the PSB by establishing the new board as an adjunct to the National Security Council. The OCB worked in tandem with the NSC to develop and implement psychological strategy. "The place for consideration of political warfare problems," a British authority consulted by the Jackson Committee observed, "is at the heart of the national strategic planning process, namely in the machinery of the NSC." In its regional and country plans, the National Security Council formulated a wide variety of "psychological" tasks to be performed by the United States. NSC policy documents commonly defined U.S. objectives with such psychologically loaded words as "convince," "persuade," "inform," "educate," "clarify," "induce," "create conditions," and "foster a sense of." Indeed, one study of U.S. information activities estimated that over one-third of the paragraphs of the national security policy documents were devoted to such psychological tasks. As the study noted, "In their totality NSC directives . . . make up an imposing array of heavy responsibilities for the persuasion of foreign peoples."[20]

Still, official statements of policy produced by the NSC were themselves too general to serve as blueprints for operations. This is where the OCB came in. Its task: to develop detailed plans of action to implement the grand strategy formulated by the NSC. Frequently the NSC planning board used generic phrasing, such as "The United States should seek . . . ," to indicate U.S. foreign policy goals. The Operations Coordinating Board translated these broad objectives into operational policy. Thus, in contrast to the PSB, which aspired to influence the policy-*making* process, the OCB focused on coordinating and implementing policies that had already been approved. As the core planning body for operations that specified the instruments and techniques to be used in implementing national security policies, the OCB had wide jurisdiction over programs designed for international persuasion.[21]

The board accomplished its work through a system of over forty ad hoc working groups. Some were carryovers from the PSB, others coordinated work in particular regions, and others managed hot problems like nuclear weapons testing. After the NSC approved a national security policy statement and assigned it to the OCB, one of the working groups prepared an outline plan of operations. In organization and design, the OCB plans closely resembled military operations orders and mirrored the psychological warfare plans used by Eisenhower in World War II. A typical outline plan provided a detailed listing of the actions agreed on, identifying the agencies responsible for each operation. The planned actions included measures in the diplomatic, economic, military, and propaganda fields. Quite frequently, they also included actions the OCB wanted private groups to perform, actions to be "stimulated" by government contacts.[22]

The OCB also served what amounted to a public relations planning function. For several of the administration's large information campaigns, the board directed the public affairs offices of various agencies and departments to adopt the administration's line in their output. The agencies and departments issued press releases, held press conferences, and delivered prepared remarks designed to place this line in the international and domestic press. Through OCB guidance, public affairs officers wrote the 1950s equivalent of sound bites into public pronouncements, choreographed appearances by department heads, and staged media events to generate favorable news. Foreign opinion was the board's foremost concern, but the OCB clearly played a role in shaping domestic PR strategies as well. For major Cold War initiatives, the OCB synchronized the public relations themes for such varying agencies as the Atomic Energy Commission, the Department of Labor, and the Department of Health, Education, and Welfare.[23]

The OCB's most important work took place during regularly scheduled Wednesday luncheons, when the high-level board members met without their staffs to discuss sensitive policy matters and debate contentious issues. OCB executive secretary Elmer Staats stated that "psychological warfare and information programs always played a prominent role" in these discussions. In addition, Staats recalled, board members used these meetings to plan and coordinate covert operations. CIA Director Allen Dulles regularly briefed board members on activities being planned or considered by his organization. The State Department's representative, Walter Bedell Smith (himself a former CIA director), worked closely with the OCB and Allen Dulles in coordinating covert projects. In a public forum in 1984, Karl Harr recalled that the OCB was involved in "things so diplomatically sensitive that I don't want to talk about them even twenty-five years later."[24]

Establishment of the United States Information Agency

The Jackson Committee made over forty recommendations for improving U.S. psychological warfare efforts, some of which remain classified to this day. Eisenhower approved almost all of them, with a particularly notable exception. He rejected the committee's recommendation that the information program remain under State Department jurisdiction. Instead he sided with Rockefeller's Committee on Government Organization and the Advisory Commission on Information, which had disagreed with the Jackson committee and recommended replacing the IIA with an independent information agency. These committees argued that information programs operated by the State Department would be "timid and unimaginative because diplomacy operates primarily through contact between governments, whereas propaganda must involve large-scale operations directed at whole peoples." The Jackson Committee argued forcefully against their recommendations. Separating the information program from the State Department would undermine Eisenhower's broad concept of psychological strategy, the committee argued: "it would strongly imply the belief that propaganda is a separate element of policy, rather than a subsidiary instrument thereof."[25]

In this debate over the future of the information program, John Foster Dulles cast the deciding vote. He wanted the IIA removed from his department. He hoped to be relieved, as much as possible, from operational and administrative responsibilities so he could focus on making policy. More importantly, he wanted to insulate himself from the never-ending controversies associated with the foreign information program, identified in the minds of many conservatives with the Office of War Information and the hated New Deal. Many Republicans believed that "Communists, left-wingers, New Dealers, radicals, and pinkos" staffed the VOA and IIA. Indeed, while the Eisenhower administration was reevaluating the information program, Senator McCarthy was holding sensational public hearings on the supposed communist infiltration of the VOA. Dulles preferred to rid himself of this lightning rod, and Eisenhower deferred to his wishes. On August 1, 1953, the United States Information Agency (USIA) was officially established.[26]

The creation of the USIA produced a confusing dual nomenclature, with the organization assuming different names at home and abroad. Within the United States, the headquarters in Washington was called the U.S. Information Agency. Abroad, information posts retained their wartime designation, U.S. Information Service. The acronyms USIA and USIS effectively denoted the same general organization, but officials used the term USIS overseas because "Information Agency" had an intelligence connotation in many languages.[27]

As initially conceived, the USIA was primarily an operational agency rather than a policy-making one—closer in administrative terms to the CIA than the State or Defense Departments. In deference to Dulles's wishes, Eisenhower required that the USIA act according to policy guidance developed by the State Department. The USIA was subordinate to the Department of State and was initially excluded from permanent representation on both the NSC and the OCB. Gradually, as the USIA rebuilt the reputation of the information services (badly tarnished by McCarthy's investigations), it was accorded a greater, albeit subordinate, role in policy making. By 1957 the USIA director had become a full-fledged member of the NSC and OCB.

Although the USIA appropriated a wide range of functions, it was by no means the only government body concerned with propaganda and information. The new Foreign Operations Administration (formerly the Mutual

Security Agency and a precursor to the Agency for International Development), Defense Department (including the Army and the Air Force), and Central Intelligence Agency continued to operate extensive programs to influence world opinion. A 1959 study estimated that the combined expenditures to influence world opinion, excluding those of the CIA, reached a half-billion dollars in that year. Significantly, the USIA's budget accounted for only a fifth of the total expenditures. Several organizations that engaged in domestic information complemented these primarily foreign-directed persuasive campaigns. The Federal Civil Defense Administration conducted extensive propaganda operations in the United States, using the imperative of civil defense to prepare Americans psychologically for a prolonged cold war and armaments race. Eisenhower's press secretary, James Hagerty, used his office to "educate" the American people about the communist danger. Most cabinet-level departments, such as Defense and State, worked to influence media presentations of their activities and of the Cold War in general through their public affairs offices. The rhetoric of U.S. policy makers, including Eisenhower and Dulles, was consciously crafted to raise public awareness of the communist peril. In addition, the voluntary and privately operated Advertising Council cooperated with the government in several Cold War campaigns promoting civil defense, the "ground observer corps," and the Crusade for Freedom. As with the war bond drives of World War II, these campaigns performed two functions: the declared function of promoting a particular cause and an unacknowledged function of stimulating morale in the United States.[28]

The USIA's Global Mission

In presenting the new agency to the public, administration officials declared over and over again that the USIA did not engage in propaganda. The refrain was familiar: the enemy disseminates propaganda; the United States conveys information. While "they" tell lies, "we" tell the truth. The first director of the USIA, Theodore Streibert, picked up the theme. In a public letter, he explained that his agency would stick to "facts, and comment associated with facts." He publicly maintained that the USIA's mission was not propaganda; it was news: "We shall . . . concentrate on objective, factual news reporting and appropriate commentaries, designed to present a full exposition of important United States actions and policies, especially as they affect individual countries and areas. In presenting facts we shall see to it that they are not distorted and that their selection does

Free World

TRACK TRAINING IN TAIWAN

Volume IX   Number 4

Free World Magazine, an unattributed publication of the U.S. Information Agency, was disseminated in Southeast Asia. It featured human interest stories and positive themes to highlight the shared interests that linked the United States to the region. Source: Eisenhower Library.

not misrepresent a given situation." Eisenhower issued a public statement of his own defining the mission of the USIA in similar terms. The purpose of the Information Agency, he announced, "shall be to submit evidence to peoples of other nations by means of communication techniques that the objectives and policies of the United States are in harmony with and will advance their legitimate aspirations for freedom, progress and peace."[29]

The statements failed to mention either psychological warfare against the Soviet bloc or propaganda in the "free world" as USIA activities. Both were central to the agency's mission. Streibert prepared a secret classified exposition of USIA objectives that was more forthright. The statement outlined the "strategic principles" that would govern the agency's operations. Psychological warfare lay at the heart of the USIA's mission, it admitted: "Although our publicly assigned mission does not explicitly point to our role as a weapon of political warfare, the current conflict of interests between the United States and the Soviet Union, in which each seeks its aims by methods other than the use of armed force, constitutes political warfare. . . . The content of our operations—our message—must serve our special political warfare needs as well as our generalized long-term mission directly and concurrently." While publicly presenting the USIA as a news agency, behind closed doors, officials frankly acknowledged it to be an instrument of political warfare.[30]

Audiences outside the Soviet bloc comprised the main targets of the USIA's global political warfare offensive. According to Streibert's statement of strategic principles, "We are in competition with Soviet Communism primarily for the opinion of the free world. We are (especially) concerned with the uncommitted, the wavering, the confused, the apathetic, or the doubtful within the free world." By the end of the decade, the Information Agency oversaw more than 208 posts in 91 countries, none of them behind the iron or bamboo "curtains." Roughly 50 of these posts were in Europe; 34 in the Middle East and South Asia; 40 in Latin America; 34 in Africa; and 50 in the Far East. The British Foreign Office astutely noted the global scope of the USIA: "The target of the USIA is not simply, or even preponderantly, public opinion within the Soviet Union or the Soviet Orbit," one Foreign Office official observed, "The main target is public opinion in the non-Soviet world and particularly public opinion among those who are not fully committed to opposition to Soviet communism."[31]

Judging from the agency's allocation of personnel in middecade, the top ten largest programs were (in order, beginning with the largest) Germany, India, Japan, Pakistan, France, Italy, Thailand, Austria, South Vietnam, and Korea. These employed between 230 and 1,659 individuals, operating on budgets ranging from $300 million to nearly $7 million. The work of these information posts abroad was divided between American employees of the USIA and foreign nationals hired by the agency to assist in the preparation and dissemination of propaganda materials. Foreign nationals, or "locals," possessed the cultural and linguistic knowledge necessary to translate USIA directives into intelligible content. They were also less expensive than American employees, and they helped reduce the public profile of U.S. information operations. In Germany in 1956, the USIS employed 132 Americans and 1,527 locals on a budget of $6.8 million. In India, 62 Americans and 465 locals were employed on a $1.2 million budget. In Japan, 54 Americans, 367 locals on a $1.1 million budget. In South Vietnam, 28 Americans, 202 locals on a $900,000 budget. The agency operated other large programs employing at least 100 persons in Indonesia, Brazil, Burma, Philippines, Iran, Mexico, Great Britain, Egypt, Greece, Spain, Yugoslavia, Cambodia, and Hong Kong. All told, the USIA in 1956 stationed 867 Americans abroad and employed 5,716 foreign nationals at posts in 80 countries. [32]

Covert Propaganda Strategies

A distinction between overt and covert propaganda guided the new information agency's approach to psychological warfare. In the language of the psychological warfare expert, propaganda comes in three general varieties: white, gray, and black. "White" propaganda is official propaganda and acknowledges the U.S. government as the source. "Black" propaganda, on the other hand, is covert and subversive. It is designed to appear as emanating from a hostile source (during the Cold War, this usually meant a communist source) in order to cause that source embarrassment, damage its prestige, or undermine its credibility. "Gray" propaganda falls somewhere in between. Purporting to originate neither from the United States nor its enemies, gray propaganda is designed to appear as if it emanated from a nonofficial or indigenous source—a second party of one kind or another. The objective of gray propaganda is to advance viewpoints that serve U.S. interests, but to do so covertly, in a way that would make messages more acceptable to target audiences than official statements. [33]

At Eisenhower's insistence, most U.S. propaganda went forth in darker hues. "Propaganda must be unattributed," Arthur Larson recorded him saying: "He said he'd been trying to get this idea across for a long time."

This unattributed anticommunist leaflet distributed in Latin America asks, in a style reminiscent of the *Where's Waldo* children's books, "Which one is the communist?" Source: National Archives.

Explaining the importance of covert propaganda, Eisenhower wrote on another occasion:

> A great deal of this particular type of thing would be done through arrangements with all sorts of privately operated enterprises in the field of entertainment, dramatics, music, and so on and so on. Another part of it would be done through clandestine arrangements with magazines, newspapers and other periodicals, and book publishers, in some countries. This entire part must be carefully segregated, in my opinion, from the official statement of [the] American position before the world.

There should "be no real or apparent connection" between the overt and covert propaganda activities, he continued. It was "hopeless" to tell America's story by "lecturing and pontification." In most cases, the "hidden-hand president" explained, the hand of government must be carefully concealed from target audiences receiving American propaganda. Eisenhower instructed USIA Director George Allen in 1959 that American propaganda should be "infiltrated into local radio stations and performed by people other than Americans. . . . In order to insert such propaganda, of course, such matters as bribery, etc., would have to be indulged in."[34]

The Jackson Committee concurred with Eisenhower. It recommended that most propaganda materials avoid U.S. government attribution. The committee observed that under the Campaign of Truth, "the sheer volume of material bearing the American label" was harmful. "Audiences often do not believe information provided by any foreigner and are particularly quick to take offense at advice and exhortation received from abroad." Accordingly, the committee admonished, "propaganda or information should be attributed to the United States only when such attribution is an asset." Radio, with its overt connection to the U.S. government, was deemed the least desirable means of influencing free world attitudes. The committee recommended cutting back "white" broadcasting as much as possible. Radio programs "should be used for exhortation in the free world only on a non-attributed basis," it advised, urging "maximum use . . . of local broadcasting facilities." Consequently, the VOA reduced its broadcasts to free world countries where suitable private media existed. The VOA focused instead on the USSR and China because radio was virtually the only medium for reaching these closed countries. Outside the communist bloc, the Voice was the least important—and arguably the least effective—weapon of psychological warfare. To the extent that the USIA used radio as a communications medium, it preferred the placement of unattributed programs on independent local broadcasting stations.[35]

The Jackson Committee also recommended expanding the use of private groups, front organizations, and ordinary Americans as vehicles for transmitting propaganda messages. The committee's report stated that "far greater effort should be made to utilize private American organizations for the advancement of United States objectives. The gain in dissemination and credibility through the use of such channels will more than offset the loss by the Government of some control over content." It encouraged efforts to solicit Hollywood to contribute to the national information program. The committee further advocated close cooperation with the commercial publishing industry: "subsidize its efforts when necessary to combat the flood of inexpensive communist books in the free world." Streibert's "strategic principles" explained that the agency should "ensure that non-official, private channels for carrying the impact of the U.S. abroad are given the fullest opportunity to cooperate in making that impact serve our purposes. This cooperation must be conducted under safeguards that will protect us from charges that private U.S. activities abroad are controlled or sponsored by the Government and serve official propaganda purposes."[36]

Such "unattributed" activities also included CIA covert operations. An entire chapter of the Jackson Committee report addressed "Covert Operations in the Free World," but most of it still remains classified. The declassified sections suggest that the committee advocated a reduction in U.S. covert operations in Europe. The committee cited concerns about exposure: the size and complexity of many covert projects meant that large numbers of people were familiar with them, and many European governments were becoming increasingly unwilling to tolerate American meddling in their affairs. The committee also acknowledged that continued covert operations in Europe were less necessary than they had been in the immediate postwar years, as most European governments had staved off the threat of communist takeover and many were assuming "some of the political action programs" that the CIA had conducted earlier. The committee did believe, however, that the United States should devote more resources to planning covert operations in the third world. It particularly advocated greater attention to the countries of South and Southeast Asia and of Africa: "these areas should receive a higher priority than they now enjoy with respect to assignment of key personnel and development of expanded covert capabilities for future activities."[37]

Although the rest of the Jackson Committee's recommendations for covert operations remain classified, it is clear that propaganda and psychological warfare were among the most important techniques used by the CIA. Christopher Andrew, a leading expert on intelligence and covert operations, writes that "Eisenhower placed exaggerated reliance on covert action" because he believed it offered "an apparently effective alternative to the unacceptable risks and costs of open military intervention." Eisenhower gave the CIA virtually carte blanche to engage in covert operations and propaganda. NSC 5412/2, which provided guidance for CIA covert operations, authorized the CIA to engage in "propaganda, political action; economic warfare; preventive direct action . . . ; subversion against hostile states or groups . . . ; support of indigenous and anti-communist elements in threatened countries of the free world; [and] deception plans and operations." Such tactics were to be used to create and exploit troublesome problems for the Soviet Union and communist China, to discredit communist ideology, and to undermine communist influence around the world.[38]

NSC 5412/2 gave the CIA a sweeping mandate to manipulate popular opinion in the free world. It authorized the agency to use covert operations to "strengthen the orientation toward the United States of the peoples and nations of the free world, accentuate, wherever possible, the identity of interest between such peoples and nations and the United States as well as favoring, where appropriate, those groups genuinely advocating or believing in the advancement of such mutual interests, and increase the capacity and will of such peoples and nations to resist International Communism." Eisenhower appointed Allen Dulles as director of the CIA in part because he shared the president's faith in the efficacy of covert action and psychological warfare. "Psychological Warfare," Allen Dulles explained to the Council on Foreign Relations, "may be a more powerful weapon than you suggest. It is true they (the Communists) are using it as an *indirect* weapon, but I believe it is their *major* weapon in this period."[39]

Political warfare was also the major weapon of the CIA, he might have added. William Colby later estimated that between 40 and 50 percent of the CIA's budget in the 1950s went to propaganda, political action, and paramilitary activities. According to Colby, "While some black propaganda was indeed produced by CIA and circulated abroad, by far the largest part of its effort fell in the so-called gray area." Covert operations supported a large number of foreign political organizations throughout the free world, and, as the Jackson Committee noted, attempts were made to "establish covert influence directly over key individuals and groups in foreign governments."[40]

Considering what is now known about the Cold War activities of the CIA, the fact that it used covert means to manipulate the international news media is no longer surprising. Covert manipulation of the media, however, was not the exclusive province of the CIA. The USIA relied heavily on gray propaganda for its mission, even though it was not, strictly speaking, authorized to engage in covert operations. In many countries the Information Agency became heavily involved in "media control projects" resembling those of the CIA. A board of consultants on intelligence activities reported in 1956 that "a disproportionately high percentage" of USIA propaganda was "of the 'unattributed' or 'gray' variety." In fact, USIA's operations relied on gray propaganda so much that the board "found some of the programs to be almost indistinguishable in their operational aspects from programs of the Central Intelligence Agency—and vice versa." Competition between the CIA and USIA in the gray propaganda field reached such a level that an OCB document mandating principles for coordinating these activities earned the nickname the "Gray Treaty." The only distinction between the unattributed activities of the two agencies was that the USIA carried out gray operations that could be attributed to the United States if necessary, but the CIA carried out gray operations that would be embarrassing if American sponsorship were revealed. CIA gray propaganda needed to be "plausibly deniable."[41]

Overt Propaganda Strategies

While stepping up the covert and unattributed propaganda of the government, the Eisenhower administration softened the tone of its overt white propaganda. Eisenhower believed that overt media, such as the VOA, should be "factual and *non-propaganda*." He assigned the Voice of America and official U.S. information materials a very narrow role. He wanted the VOA to "build a reputation for straight educational reporting." It should be a "strictly official" instrument, charged with presenting an "accurate statement of the American position on great questions and problems." Accordingly, Streibert instructed USIA personnel to emphasize factual analysis, to avoid exaggeration, to acknowledge U.S. weaknesses, to qualify statements, and to refrain from boasting and excessive emphasis on anti-Soviet material.[42]

Overt programs administered by the USIA did focus on news and information that were for the most part factually correct. But the USIA did not operate in a no-spin zone. As Streibert admitted, "We are no less engaged in propaganda because we are to minimize the propagandistic." The agency struck a balance between the posture of objectivity necessary to enhance the agency's credibility and the selectivity and manipulation of information needed to further U.S. objectives. The Jackson Committee explained that the factual emphasis should not override the importance of persuasion to the information program's objectives. Its primary purpose remained "to persuade foreign peoples that it lies in their own interests to take action consistent with the national objectives of the United States." The USIA was "under no compulsion to provide all the facts, to disseminate all the news, or to report events merely because they command public attention." Materials were selected and developed to convey certain messages. Evidence was highlighted, deemphasized, or omitted, depending on U.S. propaganda needs.[43]

The USIA projected an image of objectivity that served a propaganda function. USIA officials hoped that their emphasis on straightforward news and information would contrast with the shrill ranting and raving of Soviet propaganda, thereby enhancing the agency's credibility. The USIA therefore released a large body of propaganda that took the form of news, but that was used in the service of persuasion. Although propaganda is commonly associated with lies, the most effective propaganda is believable, and often the most believable propaganda is rooted in truth. Credible news and information can have an immediate effect on public attitudes; it can be deliberately created or manufactured to convey a specific message; and it can be manipulated by timing and emphasis. Moreover, news is perceived by most as a neutral, objective activity. Propaganda in the form of news, Parry-Giles explains, is itself a form of "camouflaged" propaganda: "Because of the generic form of news, material portrayed in that format is less likely to call forth a critical response from an audience conditioned to view its press as freely functioning. This 'news' form, combined with the ideology of the free press, elevates the force of democratic propaganda, making it distinct from totalitarian propaganda, where the audience is more acutely aware of its propagandistic nature."[44]

The Operational Importance of Goals that Promise Hope

The tone and style of American propaganda during the Eisenhower years departed from the strident, anticommunist polemics of the Campaign of Truth. Under Truman, the U.S. information program had become brazenly propagandistic, earning a reputation abroad for "unashamedly plug[ging] anti-Communism." As a result of the Jackson Committee's investigation, the information services backed away from this heavy-handed approach, adopting a more positive message. The Jackson Committee cautioned that anticommunist propaganda antagonized more foreigners than it convinced. Greater progress could be made by identifying and developing the positive virtues of the free world, rather than by harping on the vices of communism. A USIA research report explained this logic in 1953:

> Denunciation of Communism is especially likely to be condemned as "propaganda." At least certain kinds of all-out denunciation are likely to make many of our Free World listeners shudder and think of

the atom bomb. They see such denunciation as additional evidence that we are wholly preoccupied with our cold war against the USSR, that we are fanatical in our crusade against Communism, that the spirit of the American Government is not very different from that of the notorious witch-hunters within it, and that we are therefore dangerous to them. They actually fear that they may be drawn into an unnecessary war, which for them would be catastrophic, because of American "belligerence." They interpret forceful denunciation as belligerence, and the more belligerent they think we are, the more they retreat into neutralism.[45]

In line with this reasoning, overt information programs focused on positive themes and generally avoided emotional anticommunist polemics. The USIA "attempted generally to offer audiences more positive concepts in its output, showing that the U.S. is not merely or even primarily concerned with opposing Communism but stands for things which humanity values, and devotes itself to human progress." It devoted particular attention to conveying "the deep morality characteristic of the U.S." and showing that "America stands for positive values, including the positive freedoms—freedom to learn, to debate, to work, to live and to serve."[46]

At the end of the decade, Karl Harr reiterated the importance of positive messages in a memorandum to the president. Harr titled his memorandum "The Operational Importance of Goals that Promise Hope." According to Harr, the morale of the free world was stagnating against the feeling that historical trends favored the communists. Harr explained that the threat of mutual destruction posed by the burgeoning nuclear arsenals was "taking a serious toll in terms of loss of morale, tendencies toward escapism and paralysis of effort." Strategic concepts like "containment" and "massive retaliation" offered nothing for the maintenance of free world morale—they were "inspirationally sterile," according to Harr. They offered "only a cheerless negative future prospect in terms of human hopes and aspirations." Taking a cue from Lasswell, Harr reminded the president that in order to sustain morale domestic populations must be able to see a light at the end of the tunnel: "A worthy goal must be coupled with a conviction that there is a plausible chance of attainment if morale is to be sustained. . . . We must 'plant a flag' in terms of a long-range political objective that is credible both in terms of offering real hope for a solution of the dilemma and in terms of possibility of achievement."[47]

This emphasis on positive themes was absolutely central to U.S. psychological strategy. Although anticommunist propaganda continued to be

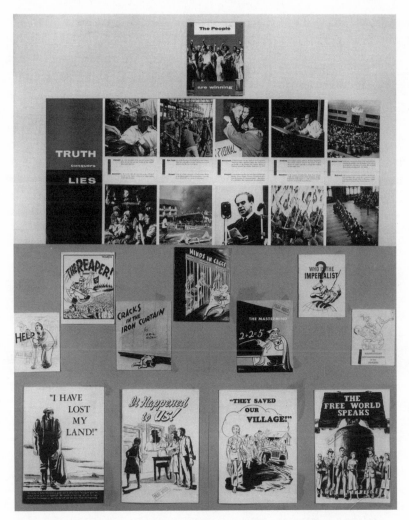

Leaflets from the Truman administration's Campaign of Truth emphasized anticommunist themes. Eisenhower believed that such materials were too brazenly propagandistic to be effective. Source: National Archives.

an important component of American psychological warfare, positive messages and goals were vital elements in the effort to win hearts and minds. This played out not just in the propaganda of USIA, but also as an element in the overall national security strategy of the administration. NSC 162/2 admonished, "Many consider U.S. attitudes toward the Soviets as too rigid and unyielding and . . . reflect too great a preoccupation with anti-communism. Important sectors of allied opinion are also concerned over developments within the United States which seem to them inconsistent with our assumed role of leader in the cause of freedom. These allied attitudes materially impair cooperation and, if not overcome, could imperil the coalition." Consequently, NSC 162/2 advised, constructive U.S. policies not related solely to anticommunism were needed: "The United States and its allies must always seek to create and sustain the hope and confidence of the free world in the ability of its basic ideas and institutions, not merely to oppose the communist threat, but to provide a way of life superior to Communism." Policies rooted in positive goals were more likely to gain acceptance from world audiences than policies rooted solely in anticommunism.[48]

The stress on positive policies and themes complemented the other aspects of psychological warfare's new look. As part of the battle for hearts and minds, Eisenhower sought to make psychological strategy an integral part of the policy-making process. Although psychological considerations did not always exert a decisive influence in foreign policy decision making, Eisenhower insisted that they be factored in at all levels, from the development of the nation's overall national security policies to their implementation in operations.

Additionally, Eisenhower sought to make U.S. propaganda more effective by making it more credible. He would not have disagreed with the Foreign Office's assessment that the USIA was primarily a "weapon of political warfare in the struggle against Soviet communism." Yet the propaganda tactics he institutionalized in the new Information Agency were more sophisticated than those of his predecessor and were far removed from the bludgeoning "big lie" approach characteristically associated with totalitarian propaganda. The Eisenhower administration camouflaged its propaganda through its overt and covert strategies. A facade of objective news reporting masked the propagandistic nature of the overt attributed materials, and covert tactics obscured the very source of the material itself. Operating on the concept that private organizations could accomplish more than official agencies, covert channels presupposed that the more they obscured the American label, the more effective they would be. Covert channels also made it possible to widely spread condemnation of communist actions without placing the U.S. government in the position of being principally responsible for this condemnation.[49]

In 1953, when the Eisenhower presidency began, a Senate investigation of the information program noted that "most countries are 'almost saturated with American propaganda' and 'suspicious of any propaganda they can detect.'" Seven years later, another study of the information program came to a different conclusion: sufficient progress had been made improving U.S. propaganda tactics that "Soviet propagandists find it desirable to borrow openly from U.S. techniques."[50]

Chapter 4

Secret Empire
Psychological Operations and the
Worldwide Anticommunist Crusade

It becomes overwhelmingly obvious that we are deeply concerned with the internal affairs
of other nations and that, insofar as we make any effort to encourage the evolution of
the world community in accord with our values, we will be endeavoring purposefully to
influence these affairs. The argument then turns out to be not about whether to influence
the internal affairs of others, but about how.

— Richard M. Bissell Jr., *Reflections of a Cold Warrior* (1996)

The officials in the new USIA saw themselves as foot soldiers in the battle for hearts and minds, ideological shock troops on the front lines of the Cold War. Typical USIA operatives were public relations counselors, journalists, secret agents, educators, missionaries, diplomats, and empire builders all wrapped up in one. They were starry-eyed idealists who sold the American way of life with missionary fervor, and at the same time hard-nosed realists, cold warriors who intervened in local politics, manipulated indigenous media, disseminated anticommunist propaganda, and did whatever else they could to advance U.S. interests.

USIA reading rooms and libraries were the most obvious signs of their presence, but they usually operated discreetly, quietly manipulating local circumstances to serve American purposes. Based at U.S. embassies and consulates in the commercial and political centers of almost every country in the world, USIA operatives worked to influence how local media presented American life, communism, and the Cold War. Armed with sound trucks, mobile film projectors, radio receivers, posters, and leaflets, they brought the communications revolution to remote parts of the developing world still largely cut off from the mass media revolution that had swept the industrialized world decades before. They taught classes in English in local schools. They printed books for teachers, journalists, government officials,

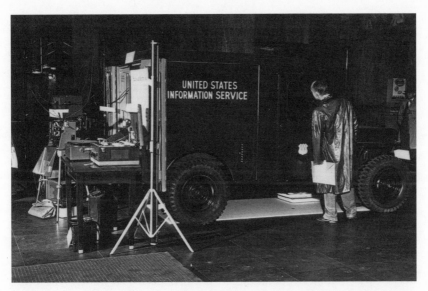

The Information Agency used trucks like this "mobile unit" to bring audiovisual programs to remote locations. A typical truck was equipped with a 16-millimeter projector, 35-millimeter filmstrip projector, tape recorder, radio receiver, record player, screen, transformer, cables, and films. The equipment was powered by the truck's engine. Source: National Archives.

and military officers. They prepared, in dozens of languages, thousands of hours of radio shows, newsreels, and films. They supplied millions of articles, photographs, and cartoons to newspapers and magazines. To radio stations, movie houses, and television studios, they brought free programming—propaganda camouflaged as education, news, and entertainment.

In these days, most U.S. propaganda themes—the themes explored in the remaining chapters—were promoted more or less uniformly around the world. Global themes were developed by the USIA, the State Department, the Operations Coordinating Board, and the National Security Council. The Department of State provided "information policy" guidance to the USIA, telling public affairs officers how to present the United States to the world. The OCB devised operations plans prescribing actions in the political, economic, military, and informational fields to expand U.S. strategic influence. In countries with a USIS post, information officers prepared "country plans" for their operations, adapting these centralized directives from Washington to fit local circumstances. To the extent USIS programs in individual countries varied, it was more a matter of emphasis than content. For the most part, the United States in the 1950s disseminated the same propaganda line across the globe.

There were some regional and national variations in programming, however, and this chapter seeks to identify the varying priorities, trends, and objectives across the "free world." American operations differed in emphasis from anticommunist indoctrination, to the promotion of free enterprise and democracy, to propaganda in support of right-wing dictatorships aligned with the United States, to public relations for American business. Information campaigns provided media support to favored candidates in democratic elections. American officials endeavored to shape the political and ideological leanings of intellectuals, teachers, students, bureaucrats, military officers, journalists, religious leaders, and other influential persons abroad. In places seemingly removed from the Cold War's front lines, countries like Iceland, Morocco, Spain, and Libya became targets of psychological operations to preserve access to military facilities. The United States also supported friendly authoritarian regimes by helping them develop mechanisms for consolidating their power, whether by training military and police forces, providing economic and military aid, assisting in campaigns against indigenous insurgencies, or disseminating propaganda. Military operations, covert action programs, and information campaigns worked to defeat communist insurgencies or undermine local communist parties. Left-leaning regimes with real or imagined communist sympathies became targets of covert political action programs to force them from power.

This global political warfare effort focused on tying countries to the "free world" coalition and isolating communist regimes according to American grand strategy. U.S. psychological operations abroad worked to foster conditions that would bind countries to the United States, thereby denying communist regimes access to critical resources, outposts for the projection of military power, and avenues for economic and political influence. In a sense, it was a program of psychologically waged containment. The term "containment" is misleading, however, because it connotes passivity and implies a reactionary response to external stimulus. In practice, U.S. operations abroad comprised a proactive effort to preserve and extend American power by manipulating political, economic, and cultural developments in other societies.

Propaganda alone did not encompass this effort. American actions to influence perceptions and politics included diplomatic, military, economic, and informational measures. Most such actions were camouflaged to mini-mize the visibility of American intervention, lest they arouse a nationalist backlash, damage American credibility, or provoke opposition at home. In these operations, one can perceive the defining feature of the new type of diplomacy that had taken root in the postwar era. Diplomatic officials were not just observing, reporting, and negotiating with other diplomats; they were working tirelessly within the limits of available resources to influence the internal affairs of the countries that hosted them. Although officials were often motivated by a desire to stop the spread of communism, their actions added up to a form of secret empire building that used covert forms of coercion and manipulation to draw countries into the American orbit.

Marketing the Atlantic Community in Europe

The new Information Agency assumed control of an existing psychologi-cal program heavily weighted toward Europe and the Soviet Union. These were the high-priority areas targeted by the Campaign of Truth, the Mar-shall Plan information program, and the CIA. Through propaganda, covert operations, economic aid, and cultural diplomacy, the United States under the Truman administration initiated a massive intervention in European affairs to prevent communist domination of the continent. This interven-tion has spawned a large body of scholarship on U.S. cultural relations with Europe, with the result that we know more about U.S. propaganda and cultural diplomacy there than anywhere else. Much of this literature has focused on the issue of "cultural imperialism" and the extent to which Euro-peans have assimilated, modified, or rejected American culture. This debate is itself a reflection of the fact that the United States had more or less won the European battle for hearts and minds by the early 1950s. The Truman administration had helped consolidate a liberal-democratic order in Western Europe that was firmly tied to the United States. Western Europe, in con-trast to other regions of the world, would not be threatened seriously with communist revolution for the remainder of the Cold War.[1]

Nevertheless, American officials had grave reservations about the strength of Europe's commitment to the worldwide anticommunist crusade. They perceived a sizable gap separating European governments, which generally supported U.S. foreign policy, and European public opinion. Officials ex-pressed particular concern about the public's "emotional neutralism," an attitude defined by its "disinclination to cooperate with U.S. objectives in the Cold War." American policy makers routinely characterized European

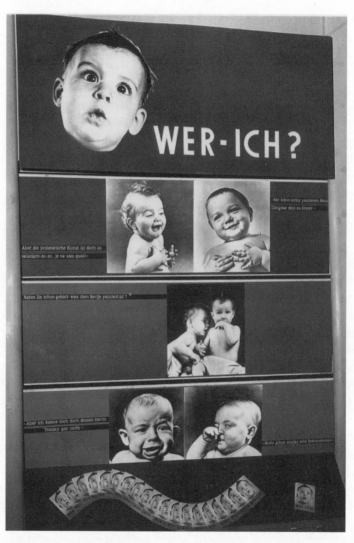

This humorous anticommunist display at the USIS office in Vienna reads, left to right, from the top: "Who Me?. . . . But proletarian art is so realistic—so—so . . . *je ne sais quoi* . . . Nothing can happen to me—my youngest is in the Kremlin . . . Have you heard what happened to that Beria? . . . But I don't even know this Mr. Trotsky . . . Not another demonstration." Source: National Archives.

public opinion as apathetic, frightened, confused, and dominated by such neutralist sentiment. "European electorates do not at present thoroughly see the realities of current international relationships in the same light as the United States," the PSB reported in a March 1953 review of U.S. information operations, citing mounting criticism of U.S. foreign policy and an alarming ignorance of the purposes of the NATO alliance. American officials feared that popular attitudes favoring neutralism and disengagement might crystallize, generating pressure on European governments to modify their foreign policies in ways inimical to U.S. interests. Making matters worse, in the eyes of American officials, Soviet propaganda was working to promote a rupture in the western alliance, striving to stimulate disagreements between the United States and Europe that would nullify U.S. leadership.[2]

Because European governments were functioning democracies with highly developed systems of mass communication, American officials concluded that the achievement of their objectives largely depended on influencing mass attitudes through the media. American psychological operations toward Europe became exercises in alliance management. U.S. information programs worked to instill a common sense of purpose and a greater awareness of the communist danger to unite NATO countries behind American leadership. U.S. psychological strategy also reflected the Eisenhower administration's priorities for the region. As part of the "New Look" national security policy, Eisenhower sought to get West European countries to bear a greater share of the defense burden. He focused American foreign policy on securing German rearmament and integration into the Atlantic system of alliances, initially through the European Defense Community, and eventually through NATO. Thus, three paramount objectives governed U.S. psychological strategy toward Western Europe: generate popular support for the Atlantic alliance and European integration; prevent the formation of neutralist attitudes toward the Cold War; and insulate American military installations from popular pressure for disengagement.[3]

To accomplish these goals, the first few years of the Eisenhower administration saw continued emphasis on propaganda in Western Europe. In 1953 and 1954, over half of all USIS overseas personnel were stationed in European countries. Of the six largest information programs in the world, four of them were in Europe: Germany, Austria, France, and Italy. The PSB concluded that U.S. infiltration of European media was of such a massive scale that even "unattributed" propaganda was provoking an anti-American backlash: "The fact that the United States has engaged in a wide array of non-attributable activities, directed to mass audiences, has resulted in a situation wherein certain of these programs promise to achieve quite the opposite effect from that intended. . . . [T]he very scale of the effort permits a suspicious and cynical public to recognize the concealed output as U.S.-sponsored. As a result, the blatancy of such propaganda contributes to the rise in anti-American attitudes and sentiments." According to the PSB, the United States needed to develop new ways and means of influencing the minds and attitudes of West Europeans while avoiding the appearance of overt intervention in European affairs. "In order to minimize the psychological disadvantages generated by . . . impressions of U.S. hegemony in Western Europe," the PSB advised, "the United States, wherever possible, should reduce the apparent scale of its intervention." Camouflaged modes of influence were imperative.[4]

USIA propaganda in Europe worked to foster the notion of an "Atlantic community," linked by a common heritage and culture and bound by shared security needs. The USIA actively promoted the administration's goal of European integration by portraying individual instances of European cooperation as steps toward European unity. It also provided unattributed media support to groups favoring European integration. The USIA stationed a public affairs officer in Paris to coordinate information support for NGOs that promoted European unification. U.S. information media likewise "played an indirect but positive role" in furthering Franco-German reconciliation. Particular emphasis was attached to increasing public understanding of the importance of West Germany's role in the mutual defense of Europe. Where large contingents of U.S. military personnel were stationed, "troop acceptance activities" were organized to reduce anti-U.S. attitudes. Above all else, U.S. information programs promoted NATO. They stressed the alliance's contribution to *collective* security: the alliance and the related network of U.S. military facilities were not there for American benefit, but to protect Europe.[5]

The USIA emphasized covert means of influencing the region's media. The tactic of cross-reporting—taking a news item from one country and publicizing it in another—became a staple of the agency's operations. Typically, information officers would "inspire" or plant a news story in one country's media, and then widely disseminate the resulting publication in neighboring countries to suggest that the ideas originated from European sources rather than American ones. The USIA also worked to increase the participation of foreign nationals in psychological operations and provided discreet financial assistance to European media. The State Department urged NATO governments to develop information programs of their own to sell the alliance to their publics and provide assistance when possible. For example, in Italy, the USIA developed a comprehensive psychological campaign in which the Italian government supplied the money and manpower for a "grassroots impact program" using propaganda materials prepared by the United States.

Similarly, in France, USIS officials labored to keep a low profile. The sensitivity of the French public to indications of manipulation by the United States reinforced the importance of acting subtly. "U.S. influence can most usefully be brought to bear by indirection and discreet persuasion," USIS Paris reported. Through contractual arrangements with five news and feature agencies, the post placed a large volume of unattributed USIS material in influential newspapers and magazines. Information officers in France courted journalists and opinion makers through regular press luncheons, off-the-record briefings, and evening sessions at the homes of public affairs officers. USIS also provided French organizations and individuals with media materials and funds to stress agency themes such as the Atlantic community and European integration. It furnished private groups with materials and equipment to produce cultural exhibits, concerts, lectures, and commemorative events that reinforced Franco-American ties.[6]

In Germany, the USIA operated with an unusual degree of visibility relative to information programs in the rest of the world. The agency stationed nearly one-third of all its overseas personnel in the country, many of them working on psychological operations toward Eastern Europe and the Soviet Union. Two media stand out as being critical to the USIS program there. First was the Radio in the American Sector (RIAS), a powerful broadcasting station aimed at Berlin and East Germany that sought to "maintain in the Soviet Zone a preference for the ideals and ideas of the free world." Second were the Amerika Häuser. Located in most of the principal cities of the Federal Republic and in Berlin, these were cultural centers and libraries used by the USIS as stations for attracting and influencing "talented young people likely to emerge as leaders in the years to come." The USIS so valued the Amerika Häuser that they consumed over half of USIS operating expenditures for the country. Unattributed media control projects sought to bolster West German morale. While conceding that West Germans were "largely if not unanimously immunized against Communist ideology," USIS officials nevertheless feared that West

Germany might drift toward neutralism, push for acceptance of disarmament schemes, or adopt a pro-Soviet foreign policy. The USIS especially targeted the Social Democratic Party and its labor constituents because they were perceived as "especially susceptible" to disengagement and disarmament schemes. One USIS program demonstrated "why armaments are necessary in today's world" in order to counter the appeal of proposals for German neutrality and disarmament.[7]

The volatile political situation in Iceland highlighted to American officials the ongoing importance of propaganda in support of the Atlantic alliance. Although Iceland was hardly a country on the front lines of the Cold War, it attracted the high-level attention of the Operations Coordinating Board when public opinion pressures threatened the American air base at Keflavik. Located some 2,000 miles from Moscow, Keflavik served as a radar and interceptor base protecting transpolar approaches to the United States. The United States faced mounting opposition to the presence of American troops in Iceland from the moment it joined NATO. Opponents of the American military presence called U.S. forces an "occupation army." They protested with petition drives, marches, and riots. The situation was "explosive," the USIS post in Reykjavik cabled Washington in January 1956. Icelandic communists and National Defense Parties were carrying on "a prolonged and intemperate campaign, as intensive and violent as they know how to make it" against the Keflavik base. American officials feared the consequences of this opposition on the country's domestic politics. "No party expecting to command extensive support at the polls can support the presence of foreign armed forces in Iceland on a permanent basis," USIS Reykjavik reported, expressing concern that Icelandic political parties, in a hotly contested election, might stake their political fortunes on the expulsion of American forces and withdrawal from NATO.[8]

To counter such trends, the USIA conducted "an all out and sustained NATO information program." Concerned that Icelanders looked at Keflavik as an American outpost, rather than a NATO installation integral to regional security, information materials stressed the importance of the base for the defense of all of Europe. Press articles attesting to Iceland's importance to NATO were placed in Scandinavian and British media, and then reprinted or cited in Icelandic newspapers. The State Department sent six leading Icelandic journalists on a tour of NATO capitals, providing them with briefings from NATO officials on Iceland's role in European security. American officials also discreetly instigated a letter-writing campaign by leading Canadians and Americans of Icelandic origin, with the expatriates alerting friends and family to the communist danger. USIS officials also helped to create a Society for Culture and Freedom in Iceland, with a roster of prestigious authors to sponsor literary contests on the Atlantic community. They flew in Danish author Hans Jorgen Lembourn to give the keynote lecture at the society's founding. Lembourn, known for his book *The Treason of the Intellectuals*, discussed the spiritual emptiness of communism and the perils of neutralism, urging Icelanders to "stand firmly and defend their ideal of democracy." In Iceland, as in the rest of Europe, American officials approached neutralism as a public opinion problem requiring a forceful counteroffensive.[9]

The Third World Challenge

Although European audiences remained targets of U.S. psychological operations throughout the Cold War, the focus of the propaganda war shifted in the years after Stalin's death. Increasingly, the American-Soviet rivalry concentrated on the third world, where revolutionary and anticolonial movements gathered steam, accelerating the long process of decolonization that so dramatically transformed the world system during the twentieth century. If the dominoes were going to fall, they would not do so in Europe, but in the third world.[10]

The ideological stakes were high. The revolutionary and antiimperialist messages of Marxist-Leninism, as well as the Soviet track record of rapid industrialization and modernization, potentially held great appeal for the new states emerging from the throes of European imperialism. Even before Khrushchev's famous speech in January 1961 declaring Soviet support for "national wars of liberation," Moscow promoted itself as a genuine supporter of anticolonial movements. As part of the policy of "peaceful coexistence," the Soviets used a combination of flexible diplomacy, trade and aid offers, and cultural, educational, and technical exchanges to build closer ties to the developing world. This shift in Soviet policy was revealed most dramatically in the fall of 1955, when a high-level Soviet delegation led by Khrushchev and Nikolai Bulganin toured India, Indonesia, Burma, and Afghanistan, publicly challenging the West for influence in these countries.[11]

The Eisenhower administration found itself embroiled in an intense political and psychological competition for the loyalties of the newly emerging states, many of them still fighting for independence and struggling through turbulent political revolutions. The U.S. Information Agency in

Officials of the U.S. Information Service traveled to a Laotian village to show American propaganda films. Here, Laotian children examine a speaker box that is playing music while the officials set up the projector and screen for the film. USIS operatives traveled great distances to bring films, screens, and projectors to towns and villages that did not have movie theaters. Source: National Archives.

middecade reduced its spending on European programs and concentrated its efforts on countries in Asia, the Middle East, Africa, and Latin America. In most of these countries, elites constituted the most important targets of U.S. psychological operations. The goal of influencing the masses received a lesser priority than winning over what the USIA termed "leadership groups": the elites who wielded most of the power in the predominantly undemocratic countries of the third world. American psychological strategists worked to inspire a form of "ideational integration" connecting influential segments of foreign societies to the United States through common intellectual, cultural, and social ties. Information programs sought to foster long-term intellectual and attitudinal developments that would enhance U.S. influence and create a positive climate for the implementation of U.S. foreign policies.[12]

Anticommunism in Asia

American propaganda in Asia, more so than in any other part of the world, prioritized expressly anticommunist themes. The region emerged as a major target of U.S. psychological warfare after the communist revolution in China and the outbreak of the Korean War in 1949–1950. Many countries were engulfed in turmoil: European colonial rule was breaking down, and new independent states were emerging amid a wash of revolutionary activity. American officials, complaining that "Red China" was acting as an effective spokesman for Asian anticolonialist sentiment, feared that communist revolutions would sweep across the region. The deteriorating position of the French imperial holdings in Indochina, along with insurgencies in Burma and Malaya, made Southeast Asia area appear especially vulnerable to communist subversion. The end of the American occupation of Japan provided another source of concern for American propaganda experts, who worried that the newly liberated Japanese might adopt a neutralist, or even procommunist, foreign policy. The Eisenhower administration embarked on a broad expansion of psychological operations in Asia. By 1960, over a third of the USIA's budget was spent on propaganda and cultural diplomacy in the region, particularly in Vietnam, Laos, Thailand, and Japan.[13]

Japan's economic potential, strategic location on the eastern flank of the Soviet Union, and ability to act as a counterweight to China made the country central to U.S. policy in Asia. Japan was also critical to U.S. strategy for denying communist regimes access to raw materials and strategic resources of the region and for integrating less developed countries in Southeast Asia into the American-led liberal-capitalist economic order. Cold War concerns led the United States and Japan to develop a close, although troubled, security relationship. Many Japanese wanted their country to pursue a more neutral foreign policy and agitated against the American military presence on Okinawa. Segments of the Japanese public resented the U.S.-Japanese security treaty, bristled at American nuclear policies and brinksmanship, and feared becoming embroiled in a new conflict. Recurring demonstrations against the security treaty culminated in demonstrations and riots that compelled Eisenhower to cancel a planned visit to Japan in 1960. American officials were certain such demonstrations were the work of communist agitators, and they were equally disturbed by the large number of Japanese intellectuals who were sympathetic to socialism or advocated closer relations with China and the Soviet Union.[14]

In an effort to counter such trends, the Eisenhower administration developed a robust psychological warfare program in Japan. From August 1945 to April 1952, U.S. occupation authorities censored Japanese media and pursued a vast array of cultural and information activities to nurture a prowestern, democratic, and capitalist mind-set among the Japanese public. But the end of the occupation raised concerns that the newly liberated Japan might pursue a neutral or procommunist foreign policy. The United States was unwilling to take that risk. In early 1952, the PSB developed a plan to govern the postoccupation phase of American psychological operations in Japan, and by 1956 the new civilian propaganda program had become the third largest U.S. information operation in the world, behind only Germany and India. The embassy reported that the USIS could "now reach large numbers of Japanese through the mass media, as well as conduct closely targeted projects directed at important audience groups." Intellectuals and students comprised the "number one target group" for U.S. psychological operations in Japan. Labor organizations and workers, perceived by American officials as leaning uncomfortably to the left, made up another critical target audience.[15]

Six years of censorship and occupation propaganda, and an even longer experience with militarist propaganda, made Japanese audiences resistant to heavy-handed indoctrination techniques. USIS officials camouflaged their propaganda by working secretly behind the scenes to manipulate the Japanese media. They disguised propagandistic intent by portraying their activities as cultural enterprises. "You're basically a cultural enterprise on the front end," one official explained, "trying to find . . . ways of doing hard propaganda things without jeopardizing your reputation and status." Thus, as in other parts of the world, USIS operations in Japan were predominantly covert and unattributed. The USIS secretly subsidized the work of journalists, scholars, and analysts with pro-American viewpoints, providing "moral and financial support to writers who wish to work cooperatively with USIS because they sincerely believe in its objectives." The USIS produced or inspired the publication of selected books at the rate of more than a hundred per year. Most were unattributed. Many reworked official American policy documents. A Japanese economist, for example, took a basic State Department document on "The Sino-Soviet Economic Offensive in Less Developed Countries," and reorganized and expanded it. *Keizai Orai*, an economic monthly, published the manuscript with USIS financial support. Copies were sent to all members of the Diet, all prefectural governors, radio and TV stations, professors, writers, and major businesses.[16]

American newsreel companies, such as Paramount and Universal, cooperated with USIS by inserting "message clips" into their regular newsreel bulletins. The USIS also contracted Japanese film production companies to distribute unattributed feature films in commercial theaters. The films incorporated anticommunist story lines, promoted Japanese defense forces, and cautioned against Sino-Japanese rapprochement "as advocated by the Communist and left wing Socialists." Five such films had been produced by 1959, all with "satisfactory distribution" and some with "very favorable press comments." The most successful was *Jet Vapor Trails in Dawn*, released in 1959 and seen by an estimated 15 million Japanese. Depicting the training of Japanese air cadets, the film differentiated between military forces under Japan's new democratic government and the old militarism of Japan. The film pointed out, for example, that jet airplanes required longer runways in order "to counter leftwing propaganda and agitation against a current U.S.-Japan program of runway extensions for military jet landings and take-offs."[17]

Hank Gosho, who directed USIS radio programs, recalled that the USIS was able to "saturate" Japanese radio with its locally produced programs in the 1950s. Because commercial radio in Japan was an emerging industry with little corporate sponsorship, Gosho and his colleagues found that radio stations would play the free USIS programs in prime time. One popular program, called *Family Album*, gave Japanese audiences "household hints" about how things were done in the United States, subtly incorporating USIS themes. Other programs were talk shows and commentaries that discussed political, economic, and military issues. Naturally, the most popular programs featured American music, which USIS officials used as leverage for getting political programming on the air. As Gosho recalled, he or one of his colleagues would travel to individual stations with a selection of pretaped programs. The music tapes, he explained, were a "good come-on, because good music programs in those early days were hard to come by in Japan. . . . In most cases, [the radio stations] knew that if they picked one or both music programs, they were, more or less, obliged to take the talk show or the commentary program." This system of informal influence, stemming from American economic largesse, provided USIS with an effective mechanism for distributing its materials so that they appeared to be private indigenous productions. Although it was known within the Japanese radio industry that the USIS engaged in the production of radio shows, the listening public was unaware of the source of these programs. The USIS used this tactic of providing free content to emerging media providers—radio, film, television, print, and newsreels—in most countries it operated in the 1950s.[18]

Southeast Asia

The perceived vulnerability of Southeast Asia to communist subversion led the National Security Council to order a strengthening of propaganda and cultural activities there in 1954. Not surprisingly, the propaganda war focused on Vietnam. Until the French withdrawal in 1954, the United States had been assisting Paris with its psychological warfare offensive against the Vietminh. After the 1954 Geneva Accords on Indochina—ending French colonial rule of the country, temporarily dividing Vietnam at the seventeenth parallel, and calling for reunification elections in 1956—the USIA initiated a "crash program" in South Vietnam to help create a viable independent government under the leadership of the American-favored strongman, Ngo Dinh Diem. The Eisenhower administration hoped that Diem's prowestern and staunchly anticommunist government would buy the United States time to strengthen defenses against the communist threat in the country and surrounding areas. The Diem regime, however, was burdened with deep-seated economic, administrative, and security problems. It was corrupt, nepotistic, dependent on American aid, and alienated from much of the population. Such factors were not immediately apparent, but building the legitimacy of the Diem government became a top priority of U.S. psychological operations in South Vietnam from the start.[19]

The USIA flooded South Vietnam with propaganda materials of all types—more than 50 million printed materials in about 180 titles in the second half of 1954 alone. The Information Agency focused its energies on public relations for Diem's government and policies. In addition to training the Vietnamese army and Ministry of Information in propaganda techniques, USIS operatives sponsored a "stepped-up village-level propaganda campaign" to build "a broader base of public support for the Diem

Vietnamese men gather in front of a USIS "boatmobile." Boats such as this one brought books, posters, and leaflets to rural areas not easily accessible by land. Most USIS operatives abroad were workers from the local population. Source: National Archives.

regime." This propaganda encouraged Vietnamese to pay taxes and trumpeted South Vietnam's social and economic development efforts. USIS also orchestrated a civic action program to provide Diem's government with "a selected, trained and disciplined body of agents who will move from village to village, seeking by the distribution of relief goods and by the organization of various propaganda efforts to counteract the effect of infiltration by Vietminh agents and to win the villages' support of the Diem government." The program intended to establish and fortify the central government's influence and control over village attitudes and activities.[20]

The OCB coordinated an "aggressive military, political, and psychological program," including covert operations, to eliminate the Vietminh forces. As part of this effort, the counterinsurgency expert Colonel Edward G. Lansdale was detailed to Vietnam in June 1954. Lansdale dispatched teams of saboteurs to North Vietnam. They targeted, among other things, a North Vietnamese printing facility. Lansdale and his agents further trained the South Vietnamese army in psychological warfare techniques.

They spread rumors of communist atrocities and organized volunteer medical teams of "free Asians" to aid populations in the South. Lansdale's most ambitious operation stimulated a mass exodus of refugees from North Vietnam. To encourage defections, Lansdale's agents warned Vietnamese Catholics of impending persecution by the communist government. They spread rumors of an imminent atomic attack on Hanoi and distributed black propaganda alleging a communist plot to deport Vietnamese workers to China. Migrants were promised "five acres and a water buffalo" in the south. The story of the American-facilitated North Vietnamese exodus became a principal theme in U.S. propaganda throughout the world. The USIA exploited the refugee story to expose the "terrorism and duplicity of the Communist Vietminh in North Vietnam" and "Communist violations of the Geneva truce." It also presented a picture of the rigors of life in North Vietnam under communism and highlighted the "moral and material assistance given Vietnam by the U.S. and other free nations to help the Vietnamese maintain their freedom and achieve national aspirations."[21]

Anticommunist and pro-Diem propaganda comprised but one element of the U.S. psywar effort in South Vietnam. Numerous aid and "modernization" programs sought to strengthen Diem's government. The Americans helped build roads, bridges, airports, waterways, railroads, telecommunications systems, and electric power facilities. They worked to improve agricultural production, to control disease, to develop educational resources, and to promote business, trade, and tourism. The United States quickly supplanted France as the number one trading partner of South Vietnam. From 1955 to 1963, the United States provided the country with almost $2 billion to finance the importation of American goods, including automobiles, typewriters, and clothing. Kathryn Statler and Marc Frey, who have studied U.S. psychological programs in Southeast Asia, have seen in these operations the unmistakable characteristics of empire building, as the United States increasingly found itself replacing French colonial bureaucracies, social services, and cultural programs with nearly identical institutions, albeit ones with an American flavor.[22]

Elsewhere in the region, the United States conducted large-scale anticommunist propaganda operations in Burma, Laos, Cambodia, Malaya, and Indonesia. As the OCB noted, "a vigorous anti-communist output has been continually sustained by USIA mass media and by field programs in all countries of Southeast Asia." In a region where literacy rates were low and systems of mass communication inadequate, the USIA contrived novel tactics to reach its audiences. To reach rural areas where communications were poor, trucks labeled "USIS" brought loudspeakers, film projectors, and movie screens to show propaganda films to villagers. Teams of operatives traveled by boats and jeeps to bring leaflets, posters, magazines and books to the countryside. The State Department operated a massive printing facility in Manila that produced millions of pamphlets and posters for distribution throughout the region. As Marc Frey notes, the agency's magazines had higher circulation levels than indigenous publications. USIA printings consumed the majority of Southeast Asian imports of ink and paper. English-teaching programs, book translation and publication programs, cultural and educational exchanges, economic aid, military assistance, and technical training provided additional means of extending U.S. influence. The USIA operated 58 libraries in Asia and published roughly 33 periodicals in 17 languages totaling 24 million copies a year. In 1959 alone, over 500 American books were translated into 23 Asian languages, and some 3 million copies of the books were distributed. The same year, 17,000 Asians studied English under USIA programs.[23]

By 1958, the United States also established mutual security programs in all the countries of Southeast Asia except Malaya and Singapore. It provided technical and material assistance, as well as training programs for local police and security forces in countersubversion techniques. These programs extended beyond practical training to include anticommunist indoctrination. In Cambodia, for example, U.S. training programs indoctrinated the National Police on "the new 'subtle' communist tactics of establishing diplomatic relationships and negotiating trade and aid agreements." The Defense Department, USIA, ICA, and CIA provided quiet assistance to develop and support the information programs of several Southeast Asian governments. The USIA also contributed publicity for anticommunist politicians in democratic elections, as in Laos, where the agency intervened in the 1955 parliamentary elections to undermine the popularity and effectiveness of the Pathet Lao. The CIA intervened in a civil war in Indonesia pitting rebellious army commanders in the outer islands against the central government of President Sukarno as part of a botched covert operation to replace the neutralist president with a more solidly anticommunist regime.[24]

Large populations of expatriated Chinese in many Southeast Asian countries comprised critical targets of USIS operations. Fearing that increasing numbers of these Chinese would turn to communism, the OCB created a special working group to develop an operations plan directed toward "overseas Chinese" in mid-1956. The working group devised

The USIA maintained a Regional Production Center in Manila to produce
pamphlets, magazines, and posters for distribution throughout Southeast Asia.
Here, a Filipino worker loads boxes of propaganda leaflets on a forklift. He is
servicing an order from personnel in Vietnam for 24 million anticommunist
leaflets. Source: National Archives.

operations to encourage the overseas Chinese communities in Southeast
Asia to organize anticommunist groups within their own communities; to
increase their orientation toward the free world; and to undermine the efforts of procommunist groups and activities. U.S. propaganda also sought
to bolster the prestige of the nationalist government on Taiwan (Formosa),
presenting it as "a symbol of Chinese political resistance and as a link in the
defense against communist expansion in Asia."[25]

The Eisenhower administration devoted particular attention to Thailand, which it considered the "the 'cork in the bottle' preventing the flow
of red ink across the map." Although Thailand was "the most pronounced-
ly pro–free world [country] of any nation in Southeast Asia," U.S. officials
feared that Thailand, surrounded by communist movements and governments, would moderate its staunchly pro-American foreign policy and
adopt a neutral, if not procommunist, position. Communist activities in
neighboring areas of Laos, Malaya, Burma, and Vietnam worried embassy
officials in Bangkok. They alerted the State Department in April 1953 of

"the urgent necessity to undertake stepped-up political and psychological action designed to maintain and fortify attitudes supporting the present pro-U.S. orientation of Thailand."[26]

In response to such concerns, the USIA developed an "anti-communist indoctrination program" to publicize the communist danger in Thailand. This included "intensive indoctrination" of government officials and military personnel, as well as a "saturation campaign" that bombarded remote villages with anticommunist propaganda materials. Because power in Thailand was not democratically based, the USIA focused on the "prime movers" within Thai society, targeting senior- and midlevel military officers and government officials; opposition political elements, intellectuals, teachers, and youth; and populations living close to insurgents in Indochina, Burma, and Malaya. The Thai government, which used anticommunist rhetoric and policies to legitimize its authority, actively cooperated with the USIA in these efforts. Thailand was the only country in Southeast Asia where "American information operations could be carried out from border to border under unobstructed political and receptive psychological conditions." A major activity of USIS-Thailand involved getting editors of local newspapers and magazines to publish articles developing USIS program themes. The articles—attributed to the publications in which they appeared—were then purchased by USIS, reprinted, and distributed throughout the country. This tactic allowed the USIS to disseminate materials appearing to originate from indigenous sources, and it enabled the USIS to subsidize independent editors who adopted pro-American and anticommunist positions. By 1956, the OCB was reporting that its anticommunist program had achieved "penetration" to the village level and it was commencing a scaled-down follow-up program to keep its anticommunist message alive.[27]

Exposing Red Colonialism

U.S. psychological operations in Asia were laced with a heavy dose of anticommunist indoctrination to counter the apparent appeal of communist ideas among nationalists and opposition groups. Anticommunist propaganda also provided a legitimizing ideology for pro-American regimes in Thailand and South Vietnam. Nevertheless, an equally serious psychological obstacle to the expansion of U.S. influence was the perception that the United States was in cahoots with the European imperialist powers that had carved up Asia and Africa at the end of the nineteenth century. So-

A USIS officer talks with Burmese author U Ba Thein in Mandalay. In the
background is a famous Buddhist shrine. Source: National Archives.

viet and Chinese propaganda routinely denounced the United States as a
neocolonialist power, portraying themselves as "the only sincere friends
of people seeking independence." Such propaganda resonated with Asian
populations, a fact that perplexed American officials. As a study of U.S.
information programs in Asia noted, "Despite our massive economic aid
and military assistance, . . . our anti-colonial record, our recognized good
intentions, our free and diverse society, we seem to be becoming more
identified with the negative aspects of the past and the status quo, particularly among younger people."[28]

The problem of identifying the United States with European imperialism was hardly unique to Asia. The threat of European domination still
loomed large in the minds of most people in the developing world—much
larger than that of the seemingly distant communist menace. To these people, imperialism, not communism, represented the greatest threat to their
peace and security, and the United States too often appeared to be aligned
with the imperialist powers struggling to maintain their empires. "It is universally admitted that the colonial era is dead," the State Department's
Policy Planning Staff observed in January 1956, "yet in the current phase

of the Cold War we [are] saddled, in the minds of millions, with the onus of colonialism."[29]

On the issue of colonialism, the United States in its propaganda work was placed in a difficult position, walking a delicate line between supporting anticolonial movements in the third world and alienating allies in Europe. Psywarriors wanted to identify the U.S. with antiimperialism, but such an identification markedly clashed with the objectives of European partners seeking to maintain an imperial presence abroad. Because revolutionary nationalism often commingled with communism and Moscow supported anticolonial movements as a way of challenging U.S. influence, the United States more often than not leaned to the side of European colonialism in the third world. This dilemma for U.S. psychological strategy made it exceedingly difficult for the administration to craft foreign policies and propaganda themes that could win the hearts and minds of peoples escaping from the yoke of colonialism. For the USIA, it was an impossible position. Year after year, the agency identified the issue of colonialism as an intractable propaganda problem, at the top of its list of global concerns.[30]

The Eisenhower administration tried to cope with these challenges by branding the Soviet Union and China as imperialist powers. Propaganda specialists developed the theme of "exposing Red colonialism" to persuade third world peoples that the expansion of communism represented a new and more powerful form of colonialism. Using an anticommunist message to deflect the onus of colonialism from the United States and its allies to the Soviet Union and China, the theme of "Red colonialism" permeated USIA media output in the third world throughout the 1950s. At the same time, the Eisenhower administration adopted a middle-of-the-road policy on colonial issues. Eisenhower and Dulles recognized that decolonization was inevitable, but they feared that a rush to independence would leave the new states vulnerable to communist subversion. Thus they expressed American support for the decolonization process, but they qualified that support by cautioning against "premature independence." On colonialism, U.S. information programs conveyed an uninspiring message that reflected this middle-of-the-road policy: the United States was for "peaceful change" and the "orderly evolution" to independence, but it was against "resistance to change" and "undue haste." By the USIA's own accounting, such efforts generally failed. Not once during the decade did the agency submit a positive report on the issue of colonialism. The agency admitted that "Western attempts to picture Soviet Russia as a colonial power itself have simply not been believed."[31]

The inextricable link between colonialism and racism further worked against U.S. efforts to promote itself as a supporter of nationalist aspirations in the third world. Segregation and discrimination in the United States called into question proclamations of the American commitment to freedom and equality, as well as the anticolonial message of American officials. In its international propaganda, the Eisenhower administration therefore sought to put civil rights issues "in perspective" by emphasizing the progress minority groups had made over the course of American history, a theme explored in Chapters 7 and 8.

Confronting Decolonization in Africa

Nowhere were the pitfalls of promoting an anticolonial message while supporting a colonial power more evident than in North Africa. The region descended into violent crisis in the 1950s, as Arab nationalists, mass protests, and terrorism challenged French imperial rule in Morocco, Tunisia, and Algeria. France tried to curb the nationalist insurgencies through a combination of token reform and brutal suppression. The Eisenhower administration struggled in vain to reconcile its desire to lend moral support for colonial self-determination with its need to cooperate with its French ally.[32]

The American interest in the North African situation was both strategic and psychological. Morocco, situated at the gateway to the Mediterranean and with access ports to the Atlantic, possessed clear strategic value. The United States had invested nearly $500 million to construct military installations there. It also had plans to build a relay station for the Voice of America. These installations were the products of agreements reached with the French government. In the likely event that Morocco achieved independence, the United States would not countenance their falling into the hands of an unfriendly regime.[33]

Psychologically, the North African crisis possessed a broader significance. The mere existence of the conflict served Soviet interests because it simultaneously fostered divisions between the United States, its European allies, and the third world. Although communist influence in the predominantly Muslim countries of North Africa was negligible (despite French claims), American officials feared the emergence of a pan-Arab, antiwestern, or neutralist movement that would jeopardize U.S. interests throughout the Middle East, providing inroads for expanded Soviet influence. A psychological strategy paper prepared in 1955 predicted a steady erosion of U.S. prestige regardless of how events developed: "Whether or not the conflict is manipulated by world Communism—and it needs not be, for, such as it is, it serves Communist aims directly—the combined violence of nationalism, settlers resistance and racial hatreds will henceforth divide the free world, whose sympathies cannot fail to be engaged simultaneously on opposite sides." Within the NATO alliance, the conflict seriously strained Franco-American relations. French efforts to suppress the nationalist insurgencies failed to gain U.S. support, because, as Martin Thomas notes, North Africa "was simply not credible as a Cold War front line." Americans grew increasingly exasperated with the brutality of French rule, and U.S. officials sought to encourage France to cut its losses and withdraw peacefully. Many French suspected that the United States was siding with the nationalists as a way of extending its hegemony over the region.[34]

The United States was trapped in a fundamental dilemma: it could not support the aspirations of Arab nationalism without increasing French suspicions of American motives. U.S. officials refrained from making bold statements in support of either the French or the Arabs, pursuing instead a middle-of-the-road policy stressing moderation and encouraging a fair and peaceful solution to the disputes. It was a recipe for paralysis. This middle-of-the-road policy created a "chronic condition," U.S. Ambassador to the United Nations Henry Cabot Lodge wrote to Eisenhower in 1956: "Because of our desire in past years not to offend either the colonial powers or the Afro-Asian powers, the United States is today in the position in which its policy receives the approval of neither."[35]

Indeed, neither did approve. Despite U.S. efforts to appear neutral and even-handed, American reluctance to support French colonial rule exacerbated anti-American sentiment in France. At the same time, USIS officials in Morocco reported working in an "atmosphere of tension and latent physical danger." In Tunisia, USIS offices were bombed one year, then raided by a French mob the next. The perpetrators of these acts may have perceived an anticolonial bias in U.S. policy, but the Arabs remained unconvinced, according to the USIA. It pleaded for "new approaches and techniques . . . to persuade Arabs that the United States is not in fact the chief support and reliance of the ex-colonial powers." It was a lose-lose situation. The North Africans suspected Americans of supporting French imperialism; the French perceived the Americans as neocolonialists out to take over "their" colonies.[36]

In such an atmosphere, winning the hearts and minds of any of the parties appeared impossible. Damage control was the more realistic—albeit unsatisfactory—goal of U.S. information programs. The OCB conceded that its psychological strategy primarily sought to "keep within bounds the damage to our standing with the Arabs." In Algeria, the goal was even more modest: "to keep our public profile to a minimum." These were hardly inspirational formulas. So long as the conflict continued, U.S. information activities remained limited, with USIS personnel confined to the minimum necessary for a "holding operation." USIS sought to avoid actions that might provoke criticism from either the French or the nationalists, pursuing instead a modest press program that kept French and Arab newspapers supplied with official U.S. statements and unofficial commentaries. Aside from this, U.S. information media strove to avoid "political and controversial subjects" altogether.[37]

The OCB closely monitored developments in North Africa. As the French withdrawal from the region became inevitable, U.S. psychological planners on the board appealed to nationalist groups to secure American influence in the resulting independent states. The OCB sought to expand ties to leadership groups in Moroccan society through cultural and educational exchanges. Such efforts were severely constrained, however, by the resistance of the French government, which remained suspicious of American motives. The Eisenhower administration tried to get around French restrictions through covert action. Jay Lovestone, a CIA operative who directed the American Federation of Labor's overseas operations, worked to build ties with moderate nationalists in the region. He nurtured contacts with free trade unions and supported lobbying efforts by Arab nationalists at the U.N. in hopes of winning the sympathies of groups likely to be influential in a resulting independent state. This "shadow foreign policy" amounted to a "deniable policy of support for moderate nationalism." While quietly courting nationalist leaders revolting against French rule, psychological strategists also sought to reassure France that the United States did indeed recognize its preeminent position in North Africa. Thus, so long as the North African crisis remained unresolved, Washington talked out of both sides of its mouth: it reassured France of its support while simultaneously courting Arab nationalists.[38]

Although France continued to fight against Algerian independence until 1962, it reluctantly signed treaties of independence with Morocco and Tunisia in March 1956. As in other postcolonial countries, independence brought with it American diplomats, economic and technical assistance workers, and information officers. In the postindependence phase of U.S. policy toward Morocco and Tunisia, U.S. psychological operations sought to consolidate the position of moderate prowestern groups and to discourage the spread of Soviet and Egyptian influence. The most urgent objective of U.S. psychological strategy for North Africa was to build public support for U.S. military bases in Morocco. The OCB's outline plan of operations for North Africa—approved days after Moroccan independence—directed the USIA to explain through press and publications the purpose of U.S. military installations. It also called for programs to improve base-community relations, tours by prominent Americans, and an expanded cultural exchange program to cement ties between leadership groups and the United States. USIA programming expanded. Its staff increased to twenty or thirty people in both Tunisia and Morocco. It opened reading rooms and libraries, and pursued the standard range of information activities: placing "packaged programs" on Radio du Maroc and Radio Tangier, distributing unattributed propaganda films, providing footage to local newsreels, and developing English-teaching programs. All were pursued with the objective of using American "soft power" to guide the new independent states into the free world camp.[39]

With the achievement of Moroccan and Tunisian independence, the pace of decolonization in Africa increased. Ghana achieved its independence in 1957, and a wave followed. Sixteen states obtained independence in 1960, bringing the total number of new independent states on the continent to twenty-seven. The drive of African territories for independence confronted the USIA with significantly enlarged opportunities and increased responsibilities. For years, the USIA had neglected the continent as the most peripheral part of the periphery—an area of primarily European concern. In most countries, USIS staff had numbered in the single digits. Improving the African program became a "major Agency task during 1957." The agency expanded its presence, creating a special office in USIA headquarters to oversee African affairs.[40]

Psychological strategists identified communism as only one small element of a broad range of psychological forces imperiling U.S. influence in Africa. As a study of U.S. information programs explained, "The African's mind is not made up, and he is being subjected to a number of contradictory forces: Xenophobic nationalism, Egyptian 'islamic' propaganda, Pan-Africanism, Afro-Asian unity, tribal rivalry, federation, sectionalism, Communism, anti-economic imperialism, and Western appeals for orderly development." Anticolonialism, however, reigned supreme. "The demand for independence in Africa today over-shadows all other issues," the study continued. "The African, whether a leader or one of the people, is comparatively disinterested and unconcerned with the issues which divide the world today, and he can be expected to resist any efforts to align him formally with either side."[41]

As in other postcolonial parts of the world, the USIA followed an often difficult path between African suspicion of U.S. support for the colonial powers and European suspicion of U.S. support for African aspirations. U.S. information policy sought to "avoid the development of a situation where thwarted nationalist and self-determinist aspirations are turned to the advantage of extremist elements, particularly Communist." Psychological operations on the continent targeted the leadership groups of the new African states. Rather than emphasizing direct appeals to mass audiences through overt forms of propaganda, the USIA prioritized programs for influencing the attitudes of military officers, students, labor leaders, intellectuals, and educators. Priority activities included English-teaching programs, cultural and educational exchanges, book and translation programs, and military and technical training programs. Such efforts endeavored to expand U.S. political influence among segments of the population most likely to exercise power. The USIA sought to win the hearts and minds of the relatively small group of educated elites in the emerging African states by forging "a cultural link between the U.S. and the influential leaders of the new states." The agency aimed at "achieving an expansion of American points of contact and influence in Africa as a means of supporting positive U.S. interests while at the same time acting as a balance to the intensified communist effort in that continent." Winning the hearts and minds of prowestern elites, it seemed, was the key to expanding U.S. influence.[42]

The Middle East and South Asia

The propaganda challenges associated with colonialism extended to the Middle East,[43] where the United States battled the perception that it was becoming the de facto successor to the British Empire that had once controlled vast swaths of territory in the region. Passions arising from the Arab-Israeli dispute and the rising tide of Arab nationalism complicated matters exponentially. In the postwar era, a broad-based nationalist movement challenged the legitimacy of the prowestern regimes created by the imperial powers as they relinquished formal empire in the Middle East. Arab nationalists opposed foreign military establishments and exploitive economic practices, resented the conservative prowestern regimes that dominated the region's politics, and urged a neutralist path between the

Cold War's opposing power blocs. Anglo-American support for the creation of a Jewish state in Palestine further inflamed Arab public opinion, providing additional fuel for the nationalist message. To American officials, the "alarming emotional drive" of Arab nationalism posed a more immediate threat to U.S. interests in the region than communism because it directly imperiled western hegemony in the Middle East. More to the point, it threatened access to the region's oil—the ultimate strategic resource and the lifeblood of Western Europe's economy. U.S. strategy toward the Middle East therefore evolved into a form of dual containment directed at preventing communist encroachment while simultaneously limiting the appeal of Arab nationalism and neutralism.[44]

The Eisenhower administration's approach to the Middle East illustrates the complex interplay between military, economic, diplomatic, and unconventional elements of American national security policies. Strategically, the overarching objective of U.S. national security strategy for the Middle East can be stated simply: preserve western access to the region's oil resources and prevent them from falling under communist control.[45] Beyond this strategic-economic imperative, the administration identified its policies and objectives for the Middle East in psychological terms. The National Security Council's 1954 policy statement for the region, NSC 5428, forecasted that the greatest danger for U.S. policy lay in the steady worsening of existing psychological trends: intensification of antiwestern sentiment stemming from the region's imperial past; hostility to the West arising from the Arab-Israeli dispute and U.S. support for Israel; and the development of sympathetic attitudes toward the USSR. The Soviet Union exacerbated these trends. It had everything to gain and nothing to lose by encouraging nationalist movements that could dislodge the United States and its allies from a region where western power was dominant. Through anticolonial rhetoric, promises of economic assistance, and cultural diplomacy, the USSR presented itself as a genuine supporter of Arab nationalism. American officials, however, privately conceded that communism per se was a distant concern. The strength of Muslim religious convictions prevented "atheistic communism" from garnering a mass following; American analysts generally acknowledged a wide gulf between the opposing ideologies of Arab nationalism and communism. Anglo-American hegemony in the region focused U.S. foreign policy on preserving the status quo and forestalling radical movements—communist, nationalist, or otherwise—that would imperil western economic and strategic interests. In the turbulent postcolonial environment, the most immediate task for U.S.

Horse and cart carry a USIS generator and film projector to remote settlements in Pakistan. Source: National Archives.

policy was to consolidate prowestern elements in Middle Eastern countries to ensure that they remained firmly tied to the free world.[46]

NSC 5428 directed that U.S. operations work to "guide the revolutionary and nationalistic pressures throughout the area into orderly channels not antagonistic to the West." Toward this end, U.S. programs focused to an extraordinary extent on elites, working to strengthen the political positions of those that offered "the best prospect of orderly progress in internal societal and economic reform and development and a prowestern orientation in foreign relations." American plans to create a regional defensive alliance reflected this objective of tying the new leadership of the new independent countries of the Middle East to the West. The Anglo-American "northern tier" strategy sought to create a defensive line along the Soviet flank consisting of Turkey, Pakistan, Iran, and Iraq. It materialized in 1955 with the signing of the Baghdad Pact, linking these countries to Great Britain. American officials had few illusions about the fighting strength of this alliance. The NSC noted that the main purpose was "political and psychological rather than military." The pact would strengthen western-oriented elements, bring about greater awareness of the Soviet threat, and encourage cooperation between local elites and the West. More importantly, the pact and western military assistance programs cultivated support from the armies in these states, which were critical in determining the extent and nature of political change. The NSC reasoned that the alliance and military aid programs "would help to induce internal stability and political orientation towards the West."[47]

Information programs comprised another element of the American strategy for nurturing ties between the postcolonial governments and the West. In January 1953, the Eisenhower administration implemented PSB D-22, the Psychological Strategy Board's program for the Middle East. The paper enunciated a program of dual psychological containment to "combat the twin extremes of Communism and anti-Western politico-religious fanaticism." Although U.S. officials recognized the potency of Arab nationalism, the psychological strategy plan developed few initiatives to counter the widespread nationalist appeal on what would later be called the "Arab street." Instead, information efforts focused on winning over local elites, concentrating to an inordinate extent on government officials, military officers, and other leadership groups. The PSB advised that U.S. programs should be designed to convince these groups that their security needs would best be met through cooperation with the West. To highlight the foreign danger, the PSB advocated strong emphasis on the "anti-religious" nature and "godlessness" of the USSR, and the deployment of every possible technique to "identify communism as a cloak for Soviet national imperialism, and to identify local communist elements as tools of a foreign power." Anticommunist messages continued to permeate U.S. information programs, despite recognition that communist ideology held little appeal in the region. The USIA stoked fears of communism to drive local elites into western arms and to prevent the Soviets from winning friends in the region.[48]

Serious limitations constrained U.S. psychological operations in the Middle East, however. Most fundamentally, the Arab-Israeli conflict worked against U.S. efforts to win the hearts and minds of the Arab world. The USIS post in Libya pointed out, "We should not delude ourselves into believing that an official information program is the be all and the end all for the U.S. position in Libya. The political picture in Libya is governed almost entirely by the U.S. position vis-à-vis Israel and Algeria." The USIS post in Syria was even less optimistic. Despite an increase in antiwestern views in local publications, it advised against a full-blown information effort there, noting that psychological operations would meet with little success so long as U.S. policy over Palestine remained unchanged. Eisenhower appeared to recognize that the United States could not hope to win over the Arab world while retaining the Truman administration's pro-Israel foreign policy bias. He endeavored to pursue a more even-handed approach to the Arab-Israeli dispute, reasoning that U.S. interests in the Middle East would be better served if the United States were perceived as a neutral broker, rather than biased interlocutor. Accordingly, U.S. policy advocated "both the fact and appearance of impartial friendship with the Arab states and Israel." In line with this approach, the USIA worked to create a favorable climate for negotiations. It played up U.S. even-handedness, publicizing official statements and diplomatic maneuvers indicating the American desire to see the conflict resolved in a fair and reasonable manner. It also worked to alleviate concerns over U.S. economic and military aid policies, focusing especially on creating "better understanding" among Israelis about the purpose of military aid to Arab governments.[49]

Operational difficulties also confronted U.S. information programs in the Middle East. Although NSC 5428 advocated a 25 percent expansion of information and cultural programs, USIS operations in the region did not receive a particularly high priority. The Voice of America broadcast for only seven hours a week in Arabic in 1953, and increased only modestly thereafter. USIS posts across the Middle East agreed that the official U.S. presence on the airwaves had a negligible impact on public opinion; most VOA programs were simultaneously too propagandistic and too dull to capture a wide audience. The paucity of indigenous, independent media outlets that could provide vehicles for unattributed propaganda further constrained USIS operations there. The governments of the region owned or controlled most media outlets in their countries. Where independent media existed, editors were often reluctant to cooperate with U.S. agents. Most materials distributed by the USIS carried the stamp of the American government, and with it the stigma of foreign propaganda that mitigated against their credibility and effectiveness.[50]

These conditions meant that USIS propaganda work relied on the active cooperation of Middle Eastern governments, cooperation that came half-heartedly, if at all. The USIA and British Foreign Office sought to prod governments in Jordan, Egypt, Iraq, Turkey, Lebanon, and elsewhere along by helping them develop propaganda capabilities. American and British officials reasoned that such aid would lead to opportunities for the discreet influence of local attitudes. At the very least, the state-controlled media outlets would provide channels for unattributed propaganda that would not be tainted by the suspicion of "plots" or "imperialism" so often associated with foreign information programs. Still, these undertakings depended heavily on the attitude of the governments, which could curtail operations or push them in directions inimical to U.S. interests.[51]

Both the British and the Americans worked to develop the state-controlled broadcasting outlets in Iraq. USIS officials in Baghdad pressed the Iraqi government to undertake positive propaganda steps against communism, hoping that the government would develop media resources that the USIS could use to disseminate unattributed anticommunist material. The Iraqi government cooperated modestly. It permitted the USIS to subsidize a few newspapers and editors and to place anticommunist programs on Radio Baghdad. From the perspective of the USIS, these were positive, but ultimately unimpressive, first steps. Neither the Americans nor the British were as successful as they would have liked in persuading Iraq to play a leading role in the region's propaganda wars.[52]

Foreign governments, of course, had their own motives for working with USIS, whether to enhance their own prestige, subvert domestic opposition, or achieve other purposes. The Iraqi case illustrates how this dynamic could produce unsavory outcomes that worked at cross purposes with American foreign policy. In December 1953, the Iraqi Director General of Propaganda, Tahsin Ibrahim, approached the American public affairs officer in Baghdad for assistance in supplying material on the "international aspects of Communism." USIS provided him with samples of unattributed anticommunist pamphlets and background studies, which Ibrahim planned to feed to local newspapers, magazines, and radio stations. The resulting campaign took on anti-Semitic overtones. The Iraqi government expressly linked communism and Zionism in a campaign it pursued with the uncomfortable assistance of USIS Baghdad. As part of this effort, the USIS and Iraqi government distributed an unattributed pamphlet entitled "We Are in Danger." Prepared by Iraqi officials, the pamphlet reprinted three newspaper articles tying communism to Zionist causes. One of the articles, titled "I Am a Zionist," purported to be a confession of a student in a local college. In the article, the student admitted that he had been procommunist and implied a Jewish-communist conspiracy. USIS Baghdad realized that such propaganda cut both ways, since "Zionism" in the Arab world was associated with the United States. Despite such an obvious disadvantage, USIS expressed satisfaction that at least the Iraqi government was attempting to attack the communists on the propaganda level, and was providing the USIS with an indigenous channel to disseminate nonattributed anticommunist material. It also acknowledged that one of the only ways to get independent editors to cooperate in anticommunist propaganda was by tying it to anti-Israeli sentiment. Eventually, the USIS and Iraqi government shifted tactics, appealing instead to nationalist sentiment when attacking the communists.[53]

Despite the many difficulties facing information operations in the Middle East, the USIA operated throughout the region and related areas of South Asia and North Africa. In Libya, the USIS labored to maintain a favorable political climate in support of the Wheelus Air Base, which performed tests of U.S. guided missiles. The Americans operated only a modest program in Saudi Arabia because of governmental resistance to USIS operations, but U.S. oil companies such as Caltex, Socony, and Aramco funneled USIS materials to their contacts. In addition, posts elsewhere in the Middle East occasionally labored to enhance the regional prestige of King Saud, whom Eisenhower hoped to promote as a prowestern "spiritual leader" on the basis of his guardianship of the holy sites of Islam.[54] Pakistan's role in the "northern tier" strategy and its participation in SEATO led the OCB to revamp U.S. psychological operations there in early 1954. The USIA created a "special psychological program" to publicize the dangers of communism and advertise "the positive contributions being made by the government, with U.S. aid, to the welfare of the people and the economic stability of the country." These efforts intensified as American officials acknowledged the narrow base of popular support in Pakistan for its alliances with the western powers, dramatically revealed by the riotous disturbances that enveloped the country in March 1956. Between 1954 and 1956, USIS operations in Pakistan increased 80 percent, making the operation there the fourth largest in the world. In neighboring India, the USIS maintained its second largest operation, employing 527 persons in 1956. There the USIS worked to alleviate the irritation created by U.S. aid to Pakistan, while simultaneously trying to limit the influence and appeal of the neutralist message of Jawaharlal Nehru.[55]

In Iran, meanwhile, aggressive psychological warfare by both the CIA and USIA played an important role in the covert operation that toppled Mohammed Mossadeq's government in August 1953. Briefly put, Operation TPAJAX (aka AJAX) traced its origins to the spring of 1951, when Mossadeq nationalized the oil industry, outraging the British who controlled half the stock of the Anglo-Iranian Oil Company. After efforts to resolve the oil issue failed, British intelligence began preparing a covert operation to restore the power of the shah of Iran and replace Mossadeq with the royalist General Fazlollah Zahedi. Playing up Mossadeq's communist ties and stressing the danger of a communist takeover in Iran, the British received a green light from the incoming Eisenhower administration in February 1953. The CIA took over planning for the coup. The agency had been conducting covert anticommunist propaganda operations, codenamed BEDAMN, since the late 1940s. It now embarked on an "all-out effort" to influence public opinion against Mossadeq and destabilize his government through psychological operations. A "mass propaganda campaign" was planned to "create, extend, and enhance public hostility and distrust and fear of Mossadeq and his government." The intense anti-Mossadeq propaganda included black operations replete with false rumors, phony currency, and forged documents proving his communist ties. Various clerical leaders and Islamic organizations played a major role by issuing condemnations of Mossadeq and statements of support for the shah. Calibrated threats by a local terrorist group and allegedly "spontaneous" demonstrations ratcheted up the pressure. Information officers in the embassy escalated propaganda attacks on Mossadeq's government, transmitting critical editorials from the American press on the VOA. Such efforts, Mark Gasiorowski argues, "played a key role in preparing the groundwork for [the coup] by undermining Mossadeq's base of support." The dramatic events of August 18–19, with crowds of CIA-backed demonstrators swarming the streets to provoke a showdown, provided the final push that led to Mossadeq's fall and the return of the shah.[56]

U.S. propaganda regarding the Iranian situation brazenly crossed the line between domestic and foreign propaganda. Both before and after the coup, the State Department worked to inspire editorials in American publications to convey "certain points of view" for the "benefit" of the American public. To illustrate the popular support enjoyed by the shah and to publicize U.S. aid efforts in Iran, the State Department took propaganda materials originally prepared by the USIS for distribution in Iran, reworked them, and distributed them to sympathetic American journalists. Such cross-reporting went in the other direction as well. The American ambassador to Iran, Loy Henderson, asked the State Department to place an editorial written by his embassy in a prominent American publication, such as the *New York Times, Time Magazine,* or *Newsweek.* Henderson wanted the editorial to appear in the U.S. press so that it could be copied and distributed in Iran without the stigma of government sponsorship. Entitled "Dilemma in Iran," the editorial argued that "Wily Dr. Mossadeq" was deposed because he had allowed his country to be infiltrated by communists. The article went on to plant the idea Henderson wanted Iranian leaders to understand: American aid might be withheld if the Iranian-government did not settle its oil dispute with Britain; time was running out.[57]

One consequence of the covertly engineered coup was an improvement in the propaganda avenues open to U.S. agencies in Iran. The United States could now count on the "complete cooperation" of the new government's division of propaganda. The USIA dispatched its operatives to various Iranian cities to seek avenues for distributing unattributed materials in local newspapers, magazines, and radio programs. When they were not disseminating hard-hitting anticommunist films and pamphlets, U.S. information officers stressed the "progressive social programs" of the new government, as well as the regime's popular support among Iranians. Despite such efforts, the shah's government developed into a propaganda liability for the United States. Its dependency on American aid and the escalating brutality of the shah's rule tainted the regime as an American client state and a symbol of neocolonialism. Even before the end of the decade, a study of psychological programs admitted that the "U.S. identification with regimes that seem to have decreasing public support" was "a very real problem in Iran."[58]

Blowback in Egypt

For all the propaganda challenges facing the United States in the Middle East, none so vexed American officials as the crusading Arab nationalist-neutralist message of the Egyptian leader Gamal Abdul Nasser. In the first few years after the 1952 coup that brought Nasser to power, the Eisenhower administration courted Nasser. It hoped to enlist him in an anticommunist coalition and secure his participation in an Arab-Israeli settlement. Recognizing Egypt's potential for shaping public opinion in the Arab world, the Eisenhower administration also helped Nasser develop Egypt's propaganda capabilities. A close relationship between American and Egyptian propaganda experts developed. The CIA provided the Egyptian government with the region's most powerful broadcasting equipment, helping to put the Voice of the Arabs on the air on July 4, 1953. The CIA furnished Nasser with analyses of Egyptian public opinion and supplied him with numerous psychological warfare experts to help him consolidate his rule through effective propaganda appeals. A former advertising executive turned CIA agent, James Eichelberger, functioned as Nasser's public relations counselor, "dreaming up ways to popularize Nasser's government in Egypt and the Arab world." Renowned psychological warfare expert Paul Linebarger—described by one CIA agent as "perhaps the leading practitioner of 'black' and 'gray' propaganda in the Western world"— secretly advised Egyptian officials on psychological operations. The CIA provided additional expertise in propaganda and internal security activities. It may have assisted in sending to Egypt about 100 German ex-Nazis and former S.S. agents, including Franz Bünsch, who had worked for Goebbels producing such anti-Semitic propaganda as *The Sexual Habits of Jews*. American specialists also taught propaganda tactics to Egyptian broadcasters and trained radio technicians and engineers. Egyptian officials involved in information activities were given "leader grants" to come to the United States, where they toured broadcasting facilities and learned media techniques.[59]

Because USIS operations required the consent of the Egyptian government, an implicit arrangement emerged whereby the USIS promoted Nasser's prestige in return for access to Egyptian media. Egyptian authorities worked closely with American information officers in exploiting such themes as "Red colonialism," while the USIS conducted PR campaigns to encourage public confidence in the Egyptian government by highlighting Nasser's "plans to improve the lot of the Egyptian people." This relationship produced operational results. The USIA maintained a production facility known as the Cairo Packaging Center that supplied Radio Cairo and other regional stations with prerecorded programs and material. This became the clearinghouse for unattributed programming. Until U.S.-Egyptian relations deteriorated in the middle of the decade, approximately half of all foreign language broadcasts on Egyptian radio were furnished by the USIA. The agency also experienced some success in placing material in Egyptian publications. USIS Cairo reported that it supplied a total of 3,500 column-inches of material per month to 110 newspapers and 40 magazines and other periodicals. The USIS also published a twelve-page Arabic weekly, *Al Sadaka* (Friendship), that was distributed to a hand-picked group of nearly 48,000 government officials, business professionals, religious leaders, and educators. It contained simple interpretations of USIA themes that were sugar-coated with entertainment features and pictures. To make the publication more acceptable to Egyptian authorities, the publication also emphasized Egypt's "economic and social rebirth" under Nasser's leadership. The USIS was promoting both Nasser and U.S.-Egyptian friendship.[60]

Providing Nasser's government with broadcasting equipment and propaganda expertise quickly became an extraordinary example of Cold War blowback. Nasser turned this equipment to waging a full-fledged propaganda assault on western imperialism and Zionism that directly contravened Washington's goals for the region and put Nasser on a collision course with the United States. Radio Cairo attacked America's British ally relentlessly, encouraging insurrection against the "British imperialists" in Sudan, Cyprus, and Kenya. Broadcasts in Jordan against the British General John Bagot Glubb, commander of the Arab Legion, were so unremitting that King Hussein felt compelled to dismiss him. Nasser vociferously opposed the Baghdad Pact as an instrument of western imperialism. Seeking to mobilize popular opinion against the Pact, he declared on the airwaves that "no Arab country should join the alliance" and charged that "the West wants to remain the master of the world so it may colonize, enslave, and exploit it." By early 1956, U.S. policy toward Nasser's Egypt had hardened. Several developments provoked the change. Nasser alienated the United States by refusing to go along with its Middle East peace plan, by reaching an arms agreement with Czechoslovakia, and by becoming increasingly vocal in his advocacy of neutralism. Nasser precipitated a full-on crisis in the western alliance when he nationalized the Suez Canal on July 26, 1956, a move that subsequently led French, British, and Israeli forces, operating under an elaborate pretext, to stage an invasion of the Sinai in October.[61]

Concerns about American standing in the Arab Middle East and the wider third world colored the Eisenhower administration's reaction to the Suez crisis. In the months leading up to the invasion, Eisenhower and Dulles attempted to convince their British ally to find a peaceful solution to the Suez situation. They argued that resorting to arms would inflame Arab nationalism, corroborate the colonialist charges of the communists, antagonize the entire Muslim world, and create a broad third world alliance against the West. Eisenhower was especially concerned about the impact of the Suez intervention in the postcolonial regions of Africa and Asia, where the crisis deeply aggravated anticolonial sentiment. He warned that the use of force "might well array the world from Dakar to the Philippine islands against us." After the tripartite invasion, he acted swiftly to distance the United States from the aggression of its allies. Determined to exert the best "psychological effect" from an unpleasant situation, Eisenhower rushed Henry Cabot Lodge to put the issue before the United Nations "first thing in the morning—when the doors open, before the U.S.S.R. gets there." Administration officials believed that Eisenhower's decisive action to rein in his allies improved U.S. standing in the Middle East and the third world. Lodge reported that "never has there been such a tremendous acclaim for your policy. Absolutely spectacular."[62]

There was some truth to this. Eisenhower's opposition to the invasion ultimately compelled the French and British to accept a U.N. cease-fire resolution. Still, Nasser was the real victor in the crisis. He emerged a hero for the cause of nationalism. Buoyed by his success, Nasser intensified his bid to unify the Arab world under his leadership, all the while declaring his neutrality in the Cold War. Nasser turned his post-Suez popularity, and his now formidable propaganda machine, to spreading his nationalist-neutralist message. The USIA identified Nasser's nationalist propaganda drive as a greater threat to American interests in the Middle East than communism: "While Communist propaganda continued to attack the U.S. as imperialistic, of more concern to USIS was nationalistic propaganda from the area which carried the same line. This applied particularly to the Pan-Arabic movement led by President Nasser . . . This propaganda machine professed . . . 'neutralism' but, for whatever motive, the propaganda was aimed primarily against the United States and its Western allies." The main task for U.S. information policy became showing "Arab peoples that, contrary to Egyptian and Soviet propaganda charges, the United States does indeed favor Arab nationalism and the drive for Arab unity."[63]

Support for Arab unity may have been the propaganda line, but it hardly described U.S. policy. The Eisenhower administration actively intervened in what Malcolm Kerr described as the "Arab Cold War," the competition between Nasser and his rivals for regional leadership. While professing support for Arab unity, the Eisenhower administration worked strenuously against Nasser's effort to achieve it by pursuing a "divide and rule" strategy to undermine Nasser's prestige and support the opposition by his rivals. John Foster Dulles began implementing this strategy in March 1956. In cooperation with the British, he developed Operation Omega, a covert psychological-political strategy for undermining Nasser's prestige, containing his influence, and compelling him to adopt a prowestern orientation. On Dulles's instructions, U.S. aid deliveries to Egypt were reduced, halted, or delayed. At the same time, Dulles directed U.S. agencies to build up Nasser's regional rivals, lending additional support to the Baghdad Pact and to prowestern elements in Lebanon, Jordan, and Saudi Arabia.[64] Covert operations were developed to undermine and discredit pro-Nasser nationalist groups in Syria and Lebanon. In Syria, the American and British intelligence services developed Operation Straggle to destabilize the government

and undermine the socialist and pan-Arabist Ba'th Party, which was nurturing close ties to Egypt. On the eve of Syria's merger with Egypt to form the
United Arab Republic, the USIS employed "indirect propaganda methods" on a scale "similar to that undertaken in countering the Communist
drive in Guatemala." In Lebanon, CIA operatives distributed briefcases
full of cash as "campaign contributions" to help prowestern politicians defeat the nationalist candidates supported by Nasser. The enunciation of the
Eisenhower Doctrine in the spring of 1957 and subsequent deployment of
troops to Lebanon were extensions of these efforts to thwart Nasser's drive
for regional hegemony. Although couched in anticommunist terms, they
were ultimately more about containing Nasser than communism. Both
moves were laden with psychological significance: to stem the tide of Nas-
serist nationalism and to reassure prowestern governments of American
resolve, thereby preventing a global slide into neutralism.[65]

At the propaganda level, the USIA contributed to this policy by changing its approach toward Nasser and Egypt. The agency sharply curtailed
cooperation with Egyptian media channels, halting all programming that
contributed to the prestige of Nasser and Egypt. It canceled a series of information projects that were in the works, including programming on the
Aswan Dam, Egypt's land reform programs, and Nasser himself. Egyptian
commentators on USIS radio programs were replaced with non-Egyptians.
U.S. propaganda experts that had once helped Nasser develop his regional
propaganda network now began exploring measures to block Radio Cairo's
antiwestern broadcasts by providing jamming facilities to Egypt's neighbors.
They also reversed their pro-Nasser propaganda line and began promoting
the prestige and power of Iraq and the Hashemite King Faisall II to serve as
a counterweight to Nasser. Unattributed stories about Iraq were prioritized
throughout the region, especially in Lebanon, Syria, and Jordan. These
efforts to build up Iraq continued until a coup in 1958 overthrew the king
and his government, a development that led to an improvement in U.S. relations with Egypt and another turn in U.S. psychological strategy for the
region: American officials turned to Egypt as a counterweight to the new
left-leaning Iraqi government of Colonel Abd al-Karim Kassem and began
contemplating covert action to overthrow the regime in Baghdad.[66]

In the short term, Washington's "divide and rule" strategy may have
succeeded in preventing pan-Arabism from producing a unified Arab state
that could challenge western interests, but it failed to counter the ideological appeal of Arab nationalism and the accompanying strains of antiwestern and anti-American views. While working to buy the loyalties of ruling

elites, Washington had virtually no strategy for winning the hearts and minds of the masses. The United States focused its psychological efforts on boosting the prestige of western-backed monarchs and politicians lacking popular appeal and legitimacy to most people of the region. American officials focused so much on elites and regimes that they underestimated the appeal and influence of popular nationalism on regional politics, a remarkable oversight for an administration that prided itself on psychological strategy. Nasser derived much of his influence from the popular support his nationalist-neutralist message received on the "Arab street." His charged attacks on Israel and western imperialism, his emotional plea for Arab unity, and his apparent willingness to stand up to the West during the Suez crisis earned Nasser popular support among the Arab masses and gave him leverage over regional politics that Washington could not counter effectively, except through reliance on reactionary allies, covert action, and force. Washington's elite-focused strategy, moreover, fundamentally worked against its efforts to demonstrate American anticolonial credentials. The strategy, after all, closely resembled that followed by the imperial powers that had, under the "mandate" system, placed prowestern leaders in power throughout the Middle East, providing increased local autonomy for conservative rulers who guaranteed the imperial interests of their former colonial masters. For the United States, the cultivation of local elites was a pragmatic policy that ensured stability and protected western strategic and economic interests. But it was also a policy that lent credence to nationalist charges of neocolonialism and contributed to the radicalization of Middle Eastern politics.[67]

Latin America as Public Relations for American Industry

American officials also expressed concern about nationalistic sentiment in Latin America, but in this area of long-standing American hegemony their concerns revolved around a peculiar form of economic nationalism that threatened U.S. commercial interests. Psychological programs toward the region focused on overcoming resistance to U.S. economic policies as well as on anticommunist themes. Economic issues dominated USIA output toward Latin American countries. The agency stressed the benefits of free enterprise, trade unionism, and inter-American cooperation in its regional information programs.[68]

Although the United States maintained good relations with most Latin American governments, the region's authoritarian regimes retained their power through nationalistic appeals to the masses and remained susceptible to domestic political influence. The Eisenhower administration feared that runaway nationalist sentiments might create irresistible pressure to adopt anti-American positions on critical issues or to pursue radical economic policies inimical to U.S. interests. Pointedly highlighting the economic stakes involved, the NSC noted that "in some countries . . . nationalism expresses itself strongly against proposals for the development of natural resources, especially petroleum, by U.S. private capital." Although American officials conceded that there was no danger of overt communist attack in the region, they nevertheless perceived that Latin American discontent, nationalism, and anti-Americanism created a volatile public opinion climate susceptible to communist manipulation and Marxist appeals. Communists might find in regional disaffection avenues for exerting greater influence on regional political and economic affairs, thereby threatening U.S. economic interests and destabilizing the region.[69]

The Eisenhower administration attached a high priority to improving the climate of opinion in the hemisphere. Dr. Milton Eisenhower, the president's brother, toured ten Latin American countries in June 1953, returning with recommendations for improving U.S. relations with the region including an increase in USIA operations to "strengthen understanding and mutual respect." The NSC, in its 1954 policy statement for Latin America, also called for enlarging informational, cultural, and exchange programs. The OCB followed this policy statement with an urgent plea for expanded psychological operations. Winning the "support of the peoples as well as the governments for our major policy objectives is a task of enormous and long-range proportions," the OCB advised. "The principal problem will be to find the means of associating the United States with the aspirations of the peoples of Latin America, thus counteracting communist propaganda which consistently and often effectively portrays the United States as the defender of 'so called exploiters' in opposition to the interests of the common people."[70]

Heeding these warnings, the USIA program in Latin America expanded by 20 percent in 1955. It concentrated most of its regional resources on seven priority countries: Brazil, Chile, Bolivia, Mexico, Guatemala, Ecuador, and Argentina. The foremost objective of the USIA's operation was countering nationalist sentiment targeting American industry. The USIA created a special "economic information program" for Latin America that amounted to public relations work for American business. It included books extolling the virtues of free enterprise, radio programs on economic

An audience in San Andres, Mexico, views an outdoor showing of a USIS film.
Source: National Archives.

themes, and films on hemispheric interdependence. The agency also developed a regional theme, "partners in progress," to market the hemisphere's economic interdependence and to publicize U.S. economic and technical assistance programs. USIS Mexico formed the hub of information operations in the region, producing films, pamphlets, press materials, and radio scripts for use by other USIS posts throughout the hemisphere. In Brazil, Chile, and Argentina, special work was carried on "to reduce resistance to U.S. economic policy." In Brazil, this took the form of carefully guarded activity to build opinion favoring the development of local petroleum resources, whereas in Chile, USIS efforts focused on overcoming the "unreasonable resentment of the U.S. as the author of Chile's economic ills." Throughout Latin America, labor groups comprised a critical target audience for U.S. operations. The USIA expanded its labor educational program through publications, translations, moving pictures and broadcasts, urging the major labor federations to take "aggressive anti-communist action." Officials in the State Department and FOA worked to develop "responsible" labor organizations that were "conscious of the communist menace." American officials worked with ORIT, a regional trade union organization, to promote labor organizations that would challenge communist-dominated unions. A labor leader training program brought 118 Latin

Americans to the United States for orientation and training. In Chile, the USIS focused most of its media activities on strengthening labor organizations. Its motion picture program was directed exclusively at labor unions and a radio soap opera, *José Strong Obrero*, targeted labor audiences. USIS officials in Santiago also developed a close relationship with the public relations representatives of large American companies operating in Chile, encouraging those companies to utilize USIA propaganda themes in their PR programs.[71]

The potential appeal of communism and Marxism to workers and other groups in Latin America produced an additional emphasis on anticommunism in U.S. propaganda toward the region. Developments in Guatemala provided a warning for how economic discontent could fuel political revolution with disruptive consequences for U.S. economic and political goals in the hemisphere. Economic discontent formed the backdrop to the popular demonstrations that helped propel the leftist junior officer Jacobo Arbenz Guzmán to power in 1951. Arbenz embarked on a land reform program, legalized the communist party, accepted arms from Czechoslovakia, and nationalized the United Fruit Company, an American conglomerate controlling vast resources in the country. Such measures signaled to American officials that Arbenz was a tool of the communists. They feared that communism in Guatemala might spill over into neighboring states, touching off a wave of communist activity. An extensive public relations effort by the United Fruit company, organized by Edward Bernays, stoked such fears by painting Arbenz as a communist agitator threatening the stability of the entire region. In the summer of 1953, the NSC authorized the CIA to conduct an intensive psychological and paramilitary campaign to overthrow the popularly elected government of Jacobo Arbenz. In this well-known covert operation, analyzed at length elsewhere, an intensive clandestine psychological warfare operation destabilized the Arbenz regime and prepared the way for a military coup led by Colonel Castillo Armas and his CIA-backed army in June 1954. Nick Cullather's superb internal, classified history of the coup reveals that the operation rested fundamentally on economic, political, paramilitary, and informational measures used for their psychological effects: to erode support for Arbenz among the army and general public, to sow fear, confusion, and panic in his government, and to isolate Guatemala in the region.[72]

Developments in Guatemala reverberated throughout the region in U.S. propaganda programs. In the months before the coup, the USIA used the Guatemalan story to dramatize the danger of communist penetration of Latin America. The agency worked to spread awareness of "the real threat to peace and security posed by the verifiably communist penetration of the Guatemalan government." Information officers planted unattributed articles in regional publications labeling certain Guatemalan officials as communists and identifying Arbenz's policies as communist inspired. Information officers around the world labored to dispel the notion that the Arbenz regime was a "'homegrown' revolutionary movement dedicated to improving the lot of the exploited Guatemalans." They also sought to deflate the notion that economic imperialism and United Fruit interests were behind American opposition to Arbenz.[73]

Following the Czech arms shipment to Guatemala in May, the United States shifted its Guatemalan propaganda program into high gear. John Foster Dulles exaggerated the size of the arms shipment, suggesting to reporters that it would enable Guatemala to develop a massive army capable of overwhelming its neighbors. CIA operatives provided further "evidence" of communist military involvement by planting a cache of Soviet arms and then publicizing their "discovery." The USIA, for its part, embarked on an aggressive information effort, utilizing all available resources, to expose and discredit the Arbenz regime as communist-dominated and to dramatize the threat Guatemala posed to the peace and security of the hemisphere. Using CIA and State Department intelligence materials, the USIA prepared more than 200 articles and scripts for press and radio placement. It distributed 27,000 anticommunist cartoons and posters and 100,000 copies of the pamphlet "Chronology of Communism in Guatemala" throughout Latin America. In Cuba, the USIA used a local radio network to broadcast hard-hitting anticommunist commentaries on Guatemala at peak hours, without attribution.[74]

The emphasis on Guatemala in regional programming continued after the coup. The CIA developed Operation PBHISTORY to demonstrate Soviet machinations in the hemisphere by unearthing evidence of communist penetration of Guatemala. The CIA cabled its Guatemalan station advising it of the "extreme importance" of documentary evidence to counter allegations that Guatemalan communism was a "purely indigenous affair, not directed, controlled or guided by world communist [headquarters]." The PBHISTORY team reviewed more than 500,000 captured documents but discovered little evidence to support the U.S. claim that Arbenz was a Kremlin stooge. Nevertheless, some documents, including letters revealing Arbenz's "procommunist bias" and photographs of Marxist literature in his personal library, were released to the press and distributed to select individuals. The USIA joined the effort, dispatching two cameramen to Guatemala to document communist atrocities. The agency gave the resulting short film on Guatemalan communism worldwide distribution. Within Guatemala, the USIS sought to help the new government consolidate its control by hyping all newsworthy events that could be used "to recall the damage done there by the previous pro-Communist regime." Radio programs, traveling exhibits, press, and pamphlets were used to "re-educate those sectors formerly most exposed to Communist propaganda" and to "encourage confidence in the present, middle-of-the-road government." Shortly after the coup, Richard Nixon toured the country as part of a Latin American tour. The USIS commemorated his visit with a film entitled *Guatemala Makes a Friend.* Two years later, Armas was assassinated. The USIA portrayed it as a communist plot.[75]

Soviet efforts to expand trade and cultural contacts with Latin America in middecade set off alarms about the danger of communist penetration. In April 1956, the OCB developed an "Outline Plan of Operations against Communism in Latin America" calling for a "maximum effort to associate communism with subversion." The board ordered a wide range of activities to discourage diplomatic, military, and other contacts between Latin America and communist countries. Diplomatic officials exerted pressure on Latin American governments to outlaw communism and communist parties, to exclude communists from political and military offices, and to implement other restrictions on communist activities. U.S. diplomats also prodded regional governments to adopt travel restrictions that discouraged private contacts with people from communist countries, and USIA programs sought to intensify "psychological deterrents to travel, especially on the part of youth, to the Soviet Bloc governments." The OCB further directed that "all attributed and non-attributed actions" educate Latin Americans on the "subversive, conspiratorial, fraudulent and brutal nature of communist action, and of its overriding ulterior purpose to serve Soviet Bloc intervention at the sacrifice of the welfare of the people of the country." The USIA produced scores of new anticommunist posters and booklets. It distributed 800 books on anticommunism and 90,000 copies of an anticommunist cartoon book. Many additional unattributed political cartoons by local artists were disseminated through the regular media. The United States later initiated covert psychological operations to undermine the communist government of Fidel Castro, which seized power in January 1959. The CIA developed a covert action program to replace the Castro regime with "one more acceptable to the U.S." including an intensive program of clandestine anti-Castro propaganda. American, Caribbean, and Latin American media resources were used to incite opposition to Castro both inside and outside of Cuba.[76]

American psychological operations abroad functioned both as weapons of political warfare and tools of empire. Reflecting Eisenhower's preoccupation with containing or rolling back communism wherever it appeared, anticommunist themes pervaded American propaganda in all regions of the world. Stemming the spread of communism determined many operations abroad. Yet it is notable that in the global battle for hearts and minds Eisenhower's psychological strategists often identified ideological currents other than communism as greater obstacles to the achievement of U.S. objectives. Psychological operations were pursued not just in areas susceptible to communist revolution, but in virtually every country of the world. Some of the largest information programs were conducted in countries with strong governments possessing negligible ties to communism and pursuing pro-American or neutral foreign policies. In many countries, officials perceived runaway nationalist sentiment, emotional neutralism, or passionate anticolonialism as more immediate psychological threats to American interests than communism. It was a matter of faith that political and ideological movements that challenged American economic hegemony, strategic interests, or political objectives by their very existence served the presumed Soviet aim of world domination. Covert media control projects and psychological programs endeavored to neutralize these oppositional elements, while at the same time lending support and inspiration to those that shared U.S. interests and aspirations.

The perception that the Cold War was a total war, albeit a war by other means, meant that political warfare—perhaps more than shifting strategies of containment—became the strategic concept underpinning U.S. programs for waging the Cold War. In the global anticommunist crusade, political warfare came to mean the use of all available means to influence the political, economic, strategic, and psychological orientation of foreign countries. The objective was to bind them to the American-led "free world" coalition. The United States became deeply involved in the internal affairs of other nations, not just through the covert operations wing of the Central Intelligence Agency, but through a wide range of programs designed to extend American influence abroad. Operations to manipulate perceptions and politics became normative features of American diplomacy. Intervention in foreign internal affairs was standard operating procedure.

This did not mean that American operations always met with success, or that the extensive penetration of foreign media necessarily translated into political influence. Nevertheless, it does highlight a key feature of American empire building after World War II. This was an empire built not just by "invitation," not just through persuasion, not just by economic expansion, and not just through mutual recognition of shared values, interests, or security needs. The American empire was also a covert empire built on subtle manipulation. It rested not on military conquest and absolute control, but on informal modes of dominance camouflaged to reduce the apparent size of intervention. The United States projected its power through the more sophisticated and secretive means of media and political manipulation, rather than through the gunboat diplomacy and political residents that characterized more formal types of empire. As Niall Ferguson has suggested, the American empire was an "empire in denial," constructed without imperial pretensions and in fact justified by an expressly anticolonial ideology. Covert modes of dominance, along with the total Cold War mind-set, perpetuated that denial. If foreigners occasionally cried neocolonialism, it was because they perceived more fully than Americans the full impact of the anticommunist crusade.[77]

Part II
Global Themes and Campaigns

Chapter 5

Spinning the Friendly Atom
The Atoms for Peace Campaign

We are trying to convince the world that we are working for peace and not trying to blow
them to kingdom come with our atom and thermonuclear bombs.

—Dwight D. Eisenhower, Hagerty Diary, March 22, 1955

Of all the images generated by the half-century of superpower conflict, none better symbolized the Cold War than the mushroom cloud. The billowing plume of smoke and ash, burned into world consciousness with the atomic bombings of Hiroshima and Nagasaki, became the preeminent symbol of the nuclear arms race, the most frightening aspect of the superpower rivalry. Fears of the atomic bomb penetrated deep into the human psyche, generating for the first time in history the awareness that human beings, not some supernatural force, possessed the capability for bringing about their own extinction. The terror induced by the nuclear arms race intensified after the first test of a hydrogen bomb in November 1952. The first thermonuclear device, detonated in the middle of the Pacific Ocean as part of the Operation Ivy test series, was 500 times more powerful than the first atomic bombs. It left a crater in the seabed a mile long and 175 feet deep, virtually eliminating one of the test islands. Thermonuclear weapons stoked fears of atomic death by the slow poison of radioactive fallout and made possible the new horror of instantaneous annihilation via ballistic missile attack. "There is an immense gulf between the atomic and hydrogen bombs," Winston Churchill later remarked. "The atomic bomb, with all its terrors, did not carry us outside the scope of human control . . . [But with the hydrogen bomb], the entire foundation of human affairs was revolutionized, and mankind placed in a situation both measureless and laden with doom."[1]

The psychological impact of nuclear weapons deeply concerned American national security planners. Atomic fears, they acknowledged, were useful for reminding the world of the threat posed by Soviet military might, yet dangerous for their potential to induce terror, panic, and apathy.

Officials perceived that a certain level of fear was necessary to sustain public support for high defense expenditures, entangling alliances, and international engagement. But excessive attention to the horrific consequences of nuclear warfare would induce paralysis. It would undermine public morale, erode support for weapons programs, weaken allied resolve, and exacerbate neutralist tendencies abroad. Atomic terror encouraged "peace at any price" attitudes and increased public pressure for disarmament. Officials identified nuclear fear as a leading cause of neutralism because it spawned feelings of hopelessness and worked against effective Cold War collaboration. Peace activists, disarmament advocates, and communist propagandists consciously played to atomic fears in their campaigns to rally public opinion behind the abolition of nuclear weapons. To national security strategists, all of these factors pointed to the necessity of carefully managing domestic and international perceptions of the nuclear danger.[2]

From the dawn of the atomic age, U.S. officials used numerous PR strategies to present "the atom" to the international community. The atomic bombings of August 1945 provided much of the world its first introduction to nuclear science, inextricably linking the words "atom" and "bomb" in public discourse. In the late 1940s, the Atomic Energy Commission (AEC) worked to soften this association by evoking a more positive image of the atom—one associated with health and prosperity rather than the mushroom cloud. The Eisenhower administration intensified these efforts to manage worldwide fears of nuclear annihilation through a systematic and sustained effort to publicize the peaceful applications of atomic science and industry. The campaign began with the celebrated Atoms for Peace proposal of December 1953. In a dramatic speech before the United Nations, Eisenhower suggested that the superpowers contribute fissionable materials from their stockpiles to an International Atomic Energy Agency to be put to peaceful use in agriculture, medicine, and electric power production. Eisenhower presented the proposal as a new approach to arms control. Sidestepping the intractable problem of inspection that had derailed disarmament negotiations since the end of World War II, he offered Atoms for Peace as "a new channel for peaceful discussion" that would lead to more substantive agreements by enhancing superpower trust.[3]

Contrary to Eisenhower's claims at the time—and those of many historians since—Atoms for Peace was not premised on breaking the disarmament deadlock. Internal planning documents reveal that the imperative of shaping and manipulating domestic and international perceptions determined the development, planning, and exploitation of Atoms for Peace.

On one level, the proposal was a political warfare tactic to discredit Soviet peace overtures by offering a seemingly realistic proposal that the Soviet leadership would likely refuse. On another level, the initiative was part of a broader effort to mold public perceptions in the thermonuclear age. Targeting U.S. allies, neutral nations, and domestic audiences, Atoms for Peace sought to manage fears of nuclear annihilation by cultivating the image of the "friendly" atom. By flooding the media with talk of the peaceful applications of atomic energy, the administration hoped to divert attention from the nuclear buildup taking place under the doctrine of massive retaliation.[4]

Atoms for Peace thus sought to help world public opinion adapt psychologically to the presence of nuclear weapons in their everyday lives. As Paul Boyer, Allan Winkler, and Margot Henriksen have shown, the atomic bomb figured prominently in the popular culture of the early Cold War. Deeply rooted fears of nuclear destruction seeped into the consciousness of ordinary Americans through popular music, comic books, movies, and novels—not to mention the almost daily barrage of news stories pertaining in various ways to the atomic menace. Government efforts to ease these atomic fears, however, have been largely ignored by the sizable literature on Cold War culture.[5] Considering the tight control of information on the effects of atomic and thermonuclear weapons, especially in the USSR, people at home and abroad relied to an extraordinary degree on the American government for information on the meaning and consequences of developments in atomic weaponry. The government had numerous ways to spin its policies for public consumption, and Atoms for Peace represented a sophisticated public relations strategy for doing so.

Eisenhower's speech before the United Nations was merely the first round of an intensive information campaign that lasted through his eight years in office and beyond. The historical record pertaining to this effort is extraordinarily detailed, making Atoms for Peace a unique case study into the strategies and tactics—the "sources and methods," so to speak—of the American battle for hearts and minds. Atoms for Peace fulfilled Eisenhower's vision of an "ideal type" of psychological warfare campaign. It was a high-level initiative originating from the White House, backed up with concrete actions in the diplomatic and policy arenas, publicized endlessly by the USIA, exploited by every arm of the executive branch, and assisted by numerous private individuals and organizations. This carefully orchestrated psychological warfare campaign was, in every sense of the word, massive in scope. Beginning with the positive, substantive proposal on the part of the president and continuing to the coordinated exploitation of the "friendly atom" by governmental and private resources, Atoms for Peace was quite possibly the largest single propaganda campaign ever conducted by the American government.[6]

Operation Candor and The Age of Peril

Atoms for Peace developed from discussions about how to present the nation's new thermonuclear capability to the American people and to the world. A few weeks before the Ivy test series in October 1952, Stefan Possony, a Defense Department consultant to the Psychological Strategy Board (PSB), urged the PSB to begin planning a new psychological effort in the field of atomic energy to take the sting off the advent of the thermonuclear age. The forthcoming hydrogen bomb test demanded the proclamation of "a new and truly attractive atomic program," he argued, because fear of the bomb was being effectively used against the United States: "Our enemies contend that reliance on this 'weapon of mass destruction' reveals the 'barbarous' character of American 'imperialism.' Moreover, our preoccupation with the atomic bomb rather than with atomic energy allegedly is indicative of the warlike character of present American policies." Possony continued to explain that U.S. reliance on atomic power imposed a psychological constraint on American foreign policy. It stimulated the pacific inclinations of allies who feared that atomic warfare would endanger their national survival. The United States needed to create psychological space for continued nuclear weapons development by exploiting the nonmilitary applications of atomic energy to the fullest. "It must indeed be realized that the atom as a peace and prosperity maker will be more acceptable to the world than the atom as a war maker . . . even the atomic bomb will be accepted far more readily if at the same time atomic energy is being used for constructive ends." In other words, the development of atoms for war required the cultivation of atoms for peace.[7]

A long report on disarmament, submitted to Eisenhower when he took office, also addressed the issue of presenting the thermonuclear age to the public. The report on "Armaments and American Policy" was prepared by a distinguished panel of consultants appointed months earlier by Dean Acheson and chaired by nuclear physicist Robert Oppenheimer. Although the panel eloquently articulated the importance of disarmament in the nuclear age, so bleak were those prospects that the panel's five specific recommendations included no disarmament proposals. It even advised abandoning negotiations in the United Nations. Instead the report proposed measures to strengthen American continental defense, enhance the solidarity of U.S. alliances, and reduce the danger of accidental nuclear war. The report also recommended greater "candor" to the American people about the awesome destructive power of thermonuclear weapons and more openness about the growing stockpile of nuclear materials. The Oppenheimer report brought the issue of public presentation of atomic age realities to the attention of the president and the National Security Council. At an NSC meeting convened to discuss the report, Eisenhower balked at the panel's recommendation that the United States adopt a "policy of candor toward the American people" on atomic energy matters. He opposed "indicating to the American people anything about the size of our stockpile," concurring with Defense Secretary Charles Wilson's comment that it "seemed foolish to scare our people to death."[8]

Although the development of hydrogen bombs was at least as significant as the earlier step from conventional to atomic bombs, the United States entered the thermonuclear age under a veil of secrecy. Eisenhower preferred to keep the public guessing about the leap from fission to fusion. He ordered the Atomic Energy Commission to leave the word "thermonuclear" out of its press releases and speeches and advised AEC Commissioner Lewis Strauss to keep the public "confused as to 'fission' and 'fusion.'" Eisenhower himself avoided discussing the hydrogen bomb in public. When queried by reporters, he changed the subject, confused them, or just plain refused to discuss it. Throughout the year scientists, politicians, and journalists pressed the administration for greater openness about the hydrogen bomb, particularly after the Soviet Union tested its first thermonuclear device in August. Murray S. Levine characterized the sentiment well when he noted in the *Bulletin of Atomic Scientists* that "Secrecy has become a mania with us; and it is foolish to carry it to the extremes to which we have gone. [We have] arranged things so that our own people stumble blindly into the hydrogen age."[9] Eisenhower resisted this pressure. Details about the first thermonuclear test of November 1952 were not released until April 1954. Even then, the Operations Coordinating Board carefully manipulated and sanitized information on the Operation Ivy test series so as to play down the significance of the revolutionary first thermonuclear explosion.[10]

Absolute secrecy about thermonuclear developments was impractical, however, and in May 1953 the Eisenhower administration began planning "Operation Candor," a public information campaign centering around a major presidential speech on the atomic age. The planning of this operation, the genesis of Atoms for Peace, went through two phases involving nearly a year of intense study and debate. During the first phase, the administration approached Operation Candor as a morale-building exercise, an effort to alert the American people of the dangers of communism and the threat posed by Soviet atomic weapons. It would explain the destructive force of Soviet and American nuclear arsenals with an eye to stimulating public support for prolonged defense expenditures. During the second phase, when the operation changed its name from Candor to "Wheaties" (after a series of breakfast meetings), the administration shifted from an emphasis on raising public awareness of nuclear age developments to easing their anxieties about them.

Despite the name given to the operation, Candor's planners were more concerned about domestic mobilization than disclosure. The operation's advocates hoped that explaining the nuclear danger "would have an important effect in securing support from the Congress and people for taking whatever practical steps we could to build a defense of the continent." The early phases of Operation Candor reflected this desire to prepare the American public for a protracted Cold War. James Lambie, Eisenhower's liaison with the Advertising Council, explained that the operation intended to stimulate the war mentality necessary to maintain continued sacrifices for national security. Lambie, fearing that a truce in Korea might spark a movement for disarmament, declared that "a way must be found somehow, short of war, to energize the people." The operation should stimulate the same "total energy" that mobilizes a citizenry in total war. Lambie forthrightly conceded that "candor" was not the essence of the operation and noted that it was better understood as "fiber-toughening for the long-pull" because it suggested the project's goal of generating the same kind of national discipline that exists during a shooting war.[11]

Operation Candor initially called for a series of nationwide radio and TV talks by the president and other high-ranking officials. Entitled *The Age of Peril*, the radio series would educate Americans in the duties of Cold War citizenship to overcome "current public apathy to Civilian Defense, blood donating, Savings Bonds, etc." Talks by Eisenhower and Dulles on the "Nature of Communism" and "What Good Citizens Can Do" would headline the series. Broadcasts by other officials would follow, addressing such subjects as the military capabilities of the USSR, the communist threat to the United States, and the importance of the United Nations. These pep talks planned to emphasize that the Cold War would be a long war of endurance; the American people needed to be patient, and they needed to sacrifice. The proposed radio talks were to be followed by a vigorous information campaign managed by the Advertising Council involving "billions" of advertisements in newspapers and magazines, and on radio and TV. The president endorsed the operation, but he dictated that it should avoid "Psychological Strategy Board overtones" by concealing C. D. Jackson's involvement. Consistent with Eisenhower's squeamishness about discussing the thermonuclear age, he also indicated that the H-bomb should be "alluded to but not elaborated on." Far from seeking candid discussion of thermonuclear developments, and farther still from seeking a way to brake the arms race, Operation Candor proposed to stimulate a war mentality conducive to prolonged national security expenditures.[12]

Just four days after Eisenhower officially approved Operation Candor, Soviet Premier Georgii Malenkov announced that his government also possessed a thermonuclear capability. For the next four months, the American and international media ran story after story addressing the possible consequences of nuclear war. Planners began to fear that too much "candor" might lead to complacency and apathy, rather than sacrifice and endurance. As C. D. Jackson put it, "bang-bang, no hope, no way out at the end." Meanwhile, the Soviet leadership continued to express its commitment to peaceful coexistence and its desire to reach some form of accommodation on arms limitation and other divisive issues, appeals made credible by the signing of the Korean armistice in July. These overtures reached a receptive ear in Europe, where anxiety over the probable consequences of a nuclear exchange mixed with a tentative hope that a thaw in the Cold War might be on the horizon. Concerned by these developments, Eisenhower and his advisors shifted the focus of Operation Candor from an exercise in domestic morale building to a peace counteroffensive on a much grander scale than the earlier "Chance for Peace" address.[13]

The Atoms for Peace Speech

The administration still searched for a way to explain atomic age developments to the American people but it increasingly sought to place such a discussion in a context that offered hope to U.S. allies. The Jackson Committee report, completed that same summer, observed: "This presents a delicate problem, but a balance can be struck between providing the American people with information that will permit them to grasp one of the basic realities of their world, and driving more vulnerable and therefore more nervous allies into neutralism." While educating the American people of the danger before them, Eisenhower could not in the process reinforce European fears of being embroiled in an atomic holocaust. He needed to affirm to the world America's commitment to peace, not prolonged Cold War. Accordingly, the planned radio talks were canceled, and Operation Candor evolved from *The Age of Peril* scare campaign into a PR blitz for the atom.[14]

Against this background, Eisenhower developed the idea that eventually became Atoms for Peace. "Suppose the United States and the Soviets were each to turn over to the United Nations, for peaceful use, X kilograms of fissionable material," he mused in a memorandum forwarded to Lewis Strauss. Strauss responded coolly. "The proposal is novel and might have value for propaganda purposes [but] has doubtful value as a practical move," he replied, pointing out that the proposal was unlikely to achieve results in the disarmament field. Despite these reservations, Eisenhower referred the matter back to Strauss for further study, indicating his general approval of the psychological warfare approach.[15]

Eisenhower's idea had far-reaching implications for U.S.-Soviet negotiations, and it was bitterly debated within the administration. For months, Eisenhower's national security advisors battled the professional diplomats in the State Department over the merits of using a public speech to advance a disarmament initiative. John Foster Dulles objected to the use of a public speech as the basis for negotiations. The Soviets were sure to interpret a public speech as propaganda, he argued, noting that they would delay or reject implementation of the proposal. Robert Cutler and C. D. Jackson had a different view. They defended the proposal as a valuable public relations device. It would highlight the American commitment to peace, dramatize Soviet intransigence, and stimulate public acceptance of the New Look's doctrine of massive retaliation. As Cutler explained in a memorandum to Eisenhower, "The virtue of making proposals lies not so much in the likelihood of their acceptability by the other side, but in the opportunity provided by the United States—once the proposals have been made and not accepted—to put into effect a new and better (for the long run) basic policy than that we now have." The proposal would not be accepted, Cutler implied, but that was not really the point. Public opinion would blame Soviet intransigence, rather than the administration's national security policies, for the New Look's atomic buildup. Eisenhower indicated his agreement with this logic by placing the speech in the hands of psychological warfare enthusiast C. D. Jackson. Working together with Strauss, Jackson developed Eisenhower's "atomic pool" idea.[16]

After eleven major revisions, and after discussing the proposal with U.S. allies, Eisenhower delivered his Atoms for Peace speech to the United Nations on December 8. Using generic imagery, Eisenhower addressed the terrible destructive power of the burgeoning nuclear arsenals. The American stockpile "exceeds by many times the explosive equivalent of the total of all bombs and all shells that came from every plane and every gun in every theatre of war in all of the years of World War II," he announced, in keeping with Operation Candor's goal of informing Americans about the country's stockpile. Atomic public relations, however, overshadowed this half-hearted attempt at atomic candor. The president referred to hydrogen bombs only obliquely, briefly mentioning their explosive force and drawing attention to the Soviet thermonuclear capability. Four times Eisenhower spoke of freeing the world from the grip of atomic fear, of hastening "the day when fear of the atom will begin to disappear from the minds of the people." Speaking eloquently of the need to "solve the fearful atomic dilemma," Eisenhower stressed that nuclear materials should be used to serve the needs rather than the fears of humanity. "My country wants to be constructive, not destructive," he emphasized. "This greatest of destructive forces can be developed into a great boon, for the benefit of all mankind." Uttering a phrase that would soon adorn U.S. propaganda materials around the world, he pledged to find a way "by which the miraculous inventiveness of man shall not be dedicated to his death, but consecrated to his life." The United States would strip the atom of its military casing and adapt it to the arts of peace. The dreaded atom would become a force for peace, a source of life.[17]

Eisenhower's eloquent words received thunderous applause by the United Nations General Assembly. Despite the enthusiastic reception, the speech did little to cope with the need for candor or to advance progress in arms control. As McGeorge Bundy observed disapprovingly, "Instead of awakening his countrymen to the realities of the thermonuclear world, Eisenhower's speech allowed them to believe that his proposal offered a way out." This was no accident, as the administration's follow-up campaign to exploit Atoms for Peace made perfectly clear.[18]

The Follow-up Campaign

Eisenhower's speech before the United Nations kicked off a massive propaganda campaign involving virtually every arm of government. To plan this effort, the OCB created an atomic energy working group, consisting of senior representatives from State, Defense, AEC, CIA, USIA, and the Federal Civil Defense Administration (FCDA). With C. D. Jackson at the helm, the working group drafted a comprehensive operations plan to marshal all the government's resources to promote Atoms for Peace at home and abroad. Technically, the board's plan differentiated between domestic and international information activities. In practice, the lines between international and domestic propaganda were blurred.[19]

The OCB's exploitation program focused on three categories of activity. First came organized publicity of the speech itself. The board sought to make sure people were aware of Eisenhower's speech and could identify the core elements of the Atoms for Peace concept. Toward this end, the OCB enlisted the services of all of the principal agencies and departments of the executive branch and it rallied extensive cooperation from the private sector. The second aspect of the OCB program focused on diplomacy and economic policy. The administration exerted pressure in the diplomatic arena to convince domestic and world opinion that the United States was seriously committed to disarmament, and to affix blame on the Soviets for failing to implement Eisenhower's plan. The campaign went beyond mere publicity to intersect with development and aid projects. The Eisenhower administration positioned itself as the leading supporter of atomic power, offering technical assistance and fissionable material for the construction of power reactors abroad. The third objective, which came to dominate U.S. propaganda efforts for years to come, promoted an image of the United States as the world leader in peaceful applications of atomic energy. The OCB sought to drown out media treatment of weapons tests and radioactive fallout in a deluge of stories highlighting U.S. progress in medical, agricultural, and industrial applications of atomic science. Emphasizing that the forces unleashed by fission and fusion could be harnessed for constructive rather than destructive purposes, the OCB hoped to foster a public atmosphere conducive to the implementation of the New Look by managing public fears of atomic weaponry.[20]

Publicizing the Speech

USIA Director Theodore Streibert was not exaggerating when he observed that "never before in history have the words of the President of the United States been so widely disseminated to all peoples of the earth." His agency assigned atomic energy propaganda its highest priority. The USIA sent the original text of Eisenhower's speech to newspapers around the world. Ma-

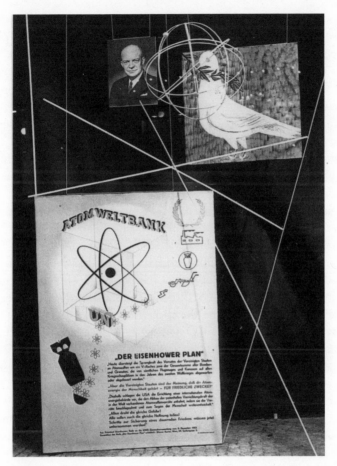

A USIA exhibit in Vienna explains Eisenhower's Atoms for Peace plan. Source: National Archives.

jor papers in twenty-five countries reprinted the speech in full. The agency published pamphlets on the speech in seventeen languages. It distributed over 16 million posters and booklets drawing attention to Eisenhower's address. The Voice of America carried the speech live in over thirty languages. Films of the speech were dispatched to thirty-five countries.[21]

As extensive as these efforts were, the USIA's activities were but the tip of the publicity iceberg. Within the United States, the speech was a major news story, hyped to the fullest by government officials. Public affairs officers at the United Nations developed a "top priority" crash program to ensure that Atoms for Peace received widespread foreign and domestic coverage. They sent advance copies of the speech to sympathetic reporters at *Newsweek, Time,* and the *New York Herald Tribune.* An additional 10,000 copies went to other publications immediately after its delivery. According to the OCB, Eisenhower's address attracted the "largest . . . motion picture coverage of a UN speech in recent years." Five American TV networks and four radio chains carried the speech live, sixteen commercial newsreels covered the entire speech, and all network news shows presented excerpts. Edward R. Murrow used the speech on the popular television program *See It Now.* Universal Pictures produced a short documentary on "Atomic Power for Peace" in forty-one languages for dissemination abroad.[22]

Such publicity developed in part because of OCB planning. The board's "National Operations Plan to Exploit the President's UN Speech" coordinated a governmentwide effort to keep Atoms for Peace in the news. "Full understanding of the President's speech by the American people will depend upon maximum repetition of the principal points over a period of months," the board explained, directing public affairs offices throughout the executive branch to hype the speech at every opportunity. The State Department and Federal Civil Defense Administration led this effort. The State Department distributed leaflets across the country through federal, state, and local agencies. The department took precautions to ensure that its publications did not physically resemble those of the USIA in order to ward off charges of engaging in domestic propaganda, but its publications closely mirrored those produced by the USIA for dissemination abroad. The FCDA contributed to the effort by sending 50,000 copies of the speech to governors, mayors, and civil defense leaders. It distributed an additional 5 million leaflets to its civilian volunteers. Local and state civil defense officials made tie-ins with the president's speech in their public pronouncements, and they arranged radio and TV discussions of Eisenhower's proposal. Private groups, working with FCDA, distributed an official film of the president's speech.[23]

Other government agencies also publicized Atoms for Peace. The OCB instructed public affairs officers throughout the government that it expected their "coordinated and harmonious" participation "without exception." Publicity efforts were conducted by the Departments of Defense, Treasury, Interior, Agriculture, Labor, and Commerce. The Attorney General's Office, the Department of Health, Education and Welfare, and even the Post Office participated. The OCB instructed these agencies to distribute copies of the speech to their employees. Groups regularly contacted by each

The U.S. Postal Service sponsored a contest to design an Atoms for Peace postage stamp to further publicize Eisenhower's proposal.

agency also received copies. Government employees discussed Atoms for Peace in their local communities and worked references to the speech into their radio and TV appearances. For an added touch of publicity, the Post Office issued an official Atoms for Peace stamp. The stamp depicted a world encircled by the symbol of atomic energy. Inscribed around the stamp were the hallmark words of the Atoms for Peace campaign: "To find the way by which the inventiveness of man shall be consecrated to his life."[24]

The OCB ensured that Americans from all walks of life heard of Atoms for Peace. The board specifically targeted agricultural workers, labor organizations, and educators—groups it believed were susceptible to communist propaganda. It also encouraged agencies to make a "special effort" to court official liaisons with national civic, religious, labor, women's, veterans', and farmers' organizations. At the OCB's direction, the Department of Agriculture prepared a short film on the uses of atomic energy in agricultural research, incorporating scenes from Eisenhower's speech. The Department of Labor sent hundreds of pamphlets and press releases to the Labor Press Service and the Associated Negro Press. The Department of Health, Education, and Welfare reprinted the speech in its publications. The Defense Department distributed a bulletin on "Atomic Energy in

Your Future" to officers in the Army and Air Force, instructing them to use the information in weekly discussions with military personnel. The Office of Defense Mobilization sent pamphlets to health professionals, business executives, and labor leaders.[25]

Perhaps the most intriguing aspect of the publicity campaign was the extent to which nongovernmental organizations joined the effort. Industrial firms with overseas operations worked with the USIA to insert leaflets in direct mailings abroad. In only the first week after the speech, various businesses requested some 100,000 leaflets. A month later, the USIA had received requests from an additional 234 private organizations for a total of 300,000 leaflets for international distribution. Westinghouse Electric Company, for example, sent leaflets to 35,000 executives, engineers, and opinion leaders in more than 125 different countries. Westinghouse attached a cover letter emphasizing the significance of Eisenhower's plan. Some companies, instead of using USIA leaflets, urged their employees to send news clippings of the speech to friends and relatives overseas. Foreign exchange students were briefed on Eisenhower's U.N. proposal; American tourists distributed leaflets on their travels. Prominent civic organizations such as Rotary International, Lions International, Kiwanis, Optimist International, and the American Legion reprinted the speech in their newsletters. Members of these groups mailed press clippings and leaflets to friends, relatives, and business associates abroad. At the USIA's request, the American Pharmaceutical Manufacturer's Association and the General Federation of Women's Clubs drafted resolutions endorsing Eisenhower's proposal. These "spontaneous" signs of support for Atoms for Peace were then widely reported as "news" stories at home and abroad. The effort to publicize Atoms for Peace was a global one, linking public and private resources in a total campaign to sell Eisenhower's plan to the world.[26]

The Diplomatic Front

The OCB's efforts to spread awareness of the Atoms for Peace speech comprised only one aspect of its exploitation program. The campaign's success required additional action beyond the transmission of the "psychological messages" embedded in the speech. The administration also wanted to convince the world that it was sincerely working for peace and disarmament. Eisenhower's psychological strategists, especially C. D. Jackson, believed that the United States needed to expose the communist peace campaign as fraudulent by challenging Soviet "words" with American "deeds." The

United States needed to appear eager to secure Soviet cooperation in the atomic pool so that the onus for the failure of negotiations could be pinned on Soviet intransigence. This required vigorous action on the diplomatic front, the second main aspect of the OCB plan.

On December 22, 1953, the Soviet government officially replied to Eisenhower's proposal in a statement given to Ambassador Charles Bohlen and broadcast to the world. The note stressed that progress in the peaceful uses of atomic energy was not nearly as important to international security as the "unconditional banning of atomic and hydrogen weapons." While repeating the familiar "ban the bomb" line that had dominated Soviet peace propaganda for years, this response nevertheless identified the diversionary aspects of Atoms for Peace. Eisenhower's plan, the Soviets charged, "would fail to assist the reaching of an agreement on the banning of atomic weapons." In fact, it would lead "to the production of more and more destructive atomic weapons" by lessening the "vigilance of the peoples regarding the problem of atomic weapons." Soviet diplomacy emphasized this theme for the rest of the year. The American vision of a "peaceful atom" was only an illusion. It would actually increase the danger of nuclear war. It would ease public pressure for disarmament and, perhaps worse, encourage nuclear proliferation by distributing atomic materials around the world.[27]

The OCB interpreted the Soviet reply as a direct propaganda challenge. It "clearly foreshadows a major effort to regain the propaganda initiative and to entangle the West psychologically and diplomatically," the board reported. C. D. Jackson grumbled to Eisenhower that the Soviets were surrounding his generous proposal "with a whole lot of old disarmament spinach." Noting that the Soviet reply had thrown the "ball into our court," Jackson sought to regain the initiative by redirecting public attention from the Soviet reply to the president's atomic pool proposal. He planned several activities to keep Eisenhower's proposals before the world media. Some of his suggestions were intended to convey American seriousness of purpose. Jackson instructed Ambassador Bohlen to resume conversations with Soviet foreign minister Vyacheslav Molotov, a move to signal "progress" in the negotiations. Without giving any substantive information, Jackson leaked to the press that the State Department and Atomic Energy Commission were preparing for these discussions. He also proposed that a prominent senator announce his belief that the U.S. should move unilaterally on the proposal, with or without Soviet participation. Jackson supplemented these actions with moves to suggest international support for Atoms for Peace. In addition to urging the British to publicly state their approval of the plan, Jackson suggested that the French take the issue up at the United Nations to attract attention in European newspapers.[28]

The Soviet note raised thorny issues that delayed the implementation of some of Jackson's countermeasures. It forced the administration to confront the difficult issue of whether or not the United States was willing to engage in disarmament negotiations. This raised a difficult policy question: would the United States negotiate on "atomic disarmament" separately from *total* disarmament, including conventional weapons? This would be a major reversal of U.S. policy, tearing at the foundation of the New Look. The State and Defense departments locked jaws over the issue. At an OCB meeting to consider Jackson's plan, Robert Bowie, director of the Policy Planning Staff, sparred with Stefan Possony of Defense. Bowie pressed for diplomatic action and quiet diplomacy to pursue the possibility of atomic disarmament. Possony objected. If the president had intended to suggest atomic disarmament without conventional disarmament, he argued, it would completely reverse U.S. defense posture, which depended on nuclear weapons to deter a Soviet conventional attack. Jackson had no patience for the State-Defense brouhaha. He complained that their "harassing tactics" were interrupting the propaganda offensive. He pleaded with Eisenhower to get the psychological warfare ball rolling.[29]

Eisenhower complied. In a revealing memorandum, he stepped into the debate to urge that Jackson's moves "be started instantly." Expressing doubts about the ability of the United States to reach an accord with the USSR, Eisenhower noted, "The question of total, as opposed to atomic, disarmament is largely academic. Neither can be accomplished without the most rigid and complete system of inspection—this we feel perfectly certain the Soviets would never allow." He met with Dulles and representatives from the Defense Department to urge them to put the pointless "atomic vs. total disarmament" debate aside. Arguing that "effective outlawing of atomic weapons is impossible," Eisenhower concluded that the United States should therefore proceed to exploit Atoms for Peace to the hilt.[30]

Still, the Soviet note put the administration in an awkward position. Having sold Atoms for Peace as a first step to world disarmament, it now had to explain to the world that Eisenhower did not, in fact, intend to engage in comprehensive talks on nuclear disarmament. The Operations Coordinating Board devised a new public line. The United States would emphasize that Atoms for Peace was a measure for sharing with the world the blessing of atomic energy, downplaying the disarmament implications of the proposal. Articulating the administration's position, the OCB explained that Atoms for Peace was first and foremost a proposal for the development of peaceful applications of atomic energy. Only secondly was it an effort to break the disarmament deadlock. If the Soviets were going to reverse that order, beclouding the issue with nuclear disarmament schemes, "they should be censured for depriving the peoples of the world of the benefits of the President's plan."[31]

Prodded by C. D. Jackson and the OCB, Eisenhower drifted away from his original plan for an international atomic energy agency. Although he paid lip service to the goal of an international organization with Soviet participation, he prioritized an assertive program to export power reactors and supporting technology through bilateral agreements. Through a series of highly publicized statements, the administration pledged thousands of kilograms of fissionable material for the construction of reactors abroad. He offered to contribute half the cost of building reactors to select countries seeking U.S. assistance. By the end of the decade, the administration had signed thirty-eight bilateral agreements and approved the construction of thirty reactors. The OCB plan explained that the reactor program would tie recipients to the United States because they would become dependent on the United States for designing, constructing, and maintaining the reactors. Martin Medhurst describes this as a form of "industrial imperialism . . . whereby an advanced technology could be embedded in a culture not yet ready to exploit its full potential as a means of getting both a technological and economic foothold. . . . If U.S. industry could be the first to establish a nuclear presence in the various countries, those countries would almost inevitably be dependent upon the U.S. for design, construction, initial operation, educational materials, and every other aspect of the infant industry."[32] The program also had psychological advantages. According to the OCB, it would attract the interest of the "underdeveloped and power-starved areas of the world" by tapping world hopes for cheap, plentiful sources of energy. Although the reactor program initially focused on industrialized nations and uranium-exporting countries like Belgium, the OCB hoped that atomic power would serve as a counterweight to the Soviet model of development for newly independent states in the third world. By demonstrating the ability to bring "rapid, cultural, economic and social improvements through application of power reactors," the United States would offer progress through the ingenuity of American capitalism.[33]

Atomic Public Relations

While the administration pursued progress in its atomic power program, the OCB and USIA directed a worldwide campaign to hype the peaceful atom, the third aspect of the exploitation program. A top secret excerpt from the minutes of an OCB meeting explained that the administration should deflect attention away from negotiations with the Soviet Union, which would confuse the issue, and instead exploit the peaceful potential of atomic energy "without let-up." The administration would channel public attention away from "atoms for war" and toward "atoms for peace." By generating widespread media coverage of the American ability to harness the peaceful applications of atomic energy, the OCB and USIA sought to show "all peoples" that the United States was "interested primarily in human aspirations rather than building up armaments." The peaceful atom served as a communications tool to shape the attitudes of world opinion— to bolster America's image and deflate public criticism of U.S. nuclear policies.[34]

To spark media interest in peaceful atomic research, the OCB developed a concerted press campaign that publicized both past accomplishments and new discoveries. The board arranged for extensive media coverage of developments such as the use of tracer atoms to increase the efficiency of fertilizers, research on cancer treatments, and the construction of a twenty-five-billion-volt particle accelerator. The AEC acted as the clearinghouse for such information on the peaceful applications of atomic energy. It advised other government agencies on news stories, provided film clips, drafted press releases, produced still pictures, and reprinted pertinent articles for distribution in the U.S. and overseas. Lewis Strauss and former chairman Gordon Dean headlined the effort by issuing public statements and writing commentaries on atomic research.[35]

The USIA, meanwhile, developed a global theme on the peaceful uses of atomic energy. In every country of the world, the theme became a top priority. The agency produced a series of twenty-six TV programs entitled the *Magic of the Atom*. It prepared dozens of short documentary-style films like *The Atom and Biological Sciences, The Atom and Agriculture, The Atom and the Doctor,* and *The Atom and Industry.* The USIA distributed these films to information posts abroad; members of the press received private screenings. In addition, the agency churned out about fourteen news stories per week on atomic energy, each selected to convey a different message or to appeal to a different audience. Some stories attested to U.S. seriousness in

An American official at a USIS library shows a visitor a display of books on atomic energy. Source: National Archives.

seeking Soviet participation in Atoms for Peace with headlines like "Eisenhower Cites Progress in Atom Talks" and "Preliminary Atomic Talks Encourage Eisenhower." Stories of this type intended to put the Soviet leadership in an awkward position should it refuse to participate. Adding to the pressure, the USIA hyped international enthusiasm for Eisenhower's plan with stories like "Indian Group Praises Eisenhower Plan," "Twenty Five Nations Endorse Eisenhower Plan," "Japan Plans Atomic Reactor," and "Belgian Senate Votes for European Nuclear Research Organization."

Other articles functioned as educational pieces explaining the potential applications of atomic research. These stories appealed to particular opinion groups to demonstrate practical applications in areas of interest to them. Agricultural and working class groups were targeted by stories like "Radioisotope Study Saves Tobacco Farmers Money" and "Labor Says

Atomic Energy May Remedy Food Problems." Industrial leaders were targeted by "Putting the Peaceful Atom to Work in Industry." Because many
associated the atom with destruction, the medical applications of atomic
energy received particular hype. Stories like "Radioactive Isotopes Widely
Used for Research and Medicine," "Atomic Cannon Helps Cancer Patients," and "Scientists See Eisenhower Plan Raising Health Level" encouraged target audiences to associate the atom with health and safety. USIA
materials emphasized to audiences that these peaceful applications were realistic, practicable, and realizable in the near future. Articles like "New Atom
Furnaces to Aid Cheap Power Development" and "AEC Approves New
Power Reactor" sought to dispel doubts about the feasibility of atomic
power while simultaneously conveying American progress in the field.[36]

These stories were accompanied by appealing black-and-white photographs reiterating the message of the peaceful atom. One photo featured
an attractive woman admiring a model atomic energy plant. Her eyes
beamed brightly as she looked at the future of atomic power. The caption
explained, "The advancing atomic age is mirrored on the face of this young
girl. The scale model of the atomic energy plant which holds her attention
is forecast of future plants which could be constructed under the Eisenhower plan for peaceful use of atomic energy." Most images were more
journalistic, featuring real individuals and real events. One such photo,
clearly targeting labor groups, depicted a worker operating a punch press
in San Francisco. The caption explained that the hands of the mechanic,
identified as J. L. Gress, were protected from danger by a peculiar safety
device: radioactive wristbands. If his hands strayed too far, Geiger counters
mounted on the press would respond instantly, halting the punching action.
Another image showed a technician checking a group of plants containing
radioactive isotopes. The caption explained, in educational prose, that "by
tracing radioactive isotopes researchers have discovered better ways of applying fertilizer and plant nutrients." Such pictures endeavored to provoke
viewers into associating these positive images with atomic energy, rather
than the foreboding mushroom cloud. These photographs also reinforced
the campaign's message that the United States was the undisputed leader in
sharing the peaceful applications of atomic energy with the world.[37]

Considering the administration's tactic of generating news through
press releases and staged events, it is not surprising that stories similar to
those pumped out by USIA appeared in the American media. Just a few
months prior to Eisenhower's speech, the press was flooded with stories

on "When an Atomic Blast Hits Your Home or Auto," "Have Atom Bomb Tests Fouled Up the Weather?," and "Atomic Bomb Blast Waves." A few months later, these terror-inducing articles competed for attention with features describing atomic research in such friendly terms as "Those Wonderful Atoms" and the "Taming of the Atom." Other articles dramatized the exciting prospects of atomic power: "Soon the World Will Need the Atom for Energy" and "Atom Power in Homes in Five Years." It was not uncommon for accompanying graphics to blend atomic reassurance with sexual suggestion. In one paper, a young woman rested affectionately in the arms of a man wearing a suit filled with compressed air to protect him from radioactivity. The caption explained: "This 'atom factory' girl doesn't seem to mind the advances of this 'fellow.' Maybe it's because she feels safe in the arms of a plastic protective outfit—filled only with air." Many articles, such as the pieces in *Today's Health* on "Hospitals for Atomic Research" and the *New York Times Magazine* on "Forecast of the Age of Atomic Power," were reprinted and distributed by USIA as part of its strategy of cross-reporting from private media whenever possible.[38]

A *National Geographic* feature that appeared shortly after Eisenhower's U.N. speech typified the positive, optimistic tone of these articles, which often parroted the language of official information materials. In contrast to the gloom and doom of other atomic-age stories, *National Geographic* told a "brighter, happier story of how atomic energy is building a better world." It explained that "the same terrific power that explodes atomic bombs has been tamed and harnessed for countless peaceful tasks" in medicine, agriculture, industry, and power production. As if that were not enough, "Even golf balls have been made radioactive experimentally to see if they could be more easily located when lost in the rough." *National Geographic*'s signature photographs accompanied the article, depicting the friendly atom at work. One photo showed a dog undergoing therapy to clear up an eye tumor. A young boy stood beside his beloved pet as the atomic treatment took place. "Even a dog's life can be made happier through atomic energy's healing power," the caption read. Another photo depicted an attractive young woman lying submissively on a hospital bed while a massive X-ray gun "fires electron bullets at cancers." Like other articles that appeared on the subject, the photos and text of the article imitated the language, style, and images of official information materials such as USIA's "Background and Action Kit." Even the title of the article, "Man's New Servant, the Friendly Atom," reworded an extensively used propaganda leaflet, "The

Atom: Servant of Man."[39]

To ensure that the initial interest in the peaceful atom stimulated by Eisenhower's speech did not fade, the USIA planned traveling Atoms for Peace exhibits to keep the plan in the press and in the public imagination. Consistent with the general theme of the Atoms for Peace campaign, information officers hoped "that anyone who has seen the exhibit will no longer think of mushroom clouds and mass destruction when he hears the words atom, atomic, or atomic energy, but rather of the peaceful uses of atomic energy in the fields of industry, agriculture, and medicine."[40] Large mobile exhibits traveled in truck caravans to major cities in Europe, Asia, and Africa. Smaller exhibits went to 217 USIS posts abroad.

The exhibits "served to bring the idea of peaceful atomic energy for the first time into the immediate consciousness of the lay viewer." They explained and dramatized the peaceful applications of atomic energy using working models, colorful displays, short films, and mini lectures. The exhibits displayed model reactors, Geiger counters, and devices for handling radioactive materials. Large panels illustrated the use of radioisotopes in food preservation, insect control, and agricultural and biological research. Many of the panels focused on medical applications, such as the use of "atomic cocktails" to diagnose cancer patients. Guides were posed throughout the exhibits to give presentations and to field questions about the various peaceful applications of atomic energy. All exhibits showed *A Is for Atom*, a color cartoon made by General Electric for domestic and international use. This animated film, which explained the concepts of atomic power in simple terms, made atomic energy seem safe, amusing, and "cute." Some exhibits included six-foot models of the atom consisting of a nucleus encircled by ninety-two electrons in their orbits. For added publicity, the models were donated to local museums after the exhibits moved on. The traveling exhibits presented the harnessing of atomic power as a defining moment in human development. One of the displays portrayed a time line of human ingenuity, a march of progress depicting mankind's evolution from stone and wood tools, to the invention of the wheel, to the steam engine, to the airplane, and finally to the harnessing of the atom itself. The exhibits made atomic power appear cheap and easily attainable, and displays explained reactor technology in the simplest of terms.[41]

The exhibits acknowledged the international contributions to atomic research, but visitors were clearly intended to come away with the impression that the future of atomic energy rested with the United States. Virtually no mention was made of Russian scientific research. The exhibits also

Entrance to Atoms for Peace exhibit in Belgrade, Yugoslavia. The U.S.
Information Agency sent dozens of exhibits around the world explaining the
peaceful applications of atomic energy. Many such exhibits drew large crowds.
Source: National Archives.

minimized the hazards of atomic energy. Indeed, guides were instructed
to tell guests that "so great is the care taken, that only very rarely is any
one exposed to an overdose of radioactivity. . . . It has been said that work
in an atomic plant is safer than any other work, as attested by a remarkably low accident record." The exhibits downplayed the problem of atomic
waste, the dangers of radioactivity, and the cost of building and maintaining atomic power reactors. Concepts were simplified and washed down so
laypersons could understand them; allusions to bombs and weapons were
deliberately avoided. One display traced the history of atomic energy research while conspicuously leaving out any developments pertaining to
the development of the A-bomb. Another showed a picture of the Oak
Ridge reactor, built during the war as part of the Manhattan Project, and
explained proudly that the facility had been converted into a plant for the
production of radioisotopes for peaceful purposes. The unmistakable suggestion was that the United States was busy converting all of its "atoms for
war" into "atoms for peace."[42]

The exhibits attracted record numbers of people. Hundreds of thousands of visitors flocked to the exhibits in major cities around the world, often waiting in line for two hours to get in. Professors and schoolteachers brought their classes. Civic groups organized field trips. Prominent scientists, doctors, industrial leaders, and government officials made highly publicized tours of the exhibits. In one report after another, USIS posts cited record numbers of attendees: 188,000 in Frankfurt; 195,860 in Buenos Aires; 135,853 in Ghana; and 155,000 in Kyoto "despite severe winter snow, rain." Occasionally the posts extended the show in order to accommodate the large crowds. So great were the numbers that the USIS mission in Belgrade cabled Washington: "By now it has become virtually trite to report that the 'Atoms for Peace' exhibit achieved a greater impact . . . than any other project undertaken . . . by the U.S. Information Service."[43]

The Atoms for Peace exhibit in Rome highlights the extent to which the traveling exhibits became full-fledged media events. The Rome exhibit became a vehicle for "spreading President Eisenhower's message on peaceful uses of atomic energy throughout Italy and well beyond its borders." Five newsreel companies and major TV networks (including CBS, NBC, and the BBC) covered the exhibit's opening day. The Associated Press, United Press, and International News Service transmitted stories around the world on the exhibit with accompanying photographs. Longer features appeared in *Life* magazine, the *New York Times Sunday Magazine*, and various Italian publications. In addition, Universal Films produced a color documentary covering the construction and opening of the exhibit. According to USIS Rome, Italian media attention was overwhelmingly positive. Local newspapers expressed "admiring surprise at the marvels the atom can accomplish in peaceful applications." The four-column headline in *Il Popolo*—"Atomic Energy, Man's Ally"—was typical. The papers described atomic energy as a "new dawn on the horizon for mankind" that promised a "happy future." As the *Corriere della Sera* of Milan commented, "You don't feel any terror, there are no nightmares: tomorrow's world is filled with extraordinary marvels . . . The world of tomorrow will be easier, happier, richer than the past." Newspapers in Sweden, Belgium, Britain, and France also reported positively on the Atoms for Peace show in Rome. The conservative Parisian paper *Le Figaro* exclaimed, "One leaves the exhibit with a great desire to live—to live in peace with work made easier, in conditions of social well being unknown till this day but of which one has now caught a glimpse."[44]

Press coverage of the exhibits consistently stressed the "swords into plowshares" theme of the Atoms for Peace campaign. When one of the traveling exhibits opened at the Library of Congress in Washington, the *Evening Star* commented, "It is decidedly worth seeing as an instructive and inspirational demonstration of how wondrously the atom can be made to serve man rather than destroy him." The *Washington Post* explained that the exhibit spoke directly to those "who believe the atom brings only tidings of blood and thunder. . . . When you see how atoms for peace already are in use to trace and treat brain cancers, or have improved the use of fertilizers and markedly increased the yield of peanuts, you cannot but come away with new vision." Considering the prohibition on domestic propaganda, it is surprising that the USIA opened exhibits in Washington, New York, and San Francisco. Although communist papers derided the exhibits as propaganda ploys, few others voiced agreement. The factual approach of the exhibits masked their manipulative elements. Many visitors saw the exhibits as educational presentations. Professor Fritz Strassman, who discovered uranium fission in 1938 with Otto Hahn, described the exhibit in Frankfurt as "the best method I have ever seen to teach students and laymen the principles and development of atomic energy." He visited the exhibit several times, bringing with him students and colleagues. In a statement widely circulated by USIA, he exclaimed: "All those people who, when they hear the word 'Atom,' immediately cover their ears and see the dreaded mushroom cloud before their eyes, all these people can now convince themselves that through this powerful force a new world of peaceful activity can be discovered." American newspapers reprinted such sentiments in their articles on the exhibit, many of which derived from USIA press releases. The journalistic style of these releases masked the agency's messages, which were hidden in quotes by ordinary people, newspapers, and government officials. Most periodicals presented the exhibits, in the words of the Associated Press, as "a serious attempt to tell the people of Western Europe and the Free World about the peaceful uses of atomic energy."[45]

The parallels between the stories generated by the Information Agency and those that appeared "spontaneously" in the American press are striking. It is difficult to identify precisely which stories were planted, which ones journalists derived from government press releases, and which ones arose independently of the administration's efforts. Undoubtedly, some of the enthusiasm for the peaceful atom arose naturally, but USIA and OCB records make clear that much of the media attention resulted from government encouragement. American propagandists frequently defended their actions, of course, by stating that they were not engaging in "propaganda," but were "explaining the facts," "educating," and "informing people of the realities they faced." Unlike the communists, they followed a "strategy of truth." But these protestations obscured a sophisticated strategy of camouflaging propaganda messages by embedding them in news stories and private initiatives. The Operations Coordinating Board orchestrated a multidirectional flow of U.S. propaganda themes that distributed the administration's messages through a number of sources. More often than not, the government's hand was, by and large, hidden.[46]

Official government propaganda channels such as USIA and the Voice of America were small links in a much bigger propaganda chain. International audiences were reached primarily by local media outlets, which reported U.S. government, international, and private activities—many of which were stimulated, encouraged, or coordinated behind the scenes by the OCB. Likewise, American audiences were reached by local media sources. They were also reached indirectly through a process known as "feedback": stories planted covertly by U.S. propagandists in the international media were picked up by U.S. media outlets and reported at home. Ordinary Americans also were encouraged to actively contribute to the propaganda campaign, making them both targets and participants. The Eisenhower administration's sophisticated approach to information management made the subtle manipulation virtually invisible to target audiences and to journalists who were (actively or unconsciously) agents of domestic propaganda.[47]

    All of this is not to say that the administration fabricated news out of thin air, coerced journalists to write stories against their will, or "invented" the peaceful atom. Research on the peaceful applications of atomic energy predated Eisenhower's proposal and would have progressed without Atoms for Peace. The administration's efforts were freely supported by scientists and AEC personnel who were eager to secure research funding and anxious about their contributions to the nation's weapons stockpiles. In addition, a "revolving door" pattern of employment between government and industry meant that the administration had an extensive network of contacts in business, advertising, and the media eager to collaborate with the government in the name of anticommunism. Most U.S. information officers began their careers in print journalism, and there was considerable movement between government positions as psychological warriors and private positions as journalists and communications specialists. These connections, plus the fact that reporters relied on public affairs bureaus in the administration for many of their stories, naturally facilitated the domestic exploitation of the administration's campaign. If government propagandists and the American news media advanced a similar message, it undoubtedly stemmed from shared interests.[48]

In any case, the administration did not have to resort to overt lies or coercion to propagate the friendly atom. For all the behind-the-scenes maneuvering of government propagandists, one reason that Atoms for Peace succeeded in getting the attention it did was because the message of the peaceful atom struck such a responsive chord in target audiences. Indeed, from today's perspective, it is hard to understand the genuine public interest and intense media attention Atoms for Peace attracted. In the U.S. press, the peaceful applications of atomic energy received extensive coverage. Abroad, the USIA believed that Atoms for Peace was its most inspired and effective propaganda theme.

Many overseas posts reported significant improvements in foreign attitudes toward atomic energy. USIS officials noted that in India, for example, Atoms for Peace had helped to alleviate the negative stigma attached to the United States from the atomic bombings of Japan: "The horror and assumed anti-Asian tinge of our Hiroshima drop still lingers but our efforts to 'civilize' the atom have moved us to the plus side of the ledger." Officials were even more optimistic about the impact of Atoms for Peace in Japan, crediting their efforts with producing a marked change in Japanese attitudes toward the atom: "The change in opinion on atomic energy from 1954 to 1955 was spectacular. Through an intensive USIS campaign, atom hysteria was almost eliminated and by the beginning of 1956, Japanese opinion was brought to popular acceptance of the peaceful uses of atomic energy. . . . Substantial progress has been made in improving Japanese opinion towards the U.S. and thereby taking some of the pressure off the Japanese Government on account of its pro-American policies." These assessments of the impact of Atoms for Peace on foreign attitudes derived from impressionistic surveys of mass media, rather than scientific evaluations of popular opinion. USIA officials in the 1950s tended to assess popular opinion according to media coverage, and they took the extensive press play that the friendly atom received as a sign that the campaign was working. Although attitudes toward the bomb would evolve with unfolding events, Atoms for Peace appeared to be a success in the short term. The Eisenhower administration had created an international news story that received positive coverage the world over.[49]

The enthusiasm greeting Atoms for Peace brings forth an appreciation of the tremendous power the atom held in the popular imagination. In seizing upon the peaceful uses of atomic energy, Eisenhower tapped into a psychological need to find something redeeming and worthwhile in this technological marvel threatening the very existence of humanity. Psychologist Robert J. Lifton has explained this process as "psychic numbing," a defense mechanism preventing "the mind from being overwhelmed and perhaps destroyed by the dreadful images confronting it." Evidence in support of psychic numbing can be found in cultural expression. Americans made light of the atomic bomb by assimilating fears of nuclear annihilation into popular songs like "Atom Bomb Baby," comic books like "Dagwood and Blondie Split the Atom," and "Atomic Fireball" candies. A popular song, recorded shortly after the atomic bombing of Japan, revealed a strong cultural current yearning to find something positive in the splitting of the atom. The verses evoked atoms for war and for peace:

> You remember two great cities
> In a distant foreign land
> When scorched from the face of earth
> The power of Japan
> Be careful my dear brother
> Don't take away the joy
> But use it for the good of man
> And never to destroy

"Atomic power, atomic power," the chorus repeated, "was given by the mighty hand of God."[50]

Such vocabulary allowed for easy acceptance of the atom and facilitated what Lifton calls "nuclearism," the acceptance of nuclear weapons as part of everyday life. The themes of the Atoms for Peace campaign aided the domestication of atomic energy. Certainly it did not put world fears of nuclear annihilation to rest. But by stressing that atomic fission was controllable, useful, and indeed desirable, the administration helped propagate a "friendly" atomic discourse to rival the apocalyptic discourse that had characterized most discussions of the atom. Even as the hazards of radioactive fallout began seeping into the press, the administration ensured that public perceptions remained focused, at least in part, on the friendly atom.[51]

Chapter 6

The Illusory Spirit of Geneva
Propaganda and the New Diplomacy

Much of what passes today for diplomacy is not diplomacy at all; it is propaganda. . . .
We are trying to use diplomacy for a task for which it has never been designed:
propaganda and psychological warfare.

— Theodore Sands, "Propaganda vs. Diplomacy," *The Nation*, May 30, 1959

We have to maintain the spirit of Geneva, whether real or illusory, and at the same time
continue the struggle against Communism by other means.

— C. D. Jackson to Nelson Rockefeller, September 22, 1955

The early years of the Cold War witnessed a transformation in how foreign policy professionals viewed their craft. The media revolution raised
awareness of the impact of popular opinion on the foreign policies of other
states, and the Cold War's ideological nature and nuclear standoff added
special significance to the symbolic dimensions of international relations.
A small but growing number of analysts perceived that the mid-twentieth
century had given rise to a new type of diplomacy. In their view, negotiations were important not just for their military and political implications,
but for the images, symbols, and psychological messages they communicated to the world at large.

Henry Kissinger was one of the most noteworthy individuals to identify the core elements of this "new diplomacy." In the fall of 1955, the
not-yet-famous diplomatist participated in a wide-ranging study on the
psychological aspects of U.S. foreign policy organized by Nelson Rockefeller, who had assumed C. D. Jackson's post in December 1954. Kissinger's contribution to the investigation drew attention to the changing
nature of international negotiations. Foreign policy could no longer be
pursued as it had been during the nineteenth century, when diplomacy was
the exclusive province of professional diplomats who used secret negotiations to reach accords based on power and interest. Things had changed,

This graphic from a classified study on the "Psychological Aspects of U.S.
Strategy" illustrates the changing nature of diplomacy. On the left, the "old"
way shows diplomats meeting secretly to conduct negotiations. On the right, the
"new" diplomacy emphasizes the use of psychological and symbolic factors to
influence world public opinion. Source: Eisenhower Library.

he argued. In the postwar world, the threat of mutual destruction posed
by thermonuclear weapons, developments in mass communications, and
the increased attentiveness of domestic audiences to foreign affairs "transformed the whole pattern of international relations." According to Kissinger, "the predominant aspect of the new diplomacy is its psychological
dimension." Negotiations now possessed a significant psychological and
symbolic dimension for the impact they exerted on international and domestic opinion. No longer were diplomatic conferences merely opportunities for resolving international disputes; they had become sounding boards
for public opinion and forums for propaganda.

Kissinger explained that the psychological stakes involved in negotiations with enemy nations were extraordinarily high because they communicated signals to the public that could affect the strength, cohesion, and
resolve of allied coalitions. A premature agreement could weaken allied
and domestic resolve in a time of crises or erode support for costly armaments programs. The wrong approach could splinter U.S. alliances, encourage neutralist sentiment, or isolate the United States in a particular
area of the world. In conducting negotiations with the Soviet Union, Kissinger advised, the U.S. government had to fight a psychological war on
three fronts: domestically, to maintain popular support necessary for high
levels of armament; toward allies, to maintain coalitions and to prevent

public pressure from blocking U.S. goals; and toward the Soviet bloc, to advance U.S. interests without conceding further gains to the opponent. In the Cold War conflict, Kissinger subsequently wrote in *Foreign Affairs*, "protagonists at the conference table address not so much one another as the world at large." Diplomatic conferences had become struggles "to capture the symbols which move humanity."[1]

Kissinger was one of many foreign policy experts to perceive that "normal" diplomatic relations had changed their meaning during the Cold War. Advocates of psychological strategy—including President Eisenhower—believed that international public opinion needed to be considered when formulating foreign policy and when conducting negotiations. Policy makers increasingly recognized that negotiations needed to take place on two levels: the diplomatic level between governments, and the popular level to win international support for policies. The State Department's Policy Planning Staff (PPS) concluded that the target of diplomacy had widened to include popular opinion as much, if not more so, than traditional diplomatic channels. The PPS explained that if the United States could get popular opinion on its side, it would put pressure on foreign governments, which would in turn create a favorable atmosphere for U.S. policies. Perceiving the impact of what later would be called "sound bites," the PPS urged diplomats to package proposals and policies with words and catch phrases that appealed to the masses. "Convincing a foreign official is often less important than carrying an issue over his head to his people, to public opinion in the country he represents," the PPS reasoned. "The people will influence the official's action more than he will influence theirs."[2]

Such concerns elevated the importance of psychological warfare in shaping U.S. diplomacy during the Cold War. This chapter explores the relationship between propaganda and diplomacy by tracing U.S.-Soviet disarmament negotiations during the Eisenhower presidency. Historians have paid little attention to the use of diplomacy as a means of influencing international public opinion. An assumption underlying most historical (and contemporary) analysis of diplomacy suggests that negotiations by their very nature are pursued to reach agreements. To explain the failure of negotiations, analysts often emphasize conflicting interests, tactical blunders, and mutual distrust. Yet in this period, reaching an accord was a secondary objective of U.S. negotiations with the Soviet Union. Far higher on the list of priorities was the use of negotiations to influence the attitudes and perceptions of domestic and international audiences.[3]

Propaganda and Disarmament

Ever since the failure of the 1946 Baruch Plan, Cold War disarmament talks had been largely ritualistic jousts for public opinion. With the on- set of the Korean War in 1950, disarmament negotiations between the United States and the Soviet Union effectively ceased. Although both su- perpowers participated in discussions in the Disarmament Subcommittee of the United Nations, the two sides mostly exchanged recriminations and rehashed old positions for appearance purposes. Beginning in 1953, with the election of Eisenhower and the death of Stalin, the whole manner of U.S.-Soviet negotiations appeared to change. Gradually, the superpowers shifted their attention from comprehensive disarmament schemes toward more limited arms control measures.[4]

On the Soviet side, the "peace offensive" initiated by Malenkov and continued by Khrushchev included a substantial effort in the disarmament field. Significantly, the Soviet government backed away from its previous insistence on an all-encompassing ban on nuclear weapons, a lopsided proposition that would have preserved Soviet superiority in conventional armaments. By May 1955, the USSR had adopted a more pragmatic ap- proach, accepting core aspects of the western position, including inspec- tion and enforcement of disarmament agreements. On the U.S. side, Presi- dent Eisenhower also adopted a more productive approach to negotiations. During his first administration, he made three major initiatives in the name of peace and disarmament. In addition to the 1953 Chance for Peace and Atoms for Peace initiatives, Eisenhower made his Open Skies proposal in the summer of 1955. In a speech at the Geneva Summit meeting, he sug- gested that the superpowers allow reconnaissance flights over each other's territories in order to provide mutual assurances against surprise attack. This proposal and Eisenhower's other initiatives were widely heralded at the time and since as evidence that the president had adopted a positive and pragmatic approach to controlling the spiraling arms race.

Historians, however, have disagreed sharply about the meaning and significance of Eisenhower's initiatives. Were they psychological warfare moves or serious attempts to reduce international tensions? This question has posed vexing problems for historians because of contradictory evidence in the declassified documentary record. On the one hand, archival sources indicate that Eisenhower's advisors devised these initiatives for political warfare purposes: to dramatize Soviet intransigence and to publicize the American commitment to peace. On the other hand, Eisenhower's personal diaries and letters, as well as his occasional remarks in private meetings, provide evidence of the president's keen understanding of the horrors of thermonuclear war and his sincere desire to halt the arms race. How do we reconcile these contradictory impulses? Revisionist historians of the Eisenhower presidency have concluded from Eisenhower's private ruminations that he prioritized the search for détente over psychological warfare. C. D. Jackson and others may have had propaganda in mind, but Eisenhower hoped his peace initiatives would foster mutual trust, paving the way for more ambitious agreements later. Postrevisionist scholars disagree. They see these proposals as nothing but clever propaganda gimmicks. Eisenhower was motivated by shrewd Cold War calculations, they argue, not an altruistic desire for world peace.[5]

All of these interpretations are weakened by an excessively narrow conception of propaganda, equating it only with "insincerity." As we have seen, propaganda need not signify lies or falsehood. A more perceptive definition identifies propaganda as words or deeds used for their impact on the perceptions and attitudes of others. In this respect, Eisenhower's peace proposals were, in a way, both sincere initiatives *and* propaganda. Recognizing the value of "propaganda of the deed," Eisenhower believed that serious proposals made the best form of propaganda. He insisted that U.S. approaches to contentious East-West issues appear practical in content lest world audiences reject them as meaningless gestures. But U.S. positions were not designed primarily, or even substantively, to secure Soviet acceptance. Allies, domestic audiences, and neutral nations were the principal targets of Eisenhower's peace initiatives. He preferred to make stunning proposals designed less for their value at the bargaining table than for their impact on the resolve and allegiance of these groups. Although Eisenhower was prepared to accept Soviet agreement with American proposals, the primary objective was influencing public opinion, not overcoming the disarmament impasse.

Disarmament 1953–1954

From the beginning of his administration, Eisenhower recognized the psychological implications of disarmament. Disarmament negotiations provided significant opportunities to win international public support and to foist blame on the USSR for the ever-spiraling arms race. Eisenhower's reaction to the Oppenheimer report on Armaments and American Policy is revealing. The report, issued in February 1953 and discussed previously, recommended abandoning disarmament negotiations under the U.N. framework because they had become dominated by propaganda maneuvering. In NSC meetings to discuss the report, Eisenhower spoke out strongly against the suggestion that the United States should not mix propaganda and diplomacy. Complaining about scientists who "moved into the realms of policy and psychology," he challenged the view that it was "bad psychologically" to continue discussing disarmament in the United Nations. Eisenhower appeared to agree with Dulles's assertion that these negotiations provided "good propaganda at least." He also approved an investigation into a dramatic propaganda move by the United States recommended by Richard Nixon. The vice president proposed making "some kind of sensational offer on the disarmament side, which the Soviets would of course not accept, and which would therefore put them on the spot. If it were possible to make such an offer . . . the effect on world opinion would be very favorable to the United States." Intrigued by this suggestion, Eisenhower instructed Dulles to explore the matter further in the Department of State. Nixon's idea informed the decision making of Eisenhower and his advisors on the disarmament question.[6]

For the next five years, the United States continued to conduct disarmament negotiations under the U.N. framework. Propaganda considerations predominated. In September 1953, the National Security Council approved NSC 112/1, a policy paper governing U.S. negotiations in the U.N. Disarmament Commission. The paper instructed negotiators to engage in discussions only to "counteract Soviet use of the Disarmament Commission as a forum for their propaganda." While avoiding "any serious attempt to break the disarmament deadlock," they should *appear* willing to negotiate. Negotiators were instructed to outline U.S. goals in broad fashion, in order to retain the tactical initiative and to convey sincerity to international observers. Detailed proposals that could become "targets" for Soviet counteraction or negotiation were to be avoided. NSC 112/1 explained the psychological rationale for these policies:

> It is advisable that the United States continue to demonstrate to the world its abiding desire for comprehensive and safe-guarded disarmament. The general desirability of such posture is heightened by the probability that the interest of our Allies in lessening international tensions and reducing armaments has been augmented by their hopes arising from the Soviet peace offensive and their fears derived from the announcement that the Soviets had exploded a hydrogen bomb.

The NSC also called for "sound" and "workable" proposals that could be developed in the unlikely event of a change in the international climate. This mandate that the United States advance political warfare proposals that were at the same time "sound" and "workable" might seem contradictory. But, as the NSC explained, serious proposals made for effective propaganda: "Our past adherence to such a course of action has convinced most of the world of our sincerity and has therefore been the best possible form of propaganda." The administration's "holding operation" in the U.N. was interrupted in December by Eisenhower's Atoms for Peace proposal—itself a "serious" initiative employed for psychological warfare purposes, as we have seen.[7]

Inflexibility and public posturing characterized subsequent negotiations in the United Nations. From April 1954 to May 1955, disarmament talks took place within a special U.N. subcommittee on disarmament consisting of five nations: the United States, the USSR, the United Kingdom, France, and Canada. In these discussions, old positions were repackaged for public consumption. Generally speaking, the western powers called for comprehensive disarmament, including both nuclear and conventional forces. They also insisted that any disarmament agreement include inspection and enforcement mechanisms to prevent "cheating." The Soviet Union sought to prevent or limit the implementation of verification schemes. It unrealistically advocated a ban on all nuclear weapons before installing a system of controls. In September 1954, however, the USSR evinced new flexibility. The Soviet representative to the United Nations, Andrei Vishinsky, announced his government's willingness to explore a proposal for a phased approach to comprehensive (nuclear and conventional) disarmament. Caroline Pruden notes that the announcement was significant because "for the first time the Soviet Union appeared to have accepted, at least in theory, the existence of an international arms control agency with powers of inspection." The Eisenhower administration chose not to explore Vishinsky's proposal seriously. Denouncing the move as propaganda, Dulles ordered the U.S. delegation to "probe Soviet proposals in order [to] deflate them." Subsequent meetings, from February to May 1955, rehashed old positions. As Henry Cabot Lodge observed, the negotiations served as a "cold war exercise." A ritualistic dance for public opinion ensued. The Soviet delegation attempted to begin every day's discussion with a concrete proposal, and the U.S. delegation countered with proposals of its own. Positions on both sides were unworkable. One American official conceded that U.S. initiatives "will help make the record but will not spell any considerable success for the conference."[8]

On May 10, 1955, the Soviets broke from this routine. Jacob Malik, the head of their delegation, tabled a stunning new proposal. Malik's proposal, Soviet specialist James Richter notes, "brought the two sides closer to nuclear and conventional disarmament than . . . at any other time during the 1950s." Since the onset of the Cold War, Soviet diplomacy had adhered to positions on disarmament that were unacceptable to the western powers: a ban on nuclear weapons, equivalent reductions of conventional forces, elimination of military bases on the territories of foreign countries, and a refusal to allow inspection to verify disarmament agreements. Malik's plan, however, represented a significant shift toward the western position. He offered to allow surveillance posts on Russian territory, and he reiterated Soviet support for an international disarmament agency to verify compliance. He also declared his government's willingness to cut conventional force levels, thus reducing an area of Soviet advantage. Although Malik's proposals included several objectionable Soviet positions, such as the elimination of foreign bases, they also accepted the core principles advanced by the West.[9]

What were Soviet intentions here? It is difficult to ascertain how open the Soviet leadership was to an agreement because the proposals never became subject to negotiation. The United States immediately dismissed Malik's proposals as mere propaganda. It is likely that the proposals, like those of the United States, were designed more to attract world opinion than to provide a basis for negotiation. Recent scholarship, however, also suggests that the initiatives reflected a genuine desire on the part of the Soviet leadership to ease the arms race so that they could focus their command economy on the production of consumer goods.[10] John Foster Dulles appeared to hold this view, acknowledging that the "Soviets generally wanted some reduction in armaments in order to be able to deal more effectively with their severe internal problems."[11]

Although U.S. officials acknowledged that the Soviets had made "tremendous concessions," they evinced little interest in exploring the proposals through negotiation. Soviet intelligence correctly concluded that the United States did not want to discuss these proposals without advancing new ones of its own. Instead of responding to the May 10 initiative, Eisenhower used the upcoming Geneva Summit as an excuse to adjourn the disarmament meetings. His reluctance to explore the May proposals undermines the revisionist claim that the president eagerly sought to relax world tensions through arms control. At the summit, the president effectively derailed the Soviet initiative with his Open Skies proposal. In so doing, he set aside whatever interest he may have had in arms limitation in favor of psychological warfare.[12]

The Geneva Summit

Few senior officials in Washington wanted a summit with the Kremlin's new leadership. Most believed that the meeting would accomplish nothing except provide grist for the communist propaganda mill. For two years, Eisenhower resisted international pressure for a face-to-face meeting with the post-Stalin leadership. Churchill had been urging the United States to agree to a summit since May 1953, but Eisenhower repeatedly put off Churchill's requests. He saw the summit as fraught with danger on the psychological front. Such a meeting would create a false sense of complacency that would doom domestic support for rearmament. Eisenhower, writing to Churchill in December 1954, argued that a summit "would merely give a false impression of accord which, in our free countries, would probably make it more difficult to get parliamentary support for needed defense appropriations."[13]

Pressure for a summit meeting increased. Other prominent figures joined Churchill in calling for a summit: the successive French premiers, Pierre Mendès-France and Edgar Faure; the chairman of the U.S. Senate Foreign Relations Committee, Walter George; and Churchill's successor, Anthony Eden. Dulles's talk of massive retaliation and threats of atomic attack during the Taiwan Straits crisis raised public apprehensions about the possibility of nuclear war and furthered the U.S. image as a "trigger-happy militaristic power." The continuing "soft line" in Soviet foreign policy ratcheted the pressure. Soviet support for the Korean armistice, the ceasefire in Indochina, and the Austrian State Treaty made the summit almost unavoidable. Reluctantly, Eisenhower bowed to public opinion. Together with France and the United Kingdom, the United States invited the USSR to a four-power conference on May 10, 1955. It is worth underlining that neither the American president nor the secretary of state desired the meeting. Eisenhower agreed to the summit, as he relates in his memoirs, because he did not wish to "appear senselessly stubborn in my attitude toward a Summit meeting—so hopefully desired by so many." Dulles recalled later that "we never wanted to go to Geneva, but the pressure of people of the world forced us to."[14]

Dulles dreaded the summit. In his view, the conference would be an exercise in damage control, an effort to contain the naive popular hope for détente that threatened to unravel the policies of strength he had worked so tirelessly to build. The western arms buildup was straining the Soviet economy, he acknowledged: continued pressure held the key to extracting political concessions. Munich too was on his mind. He did not want the conference to produce anything smacking of appeasement. "I am terribly worried about this Geneva conference," he confided to C. D. Jackson, "I might have to be the Devil at Geneva, and I dread the prospect." Dulles may have been stewing with pessimism, but C. D. Jackson was brimming with optimism. The summit, he wrote, was the "first really tremendous opportunity that has opened up in a long time." Whatever the dangers of this conference, it also presented a great opportunity for psychological warfare. Having left his post as the administration's psywar chief, Jackson sent his successor Nelson Rockefeller a call to arms. The Kremlin's flexible diplomacy demanded a bold response, he declared, suggesting that Rockefeller assemble a handful of experts and concoct some "screwy" ideas for psychological warfare at Geneva. Rockefeller needed little convincing. He and Jackson immediately organized a brainstorming session with academics, intelligence experts, and government officials at the Marine Corps base in Quantico, Virginia.[15]

The Quantico panelists approached the Geneva summit like generals preparing for a spring offensive. The summit was an opportunity for dealing the Soviet Union a psychological blow before world opinion. Dismissing optimistic predictions of détente, the final report of the Quantico panel declared: "The current disposition of the Soviet leaders to sit down at the 'summit' cannot be traced to a genuine interest on their part to ease any tensions for the sake of peace and harmony. . . . The United States would play into the hands of the Soviet Union if it were to approach the conference with the primary purpose of easing tension. It should meet the Soviet leaders with the intention to force them to retreat." Seeking a mechanism for putting the Soviets on the defensive, the Quantico panelists developed the idea of mutual aerial inspection that would soon be dubbed Open Skies. As a simple concept, easily understood by all, Open Skies would trump anything the Soviets might table at the conference. It was "almost certain" that the Soviets would reject the proposal, the panel predicted, but Open Skies would help reverse "the unfavorable image of the U.S. as a trigger-happy militaristic power, uninterested in resolving the cold war." The possibility that Open Skies would reclaim the magic of Atoms for Peace left the Quantico panelists exuberant. After the meeting, a flurry of memos expressed the view that Cold War victory was in sight.[16]

Dulles was decidedly less optimistic. From the moment the summit was proposed, he wanted to restrict the conference agenda to merely an exchange of views. He did not want the United States to table concrete proposals at the summit, especially not in disarmament. Opposing the Open Skies idea, Dulles advised Eisenhower to keep disarmament discussions confined to the U.N. framework, where they could be buried in technical details. Yet for all Dulles's opposition to Open Skies, he and Rockefeller shared one thing in common: both viewed the Geneva Summit as a propaganda battleground. But Dulles, the international lawyer, preferred Machiavellian maneuverings at diplomatic conferences to spectacular initiatives. He also did not like intrusions on his domain as the president's principle foreign policy advisor. "I have grave question as to the propriety of the President getting this kind of advice from sources outside the State Department," Dulles lectured Rockefeller. The psychological warriors should stick to propaganda and leave diplomacy to the diplomats.[17]

Undeterred by Dulles's resistance, Rockefeller brought his case to the president. On July 6 he hand-carried two memoranda to the White House arguing for Quantico's bold approach. The first, "Psychological Aspects of U.S. Position at Conference," reiterated the Quantico panel's belief that the Geneva summit would be a psychological contest for world opinion. "The Four Power Conference at Geneva will be the most important psychological-propaganda forum in the world," Rockefeller argued. "Regardless of whether the Soviets meet with us primarily for negotiations or primarily for propaganda, the propaganda stakes at Geneva will be extremely high. The psychological aspects may prove more significant than the actual conference results." Rockefeller warned that the Soviets would use the conference to drive a wedge between the United States and public opinion in allied countries by raising hopes for an early end to the Cold War and by making the U.S. appear intransigent and militaristic. Rockefeller further cautioned against Dulles's uninspiring and legalistic approach. The United States "would lose ground in terms of world opinion" if Eisenhower did not use the Geneva opportunity to do something spectacular.

Public opinion trends demanded a new and striking proposal in disarmament, Rockefeller continued. Recent public opinion surveys in Europe revealed that most people favored an all-out ban on the use of nuclear weapons, including both strategic and tactical weapons. "This trend in European public opinion," Rockefeller advised, "may eventually reduce both the strength of our alliance and our freedom to use atomic weapons. Unless we do our utmost to work for disarmament—an aspiration widely cherished by the people of Western Europe—it is very likely that there may be a significant increase in neutralist sentiment both on the Continent and in Britain, together with a growth in pressure for abandoning the use of atomic weapons." Rockefeller hinted at a significant vulnerability in the administration's New Look security doctrine. The U.S. ability to rely overwhelmingly on nuclear weapons for its defense could be circumscribed by public anxiety about such weapons. To arrest public pressure for atomic disarmament, the United States needed to appear earnest in seeking arms limitation. That way, the United States could blame the Soviet Union for the spiraling arms race.[18]

In his second memorandum to the president, Rockefeller pressed the case for Open Skies. He articulated seven reasons for making the proposal:

1. Regains the initiative in disarmament negotiation. . . .
2. Helps break down the Iron Curtain.
3. Provides us intelligence.
4. Poses a difficult decision for the Soviets.
5. Focuses on a practical and immediate aspect of disarmament which people in general can understand.
6. Exposes the phoniness of the proposed Soviet inspection system. . . .
7. Demonstrates first hand to the Soviets our greater war potential.

Significantly, none of these reasons included paving the way for future agreements, achieving a reduction of armaments, or even testing Soviet intentions. As Rockefeller presented it, Open Skies would be good for the United States whether it were accepted or not. Soviet acceptance of Open Skies would give "the U.S. a decided intelligence advantage," whereas Soviet rejection would bring a "decided public opinion advantage." Open Skies also served the purpose of public education. By dramatizing the issue of inspection, the administration could hammer home to world opinion the importance of verifying disarmament agreements. It also would highlight the danger of surprise attack, thereby reminding audiences of the need for maintaining adequate defenses. Further, by contrasting U.S. openness with Soviet secrecy, the proposal provided an additional means of criticizing the closed nature of Soviet society, a recurring theme of American propaganda. Rockefeller concluded, "I cannot see any aspect of it—even if the Soviets accept it, which is highly doubtful—which in any way seriously jeopardizes our security." Open Skies was a risk-free proposition.[19]

Eisenhower showed some signs of being persuaded by Rockefeller's argument. He was impressed with the intelligence implications of the proposal and saw it as something that could reduce the danger of surprise attack, a concern that weighed heavily on his mind. Perhaps more importantly, because no one expected the Soviets to accept Open Skies, Eisenhower was affected by international public opinion trends. At an NSC meeting the day after Rockefeller presented his memoranda to the president, Eisenhower pointedly referred to European polls showing strong support for reducing East-West tensions. This was a convincing reason for doing something dramatic at the summit, he suggested.[20]

In the meantime, the administration developed propaganda themes to contain public pressure for disarmament. The Operations Coordinating Board (OCB) created an ad hoc working group to manage the administration's public posture for the Geneva conference. Among other OCB recommendations, the board proposed a slogan to popularize the view that continued policies of strength were needed to pressure the Soviets to make further concessions. The board initially considered "strength through peace" or "strength for peace" as possibilities, but rejected them for sounding too much like the Nazi slogan "strength through joy" campaign. The board finally settled on "peace from strength" and its variant "peace through strength," slogans that U.S. officials began uttering in press conferences and plastering in official statements. By privileging "strength" over "peace," the OCB used this rhetorical device to "vividly set forth the U.S. purposes in maintaining armed strength during a period of relaxation of tensions." That the slogans remain widely used today by advocates of a strong defense posture is testimony to the power of capsule slogans as instruments of persuasion. Under OCB guidance, the USIA set out to dash any high expectations of what the summit might accomplish. U.S. information officers abroad stressed the limited objectives of the meeting, emphasizing *"we do not wish to give rise to what* [Secretary of State Dulles] *called 'exaggerated and false hopes.'"* USIA officers reminded their audiences of the "procedural rather than substantive character" of the summit. U.S. propaganda stressed the idea of "peace through strength" to maintain a sense of vigilance and urgency among domestic populations, so that "false hopes" would not relax rearmament efforts. The agency also prepared contingency plans in the event that the Soviets violated the agreed-upon "procedural" focus of the conference and announced a surprise proposal to score a propaganda coup. Unbeknownst to the USIA, this would become precisely the strategy pursued by the United States.[21]

In the midst of the Geneva summit, on July 21, 1955, Eisenhower sprung the Open Skies proposal on the unsuspecting Soviet delegation.

The president inserted into his prepared statement the seemingly extemporaneous suggestion that the United States and the Soviet Union trade blueprints of their military establishments and allow aerial reconnaissance over each other's territories. He argued that this would lighten tensions by reassuring people against the dangers of surprise attack. He also suggested that this would prepare the way for more comprehensive agreements in the future by establishing a workable, and comparatively nonintrusive, inspection system.[22]

Like the Chance for Peace and Atoms for Peace initiatives before it, Open Skies was both a serious proposal and a political warfare move. Believing that realistic proposals made for the best propaganda, Eisenhower was prepared to enter into serious discussions on the issue. But Open Skies was clearly a one-sided proposition. It promised to open up Soviet territory to American reconnaissance without providing for arms reductions. Virtually all officials in Washington assumed the Kremlin would reject the offer. Few were surprised when Khrushchev denounced the idea as a bald espionage plot. Assuming Soviet bad faith, the Eisenhower administration designed a lopsided proposal that everyone knew the Russians would reject. When they did so, U.S. officials took it as proof of Soviet insincerity. This was not a true test of Soviet intentions, however; it was a self-fulfilling prophecy.[23]

In the wake of the summit, the international press cheered the "spirit of Geneva." A new atmosphere of international cooperation appeared on the horizon. Yet even though Eisenhower scored a PR coup with Open Skies, officials believed that the summit produced dangerous trends in international opinion. The USIA reported that public opinion trends in the aftermath of Geneva included reduced fear of Soviet aggression and increased faith in Soviet peaceful intentions. War appeared less likely. According to the agency's research, the percentage of Western Europeans indicating they had "bad" or "very bad" opinions about the USSR declined from 50 percent in October 1954 to 37 percent after the summit. The percentage of people indicating they had a "fair" opinion of the USSR almost doubled. "The Russian leaders accomplished a 'public relations' success of no mean proportions, by reducing substantially the unpopularity of the Soviet Union in Western European eyes, and the fear of Soviet aggression," the agency lamented.[24]

To the Eisenhower administration, the "spirit of Geneva" created a psychological climate fraught with danger for U.S. national security policies. International public opinion trends portended a dangerous relaxation of vigilance that would erode popular support for NATO, American Cold

War policies, and rearmament efforts. John Foster Dulles remarked that the conference "poses certain problems for us, as fear and moral superiority have been the cement which has held the free world together. Now the fear and moral superiority have been importantly dissipated, and the result already is causing a considerable amount of confusion in the Free World." Now that war appeared less likely, Dulles questioned, what "cohesive force" was going to take the place of fear in holding the alliance together?[25]

Concerned that excessive optimism for a Cold War thaw would relax defense preparations, the Eisenhower administration tried to offset these attitudes with propaganda stressing the continuing need for western strength and unity. Officials in the USIA felt that they had to "go along" with the spirit of Geneva to "dispel the impression that the U.S. is trigger happy." But the USIA's attitude toward the spirit of Geneva was skeptical, if not hostile. One official summarized the Information Agency's perspective:

> The information job in the post-Geneva period is much more difficult than before but it is more necessary. Disarmament and other plans which would weaken our defenses must be fought. The notion of banning the bomb has made much headway and must be countered. The let-down of our allies must be combated—we must hammer on collective security and oppose the resurgence of neutralism.

Above all, the USIA stressed vigilance. It credited western policies of strength for bringing about Moscow's new conciliatory demeanor. If such policies continued, the agency's propaganda line emphasized, further concessions might be in the offing—but a relaxation of vigilance might exacerbate international tension because it would encourage the aggressive designs of the Kremlin. The USIA's propaganda reminded audiences that peaceful coexistence was only a tactical change designed to weaken the West. "We do not consider that relaxation of tension and a more peaceful atmosphere permit us either to scrap programs for individual and collective self-defense, or to tolerate covert aggression and to sanctify the injustices of the *status quo*." The USIA wanted the world to see that the spirit of Geneva was illusory.[26]

The Test Ban

If the major initiatives of Eisenhower's first administration were political warfare exercises designed to secure immediate popular success at home and abroad, the test ban negotiations at the end of the decade were another

story. For the first time since the beginning of the Cold War, serious negotiations on arms control between Washington and Moscow took place.
Neither Atoms for Peace nor Open Skies produced anything resembling
genuine negotiations, but on the question of nuclear tests, as McGeorge
Bundy noted, "we find real efforts by each side to achieve its objective in
terms that may be acceptable to the other; we have a negotiation." Clearly
Eisenhower's pursuit of a test ban was not the sort of propaganda exercise
that Atoms for Peace and Open Skies had been. Yet neither was it part of a
quest for comprehensive disarmament and East-West détente, as the revisionist interpretation suggests. Eisenhower was propelled to the bargaining table by concerns about the U.S. image and the deleterious impact of
nuclear testing on international public opinion. The picture that emerges
from the documentary record is one of great reluctance to negotiate coupled with an appreciation that the United States needed to appear willing
to do so for public relations purposes. Only when it became apparent that
the U.S. would be forced by public opinion to halt nuclear testing—perhaps unilaterally—did Eisenhower buckle down for serious negotiations.[27]

In the two years after the Geneva summit, the U.S. adopted a two-
pronged approach to disarmament. Vigorous propaganda exploitation of
Eisenhower's Open Skies plan accompanied a diplomatic holding operation to scuttle Soviet counterproposals. The administration focused its resources on extracting psychological gains from Open Skies. In the United
Nations, Henry Cabot Lodge promoted Open Skies as the centerpiece of a
"new World Disarmament Policy." He worked to secure a U.N. resolution
endorsing the plan so that the United States could brand the USSR "as the
troublemaker and war-monger of the world" for refusing to implement it.
Lodge maneuvered to get the British to introduce the resolution, which
was promptly approved in the General Assembly by a vote of 56 to 7.[28]

The USIA also made the promotion of Open Skies a central theme
of its propaganda. Under the guidance of an interdepartmental "public
information working group," the agency promoted Open Skies even as it
continued to hype Atoms for Peace. Films, pamphlets, cartoons, feature
stories, and press releases were used to publicize Open Skies and to bring
public pressure on the USSR to accept the proposal. The USIA produced a
fifteen-page pamphlet entitled "Mutual Inspection for Peace" and distributed copies of it to editors and government officials around the world. The
agency also produced a ten-minute documentary explaining how a typical
Open Skies flight would work. It released a picture story on Open Skies
and the "new science of 'photo interpretation'" for publication in illus-

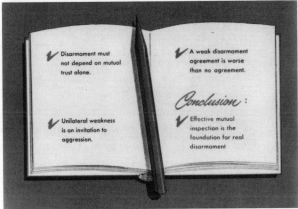

The USIA used pamphlets, films, cartoons, radio programs, and exhibits to publicize U.S. disarmament positions and proposals. This pamphlet, "Pathway to Peace," explains Eisenhower's Open Skies plan in simple, easy-to-understand terms. Source: Eisenhower Library.

trated magazines. According to the final caption in the picture story, "The United States is ready to cross out the 'SECRET' stamp on its military installations, and hold them up for the world to see."[29]

The USIA campaign also included an exhibit illustrating the Open Skies plan. The exhibit, developed in cooperation with the Advertising Council and entitled Aerial Inspection for Peace, sought both to prove America's commitment to peace and to demonstrate the practical feasibility of mutual inspection. USIA public opinion research indicated that few people at home or abroad comprehended the idea that photographs taken from airplanes could provide sufficient detail to verify compliance with arms control accords. In the days before satellite imagery, the agency needed to convince people that aerial reconnaissance was in fact feasible. The exhibit therefore focused on the technical gadgets that made the Open Skies proposal possible, an approach that also made the exhibit seem more educational than propagandistic. The agency displayed the radar equipment, infrared photography, cameras, and airplanes that would enforce an aerial inspection agreement. The USIA reasoned, "As we see it, the visitor would come to the exhibit to find out about the latest gadgets but he would leave with a complete understanding and appreciation of present American efforts toward peace." The exhibit opened at the Carnegie International Center in New York City. Ambassador Lodge personally escorted fifty-nine representatives of member nations through the center, and the USIA attributed the General Assembly's subsequent endorsement of the aerial inspection proposal to the exhibit. After the New York showing, the USIA sent the exhibit and smaller versions of it around the world to increase popular support for Open Skies.[30]

While promoting Open Skies with American propaganda, the United States conducted a holding action on the disarmament question in the United Nations. In March 1955, Eisenhower had appointed former Minnesota governor Harold Stassen to a new post as special assistant for disarmament policy. The press soon dubbed Stassen "secretary of peace," a designation that greatly irritated John Foster Dulles but that served a useful public relations purpose. Stassen became the administration's most outspoken advocate of disarmament. Encouraged by Soviet proposals, he reported at one point that the Soviets were engaged in "plain talking" rather than "propaganda techniques."[31]

Nevertheless, through most of 1956 and into 1957 the Eisenhower administration evinced little interest in concluding an agreement with the Soviet leadership. Rather, it worked to *appear* willing to negotiate, an entirely different proposition. Eisenhower's main concern, shared by Dulles, was in submitting proposals that had "some real appeal, both to our own people and to the people of the world." This strategy, Eisenhower wrote, consisted "of wrapping U.S. proposals in different packages . . . and tying them up in different colored ribbons." It fit nicely with the John Foster Dulles approach to diplomacy: complicated proposals and counterproposals designed to outmaneuver the opposition and to establish a diplomatic record that the United States could point to as evidence of its commitment to peace. The approach delayed serious exploration of Soviet proposals while "making the record" for public opinion purposes. Dulles candidly admitted to the NSC that disarmament negotiations were primarily an "operation in public relations," stressing that the United States needed to appear willing to negotiate, or else risk isolation in the world. To deflate public expectations for what such negotiations might accomplish, State Department officials urged substituting the term "controlled disarmament"—and later "arms control"—for "disarmament" to suggest more limited aspirations. They purged the word "disarmament" from briefings, speeches, and other public statements by officials.[32]

Throughout this period, the United States and the Soviet Union remained deadlocked in their opposing positions. The U.S. would negotiate on nuclear tests only as part of a big indivisible package of disarmament measures, including a cutoff of fissionable materials production and a comprehensive system for inspection and verification. The USSR, on the other hand, rejected the cutoff because it possessed much less fissionable material. The USSR also continued to link a test ban with a ban on all nuclear weapons, an idea unacceptable to the West because of Soviet superiority in conventional forces. Ever since the May 10 proposals, the Soviet Union accepted the idea of inspection in principle, but it agreed only to a weak inspection system, insisting on the right to veto on-site inspections. A potential breakthrough in this deadlock occurred during the spring of 1957, when the Soviets declared their willingness to explore the Open Skies proposal. Eisenhower instructed Stassen to explore the Soviet position. Progress stalled when Stassen, in his zeal to produce an agreement, delivered the Soviet representative an "informal memorandum" that overstepped Eisenhower's instructions and the position of the western allies. The British were outraged at this breach of allied unity. The Soviets worsened the situation by announcing that the two sides were now close to an agreement. Eisenhower called Stassen's move "one of the most stupid things that anyone on a diplomatic mission could possibly [do]" and dismissed him seven months later. In the meantime, Stassen's memorandum became buried in proposals and counterproposals. The Soviet delegation walked out in November.[33]

We Can Get One Meaningful Agreement, 1958–1960

Ironically, a turnaround in the U.S. approach to the test ban followed the bitter failure of disarmament negotiations in the U.N. Gradually Eisenhower and Dulles reappraised their position that a test ban must be linked to an overall disarmament package. They began working in earnest for a separate test ban agreement. Above all, this change in attitude toward negotiations stemmed from two developments: the impact of the radioactive fallout issue on world opinion, and diplomatic pressure generated by the Soviet Union.

By the winter of 1958, radioactive fallout had become one of the most important issues in international affairs. The origins of the issue date back to several incidents in 1953–1954, when American nuclear bomb tests inadvertently showered radioactive waste on persons in the vicinity of the test sites. In response to these incidents, Indian Prime Minister Jawaharlal Nehru proposed that the United States and the Soviet Union halt further tests of hydrogen bombs. Others took up the cry. Many notable voices called for a test ban, including the British philosopher Bertrand Russell, *Saturday Review* editor Norman Cousins, the famed musician and philosopher Albert Schweitzer, physicist Albert Einstein, and Pope Pius XII. The 1955 Afro-Asian conference in Bandung, Indonesia, adopted a resolution urging all powers to suspend nuclear weapons "experiments." A few years later the Afro-Asian Solidarity Conference in Cairo repeated the plea. Numerous resolutions were tabled at the United Nations calling for an end to weapons tests. In 1955, the U.N. created a special commission to investigate the dangers of radioactive fallout. In January 1958, the noted scientist Linus Pauling presented an antitesting petition to the United Nations with more than 9,000 signatures from 44 countries. It included signatures of 36 Nobel laureates.

Nuclear testing emerged as a domestic political issue as well. During the 1956 election, presidential candidate Adlai Stevenson promoted a test ban as part of his campaign. Congressman Chet Holifield and Senator Hubert Humphrey conducted widely publicized hearings on radioactive fallout in 1957 and 1958. The danger of radioactive fallout was further dramatized in 1957 by the publication of Nevil Shute's best-selling novel *On the Beach*. The grim story, later made into a popular film, told of a war that unleashed so much radioactivity that all life on earth was destroyed. The following year, prominent critics of testing initiated a lawsuit in the United States to bar the AEC from testing. Beginning in late 1957, public opinion polls reported that over half of Americans considered fallout a real danger and believed that the United States should cease testing if all other nations did so. USIA public opinion surveys indicated even more powerful antitesting sentiment abroad, with clear majorities in most countries favoring a test ban. In short, by 1958 the campaign against nuclear weapons tests had become a major international movement, involving scientists, politicians, philosophers, authors, and journalists.[34]

U.S. propaganda tried to restrain the rising tide of public opinion. The United States adopted an assertive public relations stance stressing that it tested only to maintain "an atomic shield against aggression." Officials also made the perhaps unhelpful argument that continued testing was necessary to develop "clean" weapons that did not produce radioactive fallout. Stassen's disarmament staff carried out a "public information program," and the OCB created a special working group to tackle the public relations problems created by nuclear testing. The OCB working group prepared press releases, staged media events, and scrutinized public statements on nuclear testing so as to minimize adverse reactions to U.S. tests. The OCB especially publicized Soviet nuclear tests, many of which were not acknowledged by the USSR. The board's pronouncements were carefully phrased to provoke public hostility to Soviet testing without increasing the pressure for banning *all* test explosions. The board sought to attach a moral stigma on Soviet tests by characterizing them as secret, irresponsible, and provocative, whereas U.S. tests were openly acknowledged, safe, and defensive. The working group additionally sought to obtain "maximum psychological advantage" from U.S. programs promoting the peaceful uses of atomic energy, advertising the Atoms for Peace theme as a foil to antinuclear sentiment.[35]

Moscow fanned apprehensions about radioactive fallout. Soviet propaganda portrayed the U.S. as recklessly endangering public health by spreading radioactive clouds around the world. The Soviet government simultaneously pursued an assertive diplomatic strategy that made the United States appear hopelessly defensive and evasive. Spurred on by the successful launch of Sputnik in October 1957, the Soviet leadership went on the diplomatic offensive. The day after the Sputnik launch, Soviet Foreign Minister Andrei Gromyko advised Dulles that the United States and the Soviet Union should work for partial agreements. He suggested that they pursue a test ban agreement as a separate measure from other disarmament proposals. He also indicated his acceptance of the idea of control posts to monitor compliance.[36]

On December 10, 1957, Nikolai Bulganin sent Eisenhower a public letter suggesting that Great Britain, the United States, and the Soviet Union suspend nuclear tests for two or three years. He proposed a second summit conference to discuss arms limitation and other issues. Over the next nine months, the Soviets sent more than two dozen public letters advocating a summit conference and a variety of disarmament measures, including a nuclear-free zone in Central Europe (the Rapacki Plan), measures to guard against surprise attack, an East-West nonaggression pact, neutralization, prevention of nuclear proliferation, and disengagement of U.S. and Soviet forces. The letter exchanges vividly displayed the "new diplomacy" in action. Negotiations were taking place in the international media. While putting forth legitimate positions on diplomatic issues, the letters aimed primarily to influence international public opinion rather than the governments they addressed. In this, Moscow proved more adept than Washington. The Eisenhower administration was placed in the awkward position of repeatedly giving negative answers to seemingly legitimate Soviet proposals.[37]

The Soviet government accelerated the pressure on January 6, 1958. Khrushchev announced his intention to unilaterally reduce Soviet troops by 300,000. At an NSC meeting the same day, Dulles complained that the move increased the pressure on the United States to show progress on disarmament. People would be duped by the "simplified Soviet views" of disarmament, he grumbled. Even though many governments were sympathetic to U.S. policies, Dulles worried that "ill-informed public opinion" might compel foreign governments to take steps in disarmament they would prefer to avoid. Echoing this sentiment, Henry Cabot Lodge cautioned, "In the long run the effect of world public opinion is very important, and at the present time the United States is in not a very good posture with respect to world public opinion on disarmament."[38]

To address this problem, the administration initiated a major gearing up of U.S. disarmament propaganda. In September 1957, deputy director of the USIA Abbott Washburn prepared a memorandum to the NSC articulating plans for a worldwide propaganda effort. Washburn declared: "The No. 1 priority task of the Government's overseas information effort over the next 20 months must be to explain and make clear to peoples abroad the United States's (Free World) disarmament proposals. . . . All other themes must be subordinated to this." The agency allocated $4 million in its 1959 budget to cover the worldwide information campaign on U.S. disarmament proposals. The Information Agency launched an intensive, long-range program to promote western disarmament proposals and "to counter more effectively the Communist emotional slogans 'Stop the Tests' and 'Ban the Bomb.'" All communication media were used to clarify and promote the disarmament program of the West as the "surest road to durable peace," including documentaries, exhibits, pamphlets, lectures, and personal contacts. The USIA also enlisted C. D. Jackson to ghostwrite a 3,000-word "presidential monograph" articulating Eisenhower's personal feelings about disarmament, which the USIA planned to send to influential persons at home and abroad.[39]

Concerned that the United States was not making an adequate case before the court of world public opinion on disarmament, the OCB created another ad hoc committee on "disarmament information programs." The OCB committee sought to increase understanding of the U.S. position in disarmament and to explain "simply and convincingly" why Soviet proposals were unacceptable to the United States. According to the board, the U.S. had "not formulated a concise and convincing statement to the world on why we continue to rely heavily on nuclear weapons despite the Soviet proposal to ban them." The key American proposal, for a cutoff on the production of fissionable materials, sounded unconvincing when compared with the Soviets' more straightforward "ban the bomb" slogan. The OCB wanted to make the U.S. disarmament position "more understandable and more appealing" to the peoples of the world in order to bring world public opinion pressure to bear on the Soviet Union to accept American proposals. U.S. disarmament proposals, as Harold Stassen noted, needed to be of a "type easily understood by the people."[40]

Eisenhower sympathized with the idea that U.S. proposals needed to be dressed up and simplified to widen their popular appeal. On March 21, 1958, Eisenhower sent Dulles a long memorandum complaining that the Kremlin had maintained the initiative on disarmament questions. He wrote to Dulles that he wanted to make a "rather startling" new proposal to communicate to the world the American desire for peace. "The world is asking for something that is almost impossible when it insists that we should give all peoples complete assurance that we are not only peaceful and friendly, but that we shall 'hold the initiative' in striving for peace." He continued in a lengthy digression: "Our public relations problems almost defy solution. The need always for concerting our views with those of our principal allies, the seductive quality of Soviet promises and pronouncements in spite of their unreliability, the propaganda disadvantage under which we operate because of the monolithic character of Soviet news broadcasts . . . all serve to make us appear before the world as something less than persuasive in proclaiming our peaceful purposes."[41] This memo and other documents revealed a core truth about the president's view of disarmament. The arms control dilemma was not a question of building East-West trust and finding a means to reconcile divergent points of view; it was a public relations problem requiring innovative packaging.

Two days after Eisenhower sent this memo to Dulles, U.S. intelligence received warning that the Kremlin was about to unilaterally suspend testing. At a crisis meeting to discuss the U.S. response, Dulles argued that the United States must beat the Soviets to the punch by immediately announcing a test suspension of its own. He said he "felt desperately the need for some important gesture in order to gain effect on world opinion." When the Defense Department and AEC raised objections, Dulles responded strongly: "Defense was approaching the problem in terms of winning a war. State must, however, think in terms of all means of conducting the international struggle. . . . [W]e are increasingly being given a militaristic and bellicose aspect toward world public opinion, and are losing the struggle for world opinion." Eisenhower agreed. "World opinion is a fact," he admonished his advisors, "world anxiety exists over tests, and causes tension." Although "testing is not evil," he explained, "people have been brought to believe it is." He conceded that the abolition of tests might hurt the United States in a strictly military sense, but, on the other hand, he emphasized "we need some basis of hope for our own people and for world opinion. . . . [I]t is simply intolerable to remain in a position wherein the United States, seeking peace, and giving loyal partnership to our allies, is unable to achieve an advantageous impact on world opinion." Despite his concerns, Eisenhower caved in to the fierce resistance of the Defense Department and the AEC, who argued that the American military posture depended on further tests.[42]

Eisenhower came to regret this decision. As American intelligence had predicted, the Soviet government announced on March 31 that it was suspending all nuclear tests. Moscow called upon the United States and Great Britain to follow suit. In press conferences, Dulles and Eisenhower denounced the move as a "gimmick." Indeed, the test suspension clearly was a propaganda ploy because the Kremlin had just completed a massive test series of its own. Nevertheless, the Soviet announcement had an electrifying effect on public opinion. *Newsweek* commented that the Soviet test suspension won worldwide approval as a "major move toward peace." Antinuclear demonstrations took place in Great Britain, West Germany, India, and Japan. Soviet diplomatic activity of the previous five months, culminating in the unilateral conventional force reductions and testing moratorium, created the impression that the USSR was the only one acting to further the cause of international peace. *New York Times* columnist

James Reston wrote that while Moscow acted, "Washington reacted with words, and found itself once more on the defensive." The *New Republic* criticized the Eisenhower administration for "the sleep walking that passes for American diplomacy." The *Nation* commented that if the Soviet testing moratorium was a gimmick, "one can only wish to God that our statesmen could concoct such gimmicks once in a while."[43]

The Soviet announcement made it increasingly difficult for the United States to continue testing. In the wake of the announcement, Dulles and Eisenhower became convinced that the United States needed to do something to placate public opinion. Dulles, who had opposed bold U.S. ventures in disarmament at almost every turn, now insisted that the United States needed to act fast. He repeatedly insisted that failure to appear willing to negotiate would produce the "moral isolation" of the United States. "We must recognize," he remarked, "that we can lose the whole struggle with the Soviets if we fail to take into account such imponderables as world opinion, and it is in this area that we have been taking a beating." Although Dulles's views on the test ban had fluctuated markedly over the previous five years, by the spring of 1958 he was convinced that the United States needed some kind of agreement. Writing to Eisenhower on April 30, Dulles pleaded for results: "Only by concrete actions can we counteract the false picture, all too prevalent abroad, of the United States as a militaristic nation. . . . We should not enhance Soviet prestige by remaining inflexible on the nuclear testing issue."[44]

Eisenhower concurred that international public opinion demanded U.S. action. At an NSC meeting to discuss the Soviet testing moratorium, Eisenhower observed that he was "very much concerned about the psychological effects of continued nuclear tests." After a lengthy discussion of the technical aspects of test cessation, the president advised the NSC that the U.S. was "facing a psychological erosion of our position with respect to nuclear testing, and that we must take this fact into account along with the other pro's and con's respecting the cessation of nuclear testing." In meeting after meeting with his advisors on the test ban issue, Eisenhower stressed the importance of world public opinion, insisting that the United States must keep the "hope of disarmament" before the world. "Much of this public opinion is very uninformed in the area of disarmament," he complained. "What the world wants is easy answers to the disarmament dilemma, and we must be clear that this opinion on the necessity for disarmament steps is steadily growing stronger and insisting on results." "Even in confidential talks at the NATO meetings," Eisenhower continued, he "encountered very strong insistence that something must be done to advance disarmament."[45]

Although widespread fear of radioactive fallout created an enormous amount of international public opinion pressure to halt the testing of nuclear weapons, there were a number of other factors that also brought the test ban to the forefront of U.S. arms control efforts. First, the test ban was the least difficult arms control measure to monitor. It would be much easier to gain Soviet acceptance of a viable system of inspection and control than more comprehensive disarmament schemes. Second, by opening up Soviet territory to American inspection teams, a test ban would provide intelligence information and help guard against surprise attack. Third, a test ban would help prevent the proliferation of nuclear weapons to other countries. Fourth, it would lock in U.S. technological superiority in nuclear weapons design. Finally, a test ban met one of the crucial tests for disarmament in an information age: a "test ban" was a proposal easily understood by the people.[46]

Despite these advantages, the administration was deeply divided over the question of nuclear testing. It was one of the most contentious issues of the Eisenhower presidency. The AEC, Defense Department, and Joint Chiefs of Staff (JCS) vigorously opposed a test ban. The JCS argued that a test ban treaty needed to be part of a larger disarmament agreement with an effective system of inspection and verification. The AEC insisted that the United States needed to continue testing in order to maintain the U.S. technological edge and to develop "clean" bombs that did not produce radioactive fallout. John Foster Dulles and the State Department, on the other hand, pressed the case for test ban negotiations on the grounds that they were crucial for world public opinion. Dulles argued time and again that the United States must not appear reluctant to negotiate. Also advocating a test ban were Eisenhower's science advisors. After the Sputnik crisis in October 1957, Eisenhower created the President's Science Advisory Committee (PSAC), headed first by James R. Killian and later by George B. Kistiakowsky. Significantly, PSAC challenged the pessimistic technical advice Eisenhower had been receiving from the defense establishment. It demonstrated the feasibility of a detection system and argued that a test ban agreement would work to the strategic advantage of the United States.[47]

The new technical advice from PSAC facilitated the policy change that Eisenhower and Dulles had been exploring. The Eisenhower administration announced in April that it would negotiate a separate ban on nuclear tests. Dulles argued that such a measure was necessary because U.S. intransigence was giving the United States a "militaristic image." Although the announcement held out hope for an eventual agreement, Dulles believed that a gesture to capture public support was needed. He wanted the administration to propose a temporary moratorium on testing while the negotiations proceeded. The situation became urgent in August as the United Nations prepared to release a report on the effects of atomic radiation. On the eve of the report's release, Dulles argued forcefully that international opinion compelled the United States to take bold action:

> Unless we take a radical step now, our failure to do so will in effect be a step to 'go it alone' as a militaristic nation in world opinion without friends and allies. This will become apparent in the course of the year. You could count now nations that will turn from us. Many say it is irrational to turn from us; we have a powerful case for continued testing. But the Government of Japan cannot stand with us; India will not; the Governments likely to come into power in the UK and Germany will not. Few will want to be our friends and allies, want us to station our facilities on their soil and be willing to stand with us. Stopping tests, of course, exposes us to certain dangers, but the danger of being isolated, encircled and strangled is even greater than the threat of a massive atomic attack so long as we retain our retaliatory power.

Eisenhower took notice of this powerful statement of the psychological significance of disarmament negotiations. At a meeting with AEC officials, he remarked that "the new thermonuclear weapons are tremendously powerful; however they are not as powerful as is world opinion today in obliging the United States to follow certain lines of policy." Overruling Defense and AEC arguments that the United States could not afford to stop nuclear testing, Eisenhower insisted, "our world situation requires that we achieve the political benefits of this action." On August 22, Eisenhower publicly announced that the United States would suspend nuclear testing for a year beginning October 31. Negotiations for a test ban would begin on that date.[48]

In the short term, the announcement of the impending moratorium precipitated a nuclear testing frenzy. For the next several weeks, the governments of the Soviet Union, Britain, and the United States exploded bomb after bomb to maximize their military gains. The Americans and British detonated thirty-seven explosions in nine weeks of frantic testing, and the Soviet Union broke its own self-imposed moratorium with a series of fourteen nuclear weapons tests in October. The testing before the October 31 deadline was so extensive that sharp increases in radioactive fallout were detected in the early months of 1959. In the United States, revelations of alarming increases in radioactivity precipitated a major fallout scare that "bordered on panic." In March, *Consumer Reports* published a study revealing heavy doses of the radioactive strontium-90 in milk. Warning that the presence of the radioactive material in the "milk we drink" was a health hazard, the study called for an end to nuclear testing. The article ignited a firestorm of public criticism—including a campaign by a group of scientists to publicize radioactivity levels in baby teeth—and exacerbated public fears about the ramifications of nuclear testing.[49]

Alarmed at the public outcry, Eisenhower pressed the case for a test ban with unprecedented urgency. At a meeting on March 17, 1959, Eisenhower acknowledged that the fallout issue obliged the United States to halt its nuclear weapons tests. The minutes of the meeting record the president as saying: "All available evidence indicates that nuclear testing is bad. The allowable dose of strontium 90 is being approached in some foods in some areas of the country. With this development, the President feels that we would no longer test atomic weapons in the atmosphere. There is a requirement now for a system which both sides know would work." Eisenhower argued that it was no longer possible for the United States to be as "rigid in the details" on such matters as inspection as it had been before. The United States needed to work toward acceptance of a test ban, even if it was not "as good as we want," in order to demonstrate to the world that "we *can* get one meaningful agreement."[50]

This represented a major shift in the president's position. Never before had Eisenhower pressed his advisors so forcefully to accept an arms control accord with the USSR. Eisenhower had in the past stated his belief in the importance of negotiations for the U.S. image, but he always stopped short of advocating an agreement. Now, for the first time since his inauguration, the president actively pushed for an accord on a substantive matter with the Soviets—not a spectacular political warfare initiative, not mere negotiations for appearance purposes, but an *agreement.*

As international opinion became increasingly inflamed by the question of fallout, Eisenhower concluded that the United States would have to stop testing, with or without a test ban treaty. At a meeting with his disarmament advisors in May, an exasperated President Eisenhower referred to reports of high fallout levels and exclaimed, "we were going to be forced by public opinion in the United States to stop tests unilaterally." He argued that because public opinion was forcing them to stop testing anyway, they should try to get an agreement in order to secure the PR and intelligence benefits of a treaty with inspection provisions. "We must find a reasonable and decent way to do this by agreement if possible," he insisted, "even if the arrangement is not necessarily a perfect one."[51]

In the spring of 1959, the Eisenhower administration backed away from a ban on all nuclear weapons tests, advocating instead a more limited agreement banning tests in the atmosphere and underground tests above a certain threshold. This shift to a "threshold" ban resulted from new technical data indicating that monitoring compliance required more intrusive inspections than previously thought. Earlier, in August 1958, technical experts from both sides had agreed that 160 to 170 land-based control posts would comprise an adequate verification scheme. Data from the last U.S. test series in October, however, indicated that that ten times as many inspection stations were needed to distinguish underground tests from earthquakes. When American negotiators revealed these findings to their Russian counterparts, Moscow reacted with alarm. Suspicious that the United States was using the information to subvert the negotiations, the Kremlin refused to allow the data to factor into the discussions. In previous years, this would have been what the Eisenhower administration was looking for: evidence of Soviet intransigence that could be used both to block further progress in disarmament and to blame the USSR for the failure of the negotiations. Instead, Eisenhower adjusted the U.S. negotiating position to make an accord possible. At one point, Eisenhower commented that "even an agreement limited to atmospheric tests, with only as few as three or four control posts, would be better than no agreement at all." Eisenhower's willingness to compromise did not indicate his trust of the Soviets, nor did it represent his pursuit of détente. He assumed that only the West would adhere to the agreements and believed that it would be difficult to trust the "liar" Khrushchev. The most important objective was to secure an agreement banning atmospheric tests in order to alleviate mounting public pressure on the fallout issue. Although the United States could have done this by merely testing underground, Eisenhower wanted to receive the psychological and intelligence benefits of a test ban treaty that included inspection provisions.[52]

By early 1960, significant progress toward a test ban agreement had been made. The previous fall, Khrushchev had traveled to the United States on a goodwill tour. Shortly thereafter, Eisenhower agreed to meet the Soviet leader at a four-power summit conference in Paris in mid-May 1960. In the meantime, the Soviets accepted an American proposal for halting atmospheric tests and underground tests registering above 4.75 on the Richter scale, providing that the nuclear powers agree to a voluntary moratorium on tests below the 4.75 mark. The two sides were very close to an agreement. Only two major issues remained: the duration of the moratorium, and the number of permissible on-site inspections. These would be addressed at the Paris summit.[53]

Then came the U-2 incident. Two weeks before the Paris meeting, Eisenhower authorized an espionage flight by a U-2 spy plane over Soviet territory. The United States had been conducting these reconnaissance flights since 1956, but Moscow had been unable to shoot down the U-2s because they flew above the range of Soviet air defenses. On May 1, however, the Soviet military successfully shot down the aircraft and captured the pilot. When Khrushchev announced the downed aircraft, the Eisenhower administration issued a bogus cover story claiming the U-2 was a lost weather plane. To Eisenhower's embarrassment, Khrushchev exposed the American subterfuge by unveiling the wreckage, the pilot, and the photography equipment. On the first day of the Paris summit, Khrushchev asked Eisenhower to denounce the flights publicly. He demanded an apology. Eisenhower pledged to refrain from further flights but refused to apologize. Khrushchev stormed off. The summit had failed.[54]

Eisenhower wrote in his memoirs that one of the "greatest disappointments" of his presidency was the failure to reach an agreement on disarmament. For eight years, he recalled, his administration had toiled to secure some sort of agreement that would mark a first, even if only a small, step toward a satisfactory disarmament plan. Perhaps Eisenhower came under the spell of his own propaganda, for evidence supporting the claim that he pursued a long-term strategy for making gradual progress in disarmament by small steps is absent from all major discussions of disarmament within the administration. Nowhere in NSC planning documents or in memoranda of discussions do we find Eisenhower or anyone else advocating a step-by-step approach to disarmament. Eisenhower was more concerned with presenting a public posture of openness to negotiation than he was eager to negotiate. Not until the pressure of public opinion on the fallout issue became unbearable did Eisenhower give priority to negotiations over political warfare. Only then did Eisenhower press his advisors for an agree-

ment. Only then did the president give any thought to the question of finding balancing incentives to earn Soviet acceptance of U.S. proposals.[55]

Many factors propelled Eisenhower to the bargaining table at the end
of his administration. Détente was least among them. The president simply did not trust the "men in the Kremlin."[56] He saw the Cold War as a
long-haul struggle and believed it might take years to win. To him, it was
vital that the United States maintain the support of world public opinion
to prevent further communist gains. Negotiations and international agreements were desirable if they won the approval of public opinion, mitigated neutralist pressures, locked in U.S. superiority in atomic armament,
or provided the U.S. with other collateral benefits, such as intelligence.
Eisenhower publicly presented his diplomatic efforts as measures to promote world peace, but he saw diplomacy more as a psychological weapon
for waging the Cold War than as means of achieving détente.

To argue that the Eisenhower administration did not pursue negotiations to arrive at an East-West détente is not to suggest that the USSR did,
however. Nor is it to claim that the United States missed an opportunity
for achieving an accord. Although some documentation from the Kremlin's
archives supports these interpretations, more high-level material is needed
for historians to reach such conclusions with certainty. Lingering questions about Soviet intentions, however, should not obscure a reexamina-
tion of American objectives. The documentary record of the Eisenhower
presidency strongly indicates that the quest for disarmament took a back
seat to psychological warfare initiatives and the maintenance of a posture
of strength. This does not mean that propaganda considerations were the
only considerations, or that strategic, economic, and military factors did
not influence American strategy. Rather, the research presented here suggests the more modest conclusion that propaganda shaped U.S. arms control strategy at critical junctures.

Concerns about propaganda and international public opinion influenced U.S. diplomatic activity on several levels. In the first place, propaganda affected *interpretations* of Soviet intentions. U.S. officials interpreted
Soviet peace and disarmament initiatives as menacing political warfare tactics designed to divide the United States from its allies, encourage neutralist tendencies, and undermine domestic support in the United States and
its allies for rearmament. Precisely because the "soft line" in Soviet foreign
policy raised worldwide hopes that the Cold War and the arms race might
come to an early end, policy makers in the Eisenhower administration
viewed the new Soviet tactics with alarm. Public opinion pressures, fanned

by the fear of nuclear war and the hope for peace, meant that the United States simply could not dismiss Soviet gestures out of hand. The United States had to appear willing to negotiate or risk isolation in the international arena. At the same time, the perception that the USSR was using negotiations to rally public opinion against American intransigence and to pressure the United States to engage in unsafeguarded nuclear disarmament led American policy makers to discount Soviet proposals as mere propaganda ploys.

As a consequence, U.S. proposals were designed less to explore Soviet initiatives than to expose them as fraudulent. At the *planning* level, policy makers worked to devise initiatives and positions that were strategically advantageous to the U.S. and that could serve propaganda purposes. Operating on the assumption that the best propaganda was believable propaganda, U.S. policy makers designed initiatives to be sound and workable, so that they would not be dismissed as "mere propaganda" by international observers. Most major U.S. positions, however, were designed as win-win propositions; the United States would win strategically if the Soviets accepted them, and it would win politically if the Soviets rejected them. American policy makers also privileged proposals that were easily communicated through the media and readily comprehended by public opinion, so as to not lose public opinion in the technical details of the negotiations. Not coincidentally, the administration's principal proposals could be summed up through simple phrases: the Chance for Peace, Atoms for Peace, Open Skies, and Test Ban. As Henry Kissinger argued in *Foreign Affairs*, "Proposals . . . must be framed with a maximum of clarity and even simplicity, for their major utility is their symbolic content."[57]

At the level of *negotiations*, concerns about international opinion and propaganda influenced how the United States responded to Moscow's initiatives. U.S. officials believed that, if they accepted a Soviet proposal, they would add respectability to the Soviet leadership and enhance Moscow's prestige. In their view, consenting to a Soviet initiative was tantamount to receiving a propaganda defeat before world opinion. Accordingly, they assumed that the Soviets would have to accept an American proposal so that the United States would reap any propaganda benefits to be gained. The propaganda war ensued that Soviet proposals were often dead on arrival. Finally, at the level of the *presentation* of American diplomacy to the public, the United States used conventional propaganda tactics to influence public perceptions of the negotiations, to generate international pressure on the Kremlin to accept American proposals, to demonstrate the American commitment to peace and disarmament, and to expose the insincerity and impracticability of Soviet initiatives.

On all these levels, U.S. diplomacy was affected by concerns about international public opinion and a preoccupation with symbols and images. The propaganda war had acted as a significant impediment to conflict resolution. The fact that each side coupled its disarmament proposals with massive media exploitation reinforced the view that each was using negotiations for propaganda purposes. Moreover, the notion that accepting the opponent's position would be regarded as a propaganda setback prevented negotiators from exploring proposals seriously. The objective became out-maneuvering the opponent in the battle for public opinion; positions were put forward more to win public acclaim than to pave the way for compromise at the bargaining table.

Chapter 7

Every Man an Ambassador

Cultural Propaganda and the People-to-People Campaign

The United States Government . . . is not the sole, or perhaps even the most important
factor in the projection of the American image. . . . Americans traveling or living abroad
contribute heavily to the process of foreign image-making.

— President's Committee on Information Activities Abroad, May 9, 1960

The promotion of peaceful United States objectives should be facilitated by every means
at our disposal. Frequently, an international trade fair, an important cultural event, or a
scientific gathering provide opportunities to influence public sentiment, of value as great, or
even greater than more formal occasions.

— Herbert Hoover Jr., Hearings Before the Subcommittee of the Committee on
Appropriations, 84th Congress, 1st Session, 1955

In the summer of 1954, Eisenhower asked Congress for "emergency" legislation appropriating $5 million for extraordinary circumstances arising
in international affairs. This request would be less remarkable if it were
not for such an unusual emergency: to fund a crash expansion in cultural
exchange activities and to boost the American presence at international
trade fairs. It was "essential," Eisenhower declared, that the United States
"take immediate and vigorous action to demonstrate the superiority of
the products and cultural values of our system of free enterprise." Congress swiftly approved Eisenhower's request, establishing what was called
the "President's Special Emergency Fund" to subsidize trade fair presentations by private industry and tours by artistic groups traveling abroad.
The "emergency" appropriation was renewed in 1955. The following year,
Congress passed the International Cultural Exchange and Trade Fair Act,
authorizing permanent government sponsorship of these activities.[1]

It is testimony to the peculiar nature of Cold War political culture that
such a sum could be granted on an emergency basis for such a purpose
as cultural relations. In the worldwide battle for hearts and minds, how-

ever, cultural interchange and expression became weapons of ideological warfare—mechanisms for illuminating, disseminating, and demonstrating ideological principles. Publicly, Eisenhower and other American officials defended the cultural program with lofty rhetoric about furthering "international understanding." Privately, they emphasized propaganda motives. Cultural exchanges and trade fairs were used as positive, long-range programs to create a favorable atmosphere abroad for U.S. policies. "Culture for its own sake is not likely to be effective in a planned program," the USIA conceded, but "culture used selectively in the service of long-range solid opinion (and, for that matter, short range impact) is an important task of USIA."[2]

The Emergency Fund programs were parts of a much larger effort to promote a favorable view of the United States by working through private intermediaries. This strategy of "private cooperation" called for the maximum use of private groups and nongovernmental organizations (NGOs) in achieving U.S. psychological objectives. The private facade would camouflage the propaganda motives; cultural initiatives devoid of political messages were less likely to elicit hostile responses from audiences suspicious of foreign propaganda. Basic public relations principles provided the underlying rationale for government-sponsored cultural diplomacy. Its purpose was not to affect perceptions of U.S. policies per se, but rather to soften the image of the United States by "humanizing" America in the eyes of the world—a PR strategy long used to manage corporate images. Instead of viewing the United States merely as an impersonal government pursuing its own interests, so the reasoning went, people would associate the United States with positive cultural, artistic, and material accomplishments. Cultural activities served another rudimentary public relations purpose: they generated news. Psychological strategists sought ways to create news through action, pushing governmental and private groups to *do* things that provided good copy. The USIA would then put its formidable media resources to work saturating the world's news outlets with positive stories about American life and culture, balancing frightening news of Cold War confrontations with more positive representations.

State-private cooperation in the Cold War's propaganda battles developed logically from the total war mind-set that called for mobilizing the entire nation to support U.S. objectives. Drawing from wartime experience, government officials sought to mobilize private groups and individuals as they had during World War II, when Americans were called to contribute actively to victory through such acts as buying war bonds, collecting scrap,

and conserving resources. Now Americans were told that *they* were engaged in a war of ideas, and that *they* had better get involved. Such efforts
achieved greatest visibility in the People-to-People program, an ambitious
state-private cooperative venture developed in the aftermath of the Geneva Summit. The program was designed to encourage ordinary Americans,
NGOs, and businesses to engage in public relations work on behalf of the
United States. As one article distributed by the USIA explained, it sought
to make "every man an ambassador." Under its auspices, civilians engaged
in "grassroots diplomacy" by traveling abroad as unofficial ambassadors,
promoting hospitality for foreign visitors to the United States, collecting
books and magazines for international distribution, creating "sister city"
affiliations, and engaging in other activities designed to promote a favorable view of ordinary Americans. For the government planners who developed the program, People-to-People provided a way to boost morale
and remind Americans of the political and ideological stakes involved in
the fight against communism without relying on the heavy-handed and
politically controversial tactics of domestic propaganda agencies. It thus
served the purposes of both international propaganda and domestic mobilization; while promoting a favorable view of the United States internationally, it also gave Americans a sense of personal participation in the Cold
War struggle. In the total public relations contest for world opinion, every
American had a part to play.[3]

The Cultural Offensive

The Emergency Fund and the People-to-People program reflected the
American concern that the communists might conquer the world through
peaceful diplomacy and cultural infiltration, a fear that grew appreciably
with the changing Soviet tactics after Stalin's death. The Operations Coordinating Board (OCB) perceived a "vastly increased and coordinated
effort" by communist countries to use cultural exchanges and trade fairs
as vehicles for propaganda. Indeed, as part of the new Soviet leadership's
"peaceful competition for men's minds," exchange programs, tours by cultural groups, and participation in international trade fairs developed into
major forms of communication between the Soviet bloc and the outside
world. Under Stalin, the Soviet Union and its satellites had participated in
fewer than a dozen trade fairs each year. After 1953 the number of communist exhibits at trade fairs quadrupled. In 1954 communist bloc countries
participated in 60 fairs in 26 countries. The following year, communist

participation increased to approximately 170 fairs in 45 countries. Soviet cultural exchange activities likewise increased. The number of Soviet cultural delegations, including tours by theatrical, musical, and dance groups, tripled between 1953 and 1955. This outpouring of communist cultural groups traveling abroad coincided with a loosening of state restrictions on cultural expression within the USSR. In addition to granting writers and artists more (though not complete) freedom of expression, the post-Stalin leadership allowed increasing numbers of foreigners to travel to the Soviet Union. Beginning in April 1953, the Soviet government granted visas to Americans to travel to the USSR after having denied them for years. Journalists, editors, students, professors, athletes, and members of Congress traveled to the Soviet Union for the first time since the Cold War began. As these measures suggest, the Soviet leadership evinced a growing interest in public relations and in enhancing the respectability of their regime at home and abroad. In the words of one Soviet official, they intended "once and for all to liquidate the notorious Iron Curtain" in the eyes of the world.[4]

The thaw in Stalinist cultural policies made it much harder for U.S. propagandists to demonize the leaders in Moscow, who seemed to be embarking on a genuine, if limited, reform of the Soviet system. American analysts did not view these measures as hopeful signs of peaceful change. They perceived instead a "cultural offensive"—a propaganda effort to demonstrate to the world the richness of cultural expression in the USSR. The cultural offensive gave teeth to Soviet charges that American culture was sterile and polluted by corporate capitalism and consumerism. The USIA argued that such propaganda posed a direct threat to American leadership abroad. It noted with concern that many foreigners believed in the barrenness of American culture. Most appeared to agree with Soviet propaganda that the "American people are preeminently a gadget-loving people produced by an exclusively mechanical, technological and materialist civilization." A vigorous and effective cultural program was necessary to dispel such notions, the agency argued, because cultural leadership was a prerequisite of world leadership.[5]

One might expect a battle-hardened soldier and fiscally conservative politician like Eisenhower to shun such reasoning and balk at plans to spend taxpayers' money on cultural attractions. On the contrary: the president believed deeply in the value of cultural exchange activities. As early as 1947, he was arguing for an intensive program to inform foreigners about American culture and living standards. Testifying before the House, he advocated an information program that disseminated cultural information in terms "readily comprehended by the people . . . in terms of ice boxes, radios, cars, how much did [Americans] have to eat, what they wear, when they get to go to sports spectacles, and what they have available in the way of art galleries and things like that." As president, Eisenhower repeatedly commented that he was "keenly interested" and "very enthused" about the U.S. cultural program. Several administration officials later recalled Eisenhower's strong interest in these types of activities. The fact that Congress ritualistically slashed his requests for USIA appropriations provided one reason for Eisenhower's support of the cultural program. Cultural exchange programs provided a means of squeezing more money out of the legislature for activities designed for international persuasion. Eisenhower also recognized that the overall image of the United States, including its society and culture, had an impact on U.S. leadership in the world. Writing to his brother Edgar, Eisenhower argued forcefully for the importance of cultural diplomacy:

> It is possible that you do not understand how ignorant most of the world is about America and how important it is . . . that some of the misunderstandings be corrected. One of them involves our cultural standards and artistic tastes. Europeans have been taught that we are a race of materialists, whose only diversions are golf, baseball, football, horse racing, and an especially brutalized brand of boxing. Our successes are described in terms of automobiles and not in terms of worthwhile cultural works of any kind. Spiritual and intellectual values are deemed to be almost nonexistent in our country.

"This picture of their misunderstanding is not overdrawn," Eisenhower continued emphatically. "In fact, in some areas we are believed to be bombastic, jingoistic, and totally devoted to the theories of force and power as the only worth while elements in the world." It was necessary to correct these impressions, Eisenhower believed, in order to convince world audiences that true progress rested with allegiance to the United States.[6]

Eisenhower's acknowledgment that the government had an obligation to shape international perceptions of American culture was remarkable. Although NGOs, religious groups, and businesses had participated in cultural exchange and trade fair activities for years, the government had not actively promoted such activities until the late 1930s, when the Roosevelt administration created a cultural program for Latin America to counter Nazi propaganda in the region. In the early Cold War years, the Fulbright and Smith-Mundt Acts formalized the government's interest in educational and cultural exchanges, but these activities were explicitly separated from government propaganda, at least on paper. Cultural affairs were to promote "increased understanding" rather than support for U.S. foreign policy goals. When the USIA assumed control of information operations in 1953, the State Department retained jurisdiction over educational and cultural exchange programs in order to maintain the distinction between "propaganda" and cultural relations.[7]

Under the Emergency Fund, these distinctions blurred. Propaganda motives became more pronounced. The day Congress approved the Emergency Fund, August 18, 1954, Eisenhower created an interdepartmental "action group" to implement this expanded cultural effort. Citing a "pressing need for early action," he directed his advisors to make "immediate and effective" use of the funds. Consistent with the strategy of private cooperation, Eisenhower called for maximum participation from private individuals and NGOs. The president put USIA director Theodore Streibert in charge of the interdepartmental effort, with the State Department operating the cultural presentations programs and the Commerce Department implementing the trade fair programs. Both departments acted under guidance from the USIA and OCB with respect to the "propaganda aspects" of their operations. The OCB created two working groups to ensure that the projects strengthened the "climate of world opinion" and refuted "communist propaganda by demonstrating clearly the United States' dedication to peace, human well-being, and spiritual values."[8]

Administration officials and journalists occasionally referred to the cultural program in terms of "piercing the Iron Curtain," but the Emergency Fund provided for activities only in the "free world." The OCB determined the priorities of the trade fair and cultural programs. It attached greatest importance to the "political or ideological tendency" of the host countries; those places in greatest "danger of communist infiltration" received highest priority. Initially, the cultural program put emphasis on European countries, a focus that reflected the administration's goals as it entered office and sought to achieve German rearmament in a western military pact. But increasingly the focus of the cultural program shifted to the developing world. In part, this change in priorities came from the realization that Europeans were already exposed to the products of American high culture and that they could afford to pay high prices for tickets. More importantly, the American cultural blitz in the developing world reflected the administration's belief that the major battlegrounds of the Cold War were in the "periphery"—in Asia, Latin American, Africa, and the Middle East. It was more urgent to target neutralist countries than those allied with the United States. This was not culture for its own sake; it was culture put to work in the service of the Cold War.[9]

Trade Fairs and American Propaganda

Trade fairs had long been used to promote international commerce. In the context of the Cold War, however, trade fairs assumed special ideological and cultural significance. These were occasions for reaching economic leaders as well as ordinary people who attended the fairs to see a free show. They were also media events, opportunities to generate favorable publicity for participating countries and their industries. International trade fairs provided built-in opportunities for the United States and the Soviet Union to demonstrate the products of the capitalist and socialist systems. They also could tangibly illustrate the material benefits of communism and capitalism, proving in effect that they "worked." Although the private companies and organizations that participated may have been more concerned with expanding export markets for their goods, U.S. officials remained focused on using the trade fairs as vehicles to sell American culture and ideology. Many business leaders shared this concern. One sales manager sent to Italy in 1955 remarked, "Our objective was selling; selling on many levels. We were selling our government's sincerity and interest in promoting two-way trade; selling our president's over-all interest and sincerity of purpose in bringing a closer rapprochement between countries; selling the American way of life and the democratic philosophy of our government."[10]

The OCB identified the basic objective of the trade fair program. It should "demonstrate to the people of the world . . . that the United States is (1) dedicated to the basic principle of human dignity and freedom in all its aspects and, (2) is the greatest producer of peaceful goods for the service of mankind." A further operating objective sought to counteract Soviet efforts to use international trade fairs as instruments for increasing goodwill toward the communist bloc. A typical U.S. trade fair exhibit, the OCB summarized, was not a "pure trade fair project" but more in the nature of a "cultural exhibit." American officials, believing that the Soviets were "using fairs as a means of disseminating propaganda and impressing the audience with the wonders of life in the Soviet Union," assumed that they must do the same. Government planners explicitly tied the material prosperity embodied in the consumer products displayed at the trade fairs

"Fruits of Freedom" reads a large sign in Siamese, announcing the theme of the U.S. exhibit in Bangkok. To the left, a Cinerama theater shows a wide-screen movie on a curved screen. Cinerama films were often the most popular displays at the U.S. pavilions at international fairs. Source: National Archives.

to American values and ideals. They did not seek merely to demonstrate that Americans enjoyed a high standard of living, which was well known. Rather, they preached that the American system produced prosperity: the high quality of life in the United States grew directly from its capitalist economy, democratic political system, and cultural traditions.[11]

The political messages of the American trade fair exhibits were conveyed through sloganistic themes. The most common theme was "Industry in the Service of Man." Reversing Marxist claims that capitalism turned workers into cogs of the industrial apparatus, the American exhibits proclaimed that industry served the needs and aspirations of humanity. American-style capitalism, the exhibits illustrated, provided a wide variety of goods, a high standard of living, and pleasant working conditions. If communist ideology promoted the view that workers in capitalist countries were selling their labor to produce profits for exploitive industrialists, American exhibits sought to demonstrate that the fruits of free enterprise were enjoyed by all. This message was also conveyed by another theme,

People's Capitalism, which stressed the great benefits brought to all Americans by capitalist production (see Chapter 8). Other main themes of the exhibits were variously labeled "Fruits of Freedom" and "Peace, Freedom, and Progress." They explicitly connected capitalism and democracy by showing how the American productive capacity developed from its political and legal traditions. One crudely crafted exhibit, for example, plastered displays of consumer goods with quotes from the Bill of Rights to show how constitutional protections provided for American material and social progress. Such themes guided the overt propaganda emphases of the trade fair exhibits, but the most basic objective of the program was simple: to provide a positive view of the American way of life and thereby refute "as directly as possible the Soviet and anti-American propaganda line."[12]

If the exhibits were parts of a propaganda war, one goal was simply to bring more people to the American exhibits than to those of communist adversaries. In this, American marketing tactics and consumer products proved a significant advantage. According to USIA progress reports, most of the U.S. exhibits "stole the show" because of their animated exhibits with live actors, moving pictures, audio recordings, and working models. Such creative displays succeeded in drawing crowds away from "the uninspired displays which gave a museum or warehouse look to the [communist] bloc country exhibits." In Milan, for example, the U.S. exhibit presented the theme "Main Street USA." It provided a graphic look inside a model American home, school, farm, and factory, all adorned with consumer products. Similar exhibits elsewhere included actors impersonating a family in a "typical" American home and demonstrating "what the American standard of living affords the average citizen in the way of comforts, leisure, and cultural attainments." At the winter 1955 Ethiopian Silver Jubilee Fair, the U.S. exhibit included a working kitchen sponsored by General Electric in which actors made "typical American cakes" from prepackaged cake mixes. The fair also featured a large coffee-processing exhibit, From Bean to Cup. Elsewhere the exhibit displayed such products as Singer sewing machines, Dumont televisions, and a fully loaded Ford Thunderbird. In Paris, kindergarten children occupied one popular exhibit. They played with American toys, "oblivious to the thousands of folks watching them everyday." The children and the actors humanized the displays of consumer products, associating them with real people and friendly faces. The products themselves were perhaps more important in luring attendees into the American pavilions, as testified by the large quantities of Sears and Roebuck catalogs stolen by the visitors.[13]

Anticommunist themes also found their way into the American pavilions. In Bangkok, USIA officials created an elaborate exhibit highlighting communist infiltration of Asia. One display featured a Korean flag covered with hundreds of signatures scrawled in blood representing Korean soldiers pledging to defend their country from communism. Nearby lay a whip made of barbed wire that purportedly had been used by communists to abuse POWs in Korea. On the ceiling of the exhibit, communist propaganda leaflets from several neighboring countries were affixed to a map of Asia. A giant spiderweb entangled the countries on the map, and a large papier-mâché spider hung menacingly from the ceiling. An international communist conspiracy was ensnaring the region with its propaganda, the exhibit suggested. "It was corny," a USIA official involved in the project later admitted, "but at least the place was crowded with people all the time."[14]

The USIA also used the trade fairs as a venue for promoting Atoms for Peace. Together with television, displays on the peaceful uses of atomic energy were the most popular aspects of the American trade fair program. According to one report, the Atoms for Peace and television exhibits in New Delhi "drew millions of people jamming through the doors and turnstiles." The Soviet government, embarrassed by the success of these exhibits, hastily scrambled to procure its own atomic energy and TV displays. Two closed-circuit televisions and atomic energy displays were rushed to the fair. Elsewhere, the Soviets canceled their participation or withdrew early because U.S. exhibits were disproportionately popular. One Soviet atomic energy exhibit featured unsightly photographs of cancerous growths being treated by radiation therapy. It was not well received. American pavilions had considerably better luck with television, a big draw. Many people in the developing world who had never seen television before were captivated. Also drawing huge crowds were the Cinerama theaters, which used three cameras to project wide-screen movies. From a propaganda perspective, Cinerama's importance was demonstrated when the Soviet government withdrew from the Bangkok Fair to avoid competing with this spectacular product of the American entertainment industry. Cinerama was so wildly popular that the Eisenhower administration even pursued a multimillion-dollar plan to outfit a "mothballed" aircraft carrier with a huge Cinerama theater and send it on world tours. It is a measure of the fiscally conservative president's faith in such activities that Eisenhower personally endorsed the proposal. He wanted to call the ship the "S.S. *Heritage.*" Congress was less enthused by such lavish propaganda stunts. It killed the plan.[15]

Selling American Culture

Although the trade fair program whetted international appetites for American consumer products, the cultural presentations program sought to convey the sophistication, maturity, and refinement of American high culture. It primarily targeted an elite audience of opinion makers through performances of talented American artists. Such activities were necessary to refute the view that, as the *Christian Science Monitor* explained, "a mass-produced economy mass produces mass-produced people and that a high standard of living somehow leads to a low standard of the arts of living," or, as a newspaper in the Philippines put it, "Americans live in a cultural wasteland, peopled with gadgets and frankfurters and atom bombs."[16]

The fact that such views about the barrenness of American culture were trumpeted by communist propaganda made them all the more sinister to U.S. officials. Cultural presentations, therefore, were "to be so designed as to refute communist propaganda by demonstrating clearly the United States' dedication to peace, human well-being and spiritual values." As with the trade fair program, the cultural program blended government sponsorship with private initiative. Administration of the cultural presentations at the operational level rested with the State Department, which in turn relied on a private organization, the American National Theater Academy (ANTA), to set up arrangements for artists and musicians to go abroad under the President's Special Emergency Fund. The government subsidized travel expenses and guaranteed that there was no net loss on the ventures, but all the tours included at least some private sponsorship. Government funds played their biggest part in financing tours to places high on the propaganda priority list.[17]

Under the cultural program, the United States sent symphonies, operas, plays, dancers, musicians, and athletes abroad. The OCB believed that "to send abroad artists of inferior talent would negate the favorable effects of the program." The board's guidelines for selection explained that preference should be given to artists "well known for the superior quality of their performance, because of the value to the United States of their prestige." Weight lifters, synchronized swimmers, tennis players, track and field stars, and rowing teams competed before foreign audiences, while pianists, trumpet players, harpsichordists, and singers performed to capacity houses.[18]

Metropolitan symphonies from such major cities as New York, Philadelphia, and Los Angeles traveled to the four corners of the world under the program. One orchestra, NBC's Symphony of the Air, was said to be the first symphony orchestra of the western world to tour the Far East. Students stood in lines for hours to get tickets; extra shows had to be added to meet the demand. The U.S. embassy in Japan effusively reported that it was "difficult to summarize rationally the overall impact of the orchestra." The performance was so successful that the State Department approved another tour for the group. This one became a source of embarrassment, however, when members of the symphony accused their colleagues of being communists. The quarrels attracted public attention, and the tour was canceled. More successful was the Martha Graham Modern Dance Company, which received over $300,000 from the Emergency Fund for performances across Asia. The OCB reported with satisfaction that the *Manila Chronicle*, a paper often critical of the United States, wrote in connection with her visit, "The genius . . . who conceived the idea of sending out to all parts of the world the best American orchestras, the best American theater groups, and the best American dance companies to prove that the United States is not a nest of materialism . . . deserves more than routine commendation. . . . Europeans and Asians, who are inclined to regard American pretensions to culture with something akin to skepticism, now have healthy respect for the excellence of American art and artists."[19]

Although the OCB emphasized the importance of "distinctive American creations" in the arts, U.S. propagandists resisted exploiting American popular culture in the program. The USIA featured the likes of rock 'n' roll only gradually and cautiously in its programming. The agency was more preoccupied with working against those aspects of American popular culture that presented a "distorted" view of American life. The USIA, for example, maintained a liaison with the U.S. motion picture industry "to reduce the negative impact abroad of U.S. commercial films and to improve their positive impact." This "delicate and highly confidential" relationship with the Hollywood producers reportedly enabled the USIA "to exercise influence on almost all elements of the theatrical motion picture industry." (The USIA conceded, however, that its influence was greater in regard to film sequences "having foreign policy and foreign relations implications than in regard to the aspects of American life depicted.") The cultural presentations of the Emergency Fund were likewise designed to present a "sophisticated" view of American culture by demonstrating to international audiences that American tastes were more refined than Hollywood and Elvis Presley suggested.[20]

The problem, the *New York Times* pointed out in a front-page story, was that there was nothing uniquely American in the traveling performances by orchestras and ballet troupes. Jazz music, on the other hand, was America's "Secret Sonic Weapon," a truly American product with great propaganda value, "for to be interested in jazz is to be interested in America." Consequently, jazz music entered the cultural program in 1956 to widespread acclaim abroad. Long lines and scalped tickets were the order of the day as such legends as Louis Armstrong and the Dizzie Gillespie band toured the world. They played to packed houses. In Karachi, a rival Soviet performance appeared simultaneously with an American jazz concert but, it was reported, "probably not more than a thousand persons saw the Soviet singers," while Dizzie Gillespie played in sold-out movie theaters. The OCB commented, "it is plain that this type of presentation is one of the best possible for the effect desired; not only is this type of music popular throughout the world, but it also attracts the attention of serious music students and lovers of fine music who appreciate the fact that this is a distinctly American form of expression which has influenced profoundly the entire art of music."[21]

While touting American excellence in the realm of cultural achievements, the OCB and USIA also sought to use the cultural program to counter the damaging effects of racial segregation on international opinion. Communist propaganda, the *New York Times* noted, "sedulously fosters the notion that . . . American Negroes live under conditions little if any different from those described in *Uncle Tom's Cabin*." Many of the groups and artists sent abroad under the cultural program were consequently African Americans. These included pianist and conductor Scuylor and Lee, mezzo-soprano singer Betty Allen, and the Jubilee Singers, a choral group that sang African American spirituals. The USIA reported to the NSC that "the cultural attainments of these Negroes were living proof to foreign audiences of the great progress achieved by the race under the American democratic system." One of the most outstanding successes of the cultural program was the folk opera *Porgy and Bess* by George Gershwin and Dubose Heyward. With a cast of seventy African Americans, *Porgy and Bess* provided a candid but hopeful view of race relations in the United States. Performing in Europe, the Middle East, and Latin America, *Porgy and Bess* received widespread praise. In Cairo, Egyptian President Gamul Abdul Nasser summoned the cast to his office to receive his personal expression of appreciation. In Tel Aviv, tickets were scalped on the black market, and a crowd of 200 crashed through the window of the theater and occupied vacant spots and aisles. In Zagreb, Yugoslavia, the company had fourteen curtain calls and, on closing night, a half-hour ovation. Commenting on

Jazz legend Duke Ellington poses with a U.S. Information Service publicity poster in New Delhi, India. According to the *New York Times*, jazz was America's "secret sonic weapon" in the war of ideas. Source: National Archives.

the success of show, the *New York Times* wrote that audiences "responded to *Porgy and Bess* . . . with the observation that only a psychologically mature people could have placed this on stage."[22]

Members of the cast further served American public relations through numerous off-stage meetings with foreign nationals, reported as "particularly effective in dispelling erroneous beliefs regarding the cultural and educational advancement of the American negro." Other traveling groups likewise performed off-stage public relations work for the United States. The director of the Jubilee Singers, identified only as Mrs. Myers, was quoted in the Bombay *Sunday Standard* as saying, "my people have been misrepresented throughout the world because we have been made to look like ill-treated and unhappy people. Actually we are happier in America than anywhere else in the world." Such statements were widely publicized by USIS posts around the world as evidence that communist propaganda greatly exaggerated American racism.[23]

The USIA aimed to enlarge the impact of the cultural programs through press stories, newsreels, radio, and TV. Because few people saw or heard the trade fair and cultural presentations in person, the USIA publicized the events widely in order to derive the maximum benefit from them. "One of the most important aspects of the activities of the President's Emergency Fund Program is the task of pointing up the significance of these events in the minds of the public overseas," the agency reported. The events served as "the nucleus of a public information campaign to stress the cultural or industrial achievements of the United States with the objective of increasing respect and prestige." USIA media exploitation enabled the cultural presentations to reach a much larger audience. In Japan, for example, nine cultural attractions appeared from May 1955 through January 1957. In TV appearances alone, an estimated 32,900,000 persons were exposed to these attractions. A film of a performance by an American symphony orchestra opened simultaneously in 46 cities in Japan, attracting 160,000 people on the first day. Within a month, over 3,000,000 people had seen the film. Six months after the orchestra's live appearance, the film ranked among the ten most popular in the USIS lending libraries. USIS records indicate that the film had been shown to over 7 million people by early 1957.[24]

The cultural and trade fair programs provided the agency with favorable propaganda to report as "news" stories abroad. Information officers used the events as "instruments to portray many aspects of American civilization; our democratic values in education, our freedom in creative expression, American opportunities for self-development without regard to social, racial or religious origin, etc." USIA produced pamphlets, leaflets, and posters promoting American cultural achievements. It provided tickets for cultural presentations to "opinion-forming" groups and paid travel expenses to enable artists to appear before such critical target audiences as university students and workers. This exploitation served "both to increase the potential audience reached by the project and also to interpret its significance as widely as possible." USIS posts furthered the cultural propaganda by holding special "Cultural Weeks" that included live concerts of American music, radio shows, and publications, all of which stressed the cultural values of the American people. The USIA also produced documentary films of American cultural activities, such as *Symphony Across the Land*, which told the story of America's widespread interest in symphonic music and the voluntary support given to orchestras in nonmetropolitan communities. Other films included *Helen Keller in Her Story* and *Portrait of a City*, which showed how the residents of Buffalo, New York, were interested in painting.[25]

From the government's perspective, the Emergency Fund was not pernicious propaganda or cynical manipulation of popular opinion. But neither was it altruistic support for the arts. The OCB carefully orchestrated the activities of the Emergency Fund to support U.S. foreign policy aims in the areas of the world that it thought needed the support the most. The artists sent abroad under the Emergency Fund probably did not see themselves as agents in a propaganda war. They were entertainers thrilled to prove to the world their worth as icons of American culture. Historian Naima Prevots writes, "The artists who participated in the program did so not only as performers but also as individuals, whose interactions with citizens of the host country broke down barriers and cultures and ideologies." Their skills and good intentions were crucial to the cultural program's effectiveness. It was not the only time good-natured, well-meaning Americans would be called upon to engage in "grassroots diplomacy."[26]

Mobilizing Private Industry

During World War I and World War II, U.S. propaganda agencies worked to enlist the services of ordinary Americans and private groups in the wars of persuasion being waged at home and abroad. This was part of total war: mobilizing the entire citizenry to support the war effort by giving everyone a role to play. During World War I, the Committee on Public Information, aided by Edward Bernays, persuaded American corporations with overseas operations to display posters in store windows and distribute pamphlets to customers. This practice of soliciting private cooperation with U.S. propaganda campaigns continued and expanded during World War II. Private industry closely collaborated with the Office of War Information (OWI) to promote the war at home and to conduct public relations work abroad. The War Advertising Council, a consortium of private advertising firms later renamed the Advertising Council, also cooperated with OWI by promoting war-mobilization projects in the United States. It encouraged Americans to conserve fuel, to collect scrap, and to buy war bonds. Hollywood contributed its image-making power to the propaganda war, working closely with OWI to produce pictures that sold the war to Americans and sold America to the world.[27]

These state-private efforts were revived early in the Cold War. The Smith-Mundt Act, in addition to authorizing expanded government information programs, recommended that U.S. propaganda agencies cooperate with private organizations as much as possible. Accordingly, the State Department's International Information Administration (IIA)—the precursor to the USIA—created an administrative unit explicitly charged with mobilizing private industry to conduct foreign information activities. Its Office of Private Cooperation (IOC), also known as the Private Enterprise Cooperation Unit, encouraged businesses and other NGOs to cultivate a positive image of the United States abroad. With staff in Washington, New York, Chicago, San Francisco, and New Orleans, the IOC facilitated such cooperative programs as letter-writing campaigns, community and college affiliations, book and magazine donations, distribution of printed leaflets in commercial mail sent overseas, institutional advertising, and tourist indoctrination programs.[28]

The IOC did not carry out operations itself. Rather, it developed ideas for projects and then found NGOs to implement them. One of the earliest and most notable projects got Americans involved in the propaganda war by writing letters abroad. The Common Council for American Unity, an organization that worked with foreign-born American citizens, collaborated with the IOC staff in developing this "Letters from America" campaign. Under IOC guidance, the Common Council asked first- and second-generation Americans to write to their friends and relatives abroad to "tell them the truth about the United States." The IOC prepared editorials suggesting themes people might include in their letters going overseas stressing that they refute "misconceptions" about American life. The editorials were regularly published in 313 foreign-language periodicals and broadcast on 195 foreign-language radio programs in the United States. The CIA joined the act, funding women's groups that engaged in similar letter-writing activities and developing a letter-writing campaign as part of its covert effort to influence Italian elections in 1948.[29]

IOC operations expanded as Cold War tensions escalated. By July 1952, more than 500 private organizations were regularly working with IIA. Their contributions reportedly exceeded several million dollars in value. IOC officials saw the strategy of private cooperation as having several advantages. Private cooperation was a force multiplier: NGOs supplemented official information programs, thus expanding the reach of U.S. messages. Private cooperation also provided a way of camouflaging the government's role in international propaganda. Target audiences "welcome close contact with non-official Americans . . . but will shy away from close contact with the U.S. Government or officials," the IOC reasoned. "Furthermore, it is a well recognized principle of communication that people tend to be more approachable and more subject to influence by their foreign professional counterparts, or people with whom they have some other common interest . . . than by foreigners who have no common ground with them." Moreover, because many governments tended to look askance at foreign propaganda, private efforts offered politically expedient ways to promote U.S. ideas.[30]

The private cooperation strategy also provided a way to sell the information program to an American public typically skeptical of government-sponsored propaganda. This dual domestic-international objective of the private cooperation strategy is well illustrated by IOC's campaign to promote the Voice of America. The Advertising Council, working with the IOC and the Common Council for American Unity, launched a nationwide promotional campaign in November 1951 urging Americans to write "Listen to the Voice of America" on international correspondence. The Ad Council printed 900,000 leaflets in 17 languages, distributing them to 190 organizations, 200 manufacturers and export agencies, 30 town affiliations, 25 college fraternities, and foreign students in 5 colleges. The council also sent spot announcements to 3,300 radio stations and advertising agencies, 420 foreign-language radio stations, and 115 television stations. Newspaper advertisements, editorials, and press releases supporting the campaign went to 2,000 daily, 6,000 weekly, and 675 foreign-language newspapers. According to the IOC, "the publicity . . . generated for the Voice of

America through this program was undoubtedly the greatest ever assured for or by the VOA." It not only publicized the VOA abroad; it also marketed the VOA at home.[31]

The People-to-People Program

Efforts to secure private participation in the propaganda war were significantly stepped up during Eisenhower's presidency. The Jackson Committee had called for expanded use of private groups, front organizations, and ordinary Americans as vehicles for transmitting propaganda messages. The U.S. Information Agency also made "private cooperation" a key element of its propaganda strategy. Under the new USIA, the Office of Private Cooperation was more generously funded than it had been under State Department jurisdiction. It expanded its operations and it tripled its staff to three dozen members.

Theodore Streibert, the agency's first director, pushed the IOC in new directions. He wanted the IOC to shift its focus from generating ad hoc operations to creating permanent organizations that could devise and implement long-lasting programs on their own. He hoped a nongovernmental body could be created that would take on the responsibility of planning public affairs programs and securing the cooperation of American businesses in carrying them out. Such an approach potentially offered more bang for the buck. Private organizations would initiate, plan, and implement various public relations programs, freeing the IOC staff from operational planning and allowing them to focus their energy on creating more "mechanisms" for private cooperation. The IOC also could devote more resources to supervising the content—or propaganda themes—promoted under private auspices.[32]

The IOC first created the Business Council for International Understanding (BCIU). Its charge: enlist the aid of businesses in "counteracting and dissipating any unfriendly attitudes that may exist in other countries." The BCIU, an independent organization run by corporate leaders and modeled after the Ad Council, planned and carried out public relations projects abroad using USIA themes and ideas. The BCIU was sold as something that would benefit both business and the government. In addition to shaping international attitudes for the benefit of U.S. foreign policy, it would also improve the climate for American business. It was a "national business mechanism which will mobilize American commerce and industry in a nation-wide campaign to meet the communist threat in the market places of the world."[33]

A few weeks after the Geneva Summit, Streibert proposed to the president a more ambitious "program for world understanding" to marshal the participation of ordinary Americans in cultural and peace initiatives around the world. He called for a massive expansion of the entire U.S. information program to meet the communist ideological and cultural offensives, including a tripling of the information program's budget. Although Streibert's proposal did not lead to the fat USIA budget he desired, from it grew the administration's most ambitious program for stimulating private cooperation in waging the Cold War: the People-to-People program.[34]

If the cultural program under the emergency fund sought to demonstrate the achievements of American high culture, the People-to-People concept rested on more modest precepts. The basic idea was that the government would encourage ordinary Americans to develop friendly contacts with like-minded foreigners to convince them of the basic goodness of the American people. In addition, the program served the explicit purpose of stimulating interest in foreign affairs at home. Streibert saw increased private participation in the propaganda war as a way to impress upon U.S. citizens the importance of the agency's task in the Cold War. It would "get the people to understand the general information program" and shore up Cold War morale by giving ordinary Americans a personal stake in the nation's foreign policy.[35]

Both the president and the secretary of state supported Streibert's proposal for stimulating private Cold War initiatives. Dulles emphasized that he liked the program for domestic information purposes: such a program would be a valuable device to create a sense of public participation in the government's Cold War policies. Eisenhower, like the secretary, saw the program as a way of boosting civilian morale by giving Americans something to contribute to the nation's foreign policy. He did not, however, want the program to appear propagandistic. He instructed Streibert to take a "positive approach" in presenting the People-to-People program, selling it not as an anticommunist initiative but as a friendly way of encouraging ordinary people to promote a positive view of the United States through informal contacts. Americans would be goodwill ambassadors, not ideological warriors. Throughout his presidency, Eisenhower actively supported the People-to-People program. "It doesn't grab big headlines like foreign aid, reciprocal trade and other major foreign policy programs," a widely printed *United Press* story observed, but "Eisenhower seldom passes up an opportunity to plug for the program when talking off-the-cuff." The president, as Abbott Washburn recalled later, did "more for the People-to-People Foundation than virtually any other cause." Eisenhower told Sherman Adams that People-to-People was "a great and useful effort in getting ahead in the cold war."[36]

The IOC staff took charge of developing the People-to-People program. To encourage "a broad stimulation of private activity," the IOC developed a novel form of organization for People-to-People. Independent citizens' committees would mobilize particular segments of society. The IOC created more than three dozen such committees. Some were organized around various interests, such as the Fine Arts Committee, Motion Pictures Committee, and Hobbies Committee. Others targeted professions, with committees for advertising, insurance, medical and health, and public relations. Additional committees represented the handicapped, travelers, veterans, and youth. Several of the People-to-People committees acted as liaisons with the hundreds of national organizations in the United States, including civic and women's organizations. As the basic units of the People-to-People program, the committees were designed to develop goodwill projects, stimulate group action, and communicate with foreigners who shared similar interests. Many of the committees were chaired by corporate leaders from such companies as Eastman Kodak, Dupont, Chase National Bank, and American Express. Notable committee chairmen included the cartoonist Al Capp, creator of *Li'l Abner* and the *Schmoo*, who chaired the Cartoonist Committee, Nobel Prize–winning author William Faulkner, who chaired the Writer's Committee, and General "Wild Bill" Donovan, former head of the Office of Strategic Services, who chaired the Fraternal Organizations Committee.[37]

Eisenhower wrote personal appeals to the committee chairs requesting their help in supplementing the "modest" information apparatus of the United States. Evoking wartime themes of total mobilization, Eisenhower instructed committee leaders that all Americans had parts to play in the ideological struggle. "If our American ideology is eventually to win out . . . it must have the active support of thousands of independent private groups and institutions and of millions of individual Americans acting through person-to-person communication in foreign lands." Engaging in personal diplomacy was "something which every U.S. citizen—man, woman, and child—can do to help make the truth of our peaceful goals and of our respect for the rights of others known to more people overseas." To defeat communism, Eisenhower explained, the United States must wage total Cold War: "In a very real sense, to be successful we must wage peace with all the vigor and resourcefulness and universal participation of wartime."[38]

A kickoff conference in Washington launched the program on September 11, 1956. A parade of high-level officials, including Eisenhower and John Foster Dulles, addressed the committee chairs to impress upon them the importance of their task. All evoked total war imagery. USIA chief Theodore Streibert told the audience that the Cold War was the same kind of total conflict that World War II had been. There needed to be a "total effort" to mobilize "the full strength of America, private as well as public, . . . for today's emergency." Eisenhower's speech before the volunteer committee chairs similarly stressed the "total" nature of the Cold War, which "colors everything we do." He expressed his interest in rallying citizen participation in the Cold War effort and pleaded for private help in convincing world audiences of American peaceful intentions. Eisenhower explained that "Some people are taught [that we] want war: that we are warlike, that we are materialistic, that we are, in fact, hoping for cataclysms . . . so that a few may profit . . . out of the misery of the world." He called for a "great American effort," involving doctors, labor unions, and ordinary travelers to dispel such notions. In his speech, Eisenhower took pains to stress that People-to-People was not propaganda. While the communists masterminded "great propaganda program[s] all laid out in the details," Americans created "understanding between peoples" by marshaling "the forces of initiative, independent action, and independent thinking."[39]

The credibility of such claims to independence depended on a low government profile. To minimize the visibility of the USIA's involvement in People-to-People, the name "Office of Private Cooperation" rather than "U.S. Information Agency" was used in correspondence to give the program a less obvious government affiliation. Despite this ploy, the People-to-People committees became the USIA's primary mechanism for mobilizing private resources. The IOC staff focused more on People-to-People than any other program. By 1960, as a result of the prestige and importance attached to private initiatives by the People-to-People committees, the IOC had expanded dramatically: its staff doubled in size to 40 members and its budget increased from $205,000 to $573,000.[40]

To maintain the private image of People-to-People, the Eisenhower administration insisted from the start that funding for the program come from private sources. The whole objective was to create a self-sustaining organization apparently free of government influence. Toward this end, the chairmen of the various committees decided to create a nonprofit corporation known as the People-to-People Program Inc. The corporation would serve as a central fund-raising engine, seeking large donations from foundations and other sources to redistribute among the individual committees.

The creation of the foundation in February 1957 further complicated an already convoluted administrative arrangement of state and private organizations. On the private side of the equation, there were the autonomous committees charged with stimulating grassroots action. They looked to the People-to-People Program Inc. for financial support. On the government side, there were the administration officials who initially devised the program. The tenuous state-private relationship was further complicated by the creation of an OCB committee to oversee the project in the summer of 1957. The OCB created an interdepartmental working group consisting of an unusually large number of agencies to stimulate action in the private sector. This only accentuated the fiction of the completely private nature of the project. While enhancing governmental resources to oversee the People-to-People program, the OCB maintained a public posture of noninterference: "The concept of the Foundation as a private agency, free from Government control and direction, should be maintained at all times," the board decided. Disingenuously, less than two months after the creation of the OCB working group, the People-to-People Foundation issued a press release stating that the program was "now operating free of government influence."[41]

From a financial perspective, government involvement turned out to be a liability for the program. To the consternation of President Eisenhower and to the surprise of many, the Ford Foundation refused a request for a $5 million grant to fund the People-to-People Foundation. More than twenty other foundations followed suit, also declining to contribute to the program. All expressed concerns about the overt connections of People-to-People to the U.S. government. Many of the foundations quite reasonably found it difficult to understand how a program announced by the president, pushed by his staff, and financed by USIA qualified as a private activity requiring NGO assistance. Foundation directors also wondered why, if People-to-People was not a government operation, the committee chairs had to obtain security clearances. The fact that Eisenhower and other government officials lent their personal and well-publicized support to the program only convinced prospective donors that they were being asked to contribute to a government activity. As one foundation member noted, "We were doubtful about blanketing private activities into official government programs already rather explicitly related to the Cold War." Other foundation directors, such as Henry Ford II, believed that the program was ill-conceived. It was too diffuse, too amateurish. To fund People-to-People would be to throw money away to well-meaning but inexperienced do-gooders.[42]

In June 1958, after a year and a half of unsuccessful attempts to secure large-scale financial assistance, People-to-People Program Inc. was officially dissolved. The dismal fund-raising performance of the central foundation limited the size of the People-to-People program, but it did not measurably affect its over all impact. Because the committees acted independently of the foundation, most were able to secure financing for their activities from other sources. As early as 1957, many of them could point to some impressive accomplishments, amateurish or not.

People Are the Best Propaganda

The activities of the People-to-People committees resembled, in many ways, home-front mobilization efforts of World War II. They were Cold War versions of the activities promoted by the War Advertising Council and the OWI ten years earlier. Instead of collecting scrap and rubber for the production of war machines, however, Americans collected books and magazines for distribution abroad as ideological weapons. If during World War II Americans were exhorted by OWI to purchase war bonds, now they were instructed that $30 could send a portable library of ninety-nine American books to schools and libraries overseas. People-to-People committees helped organize over a hundred sister city affiliations. They marched in parades, hosted exchange students, and organized traveling "People-to-People delegations" representing their various communities. Americans exchanged letters, scrapbooks, and photographs with foreigners all over the world. Civic organizations such as Kiwanis, Rotary, and Lions organized book and magazine drives. Corporations used portions of their overseas advertising to build a "better understanding" of the United States.[43]

The private and voluntary nature of the program ensured that the public relations activities undertaken on behalf of People-to-People were creative and diverse. Many projects derived from corporate participation. Publishers, prompted by the magazine and publishing committees, donated thousands of copies of magazines and books for free distribution in foreign countries. *Woman's Day*, for example, volunteered 6,000 copies of the magazine per month for free distribution in foreign countries. Similarly, the Metropolitan Life Insurance Company was inspired by the Insurance Committee to organize a book drive. Employees collected 16,000 books and sent them, accompanied with personal greetings, to seven libraries in Sudan. To encourage more contributions from publishers, George P. Brett, president of Macmillan and the head of the People-to-People Book Committee, staged a contest. Speaking to a convention of textbook publishers, he challenged his colleagues in the industry to select from their personal libraries the best books about America. Urging them to donate copies of their picks to People-to-People book drives, he advised them to "keep away from books that might be considered straight propaganda." They should choose selections that describe "our country as it is—a *friendly place.*"[44]

Several projects linked Cold War concerns with humanitarian inclinations. Many committees provided food and other relief items to émigrés from Eastern Europe, highlighting American generosity and the privations of the captive nations in the process. The Nationalities Committee, which established connections with ethnic enclaves in the U.S., sent five tons of vitamins and ten tons of winter clothing to "Polish repatriates from Communist forced labor camps." Other projects revolved around creating personal associations between American and foreign communities. Small-town newspapers and governments organized "community salutes" to their foreign counterparts. In the Philadelphia suburb of Ambler, Pennsylvania, the local newspaper printed a special "freedom edition" to celebrate the 900th anniversary of Coburg, Germany. Copies of the special edition were printed in English and German and were distributed to readers in both cities. The Youth Committee mobilized young people to join the effort. Camp Fire Girls from over 3,000 communities launched a People-to-People project portraying America through photographs. Following the theme "This is our home. This is how we live. These are my People," the photographs were assembled in albums and sent to other girls in Africa, Asia, and the Middle East.

Clubs and civic organizations provided another means of inspiring People-to-People projects. One of the most active People-to-People groups was the Hobbies Committee. It connected Americans with such interests as photography, horticulture, and stamps to foreigners sharing these interests. Hobbyists corresponded through the mail with their foreign counterparts, exchanging stories, photographs, and items of mutual interest. Occasionally they formed People-to-People delegations that traveled on goodwill visits to meet the friends they had made through these letter exchanges.

One such People-to-People group implemented a "friendship through gardening" plan. Helen Hull, a gardening enthusiast and the author of *Wild Flowers for Your Garden*, organized the effort. At her suggestion, the New Jersey Garden Club donated fifty white-flowering dogwood to the city of Niigata, Japan, which had been ravaged by fire four years earlier. The club also sent fifty rosebushes to an amateur gardening group in Seoul for planting on the grounds of the South Korean capital. The group chose its rosebushes carefully: it included the 'President Eisenhower' and 'Peace' varietals. The group may have selected the roses, but American officials were behind the projects. Both were done at the suggestion of officials in U.S. embassies, who saw opportunities for good public relations. G. G. Wynne, a public affairs officer at the American embassy in Seoul, saw great symbolic value in the rosebush project. "Something without practical value, that serves beauty alone—this is certainly not the usual image of America, but it is one I feel that should also be projected, because it is part of the rich American fabric, and even more important, it is one on which a great value is placed in the Orient among intellectual leaders."

Helen Hull, like other People-to-People activists, believed she was furthering the cause of peace. She believed, as she told the *New York Times*, that American gardeners "can make an important contribution to peace since an appreciation of nature's beauties unites gardeners everywhere." Other gardeners appeared to agree. In March 1958, the National Council of State Garden Clubs started a letter-writing campaign. It sent 5,000 letters abroad inviting inquiries about American gardening. The letters, which included packets of flower seed donated by the Burpee company, were mailed using the newly issued Hyde Bailey Memorial postage stamp, which paid tribute to the venerable American botanist. Within a few months, the American gardeners received responses from 500 people in 83 countries. One respondent, a man from Czechoslovakia, expressed his interest in joining the "fight for peace on this front [of] exchanging ideas and experiences in gardening." Members of the garden clubs kept up the correspondence, often replying in the original languages of their new pen pals. The Burpee company—using the initiative as an opportunity for a little PR of its own—supplied translators to assist in reading and replying to the letters. When the project was completed, a bound volume of the letters and answers was presented to President Eisenhower.[45]

Press coverage of People-to-People consistently described Americans who participated in the program as "diplomats" and "ambassadors." They were spreading the truth about America, refuting enemy propaganda, and defending the United States against communist lies. They were making personal contributions to the cause of peace. "People are the best propaganda," proclaimed *Current Events*, a national student newspaper. "Here's Your Chance to be a Diplomat," added the *Baltimore News-Post*: "Every American who participates in [the People-to-People program] will be in a sense a diplomat—not a diplomat in top hat and frock coat but one in a sports shirt or a house dress who can, in his or her own way, explain to the peoples of other nations the truth about the United States and its aims." Those who participated in the program appeared to share the view that they were working for truth and international friendship. Juvenal Marchisio, an Italian American judge from Brooklyn, joined the Nationalities Committee to help prove to others the good intentions of the United States. Appalled by the fact that many Italians interpreted American aid as "imperialism," he wanted to show them that "America does not wish to interfere with their way of life." Speaking to a New Jersey newspaper, he said that recent immigrants and second-generation Americans had an obligation to refute misconceptions about the United States. They should serve as "ambassadors" to people from their homelands, providing them with "a peek into the American way of life and Democratic capitalism."[46]

As such comments suggested, the Americans who joined the People-to-People effort often interpreted their activities as small but meaningful contributions to international understanding. This was the case even when the activities seemed to have very little to do with world peace. One of the clubs, for example, brought together pet owners in the United States with animal lovers abroad. The Americans who participated in the program wrote letters to other pet owners overseas, mailed photographs of their pets, and occasionally met up with their animal-loving friends on their travels. One club member reported his group's contribution to a local newspaper. "Dogs make good ambassadors," he explained. They "are capable of hurdling the barriers of language and ideologies in the quest for peace."[47]

Probably the most dramatic initiative undertaken by the People-to-People program was developed by Dr. Robert Walsh, a heart specialist from Washington, D.C. He devised a plan whereby an old U.S. Navy hospital ship would be rescued from the scrap yard and retooled to provide health care and medical training to people in the third world. The *Consolation*, which had treated wounded soldiers during World War II and the Korean War, was about to be mothballed, but Dr. Walsh convinced the government to pay $2 million to refit the ship for People-to-People. Now

President Eisenhower meets with Dr. Robert Walsh to congratulate him for his People-to-People project: the floating hospital ship *Hope*. Source: Project Hope, Millwood, Virginia.

rechristened the *Hope*, private donations funded the ship's mission. American President Lines, a cruise company, subsidized operating costs for the vessel. Nearly five thousand other companies provided additional funds to support *Hope*'s expenses. Fund-raising efforts received a boost from a large Advertising Council campaign as well as the editorial support of *Life* magazine (helped by C. D. Jackson's position at *Time-Life International*).

The *Hope* embarked on its first voyage to Southeast Asia in the fall of 1960. It stopped first in Indonesia. Aside from briefly running aground east of Java, the voyage was a success. Volunteer doctors and nurses performed operations on Indonesian locals and instructed Indonesian medical personnel in modern medicine. In a two-week stay in Djakarta, 2,400 people visited the ship, 66 patients were treated, and 600 Indonesian doctors and nurses participated in training lectures and seminars. After its stay in Djakarta, the *Hope* traveled to Saigon before returning to San Francisco in September 1961.

As some administration officials pointed out, an expensive floating vessel was not the most "economical and efficient" way to provide medical care or administer technical assistance to the people of Southeast Asia. But it did have great visibility. C. D. Jackson noted that the cost was "more than justified" by the goodwill accrued "from a privately-endowed U.S. hospital ship riding at anchor in the steaming, disease-ridden harbors of Southeast Asia." In addition, the USIA reported, Project Hope "had an important effect on the domestic U.S. scene." It brought "into active participation in foreign affairs thousands of important people in American life." It was a "great aid in the task of making it better understood throughout America that our national security depends on aid to the struggling peoples of the less-developed countries. It should aid in obtaining support for the overall Mutual Security Program."

The people who made the *Hope* journey possible similarly emphasized the positive impact of the ship on foreign perceptions of the United States. Abbott Washburn recalled that when he first told the president about the idea, Eisenhower responded enthusiastically "not because it was the humane thing to do, but because it showed the U.S. in a good light." Eisenhower announced in a press conference that he knew "of no better way in which you could bring to many thousands of people, many millions, the concern of the United States in humanitarian things." The journey bolstered America's "good image," Dr. Walsh told the *New York Times*. Dr. James Yates, a member of the medical team from Stockton, California, agreed: *Hope*'s journey "more than anything paid off in goodwill." U.S. Customs officials were less kind. When the *Hope* returned home to San Francisco, the ship was greeted by an oversized contingent of customs officials who assessed the medical team for $200,000 worth of merchandise acquired on the journey.[48]

The individuals who participated in these activities may not have realized it, but the People-to-People campaign served the purposes of domestic as much as foreign propaganda. As one State Department official noted, People-to-People was designed to "awaken interest in foreign affairs in the ranks of the 700 national organizations represented on its committees." Indeed, many of the activities of the individual committees were as concerned with stimulating action by Americans as influencing opinions overseas.[49]

Pamphlets distributed by the People-to-People committees urged Americans to become personally involved in the waging of the Cold War by acting as public relations representatives for their country. One such

The *Hope*, a refitted naval hospital ship, sailed to Saigon on its second voyage in 1961. Source: Project Hope, Millwood, Virginia.

pamphlet, "What You Can Do in People to People," urged Americans to serve as "citizen ambassadors" by acting responsibly and courteously when traveling abroad and by welcoming overseas visitors to the United States. The pamphlet suggested writing letters to pen pals in other countries, sharing hobbies with foreigners, sending books and magazines overseas, and stimulating group action at home. One pamphlet, "Your Community in World Affairs," encouraged Americans to "help combat the distortions of Communistic propaganda" by bringing about "a better understanding of what your country is doing to bring peace and understanding in the world today." Another added, "Nothing that you may do, so far as improving the relationship between the people of the United States and the people of foreign countries, need be considered too small."[50]

Many of the campaigns undertaken by the People-to-People committees sought to teach Americans coming in contact with foreigners to "behave themselves." The Cartoonist Committee, for example, produced a 100-page booklet called "You Don't See These Sights on the Regular Tours." In it, several prominent cartoonists illustrated how Americans should *not* act on their travels. Another such campaign, keyed to American tourists and businessmen traveling overseas, emphasized the theme: "Make a Friend This Trip—For Yourself, For Your Business, For Your Country."

Posters emphasizing the theme were placed at major points of embarkation. A million passport-sized leaflets were distributed to travelers before they departed the country. Some were included in envelopes containing traveler's checks; others were handed out at travel agencies. Leaflets encouraged travelers to act modestly and responsibly, and they cautioned travelers that every person they talk to "will think more of my country, or less, because of my words and actions and attitude." The armed services stressed similar themes in their indoctrination programs for service personnel and their dependents. Service personnel were repeatedly advised of the "importance of harmonious relations" with foreign nationals. They were encouraged to give money to orphanages, study foreign languages, participate in sporting events with the "locals," and entertain foreigners in their homes. The Defense Department produced a motion picture explaining to service wives their "responsibilities as person-to-person ambassadors." Posters illustrated the People-to-People concepts through such themes as "You Are a Symbol of Liberty," "What Do They Think of You?," and "I Learned to Speak Their Language."[51]

Americans Abroad as Spokesmen for the USA

Such efforts to instill Americans with good manners were responses to the agonizing public relations problems created by Americans who were already living abroad, particularly those stationed at U.S. military bases overseas. The fact that many Americans were living in "relatively privileged positions" often sparked resentment among local populations. In addition, automobile accidents involving civilian personnel and violent crimes perpetrated by U.S. troops attracted widespread media attention around the world. One of the most notorious incidents was the Girard case, involving Army specialist William S. Girard, who shot in cold blood a woman collecting brass scrap on an American firing range in Japan. The incident provoked such a widespread outcry that ambassador Douglas MacArthur II warned from Tokyo that unless Girard was handed over to local authorities, "vital interests . . . throughout free Asia" would be jeopardized. Tensions were exacerbated by U.S. status of forces agreements granting to the United States the right to try military personnel involved in civil crimes abroad, provisions that were often compared to the old extraterritorial rights of the western colonial powers. The problems were particularly acute in Asia and the Near East where, as one official noted in 1957, "rarely has there been so much feeling against the U.S. military presence."[52]

MAKE A FRIEND
THIS TRIP

FOR YOURSELF
FOR YOUR BUSINESS
FOR YOUR COUNTRY

Leaflets such as this were distributed with new passports. They encouraged Americans to spread goodwill about the United States on their travels. Source: Eisenhower Library.

The publication of William Lederer's and Eugene Burdick's novel *The Ugly American* in 1957 further dramatized the difficulties engendered by the U.S. presence overseas. The story, set in a mythical Far Eastern country, criticized Americans for living in "golden ghettos" abroad, isolated from the locals they were supposed to help and ignorant of foreign cultures. The British Foreign Office noted *The Ugly American* became "part of the background of thought here about the image of America overseas." Indeed, this best-selling novel, Robert Schulzinger observed, "had an impact similar to that of Harriet Beecher Stowe's *Uncle Tom's Cabin* in the years before the American Civil War."[53]

The negative impact of U.S. personnel on foreign opinion alarmed the Eisenhower administration. In 1956 the NSC ordered U.S. missions to reduce the number of Americans in each country to a minimum. A year later, after the publication of *The Ugly American*, Eisenhower commented that the problem "was bigger than he had supposed." He directed the OCB to study ways to reduce the friction between Americans overseas and local populations. Beginning in July 1957, all Americans who ordered new passports received a letter from President Eisenhower asking them to act as goodwill envoys. In the pamphlet, Eisenhower's message instructed travelers that they had a duty to act responsibly. "As you travel abroad, the respect you show for foreign laws and customs, your courteous regard for other ways of life, and your speech and manner help to mold the reputation of our country." The president implored travelers to work to convince other peoples that "the United States is a friendly nation and one dedicated to the search for world peace and to the promotion of the well being and security of the community of nations." Eisenhower issued a similar message to all members of the armed forces serving outside the United States, adding that an essential part of their mission was "building good will for our country."[54]

USIA officials also concerned themselves with the behavior of Americans abroad. George Allen, director of the agency from 1957 to 1960, declared in a speech that the USIA "should make strenuous efforts" to "elevate the general cultural level of the American people. If American tourists must chew gum, they should be told at least to chew it as inconspicuously as possible." Accordingly, the agency's Office of Private Cooperation orchestrated a public relations program to teach Americans how to behave abroad. In cooperation with private industry, the IOC produced a television program, motion picture, and scores of pamphlets to educate Americans about their responsibilities as "unofficial Ambassadors." A particularly striking pamphlet was "Americans Abroad: Spokesmen for the USA." The cover of the pamphlet instructed readers, "YOU TOO SPEAK FOR AMERICA." Inside, a statement by President Eisenhower encouraged travelers to demonstrate "courteous regard [for] ways of life that differ from ours" and asked them to convince "peoples in other lands that the United States is dedicated to the cause of peace." A lecture on good behavior followed. American travelers were urged to learn a few words from the native language of the country they were visiting and to show respect for foreign customs and traditions. Americans should praise the things in the visiting country they admire, while keeping things they find "strange or disagreeable or uncomfortable" to themselves. Americans should not compare the countries they visited unfavorably with the United States, they should avoid criticizing foreigners, and they should "resist the tendency to talk down to people." Travelers were cautioned to "lay off preaching to the friends you make" because "people all over the world chide Americans about this tendency." The pamphlet lectured travelers on good manners: "Friendliness, politeness and a willingness to get along are coins that are good in any country in the world."[55]

The pamphlet then proceeded to tell travelers how to respond to questions they might be asked about the United States while on their travels. It advised readers that they had a patriotic duty to refute anti-American sentiments and communist propaganda. It explained, "One thing you want to keep in mind is this: you'll probably be asked some questions that are mighty tough to answer—yet mighty important you try to answer well. Don't be surprised if the questions asked occasionally reflect current anti-American propaganda. If *you* can give an accurate and convincing reply, *you* can blunt the edge of propaganda" [emphasis added]. The pamphlet functioned as "talking points," briefing travelers on the appropriate responses to questions they might receive from foreigners about American practices. These included queries like: "Do Americans really want peace?" and "Isn't the Soviet Union ahead of the United States in science and technology?" Other questions addressed specific foreign policy issues, such as, "Wasn't the U.S. guilty of aggression in sending its troops into Lebanon?" Some addressed American race relations: "If Americans believe in democracy, how do you explain racial discrimination in the United States?" Others focused on labor relations: "Isn't the American worker exploited by Big Business?"

The pamphlet then gave the American tourists answers that they were to recite in a presumably impromptu fashion if they happened to discuss these matters during their travels. They should deny that American big business exploits its workers, and that American serviceman "think they can flout the laws of countries where they are stationed." Should U.S. travelers encounter queries about American "colonialism," they should emphasize that Americans were "the pioneer rebels against colonialism," while stressing (presumably for the benefit of European audiences) that "we do not . . . 'stir up' native peoples in colonial areas to demand their independence as some critics say." As for segregation, travelers were to point out that "slowly, perhaps, but surely, through legislation, education and individual good will, racial discrimination is being eliminated from American life. . . . Every year the Negro in America is gaining in his march toward equality." Above all else, travelers should stress the peaceful intentions of the United

This pamphlet distributed by the People-to-People program instructs Americans that they are "ambassadors" and "spokesmen" for the United States. The forty-one-page leaflet advised tourists how they should respond to questions from foreigners about American life and foreign policy. Source: National Archives.

States. They should tell others that "peace is foremost in the minds of all Americans," but they should caution against peaceful coexistence with the communists: "Peaceful co-existence in a genuine sense—not as the Communist synonym for 'cold war'—is impossible as long as the Communists insist on trying to attain world domination by subversion, promoting armed revolution, and military aggression. If they demonstrate they have renounced such methods, and if progress is made in halting the atomic weapons race, peaceful coexistence can become real and lasting." Other questions prepared by the pamphleteers included: "Isn't U.S. talk about a 'position of strength' evidence of your aggressive intentions? . . . Why does the U.S. maintain a ring of bases around the Soviet bloc? . . . Why has the U.S. been so reluctant to stop its nuclear weapons tests?"

The prescribed answers to all of these questions predictably defended the actions of the United States and stressed the peaceful intentions of American foreign policy. On the question of whether Americans were materialistic, the pamphlet began by dictating the following answer: "If by being materialistic you mean that Americans want to live as well and comfortably as they can, we plead guilty. But a high standard of living is not proof of a materialistic outlook on life. Many of us are deeply concerned with the spiritual and cultural side of life." To the question, "Why does the U.S. Government spend so much money for military purposes?" the pamphlet answered: "No one would be more pleased than the American people to spend less money on the military. When the danger to our security recedes, we shall gladly shift such spending to more constructive purposes."

The answers were actually much longer than these illustrative quotations suggest; each explanation ran from one to two pages. How the authors of the pamphlet expected tourists and business travelers to absorb the long, detailed expositions of every foreign policy position of the United States is a mystery. One angry citizen complained about the pamphlet in a letter to the secretary of state in 1961. The pamphlet was "extremely superficial and fatuous [and gives] an over-idealized picture of the United States," he observed. "Is it not possible that such glib and inane answers to what indeed are difficult questions, may make exactly the opposite impression on the foreign interrogator than that intended." Perhaps, the writer admonished, "Americans traveling abroad should be armed with . . . more sophisticated arguments than those presented in this pamphlet." The State Department responded by pointing out that the pamphlet was not a publication of the Department of State, but the angry citizen correctly perceived the government's hand at work.[56]

A note on the back page of the pamphlet explained in tiny print that it was published by private citizens on behalf of the People-to-People Corporation. This was at best half true. The corporation had very close ties to the government, as we have seen, but technically it was private, and the pamphlet was paid for and distributed by a private organization. Declassified documents, however, reveal that the pamphlet was written by public affairs officers within the government. It was an OCB product. Every word was pored over by officials in the business of shaping public opinion. At forty-one pages, it systematically addressed all of the major propaganda themes of concern to the administration. The pamphlet was attempting to influence two different audiences, one of them overtly identified, the other camouflaged; one international, the other domestic. The obvious one is the international audiences who encountered American tourists on their travels. If the tourists did their homework, studied the pamphlet, and learned the answers, then they would be in a good position to defend U.S. policies and to convey an overall favorable impression of the United States. The camouflaged audience was the tourists themselves. The pamphlet sought to reinforce *their* confidence in American foreign policy, to remind *them* of the positive aspects of American life, and to build up *their* resistance to communist propaganda and so-called anti-American sentiments. The pamphlet served to remind Americans of the importance of the Cold War struggle, of the seriousness of the ideological challenge, and of the essentially defensive and morally superior character of American foreign policy. The officials who prepared the pamphlet and devised the People-to-People program hoped that by making Americans active participants in the Cold War abroad, they would be more likely to support it at home—going along with defense expenditures, foreign interventionism, economic aid packages, and other Cold War measures.

Although the Eisenhower administration was highly sensitive to the effects large numbers of Americans traveling and living abroad would have on foreign opinion, it was also hopeful that increased contacts between ordinary Americans and other peoples could serve U.S. foreign policy interests. The Jackson Committee's successor, the President's Committee on Information Activities Abroad, chaired by Mansfield Sprague, explained that Americans abroad were as important to U.S. propaganda as official information programs. It observed that the government was "not the sole, or perhaps even the most important, factor in the projection of the American image. . . . Americans traveling or living abroad contribute heavily to the process of foreign image-making." It expressed hope that person-to-person contacts, if properly channeled, could make a positive contribution to foreign policy objectives abroad. It advised, "the U.S. will be better understood in the world and American interests would be better served if the greatest possible volume of contacts is established and developed between individuals and groups in the U.S. and individuals and groups in other countries."[57]

Indeed, the dramatic growth in the State Department's exchange programs indicated the importance the administration attached to cultural and educational exchanges as a way of promoting U.S. interests. The sum total of all exchanges sponsored by the government approximated 20,000 persons in 1953. By 1959, that number had more than doubled. The State Department's exchange program in 1959 involved exchanges with 105 countries totaling 7,292 persons: 5,165 foreigners coming to the United States and 2,127 Americans going abroad. During the same year, the technical assistance program supported an additional 15,715 exchange visits in over 60 countries, and military exchanges contributed another 15,000 persons to the overall figure. Between 1950 and 1959, well over 300,000 foreigners were brought to the United States under official programs, and at least 50,000 Americans had gone abroad.[58]

The Eisenhower administration hoped that such exchanges, combined with its cultural propaganda efforts, would "humanize" Americans in the minds of foreign audiences. The USIA ranked the "long-range approach" of "telling America's story" as one of its most important missions. It reported to the NSC in 1958 that its "most important task lay in keeping before foreign audiences the solid and enduring features of American life: the nation's achievements in literature, in painting, and music, not to mention its spectacular advances in medicine, science, and technology." These efforts were viewed as long-term projects that would ultimately benefit the political, economic, and military climates in which U.S. interests could flourish.[59]

Private cooperation remained one of the USIA's most important tactics in this mission. At a practical level, private initiatives were cost-effective. They allowed the USIA to do more than the limited appropriations it received from Congress allowed by getting private groups to do the legwork and assume some of the costs. Contrasting the voluntary American approach with the communist approach of "forcing people into ideological warfare," private initiatives provided cover from opponents of government propaganda. In addition, the spontaneous and voluntary nature of private initiatives added an air of legitimacy to the information program as a whole. A government stamp on any form of propaganda, no matter how benign the message, immediately raised suspicions in the minds of target audiences. By concealing the hand of government, however, private initiatives enhanced the persuasive potential of U.S. propaganda themes because government involvement was hidden from target audiences. Obscuring the source of the message, it seemed, compensated for what the USIA lost in control and professionalism. Private cooperation also enabled the USIA to reach "some areas where suspicion of U.S. Government intentions exists." Private American organizations, acting on behalf of the agency or engaged in tasks complementary to USIA projects, would be welcomed in these areas because they were thought to have no political or propaganda intention.[60]

The strategy of private cooperation also supported those information media that openly acknowledged affiliation with the U.S. government, such as the USIA's press releases and the Voice of America. These overt media widely publicized activities in the private sector as part of its "news format," ensuring wide international impact. By providing the U.S. information services with factual material for their programming, private initiatives enhanced the credibility of government propaganda without sacrificing persuasive potential. Finally, as the Sprague committee noted, private initiatives enabled U.S. citizens "to acquire a greater sense of participation—and therefore a heightened interest—in world affairs by representing America to foreign individuals or groups through an organized effort." Private initiatives not only promoted morale and a sense of participation in the Cold War, they also nurtured an awareness among participants of the importance of the USIA's mission—something that would come in handy when the Eisenhower administration pursued increased appropriations for informational and cultural work from Congress. Private cooperation, in short, provided additional avenues for the administration to camouflage its propaganda—tactics that served its purposes.[61]

As instruments of U.S. policy, the People-to-People program and the Emergency Fund highlighted the extent to which propaganda concerns permeated diverse aspects of U.S. foreign relations. The ways Americans acted at home—their cultural interests and business practices, their consumer goods and household appliances, their social relationships and race relations—were considered vital ingredients in the Cold War struggle. In the "all-out public-relations battle" with the Soviets, so the reasoning went, the very picture foreigners had of American life could make a difference between victory and defeat in the struggle for the allegiance of the world's peoples. It was not enough for the U.S. government to stockpile weapons and atom bombs to prevent a communist victory; both state and private resources were rallied to combat the influence of communist ideology around the world.[62]

Chapter 8
Facts About the United States
The USIA Presents Everyday Life in America

A need which cries to heaven in this world-wide war of ideas is to make clear the triumphant
fact that a new way of living has come to pass in the Western Hemisphere.

— Theodore S. Repplier, February 29, 1956

The U.S. Information Agency widely distributed a hundred-page reference booklet titled simply *Facts About the United States*. It was, in many ways,
unremarkable. The pamphlet merely contained an assortment of facts, statistics, and basic information about life in America. It included statistics on
the birthrate and life expectancy of Americans. It detailed average rainfall
and temperatures of places in the United States. It explained the three
branches of the U.S. government and it listed the time zones that divided
the North American continent. Sections of the pamphlet explained the recreational and religious practices of ordinary Americans and described the
everyday lives of American women and children. Other sections addressed
labor, agriculture, and industry in the United States; still others provided
information about American living standards, health care, education, social
security, and finances. On the surface, the pamphlet appeared to be an unlikely product of a government propaganda agency. No loud pro-American
blandishments or crude anticommunist digs appeared in its pages. It read
more like an almanac or an encyclopedia entry than a work of propaganda.
To the ordinary observer, the pamphlet would have simply appeared to be a
compendium of information about the United States. Why, then, was this
innocuous collection of facts and figures one of the most widely distributed
pamphlets of the United States Information Agency?[1]

Facts About the United States was but one piece of a large volume of
USIA material that presented "typical citizens playing their role in daily
living" to the world. The agency's wide-ranging cultural effort included a
sustained and systematic effort to present everyday life in the United States
to the world. USIA operatives believed that wide distribution of factually
based material, such as *Facts About the United States*, would enhance the

agency's credibility and would in the long run advance American interests more than lies and vituperative anticommunist pronouncements. They also believed that basic information on life in the United States would have a positive impact on international public opinion and would help create a favorable climate for U.S. foreign policies. If foreigners could see ordinary Americans as individuals like themselves, working and struggling for a better life, so the reasoning went, they would perceive the American government likewise. A USIA planning paper explained, "Over the long pull, our effort to increase community knowledge and understanding of such things as our sense of community, our religious beliefs, our form of government, our system of education, our interest in others, [and] our contributions to the arts will create respect for us as a people capable of wise leadership in the Free World." President Eisenhower concurred with this reasoning. He instructed subordinates in the USIA to devote more attention to "telling the story of America," something he saw as one of the most important objectives of the information program. The mission statement Eisenhower approved for the USIA charged the agency with "delineating those aspects of the life and culture of the people of the United States which facilitate understanding of the policies and objectives of the Government of the United States."[2]

Although many of the materials disseminated by the USIA to defend the American way of life appear silly or superfluous today, daily life in the world's two superpowers was one of the hottest topics of Cold War propaganda. Soviet and American propagandists devoted enormous resources to demonstrating the superiority of their competing ways of life. The competition for hearts and minds, especially in the developing world, focused on proving to the international community that capitalism and communism provided the most equitable, effective, and just paths to individual and national progress. Key themes of Cold War propaganda included standards of living, availability of consumer goods, cultural products, scientific accomplishments, family relationships, educational systems, social values, and economic benefits. So too did issues pertaining to race, class, and gender figure prominently in the propaganda output of the superpower rivals. The USIA drafted elaborate instructions telling its personnel how to use race, class, and gender as themes in the service of U.S. foreign policy. The agency specifically targeted women and workers with materials designed to appeal to their values and interests. It sought to prove to all audiences that the United States was addressing the problem of civil rights for American minorities, that capitalism benefited all Americans, that workers shared in the fruits of their labor, and that women lived full, happy lives as mothers and homemakers. The story of America presented by the USIA was characterized by social mobility, spiritual vitality, rugged individualism, equality of opportunity, the rule of law, and widespread belief in the virtues of capitalism and democracy. It was a story of progress.[3]

This portrayal of American life to the world is revealing not just for what it says about American propaganda, but also for what it says about American self-perceptions and ideology. Although the USIA's factual emphasis obscured the subtle (and not so subtle) interpretation of the facts and the careful selection of facts to suit the needs of American propaganda, USIA officials generally believed they were telling the truth about the United States. They constructed a coherent picture of American life as one of progress and consensus that mirrored the assumptions underlying the white, middle-class political culture from which they came.

The USIA's Story of America

Communist propaganda portrayed America as a cultural wasteland where workers, women, and minorities were ruthlessly exploited by the ruling classes. The USIA told a happier story. It portrayed America as a land where free enterprise and democracy brought prosperity and freedom to all. Through "judicious selection of content, audiences, and emphases," the USIA used news items, human interest features, documentaries, and radio programs to carry this positive message around the world. American shortcomings were placed within a narrative of progress. The agency frankly admitted the existence of race-, class-, and gender-related conflicts in American life, but these admissions were presented as triumphal stories of a democratic society peacefully resolving problems as individuals without reliance on the state. The USIA hoped to offset the picture of America as the land of "*Tobacco Road*, gangsters, immorality, Jim Crow, juvenile delinquents, and restless Philistines" with information on U.S. church membership, art museums, community orchestras, cultural and educational projects, libraries, and publications.[4]

The agency's Office of Policy and Plans cataloged a list of impressions to be conveyed about ordinary Americans to world audiences. Americans were to be presented as "hard working," "resourceful," "willing to change," "self-critical," "responsible," and "well-rounded." They "have little class feeling," and they "constantly strive for progress and improvement in all aspects of their society." The USIA informed its audiences that many Americans owned automobiles, televisions, and radios. Americans also enjoyed free public schools, four-lane superhighways, plenty of free time for leisure activities, and wide access to medical care. The agency further stressed that freedom of thought and expression was fundamental to American democracy and that the U.S. economy was managed essentially by consumers. The USIA presented the federal government as having an important, but not excessive role, in American life. Although it was "decentralized as far as possible," the government provided social protections for workers, needy persons, and the elderly. It oversaw a minimum wage, negotiated labor disputes, and carried out a foreign policy motivated by a "desire for peace and freedom." *Facts About the United States* explained that in the American system of government "the people retain full sovereignty." The Constitution guaranteed individual rights and freedoms to everyone. Officials, regardless of position, could be ousted through recall or impeachment.[5]

In presenting the social values of ordinary Americans, the USIA stressed that individualism and diversity characterized life in the United States. While exalting the freedom of individuals to think and act as they pleased, the Information Agency emphasized the shared values that linked individual Americans to each other. Diversity was presented within a narrative framework of consensus: fundamental principles guided American society. *Facts About the United States* stated that "effective self-expression is a keynote of American social and political life," but clarified that this self-expression operated within boundaries delineated by wide acceptance of certain core values. Bold text in the pamphlet explained: "The United States is a country of great diversity . . . but in spite of many differences, certain traditions—freedom, equality, equal rights—are common to all and are taught in the home, in the church, and in the schools." Individualism was kept in check by the strong community orientation of ordinary Americans. From a very young age, boys and girls learned to "work together." "The emphasis in American family life is upon sharing," *Facts About the United States* explained.[6]

The USIA also presented America as a land of spiritual and religious vitality. Religion was a vital force in American society, the agency emphasized: in the United States, followers of many faiths lived and worked together in an atmosphere of toleration and cooperation. *Facts About the United States* conspicuously reported that the Bible continued to be the best-selling book in the United States. Representatives of "virtually every religion known to the world" could be found in the fifty states. American religions were "free of governmental influence or control," the booklet continued, because the separation of church and state was a cardinal principle of American democracy.[7]

A simple running theme tied this propaganda about everyday life: Americans were human beings who worked, played, and lived their lives in ways that international audiences could relate to. According to *Facts About the United States*, most Americans lived in small towns, liked fishing, listened to the radio, attended theatrical productions, enjoyed sports, and loved to travel. They loved to stay in hotels too, but mostly "modest and inexpensive" ones. This stress on Americans as normal people who worked and played just as others in the world did was counterbalanced by the presentation of Americans as exceptional individuals, possessed of strong convictions, noble aspirations, and good intentions. The USIA tried to convey to target audiences that Americans only wanted good things for themselves and for the world in the hopes that audiences would come away with the same impression of the U.S. government.

Women in USIA Propaganda

One "typical American" featured in the USIA's publications was Mrs. Gail Forster, a housewife living in a Philadelphia suburb. She received a visit from a photographer working for the agency's International Press Service (IPS), the division that wired news, feature stories, and photographs abroad for placement in local publications. The photographer spent a few days with her, followed her daily routine, and developed a "picture story" about her life. The USIA would showcase Mrs. Forster as a typical American housewife and mother. Eight photographs and seven short paragraphs chronicled her daily life. According to the IPS picture story, Mrs. Forster lived modestly. She loved her three children and worked hard raising them. Her husband, William, was an engineer for a radio manufacturing company. "Like most American families, they have no servants," the IPS article explained. "Mrs. Forster cooks the meals, cleans the house, washes, irons and mends the clothes, cares for the children, and works in her flower garden. Her workday begins early and ends late." In her spare time, Mrs. Forster taught at a nursery school and volunteered for a citizens' committee working to improve housing conditions in Philadelphia. On Sundays, the Forster family attended church. The photographs showed Mrs. Forster holding her children, working in the garden, taking her children to the doctor for routine examinations, playing chess, studying a community improvement plan, and attending church.

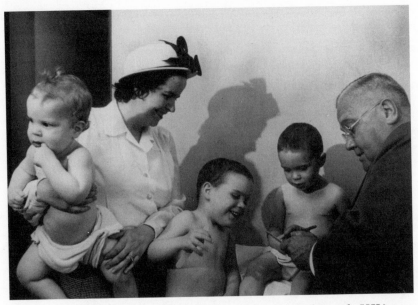

Mrs. Gail Forster, a "typical American housewife," was the subject of a USIA feature story. Source: National Archives.

Had the USIA's global audiences known more about Mrs. Forster, they might have considered her less "typical" than the IPS picture story suggested. Most of her family had attended Harvard or other Ivy League schools. Her husband was a Harvard graduate and the son of H. Walter Forster, president of a management consulting firm. In addition to the Philadelphia home mentioned in the USIA's picture story, the Forsters owned a beautiful estate on Mason's Island on the Connecticut shore. The USIA did not mention such details. It wanted its international audiences to view Mrs. Forster as a representative American woman—a happy housewife living comfortably, but not luxuriously; working hard, but with time for leisure.[8]

Mrs. Forster's story was representative of USIA propaganda about American women. The agency drafted elaborate policy guidance papers and prepared propaganda materials specifically on the subject of American women and their families. The agency used gender as a propaganda theme in appeals to world audiences and cultivated a particular image of everyday women in America. It portrayed them as it had Mrs. Forster: "thoroughly human, hard-working, [and] feminine." "The great majority of American women," a USIA planning document declared, "are characterized by devotion to family, womanliness and industriousness—qualities which foreign audiences can identify with sympathy." The agency celebrated domesticity to dispel portrayals of American women as "irresponsible glamour girl[s]" and "unfeminine, materialistic being[s]" whose main interest in life is work. It wanted to convey the impression to its international audiences that American women played an increasingly important part in political, economic, and community life. But it emphasized above all that women maintained their "primary role" within the closely knit family group. According to the USIA, most American women, like Mrs. Forster, loved their children, volunteered in their communities, and possessed religious convictions.[9]

Attacking the Soviet Union by portraying women there as sweatshop workers and manual laborers denied the joys of family life, the USIA advised that true concern for women required protecting their roles as homemakers and mothers, not exploiting them as manual laborers. According to USIA propaganda, American women cherished their domestic roles. The USIA developed such themes as "the efforts of American parents to maintain a close family relationship," "the love which the American mother, like any other mother, feels for her children," and "the extent to which American mothers share the concern of all mothers." To counteract images of American women as living extravagant, lazy lifestyles supported by kitchen gadgets, maids, and fast food, the USIA made a point of emphasizing that motherhood and homemaking involved hard work. Information officers balanced a stress on the duties of womanhood with materials on the many uses that the "American homemaker" made of her leisure time. Although the "American woman has many serious interests and responsibilities," the USIA explained, "she does like to have a good time." As Laura Belmonte notes, USIA propagandists played up the leisure opportunities available to the American women because they realized that "the freedom to dine and relax could 'sell' America as effectively as the freedom to vote and strike."[10]

The USIA also portrayed women as community activists. *Facts About the United States* declared that "there are very few women in the United States who do not belong to at least one organization, many of them devoted to religious causes, humanitarian purposes, and civic and political responsibilities." Like Mrs. Forster, most women worked to improve American society by volunteering for civic and charitable organizations. IPS feature articles highlighted the social conscience and philanthropic activities of such women as Esther Williams, a film and swimming star who devoted herself to helping handicapped children, and Alice Rohrback, who worked with the blind. One IPS picture story showcased Mr. and Mrs. Albert Dierringer, middle-aged foster parents from Enumclaw, Washington. Mrs. Dierringer was a loving and supportive wife. She explained to international readers how she coped with temper tantrums and bad behavior from her adopted children: "We just vaccinate them with love and affection, give them plenty of food and fresh air, and a large dose of understanding." The IPS story conspicuously suggested that the Dierringers loved children more than consumer goods and luxurious living. "Some people like fancy automobiles, some like long vacations—we like children," Mr. Dierringer was quoted as saying.[11]

In politics and in work, the USIA told a story of women's progress toward equality. The agency saw "special program value" in materials on "outstanding women" and on the "status which women have attained in America." It informed the world that American women participated actively in the traditionally male realm of politics. *Facts About the United States* pointed out that women comprised more than half of the potential voters in the United States and that women were permitted to run for public office. It noted approvingly that between 1916 and 1956 there had been sixty congresswomen, 304 women in state legislatures, two women governors, and two female members of presidential cabinets. The IPS churned out hundreds of feature articles highlighting women's contributions to local, national, and international welfare. Others featured the achievements and civic contributions of such women as Eleanor Roosevelt, Clare Boothe Luce, and Helen Keller. The USIA's story of progress further emphasized the gains women had made in the workplace: "Women are entering fields each year on an equal footing with men, and they may choose careers in teaching, chemistry, law, engineering, music, physics, accounting, politics, or other profession."[12]

While celebrating the accomplishments of working women, the agency also feared that the role of wage earner offered "the greatest opportunity for distortion overseas." Such an image might make American women seem materialistic or masculine. Agency planners advised that materials depicting women in the labor force should "not be presented in isolation from facts showing them first *as women.*" Women in the United States could pursue careers if they desired, the agency noted, adding that most elected to subordinate professional aspirations to homemaking. "Home and family remain the primary interest of the average American woman," the USIA instructed its audiences. A narrative of progress and freedom of choice ex-

A USIA feature story tells foreigners about the everyday lives of American women. Celebrating domesticity, the USIA explained that women in America lived full, happy lives as homemakers and mothers. The USIA's caption for this photo read: "With the feminists' battle long behind them, U.S. women are less interested in being poets and statesmen than they were 25 years ago, and more interested in domesticity." In America, it continued, "Some husbands help with the housework; some do not." Source: National Archives.

plained wage inequities between male and female workers. Although information officers frankly admitted the existence of some discrimination, they stressed the long-term trend toward increasing wage equality. Information officers explained gender-based wage discrimination as a result of the choice most American women made to prepare for marriage and motherhood rather than working life. They tended "to enter and leave the labor force periodically" and did not "plan their education with a career in mind."[13]

Information posts abroad identified women as a special target audience. Women in positions to act as "channels for political information," including leaders, members of organizations, wives of important officials, and university students, were especially targeted. USIS posts devised a number of tactics to reach women directly. They presented books and magazines to women leaders and organizations, hosted special films for women, produced radio programs for female audiences, taught English classes for women, organized seminars and discussions with "outstanding American women," and prepared exhibits and events at USIS libraries and information centers designed to appeal to women. Information officers placed articles of interest to female readers in local newspapers and magazines. Special USIS mailing lists sent information materials to women leaders of civic, political, and charitable organizations. Such efforts sought to "assure [foreign] women that American women have interests and goals in common with their own" and "to encourage on a basic human level an identification of the local women with American women." The agency hoped that if it could win the sympathy of women abroad for the American way of life, women would exert pressure on their husbands and on their governments to act in ways commensurate with U.S. interests.[14]

Reaching the World's Youth

One reason the USIA considered women an important opinion group was because of their "unusually strong influence" on the "attitude formation" of children. The agency wanted to reach young people and students because many of them would be in leadership positions later in life. Moreover, the USIA realized, young adults were the most likely group to actively oppose U.S. foreign policies. In 1958, the Policy Planning Staff recommended a "very substantially expanded" program to reach youth around the world, "particularly those of uncommitted countries and those where Communist influence is strong." It argued that "it is of the utmost importance that USIA consider youth as one of its principal, if not its primary target, that it devise major programs for developing the closest possible associations with youth groups, and that it devote a much more substantial proportion of its available resources to these programs."[15]

All USIS country plans contained special programs to reach students and young people. USIS Mexico, for example, explained that students were a "primary target group" in its programs: "the educational community, including university and advanced secondary students, [receive] priority attention from all applicable USIS resources." USIS posts sought to reach students and young people directly through teacher and student exchange programs, English classes taught by USIS personnel, and seminars and film showings at information centers. Information officers placed special features and news items in local media outlets specifically targeting youth audiences. USIS Vienna produced *Young People* magazine to introduce Austrian children to American culture. The magazine included feature stories on American children, crossword puzzles and games relating to the United States, and other fun items designed to nurture interest in the United States among Austrian youth, such as the lyrics to "My Darling Clementine." The Information Agency sent regular "youth feature packets" to its posts abroad that included news articles and feature stories on topics supposedly of interest to young people. Articles such as "Thousands of U.S. 'Little Merchant' Newsboys Conduct Own Business" and "Teen-Age Employment Agency Gets Jobs for Youngsters" positively reported the participation of America's youth in the nation's economy. Other topics, carefully selected to convey positive images of the United States, included features on the recreational activities of American schoolchildren, information on U.S. colleges and universities, and articles on the civic and volunteer activities of young Americans. The youth packets also included a steady stream of articles about American entertainers such as jazz musician Louis Armstrong and athletes such as boxing champion Rocky Marciano.[16]

Sports were a staple of the USIA output on everyday life in America because they effortlessly stirred the interest of a wide audience. Information posts abroad rarely had difficulty placing items on American athletes and sporting teams in local newspapers and magazines. The Information Agency capitalized on this natural interest in sports to spread positive messages about life in the United States. The IPS prepared monthly feature packets on sports that conveyed a world of gender equality, racial harmony, and international cooperation. Articles on Willie Mays, the legendary African American major league baseball player, and Jim Thorpe, the famous Native American athlete, highlighted the equality of opportunity available to Americans of all races. Features such as "Rosemary Jones: A Leading U.S. Woman Basketball Player," "Stella Walsh Inspires U.S. Women Athletes," "Judy Devlin: Outstanding World Badminton Player," and "Thirteen-Year-Old Girl Enters U.S. Golfing Scene" conveyed the impression that American women of all ages and races won acceptance and success as athletes in the United States. The USIA painted a picture of international harmony by highlighting athletes of international descent, such as Japanese, Pakistanis, and Cubans, who had come to the United States and achieved success in the sports world. Articles also highlighted the American interest in sports typically more popular abroad than in the United States, such as soccer, in order to promote "international goodwill" by suggesting common interests linking Americans to others. Other regular features included stories of East European émigrés who fled to the United States and achieved success as athletes. One such story told the story of Polish refugee Jan Miecznikowski, a long-distance runner at the University of Houston, "who escaped from his Communist-dominated homeland . . . and asked for political asylum in the free world." Such features were selected to convey specially targeted messages to world audiences through the medium of sports reporting, but the sports stories were also intended for the simple purpose of generating positive images about the United States in the minds of the world's peoples.[17]

A Nation of Workers

Labor was one of the most important elements of American life presented by the USIA in its global battle for hearts and minds. Soviet propaganda—rooted as it was in Marxist-Leninist ideology emphasizing class conflict—tried to appeal to the world's workers by portraying the Soviet Union as a worker's paradise and by denigrating the United States as a nation run by greedy capitalists who exploited workers for personal gain. "It has always been the aim of Communist strategy to utilize labor groups as the spearhead of the so-called proletariat in the conquest of power," a USIA planning paper warned. "No effort seems too great for the Communist apparatus to make in order to infiltrate existing unions, to create new ones, or to devise 'united fronts' as a means of manipulating labor organizations for propaganda and . . . subversion."[18]

To prove to the world the falsity of the basic concept of class warfare, the agency developed a global Labor Information Program advertising the marvelous achievements of free labor and free enterprise. In virtually every country with a USIS post, labor leaders and workers constituted a high-priority target group. "Workers must be convinced that it is in their own interest to join with free people in democratic measures for their spiritual, economic and social well-being," a USIA planning paper explained. The USIA was on the defensive, protesting vehemently that capitalism in America had developed neither as Marx had predicted nor as Soviet propaganda claimed. The agency cultivated a utopian vision of labor relations within the United States, telling a story of ceaseless progress, one in which labor and management peacefully resolved their differences and worked together to build an equitable society with all individuals sharing in material progress. The USIA portrayed the United States as a "nation of work-

The USIA frequently used pictures of parking lots, such as this one, to highlight the high standard of living enjoyed by American workers. Source: National Archives.

ers," where labor and labor unions had made great progress in achieving high living standards, collective bargaining power, and social protections. "The abuses of capitalism attacked by Marx are practically non-existent within the American system of free enterprise," the agency argued. "We have in a real sense achieved the true 'classless' society where workers and employers alike share equitably in the abundant goods which our system provides." The United States, in short, had already achieved the wonderful classless society envisioned by Marx, and it had done so peacefully, without revolution and without relying excessively on the state.[19]

Summarizing the central theme of the agency's Labor Information Program, a feature story distributed by the USIA announced: "The story of United States industrial progress can best be told in terms of the life of the average American worker—how he lives, how he works and what benefits he has." Facts and figures would tell the world of the progress workers had achieved under America's democratic-capitalist system. From the USIA's perspective, there was much progress to report. Workers in the United States earned more than they did in the past, and they could buy more with what they earned. To drive home the point of the widely shared benefits of

American production, the agency routinely published photographs of cars parked in lots outside "typical" industrial plants. A caption explained that workers in the United States earned enough money to afford their own automobiles. The USIA's publications similarly documented workers' benefits, conspicuously pointing out that American laborers had paid vacations and holidays, health and welfare protections, provisions against sickness and accidents while on the job, unemployment insurance, and retirement plans.[20]

Life was good for average working Americans, according to the USIA. Putting the buying power of American workers in terms that peoples in impoverished countries could understand, *Facts About the United States* reported that the average American annually consumed 58 pounds of oranges, 23 pounds of chicken, 405 eggs, 72 pounds of beef, 53 pounds of tomatoes, and 17 pounds of ice cream. The "average worker" purchased 1 suit, 4 shirts, 2 pairs of trousers, 5 pieces of underwear, 1 overall or coverall, 1 pair of shoes, 11 pairs of socks, 6 handkerchiefs, and 1 necktie. The "average worker's wife" bought 1 coat or jacket, 3 dresses, 1 hat, 8 pairs of stockings, 1 slip, 2 pairs of shoes, 3 pieces of underwear, and 1 nightgown. Such figures were meant to suggest that American workers lived comfortably, but not extravagantly. They sought to communicate to target audiences that a similar standard of living—high, but modest—could be theirs too under American-style capitalism.

Biographical sketches of "typical" American workers and labor leaders rounded out the picture of labor progress. One article featured African American identical twins who belonged to the United Mine Workers. It began by stating that "it is no longer news for a coal miner to retire at the age of 65 and begin receiving $100 monthly pension checks from the fund." The article related how the twins represented typical American workers receiving benefits from their union, and suggested that American labor provided benefits to all its workers, regardless of race or class.[21]

The USIA further related U.S. labor progress by contrasting the living standards of American workers with those of the Soviet Union. Statistics told the story of American achievement and Soviet decline. *Facts About the United States* explained that in the United States, one hour's average take-home pay bought twice as much food as it did thirty years ago, most of it "of a high nutritional standard." In the USSR, by contrast, the situation had worsened: the "real earnings of workers in the Soviet Union, in terms of food-buying power, are far below the level of 1928." According to the USIA, the Soviet "workers paradise" brought laborers few of the benefits

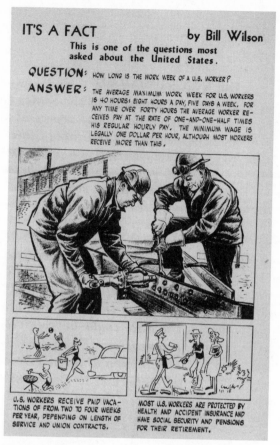

A USIA cartoon publicizes the good working conditions and benefits enjoyed by "average" workers in the United States. Source: National Archives.

that American capitalism provided for its workers. A chart distributed by the agency made the point. It compared the cost of food in "working minutes" for workers in the United States and the Soviet Union. According to the chart, it took American workers four minutes to earn enough to purchase a kilogram of potatoes, but workers in the Soviet Union had to work eighteen minutes for the same thing. A worker in the Soviet Union had to work sixteen more minutes than an American worker to purchase a kilogram of bread, and thirty-four more minutes to purchase a liter of milk. In addition to ridiculing the standard of living in communist countries, the USIA also attacked the "vast slave legions" working in forced labor camps in the USSR. It publicized communist exploitation of workers and the "perverted Communist use of labor organizations to enforce discipline and speed up production." Life for workers in the USSR was not so wonderful as Soviet propaganda claimed, the USIA charged.[22]

An underlying purpose of the USIA's Labor Information Program was to develop opposition to communist influences in international labor movements and to support anticommunist worker groups and unions. "Our efforts should constantly be to equip our friends in other countries with the facts they can use as arguments to show the advantages of free trade unionism and expose the falsity of totalitarian approaches," a USIA planning paper asserted. The USIA actively promoted trade unions abroad by publicizing the practices and benefits of trade unionism in the United States. Deriding labor unions in the "Sino-Soviet orbit" as Communist Party–controlled instruments for labor discipline, the USIA promoted American-style free trade unions as the keys to worker progress. It explained that "American workers continue to improve their earnings, working conditions, and security through militant and responsible free trade unions, which reject Communism and settle their differences with management through negotiation." A steady stream of media materials highlighted the activities of labor unions in the United States. Regular feature stories highlighted the community service activities of workers. These showed "American trade unionists acting, not as a class, but as citizens of their communities and of the nation."[23]

Labor disputes appeared in USIA materials as stories of cooperation between management and labor. The USIA explained that in spite of differences in wealth, education, and environment, the American people had little class feeling and great social mobility. While portraying conflicts between management and labor as benign and cooperative, the agency characterized "U.S. Labor's Struggle Against Communism" as a "bitter struggle against a ruthless and relentless enemy." The USIA instructed its audiences that workers in the United States rejected communism as a hostile force inimical to their interests: "American workers have examined and rejected the efforts of Communist and other totalitarian groups and fellow travelers to infiltrate and manipulate labor unions because they have found that such groups subordinate to their own political expediency the needs and wishes of the workers they profess to serve." Among U.S. workers, there was "virtually unanimous" acceptance of capitalism, the agency declared.[24]

This photograph was part of a USIA feature story on "Lunchbox Capitalists." Text accompanying the picture story explained that "virtually every man, woman or child in the country shares directly or indirectly in the ownership of American industry. In this broad ownership lies the strength of the nation's economic system." The caption to this picture read: "During his lunch hour Paul Craft reads a financial paper, the *Wall Street Journal*, to see how his stock investment is progressing. Craft is employed as an electrician at the Monsanto plant. Along with thousands of other employees, Craft buys Monsanto stock through the company's 'Stock Purchase Plan' for its workers." Source: National Archives.

The People's Capitalism Campaign

The American economy received "high priority and sustained attention" as a theme of USIA propaganda. Agency officials worried that intellectuals and others did not properly understand the American economic system. Foreigners, they complained, mistakenly perceived the U.S. economy as a "cartel-like or feudalistic" system, naively equating it with nineteenth-century laissez-faire capitalism, labor exploitation, high unemployment, and a lack of social protections. The USIA sought to change this image by declaring that the United States had evolved. It had progressed into a "new" form of capitalism, radically different from anything described by the *Communist Manifesto*. In the United States, the agency's media explained, government, industry, and labor worked together to ensure social welfare protections, high productivity, and widespread prosperity. Marxist concepts of class struggle were outdated and anachronistic. "The Communist concept of a proletarian class struggle against 'capitalism' is as destructive as it is outworn," the USIA announced. "Our free trade unionism was developed by the present generation, whereas the *Communist Manifesto* was issued in 1848."[25]

The USIA's effort to promote the U.S. economy as a new form of capitalism developed into a major propaganda campaign. It revolved around the slogan "People's Capitalism," a Madison Avenue invention created to advertise the American economy to the world. The campaign originated in 1955, when Eisenhower sent Theodore Repplier, president of the Advertising Council, on an international tour. His task: survey the climate of international opinion and recommend ways to improve U.S. information efforts. Upon his return, Repplier wrote a report to the president noting that the United States operated under a "serious propaganda handicap" because it did not hold up for the world a "counteracting inspirational concept" to communism. He complained that the communists were stealing "all our good words" by reinterpreting "democracy" and "freedom" to suit their propaganda needs. Repplier wanted to reverse the tables. He suggested that the USIA deliberately appropriate communist discourse to describe the American economic system. If the communists could have "people's democracies," Repplier reasoned, the United States should have "people's capitalism."[26]

Ideological arguments about the American economy backed up the "people's capitalism" slogan. Communist propaganda and Marxist theory came under attack. If communist ideology held that workers and "the

The entrance to the People's Capitalism exhibit. Source: USIA Historical Collection.

people" were exploited by fat-cat capitalists who controlled the means of production, the people's capitalism campaign suggested that in the United States, the people themselves were the capitalists. According to Repplier, "the *people* supply the means for expansion of prosperity, and the people share the *benefits.*" In the United States, "almost everyone" was a capitalist. A pamphlet produced by the Advertising Council explained: "American capitalism at Mid-Twentieth Century is not the capitalism of colonialism, it is not the capitalism of Karl Marx, it is not even our own capitalism of 50 years ago. It is, instead, a capitalism so widely invested in (directly or indirectly) by so many people, with the benefits in goods and wages shared in by so many people, that it is truly People's Capitalism." In the United States, the pamphlet continued, capitalism provided higher wages to workers, better products at lower prices to consumers, greater returns to investors, more goods and services to more people, and more time and money for leisure activities. "In People's Capitalism, the rewards are shared with the workers. Contrary to Karl Marx's predictions, the rich did not become richer and the poor poorer; the middle income group is the one that has grown. America is becoming classless."[27]

The People's Capitalism campaign was not just directed at foreign audiences. It developed into one the most blatant domestic propaganda campaigns since World War II, brazenly marketing the U.S. economy to Americans and foreigners alike. The Advertising Council brought the campaign to the American people in February 1956, with the opening of a People's Capitalism exhibit in Washington, D.C.'s Union Station. The exhibit pictorially presented U.S. history as a steady march of progress. Ten sequential displays told the story, taking spectators through time from the way Americans lived on the eve of the Revolution, when few had many comforts, to the widely shared prosperity of people's capitalism in the 1950s. According to the exhibit, America had developed in stages, gradually transforming from frontier life to people's capitalism. Panels and displays illustrated core ingredients of the U.S. economic system. One explained installment buying in the United States, and another illustrated "typical rooms in average American homes"—modestly furnished, but replete with the latest gadgets and attractive decorations. A panel on the "American philosophy" explained how "American society was founded on belief in God." In the United States "the state exists only to serve the individual, not the reverse." American economic prosperity, the exhibit stressed, grew logically from the limited, democratic government of the United States.[28]

The exhibit marked the beginning of a large domestic information campaign. The Advertising Council orchestrated a media blitz. Statements by union leaders and corporate presidents endorsing "people's capitalism" blanketed the American press. At the council's request, speeches on people's capitalism were given by Clarence Randall, Eisenhower's assistant for foreign economic policy, Keith Funston of the New York Stock Exchange, David J. MacDonald, president of United Steelworkers of America, Lee H. Bristol, president of Bristol-Myers Company, and dozens of other business and labor leaders. Newspapers dutifully reported the themes of the campaign, often with exclamatory approval. Several major U.S. corporations incorporated the people's capitalism theme into their advertising. General Electric, for example, placed a double-page advertisement in ten major newspapers listing the distinguishing characteristics of American capitalism.[29]

People's capitalism blended slick advertising promotionalism—the U.S. economy is "new and improved"—with sophisticated intellectual arguments. Leading economists and journalists lent their support to the program, transforming an advertising stunt into a brainy defense of the Amer-

This display in the People's Capitalism exhibit illustrates how the "average American" lives today. Source: Advertising Council Archive.

ican economy. The Advertising Council, having invented and publicized the idea of people's capitalism, feared that journalists and opinion makers might regard people's capitalism as merely an advertising stunt because it originated on Madison Avenue. It decided the people's capitalism theme needed to be considered seriously by the "intellectuals of the world—particularly the economists." To stimulate wide awareness of the academic merits of the people's capitalism idea, the Advertising Council sponsored several roundtable discussions with prominent economists, published the resulting conference papers, and distributed the publications along with press releases summarizing the desired points to newspapers, magazines, libraries, and universities. Perhaps the most notable event was a roundtable discussion at Yale University in November 1956. Yale faculty members, including propaganda specialist Harold Lasswell, met with labor and business leaders in a discussion devoted to correcting "the fantastic and often dangerous misconceptions about the American economic system that exist around the world." The Advertising Council's campaign further stimulated the authorship of academic articles on people's capitalism. After their publication in scholarly journals, the articles were reprinted as pamphlets and widely circulated in the United States and abroad. Scholars lending their intellectual weight included Yale economist Henry Wallich, who wrote about the "exportability" of American-style people's capitalism to other countries. Books published included Marcus Nadler, *People's Capitalism*, and Jacob M. Budish, *People's Capitalism: Stock Ownership and Production*.[30]

While the Advertising Council promoted the theme at home, a massive USIA blitz publicized the theme abroad. Information officers were instructed to repeat the phrase "people's capitalism" as "often as practicable" to make target audiences use the slogan in everyday discourse. The Advertising Council's exhibit toured the world as part of the administration's trade fair program, and the USIA prepared smaller exhibits on the people's capitalism theme to reach more audiences. One of these, The Changing Community—Kalamazoo, emphasized U.S. progress in social and cultural development as well as in the industrial and technological fields. Another small exhibit, the Capitalist Farmer, comprised a photo panel display demonstrating the benefits the free enterprise system provided to American farmers. The Advertising Council's PR stunts provided the USIA with volumes of news and feature items to distribute abroad. The agency distributed packages of materials illustrating the theme, many of them drawing from materials produced by the Advertising Council. Scholarly books and articles on people's capitalism went to intellectuals and journalists abroad, and the speeches by prominent Americans supporting the people's capitalism theme were reported as top news stories by the USIA's press service. Stories of ordinary Americans owning stock and insurance policies received heavy play, as did films on the role of capital investment in the development of the American economy. Pamphlets explained such subjects as profit sharing, and cartoons contrasted the "Wall Street Capitalist Fat Cat" pictured by the Soviets with the "actual American 'capitalist' as he is emerging in the U.S." A typical feature story told of a government clerk who so believed in America's future that he invested in industrial stocks during the Great Depression. His faith in capitalism paid off: his stocks eventually rose in value to nearly $1 million.[31]

President Eisenhower tours the People's Capitalism exhibit with Theodore Repplier, the president of the Advertising Council who developed the People's Capitalism theme. Source: Advertising Council Archive.

Race and U.S. Propaganda

These efforts to convince the world that all Americans enjoyed the benefits of the U.S. economy and democracy were severely handicapped by the presence of racial segregation in the South and the wide divergence in standards of living between whites and blacks in the country as a whole. Race was the issue that most interested many international observers, and segregation cast a dark shadow over the utopian picture of everyday life promulgated by the USIA. In all regions of the world, the agency reported to the NSC, racial relationships within the United States were "a matter of local human interest, often with emotional self-identification." Mary Dudziak has shown that even before the civil rights clashes of the mid-1950s and 1960s brought U.S. racism and segregation to television sets around the world, American race relations were exerting a powerful impact on international views of the United States. Throughout the postwar period, international media coverage of the "American Negro problem" was widespread and overwhelmingly critical. Reports of racial violence and segregation in the aftermath of World War II led many observers abroad to question the sincerity of the American commitment to liberty and equality. The effects of "racial discrimination in the United States on public opinion abroad is definitely adverse to our interests," a State Department study pointed out. "It clearly results to some extent in the weakening of our moral position as the champion of freedom and democracy, and in the raising or reinforcing of doubts as to the sincerity and strength of our professions of concern for the welfare of others, particularly in the nonwhite world. Moreover, it provides a solid target for anti-American propaganda."[32]

Communist propaganda used depictions of racial inequality to great effect. Factual evidence of racial discrimination was plentiful, and it provided Soviet media with a steady supply of ammunition for attacking the USIA's idealized image of America. Soviet publications routinely reported the "increased frequency of terroristic acts against negroes" and explained that in the American South "semi-slave forms of oppression and exploitation are the rule." According to the U.S. Embassy in Moscow, "The Soviet press hammers away unceasingly on such things as 'lynch law,' segregation, racial discrimination, deprivation of political rights, etc., seeking to build up a picture of America in which the Negroes are brutally downtrodden with no hope of improving their status under the existing form of government." Other communist governments followed Moscow's lead, giving heavy emphasis to American racial discrimination in their worldwide propaganda output.[33]

Even if communist propaganda had been silent on the issue, racial discrimination would have continued to attract international attention. In the developing world, American race relations received especially intense scrutiny. The fact that nonwhite foreign dignitaries who visited the United States were often subjected to the same discriminatory treatment as African Americans ensured international press coverage and diplomatic protest. In 1947, for example, a representative of the Haitian government was forced to stay in servants' quarters in a Biloxi hotel for "reasons of color." The Haitian ambassador lodged a complaint with the State Department, and editorials in the Haitian press attacked American democracy as a sham, declaring that "the Negro of Haiti understands that the word democracy in the United States has no meaning." Similar incidents occurred throughout the postwar period. The State Department reported that between 1957

Liberian students gather at the entrance to a U.S. Information Center in Liberia. News of civil rights clashes and racial conflicts in the United States received extensive attention by media outlets in Africa and other parts of the third world. The USIA sought to put segregation "in perspective" by highlighting the progress made by African Americans since emancipation. Source: National Archives.

and 1961, fourteen such incidents occurred involving diplomats from Africa, Asia, the Caribbean, and Latin America. Such incidents were the worst kind of negative PR, receiving prominent press play in their home countries and elsewhere. When a diplomat from Nigeria was refused service in a restaurant in Virginia, for example, Nigerian papers expressed their "horror and dismay" at American racial discrimination. The *Lagos Daily Times* comment was typical of international press coverage of such incidents: "By this disgraceful act of racial discrimination, the U.S. forfeits its claim to world leadership."[34]

In the new postcolonial countries in Asia, Africa, and the Middle East, perceptions of American race relations were closely tied to the issue of colonialism. The NSC noted that "the underlying distrust of the white man is an important ingredient of the basic problems with which our information programs must deal in colonial and ex-colonial areas of the world." Many countries in the developing world, which had only recently earned

their independence, not surprisingly viewed anticolonialism as a more sa-
lient issue than anticommunism. Few issues more thoroughly hampered
U.S. courtship of the developing world than American racial practices,
but race also affected European attitudes. Many European observers self-
righteously wondered how the United States could attack their imperial
practices in Asia and Africa while permitting egregious violations of civil
rights at home.[35]

The USIA's response did not smack of confidence. The agency admitted that racial discrimination was indeed a problem in American life, but
it stressed progress: despite the problem, the United States was moving
steadily down the road to equal opportunity. The Information Agency
bombarded international media outlets with press items illustrating the
progress African Americans had made since emancipation. The USIA's Office of Policy and Plans instructed information officers to identify, "but
without too much obviousness," each "newsworthy" instance of achievement by African Americans. "We should seek especially to show progress
in the fields of civil rights, employment, education, armed forces, sports,
entertainment, art, housing, business and government by individuals and
groups, with major emphasis on interracial cooperation." When stories of
racial prejudice or violence appeared in the international media, information officers should "place them in perspective" by submitting evidence of
racial progress. Stories of discrimination were balanced by materials showing the world "the social mobility of our population" and that "the income
level of Negroes and other minority groups . . . has steadily risen."[36]

The USIA utilized a variety of tactics to present to foreign audiences
evidence of racial progress in the United States. The agency's propaganda
to all countries included special programming on the accomplishments of
African Americans. Information officers arranged press conferences and
other discussions between local groups and distinguished African American personalities. Marian Anderson, for example, talked "with great impact" about the American racial picture during her Emergency Fund concert tour of the Near and Far East. The USIS post in Italy reported that
African American lecturers and Fulbright grantees were similarly used as
"living demonstrations of Negro cultural and intellectual advancements in
the US and as authoritative sources of information on race problems and
progress toward their solution in the US." One of the top country objectives of USIS Nigeria was "overtly proving American sympathy for Negroes in the U.S." The post placed features on African American leaders on
local newsreels, distributed biographies of prominent African Americans,

promoted music by African American musicians on radio programs, and presented paperback books on "the American Negro" to secondary schools and universities. USIS Ethiopia, meanwhile, placed scores of stories "on Negro advancement and the gradual breaking down of segregation" in its Addis Ababa bulletin. Behind the Iron Curtain, the Voice of America told listeners in Poland that "in the economic field there have been no restrictions or differences between the Negro and the white population" and "the existence of segregation does not in the least mean worse schools for Negro children than for the non-Negro children."[37]

USIS posts also exploited the sesquicentennial anniversary of Abraham Lincoln's birth as a device for highlighting the American commitment to equality. They commemorated the Lincoln sesquicentennial with books, pamphlets, and special programs presenting Lincoln as a symbol of freedom, justice, and racial equality. USIS Thailand linked Lincoln to his Thai contemporary King Mongut to stress the community of interests between Thailand and the United States and to remind Thais that they, too, had racial problems to contend with. (In addition to the standard tactics, USIS Thailand convinced the Thai government to produce a Mongut-Lincoln commemorative postage stamp.) Through the symbol of Lincoln, USIS posts around the world worked to "put into proper perspective the historic changes in the position of the Negro in American society."[38]

One of the most important methods of influence available to USIA personnel abroad was person-to-person contact with opinion makers and government officials. Information officers used these personal contacts to influence press coverage of the civil rights struggles in the United States. USIS Ethiopia cabled an example of this personal contact work to Washington. The post reported that the editor of a local newspaper, the *Voice of Ethiopia*, was engaging in sensational and "totally irresponsible reporting of the race question in the United States." Seeking to mute this passionate portrayal, the public affairs officer (PAO) met on several occasions with the editor of the paper to convince him to rethink his opinion of American race relations. Reiterating the standard line, the PAO stressed to the editor that "law was on the side of the Negroes" and that "the race situation in the United States is not one of murders, riots, beatings, etc., but it is a situation which both races are trying by calm but persistent evolution to correct." He supplied the editor with materials prepared by the USIA and continued to meet with him for several weeks. The PAO reported that as a result of his efforts, the *Voice of Ethiopia* stopped its publication of "sensational stories" about racial discrimination and began printing stories "about Negro advancement" in the United States. To the USIA, the real story of America was this story of progress.[39]

Did these strenuous efforts work? Public opinion surveys indicated that the world was not convinced of American racial progress. Eisenhower resisted taking the assertive stand on civil rights that American propagandists so badly needed to convince the world that the United States was indeed progressing toward racial equality. The president refused to adopt even a strong rhetorical stance opposing segregation, preferring instead to utter bland platitudes about the limits of federal authority. His self-proclaimed belief that he could not change people's views on racial matters with mere words stands in stark contrast to his avid endorsement of propaganda to mold popular attitudes on other matters. To be fair, Eisenhower did sign the 1957 Civil Rights Act in the face of massive Southern resistance, but this legislation did little to advance equal rights. Regardless, the nasty fight in the Congress over the bill's passage did more to harm than to help international perceptions of American race relations.[40]

By and large, it was the Supreme Court that provided the USIA with the hard-core facts it needed for effective propaganda on the race problem. Of particular significance was the 1954 Supreme Court decision in *Brown vs. Board of Education* that reversed the doctrine of "separate but equal." Although Eisenhower was "profoundly ambivalent" about the decision, the USIA exploited the *Brown* decision "to the fullest" as evidence of the American repudiation of racial discrimination. Articles on the decision were placed in "almost every" publication of the agency. The USIS campaign in India, where U.S. race relations were receiving widespread and sensational coverage, typified these operations. USIS India engaged in a large campaign to convince Indians that *Brown vs. Board of Education* was proof of the U.S. commitment to racial equality. The post distributed pictures depicting black and white schoolchildren playing and studying together, and it sent over twenty different news stories on the ruling to nearly a thousand newspapers around the country. The post also reviewed all of its educational films to find sequences of black and white children attending the same schools. USIS mobile teams brought the films to towns and villages across India. (Unfortunately, the post commented, "the sequences in these films were all too brief.") USIS India also prepared a special exhibit of large photographs, New Triumph for Racial Equality in U.S. The exhibit was displayed at ten colleges and business institutions around Bombay. Nearly 40,000 copies of the pamphlets "The Negro in American Life" and "American Negro—Fifty Years of Progress" were distributed to recipients on USIS mailing lists accompanied with a four-page pamphlet on the *Brown* decision.[41]

Although the *Brown* decision provided the USIA with powerful material to counter images of racial discrimination in the United States, within a year, widespread resistance to the Supreme Court's decision had begun to mar the picture. The brutal murder of Emmett Till in 1955, the mob protests that greeted Autherine Lucy's attempt to enroll in the University of Alabama in 1956, the 1955–1956 Montgomery bus boycott, the opposition to desegregation organized by white citizens' councils, and the adoption of antiintegration state statutes by many Southern states directed public attention abroad to American racism. Pictorial evidence of racial violence in the international press had an especially potent impact on international audiences, the USIA complained, because it "focused attention upon this aspect of American life and created an appetite for information on the subject." The USIA redoubled its efforts to convince people of other nations of the progress being made throughout the country. The agency argued that many of the barriers to racial equality had been removed; dramatic instances of resistance were evidence of "substantial advances rather than of retrogression."[42]

The 1957 events in Little Rock precipitated a veritable crises in international perceptions of American race relations and democracy. On September 4, nine African American students tried to enroll at Little Rock's Central High School, only to be turned away by the Arkansas National Guard. The governor of the state, Orval Faubus, had ordered the troops to Little Rock in defiance of court-ordered desegregation. His action created a national crisis and an international scandal. Stories of the armed National Guard troops forcibly turning away black schoolchildren appeared on front pages around the world. Headlines like "Armed Men Cordon Off White School" and "Troops Stop Negroes Going to School" were typical. In many countries, stories on the Little Rock crisis appeared in newspapers virtually every day for the entire month of September. U.S. embassies around the world cabled the State Department to report the damage Little Rock was inflicting on international opinion of the United States. John Foster Dulles complained that the "situation was ruining our foreign policy. The effect of this in Asia and Africa will be worse for us than Hungary was for the Russians."[43]

The crisis came to a head when a thousand paratroopers descended on Little Rock to restore order and to protect the African American schoolchildren. As Cary Fraser has argued, the international reaction to the crisis was a factor that influenced Eisenhower's decision to send in federal troops.

In his memoirs, Eisenhower rationalized the intervention by pointing to the damage being done by Soviet propagandists who were exploiting the incident. According to Eisenhower, "Overseas, the mouthpieces of Soviet propaganda in Russia and Europe were blaring out that 'anti-Negro violence' in Little Rock was being 'committed with the clear connivance of the United States government.'" In a televised address to the American people, Eisenhower argued that the Cold War struggle and international opinion virtually compelled him to take action:

> At a time when we face grave situations abroad because of the hatred Communism bears toward a system of government based on human rights, it would be difficult to exaggerate the harm that is being done to the prestige and influence, and indeed to the safety, of our nation and the world. Our enemies are gloating over this incident and using it everywhere to misrepresent our whole nation. We are portrayed as a violator of those standards of conduct which the peoples of the world united to proclaim in the Charter of the United Nations.

Eisenhower continued to state his hope that the image of America would be restored and that the "blot upon the fair name and high honor of our nation" would be removed.[44]

Eisenhower's address to the nation was also an address to the world. The USIA transmitted the full text of the speech to all its posts. To bring Little Rock "into better focus" for foreign audiences, the agency mobilized its resources to demonstrate that, despite the Arkansas episode, integration in other public schools was developing in an atmosphere of cooperation and understanding. Information officers were instructed to accord the news factual treatment but to balance "adverse sensational items" with quotes from editorials and official statements indicating "steady determined progress toward integration." Information officers were also encouraged to provide friendly editors with "constructive interpretations" of the events and to release additional information depicting racial progress. Like other posts, USIS Mexico reported that it played the Little Rock story "straight and intensively," but that it balanced its factual reporting of developments in Arkansas with materials emphasizing racial accomplishments in other schools and other fields of endeavor.[45]

The State Department advised USIS posts and PAOs "to start the long and slow job of putting these unfortunate incidents in their proper perspective." The department sent "talking points" to information officers abroad instructing them to present Little Rock as an aberration in the ongoing process of racial reform. The talking points argued that increased employment opportunities for blacks, rising wages, the desegregation of the armed forces, and Supreme Court decisions highlighted the American commitment to racial equality. Information officers were told to stress the "marked progress toward integration" and the "tremendous strides [that] have been made in removing racial barriers in the U.S." USIS posts issued a torrent of press releases providing evidence of racial progress. USIS Geneva and USIS Paris, for example, successfully placed a cover story in local magazines bearing the headline "Despite Little Rock Segregation Is Dying." Accompanied by large photographs of white and black students in school together, such articles encouraged international audiences to take a "second look" at the problem of race relations in the United States.[46]

Information officers also dealt with the crisis by reminding foreign audiences that racial conflict was an international problem, not strictly an American dilemma. USIS posts made a point of observing that racial friction in India and in slum areas of London illustrated that "the problem was not confined to the U.S." They further placed the use of force in Little Rock in perspective by contrasting it with the use of force in the Soviet Union: "In the Little Rock incident national authority has been invoked to maintain [the] equal rights of a minority. In the Soviet Union national authority has been repeatedly invoked to suppress the rights of minorities." USIS Paris asserted that "our national authority is being used to ensure the education of children, in dramatic contrast to the uses to which Soviet armed might was put last year in Hungary." Whatever the problems in the United States, the agency stressed, they paled in comparison to the brutal suppression of human rights in the Soviet bloc.[47]

The USIA's treatment of Little Rock and segregation did not go over well in the American South. An editorial in the *Charleston Post* typified the hostile reaction engendered by the agency's message that desegregation equaled progress. Expressing its disapproval of the way the Voice of America covered events in Little Rock, the editorial asked why an American propaganda station was telling people abroad about "helmeted soldiers, police, [and] crowds of angry and annoyed citizens." The *Charleston Post* further objected to the VOA's claim that the "majority of Southerners" were "beginning to accept" the Supreme Court's order for desegregating schools. To this assertion the editorial responded with hostility: "The majority of Southerners have not yet been confronted with school integration. If the federal government ever tries to force it on them there will be grievous times."[48]

Despite such heated criticism, U.S. propaganda experts continued to portray segregation as a problem that was being addressed. They adopted an unconventional approach to tackling foreign criticism of American race relations at the 1958 Brussels Universal and International Exhibition. Countries around the world poured money into their exhibits at the Brussels World's Fair, making the event an international media spectacle and a major forum for Cold War propaganda. The U.S. exhibit included many of the features common to its trade fair exhibitions, including fancy displays of consumer goods and models addressing the Atoms for Peace theme. Yet what made the American pavilion stand out was its Unfinished Business exhibition. Walt Rostow, C. D. Jackson, and others involved in planning the exhibit developed an unusual approach to the fair. Although most nations used the exhibits to tout their own successes, the Unfinished Business exhibit aired America's dirty laundry. The exhibit courageously acknowledged the existence of social problems in order to show the world how the United States wrestled with those problems as part of its ceaseless quest for progress.

The planned exhibit highlighted three problem areas: urban slums, environmental challenges, and, most controversially, segregation. Planners felt that silence on the race issue, especially in the wake of the Little Rock crisis, would hand communist propaganda a ready-made propaganda bonanza that would ruin anything else the U.S. did at the fair. "Maybe the way to 'handle' Little Rock is not to ignore it and distract the world with gadgets and appliances," wrote one of the architects of the exhibit. "Maybe the best way is to try to show the miserable events at Little Rock in the long context of the Negro's rise from slavery and his spectacular recent progress."[49]

Far from denying the existence of a race problem in the United States, the exhibit addressed the issue in a surprisingly candid manner. It began by recounting, of all things, the slave trade. It continued to inform visitors that the 17 million blacks in the United States had not yet won equal rights, a point also made by displays featuring pictures and newspaper clippings of civil rights clashes. Then came the positive message. Incorporating a rhetorical tactic that had become standard in U.S. propaganda, the exhibit went from acknowledging the problem to pointing out that progress was being made. "Not since the Civil War . . . has the Negro made such strides toward full equality as he is making now," a caption read, announcing that "the doom of the American caste system is in sight." Displays in the exhibit highlighted the growth in university enrollments, increased political participation, and rising living standards of African Americans. Photographs depicted multiracial groups of children playing and working together. The final section of the exhibit featured a large photograph of black, white, and Asian American children holding hands. The accompanying caption proclaimed that "democracy's unfinished business, already partially mastered, will get done on a national scale."[50]

This was a novel approach for a government propaganda program. Officials were taking the most widely criticized aspect of American life, putting it on display before the world, and admitting that it was a problem. The exhibit's planners hoped that such a tactic would deflate the potency of the race issue abroad and lend credibility to their larger message of progress. It seemed to be a good plan. Many visitors to the exhibit expressed admiration for the courageous way the United States dealt with the issue. Some said that they had found "new respect" for America.

Many Southerners were less impressed. Legislators from Southern states expressed outrage at the suggestion that desegregation was being presented as "unfinished business." One congressman from South Carolina described it as "colossal and unimaginable stupidity." His colleague in the Senate called it a "propaganda fiasco." A senator from Georgia wanted John Foster Dulles to explain why the United States was apologizing for racial segregation. Three congressmen traveled to Belgium to see the exhibit firsthand. They returned angry at the "anti-Southern slant" to the exhibit and its message "that segregation was a problem the United States must solve." Faced with such opposition, the Eisenhower administration quietly dismantled the exhibit.[51]

President Eisenhower, who had often sided with his propaganda experts on other matters, was silent. Although he understood that the civil rights situation was harmful to America's overseas image, he did not seem to fully grasp the dimensions of the situation. "The *real* fact that our propaganda should be able to emphasize," he wrote in 1951, "is that there is no lynching in the United States, that no worthwhile citizen is really kept from voting because of any poll tax, and that no man is kept out of employment merely because of race or religion or other factor of this kind." The year Eisenhower wrote this, there had been two lynchings and seven instances of racially motivated mob violence. Five Southern states still used poll taxes to prevent blacks from voting.[52]

Reflecting this myopic vision of civil rights, U.S. information programs most widely deviated from their professed "strategy of truth" on the subject of race relations. The USIA's statements that African Americans were not impeded from voting in the South, that segregation did not mean worse schools for black children, and that racial violence was virtually non-existent in the United States misrepresented the everyday lives of many African Americans. The distortions highlight the limits to the USIA's public claim to objectivity and its commitment to a factual approach. On this subject at least, the agency clearly privileged a factual tone over the actual facts. Yet from a propaganda perspective, it was novel for a government propaganda agency to admit that a problem existed at all. The fact that the USIA reported civil rights clashes, labor disputes, gender-based wage discrimination, and other unfavorable aspects of American life strengthened the credibility of the agency and allowed it to put a positive spin on negative images of the United States. In order to maintain its message of progress, the USIA always stressed "trends" in its propaganda and avoided listing any accomplishment as the "first" of its kind. Thus, although American women were subject to wage discrimination, the "trend" was toward greater equality in the workplace; although there was segregation in the South, African Americans were "increasingly" recognized as equals. U.S. propagandists believed the most effective approach to controversial issues was to acknowledge their existence so as to reassure the world that the problems were being addressed. According to the USIA, the United States provided mechanisms for social and economic progress to all Americans, regardless of race, class, or gender.[53]

Progress itself was an overarching theme of U.S. propaganda. Officials in the Information Agency worried about a "pronounced tendency" abroad to equate progress with socialism. According to USIA studies, important segments of international opinion, particularly in less developed areas, viewed socialism as providing a more equitable path to rapid industrialization and economic growth than capitalism. The USIA warned that "authoritarian rule that apparently succeeds in leaping from abject poverty to comparative comfort through rapid industrialization is profoundly attractive to leaders faced in their own backyards with urgent demands for improved living standards." Soviet propaganda fed these perceptions by playing up its own story of progress, claiming that the USSR had built a "classless" society freed from the rampant labor exploitation of the capitalist world. Khrushchev argued time and again that the Soviet Union would soon surpass the United States in living standards for its people.[54]

Most USIA campaigns therefore developed in some way the idea of progress. The United States was working for a better world, U.S. propagandists argued, in seeking to end the Cold War, in protecting the rights of the individual, in limiting the power of the state, in extending the benefits of capitalist production to all, and in advancing the principles of freedom and democracy. The USIA used factual information and factual-*sounding* information to submit evidence to international audiences of U.S. progress in peaceful uses of atomic energy, scientific endeavors, gender relationships, race relations, economic production, disarmament negotiations, intellectual accomplishments, and other arenas. This message in part reflected contemporary intellectual currents in an age of consensus. Yet in telling this story of progress, the Information Agency clearly subordinated its strategy of truth to its mission of international persuasion. Progress was a rhetorical device for selling the idea that the United States promised a brighter, happier future than its communist rivals. The emphasis on daily life in USIA propaganda created the simple and appealing message that American capitalism and democracy promised the good life to all who followed the U.S. example. The USIA exploited the "facts about the United States" that fit this message.

Chapter 9

A New "Magna Carta" of Freedom
The Ideological Warfare Campaign

Although the West has strengthened its military and economic resources in meeting the
Soviet challenge, insufficient attention has been accorded to the need for combating
Communism on the intellectual plane, in the sphere of ideas.

— Psychological Strategy Board, June 29, 1953

One of the most revealing representations of the communist menace, as
Americans perceived it, was *Invasion of the Body Snatchers*, the 1956 film
that depicted not communist agents, but alien "pods" out to take over the
world. Set in an archetypical small American town, the film followed Dr.
Miles Bennett as he gradually discovered that his friends and community
had been taken over by the pods. Holding out the empty promise of a perfect future, the plantlike aliens quietly infiltrated the peaceful community
of Santa Mira. Gradually they replaced a dynamic and individualistic culture with a soulless collective. Portraying the pods as "a malignant disease
spreading through the whole country," the film's dramatization of the alien
threat closely paralleled official views of communism. For the aliens in the
film did not so much snatch the bodies of their victims as take over their
minds. A dramatic expository scene explained how the pods did their work:
"Suddenly, while you're asleep, they will absorb your minds, your memo-
ries, and you'll be reborn into an untroubled world." This utopian promise, audiences were told, created a world where "everyone's the same," a
meaningless and spiritually hollow existence, devoid of love, ambition, desire, and faith. "You're in danger!" Dr. Bennett cries, frantically yelling into
the camera in a climactic final scene. "Can't you see? They're after you!
They're after all of us! . . . YOU'RE NEXT!"[1]

The mythical danger depicted in *Invasion of the Body Snatchers* could
have been culled from any number of classified national security documents that presented communist ideology as one of the most treacherous
weapons in the Soviet arsenal. These documents conceded that the communists—like the pods—did not pursue their objective of world domina-

tion by military means, but by infecting good and innocent societies like a disease, subverting them from within. The danger posed by communist ideas, as one government report put it, was everywhere: "The Communists begin their attack with ideas and intend to finish with guns if necessary. With ideas they spread their poisonous germs in every phase of American life . . . These ideas seep into American politics, American economics, American educational institutions, American neighborhoods, and American homes."[2]

To many a Cold Warrior, the communist ideological challenge demanded a forceful counteroffensive. It required Americans, presumably free of ideological restraints, to enunciate a doctrine of their own, one that rivaled Marxist-Leninist testaments. Allen Dulles, speaking before the Council on Foreign Relations in 1953, articulated a widely held sentiment when he announced a desperate need for an *American* ideology, a captivating idea, a universal doctrine that could compete head to head with communism. "The record clearly shows that [the communists] have more successfully appealed to the hopes and aspirations of other peoples than we have. . . . One thing we lack in the free world is a dynamic statement of agreed principles—a new 'Magna Carta of freedom'—that we can all rally around. We badly need some such 'universal idea.'" So all-encompassing was this battle of ideas that even professional historians took up the call for an ideological offensive. "The antidote to bad doctrine is better doctrine, not neutralized intelligence," the president of the American Historical Association, Conyers Read, declared in 1949. In this total conflict for freedom and democracy, neutrality was not an option, not even for historians. "We must clearly assume a militant attitude if we are to survive," he explained. "Total war, whether it be hot or cold, enlists everyone and calls upon everyone to assume his part. The historian is no freer from this obligation than the physicist."[3]

Psychological warfare planners sought to meet the perceived vulnerability posed by the communist ideological challenge by harnessing the creative and intellectual energies of American scholars and intellectuals. In mid-1953, the Operations Coordinating Board (OCB) approved an ambitious program to discredit communist ideology and to promote a contrasting ideology of freedom. This plan—for what the OCB variously called "doctrinal warfare," "ideological warfare," or simply the "ideological program"—targeted intellectuals and educated elites abroad. It attacked the core tenets of communist ideology and promoted the concept of freedom as a positive ideological alternative to communism. By using government resources to mobilize the private sphere in a great ideological offensive, the OCB used the government's powers of funding, access, and connections to support the production and publication of ideas that served the Cold War agenda of the national security state. Books, publications, and various educational exchange programs were the principal vehicles of the Ideological Program. Intellectuals, publishers, academics, and religious leaders were, through various covert and overt means, both wittingly and unwittingly, brought on board to support the campaign.

The actions taken to wage this ideological warfare are considered sufficiently sensitive that many records pertaining to the doctrinal warfare program remain classified to this day. It is therefore not possible to recreate the program in its entirety, and there are lamentable gaps in the pages that follow. In some areas, the ideological warfare documents raise more questions than answers. Nevertheless, enough material is now available to present a clear, if incomplete, picture of the American ideological program, an effort that highlights the extent to which communist ideology itself, as a set of ideas and as an influence, was perceived as a threat by American policy makers and as a terrain of Cold War combat.

As Much Ideological Content as Possible

The program for ideological warfare, like other key psychological warfare initiatives of the day, was developed in the waning months of the Truman administration by the Psychological Strategy Board. In September 1952, the PSB created a panel to develop a national strategy for attacking communist ideology and enhancing the appeal of the spirit and philosophy of the American system. "In recent years," the board's resulting classified plan announced in June 1953, "the largest selling book in the world—with the possible exception of the Bible—has been the *Short History of the Communist Party* which is the testament of the Communists. Its circulation exceeds 41,000,000 copies. This book is on sale in U.S. book stores for $1.00 on fine paper, beautifully printed." The wide availability of communist literature, subsidized by the Soviet Union, indicated a "massive, comprehensive, worldwide campaign of ideological indoctrination, designed to further international Communism." To meet this challenge, the board proposed a sweeping ideological crusade. The government would work with the private sphere to encourage the production of counterpropaganda, mostly books and articles, that illuminated the fallibility of communist doctrine and illustrated the spiritual vitality of the West. These would be "doctrinal

materials" that would include "all fields of intellectual and cultural interests, from anthropology and artistic creations to sociology and scientific
methodology."[4]

The PSB's program targeted the world's intellectuals, classified by the
board as educated, rational, and scientifically minded persons who would
be convinced by logical arguments substantiated by factual evidence. The
board developed an elaborate rationale for this approach. Doctrinal warfare differed from ordinary propaganda in that it did not directly seek to
influence mass behavior or popular beliefs. Rather, it aimed at the "mental
processes of the influential few": teachers, scientists, writers, and others capable of affecting the attitudes and perceptions of leaders and decision makers. Whereas propaganda presupposed an *emotional* reaction as a stimulant
for group action, doctrinal warfare operated indirectly through reason and
subtle suggestion. "Doctrinal warfare, *per se*, does not attempt to convince,
but ostensibly it develops and presents materials which would stimulate
the recipient mind to develop and to create as its own the thoughts implicit
in the doctrinal material," the PSB explained in a long exposition of its
"planned manipulation of targets." Doctrinal warfare would accomplish
its objectives by implanting in "developed minds" new facts and additional
knowledge that would change established views and mental patterns, making them "unsatisfied with their accepted ideology." The targets would be
inspired by this new discovery like converts to a new religion. They would
become motivated to transmit this enlightenment to others, thus beginning a process of ideological renewal. The PSB expected results to emerge
only gradually, after considerable time had passed. It was what propaganda
experts called a "slow media" operation.[5]

When the Operations Coordinating Board replaced the PSB in the
summer of 1953, it revised and approved the PSB's doctrinal warfare plan
as one of its first actions. It appointed an Ideological Working Group to
guide its implementation. According to the OCB, "the cold war is in considerable measure a contest between two antithetical ideologies, that of
the totalitarian Communists and that of the U.S. and the rest of the Free
World." The task for the United States, and for the doctrinal warfare program, was to find "an ideological weapon which can be used *offensively*."[6]

Reflecting the Eisenhower administration's belief that information
programs should emphasize positive messages, not simply anticommunist
polemics, the new OCB plan set forth an ambitious objective. The United
States needed to popularize a coherent statement of "free world principles"
to challenge the perceived communist monopoly on ideology. "Opposition

to Communism or other totalitarian systems of thought, particularly if based on fear alone, is not adequate to assure active support of Free World principles," the board explained, "a positive belief in the superiority of the latter is essential." Thus, in addition to attacking communism, the OCB plan also called for the "formulation of the Free World position in terms that will gain the adherence of the variety of cultural, economic, and social patterns which . . . make up the Free World." An ideology of "freedom" would be promulgated to lure intellectuals away from communist "totalitarianism."[7]

Establishing a free world position in the ideological conflict raised thorny questions for American psychological strategists. Did this mean that Americans, who saw themselves as free from ideological influences, should construct a systematic body of dogma representative of U.S. and free world principles? What were these principles anyway? American policy makers used the words "free world" to apply to all noncommunist countries, but the term was closely identified with industrialized, western democracies. This raised problems for USIS posts in the developing world, which had to translate "freedom into terms understandable to local peoples who are in varying degrees of social development." An uncomfortable fact further complicated matters: many countries in the free world camp were not particularly free. U.S. foreign policy in the third world rested on close alliances with right-wing dictatorships that adopted staunch anticommunist policies even as they suppressed freedom and human rights in their own countries. As one OCB document candidly admitted, "It must be recognized that many countries outside the Soviet orbit are indifferent to, or actively reject, our concept of freedom. At the same time not a few of these are solidly anti-Communist, are allies bound to us by treaties important to our security, and are supported by their people." In pushing the ideology of freedom, U.S. psychological strategists ran the risk of subverting established governments allied with the United States.[8]

The very premise of the program, that "freedom" constituted an ideology antithetical to communism, seemed illogical to many public affairs officers who were asked to comment on the OCB's program. The U.S. Embassy in Sweden commented that freedom "is less an ideology than a state of being." The basic weakness in the program, the embassy continued, was that it promoted the idea that "'freedom' is an ideology in the same sense that Communism is an ideology." Officials in Tehran agreed: "Freedom cannot reasonably be presented as a system of political, economic, and social thought or doctrine alternative and antithetical to communism.

The phrase 'Freedom versus Communism' is a useful slogan in dealing with undiscriminating audiences . . . if freedom is construed as 'freedom from communism'—a view which the leaders of intellectual and official life abroad, to whom the program is directed . . . would be unlikely to accept." In many countries, as the U.S. embassy in Saigon noted, "principles of freedom will meet with less response than appeals for peace, security, and economic stability." Moreover, as the embassy in Tehran added, "the freedom that each people desires is the freedom to be itself."[9]

Faced with such philosophical and bureaucratic opposition, the OCB backed away from its initial idea of creating a systematic body of dogma that could be used as a unifying free world ideology. The board had wanted to find or create a single text that could rival the testaments of Lenin and Stalin—not to mention the widely available *Short History of the Communist Party*—but this proved unworkable. So the OCB sought to keep the ideological warfare program alive by expanding its notion of ideology. "It is not suggested that there can be set forth a detailed dogmatic Free World doctrine which is applicable universally," the board conceded half-heartedly, but "there are common denominators whose presence is essential to the existence and stability of conditions in which peace and the satisfaction of basic human values can flourish." Democratic freedom, peaceful change, and individual rights were universal concepts, the OCB argued. The ideological warfare program would seek to inspire and disseminate works that could convince intellectuals of two related propositions: first, that their aspirations could "best be achieved under a free society"; and second, that these aspirations would be "endangered and destroyed by the acceptance or even the compromise with the fallacious and essentially destructive principles of Communism."[10]

The resulting OCB program expanded government involvement in intellectual spheres of activity. It disseminated officially sanctioned ideas abroad, and it provided a mechanism for integrating "ideological" themes into a wide range of government and private activities. The OCB ordered all agencies of the executive branch, even those having primarily domestic functions, to contribute to the program. It instructed officials throughout the government "to include as much ideological content as possible" in their activities. Government agencies used overt and covert means to stimulate the production and publication of books and articles meeting the psychological warfare objectives of the program. They used personal and organizational contacts to spread doctrinal materials directly to target groups and to flood the market with ideological works through a network of private publishers, nonprofit foundations, educational institutions, and front organizations.[11]

To be effective, the doctrinal warfare campaign had to be conducted in secret. American psychological warfare planners, as we have seen, preferred to disseminate propaganda materials so that target audiences perceived them as objective sources untainted by government manipulation. Accordingly, the OCB sought to avoid the perception of a "heavy handed indoctrination of other peoples in the American way of life" and "the plausible charge that it is itself undertaking any form of censorship or propaganda." The USIA stressed that in implementing the ideological program, operators should neither rely on "merely news or simple information, nor the direct frontal attack upon the Communist system." Instead, they should "aim by the use of materials, to a considerable extent non-attributable, which are rational, analytical, interpretive, discursive, [and] expository." Information officers in the field should work through local groups, such as publishers and "unemployed intellectuals," to spread the ideology of freedom. The government would camouflage its role, allowing private groups to assume center stage in the campaign of ideological warfare.[12]

Books: The Most Important Weapon of Propaganda

The ideological warfare program prioritized two methods for reaching the world's intellectuals: stimulating the publication of books, and improving mechanisms for distributing them abroad. The early PSB plan articulated the rationale for using books as weapons of propaganda: "In most parts of the world, the radio and television are still novelties; magazines have low circulation; and newspapers circulate mostly among political groups whose opinions are already formed. In the majority of countries in the free world, books—permanent literature—are by far the most powerful means of influencing the attitudes of intellectuals." More needed to be done "to promote the publication and circulation of books favorable to the free world and damaging to the Communist Party line."[13]

Books had been used as propaganda tools by the United States since World War II, when the Office of War Information (OWI) developed a large program for circulating abroad American books in English and in translation. By the end of the war, OWI had also established forty-four international libraries containing a wide range of books on the history, life, and thought of the United States. With the onset of the Cold War and the stepped-up propaganda offensive under the "Campaign of Truth," the

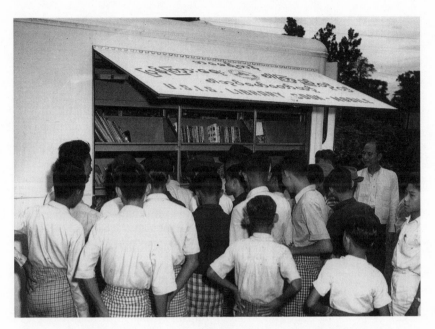

Burmese students examine a U.S. Information Service bookmobile. Traveling libraries such as this one brought American books to remote villages and small towns. Source: National Archives.

libraries were dubbed "information centers." They were charged explicitly with combating anti-Americanism, furthering U.S. foreign policy goals, and exposing "the fallacy of . . . Communist doctrine." The information centers, the largest of which housed between 10,000 and 15,000 volumes, also functioned as reading rooms, concert halls, galleries, theaters, and lecture halls. Between 1945 and 1955 the number of U.S. information centers more than quadrupled to 197 centers in 65 countries, reportedly accommodating some 23 million visitors. Sometimes referred to as "America Houses" or "propaganda shops," the U.S. information centers were described in 1961 by one sympathetic author as "the most effective single part of our entire information operation."[14]

Just as the Cold War book and library programs were gaining momentum, they fell victim to Senator McCarthy's rampages. After his attack on alleged communist sympathizers in the State Department and the Voice of America, McCarthy turned his attention to the USIS libraries. In April 1953, two of his aides, Roy Cohn and G. David Schine, embarked on a "clean-up expedition." They toured seven USIS libraries in Europe and announced that they had found 30,000 procommunist books in the stacks. In the ensuing scandal, books by communists and "fellow travelers" were purged from the libraries and in some cases burned. The State Department issued strict orders banning anything remotely controversial from the information centers. The USIA removed from its shelves books by authors who refused to testify before the House Un-American Activities Committee, writers overly critical of the United States, and members of the Communist Party. Literary victims included Thomas Paine, Albert Einstein, Helen Keller, Henry David Thoreau, Charles Beard, W. E. B. DuBois, Jean-Paul Sartre, Ernest Hemingway, Arthur Miller, and Upton Sinclair. Before McCarthy's onslaught, approximately 120,000 books had been sent to the U.S. information centers each month. This number fell to less than 400 at the height of these investigations. McCarthy destroyed, virtually overnight, the liberal image of the United States and the goodwill fostered by the information centers.[15]

The doctrinal warfare program provided a means of revitalizing these badly battered book and library programs. Its focus on "ideological" themes—prodemocracy and anticommunism—shielded it from Congressional critics, reestablishing books as critical weapons in the war of ideas. The OCB, however, took the government's book programs to a whole new level. Rather than merely establishing reading rooms and libraries abroad, government agents actively manipulated the publishing market by subsidizing the authorship and distribution of books selected for their content.

The OCB and USIA established guidelines spelling out what kinds of books were desirable, emphasizing a seemingly neutral, objective, and scientific approach to camouflage propaganda motives. Books and periodicals "couched in the characteristic methodology of the target types and accepted as reliable by them" would be more effective than emotional or brazenly propagandistic tracts. "Target audiences," the USIA explained, "must be convinced by reason, not assaulted by propaganda." This meant prioritizing scholarly output that "critically and objectively" analyzed communist and free world ideas and practices. Although objectivity in tone was desirable, objectivity in substance was another story. USIA policy gave highest priority to books that had wide "program value." The OCB explained that doctrinal warfare materials should highlight the "discrepancies between totalitarian theory and practice" and how "the application of totalitarian philosophies and practice would adversely affect the life of each individual in the target audience." In addition, works should "strengthen and intensify the belief in and support for the fundamental principles which characterize a free society." Literature conveying a favorable impression of American thought, life, and culture also had value.[16]

The OCB plan called for the imaginative use of government resources and contacts to stimulate the publication of works that filled "gaps" in the "ideological literature." Toward this end, the USIA established a special office for the express purpose of commissioning manuscripts that defended aspects of U.S. foreign policy or criticized communism and communist countries. The agency published many manuscripts, but unfortunately the declassified documentary record does not indicate which books, or how many of them, were commissioned by the agency. In 1965, the only year for which figures and details are presently available, the agency commissioned forty-three manuscripts. Undoubtedly many more were commissioned in the 1950s, as USIA correspondence files reveal that the stimulation of books was routine practice of USIS posts abroad. In many of the roughly ninety countries with a USIS post, information officers managed to stimulate the publication of a book or two every year. Within the United States, government officials from the USIA, CIA, the State Department, and the Defense Department used contacts with foundations and think tanks to encourage the production of books and articles serving "program needs." They also provided grants and other financial incentives to authors and publishers that produced materials useful to American foreign policy. The publisher Frederick Praeger, for example, admitted in 1967 to printing books at the behest of the CIA. He explained to the *New York Times* (after these activities had been exposed) that he had also published books "following the suggestion of 'somebody connected with the Pentagon' or as a result of funding by the United States Information Agency." Most of the publishing, which took place during the late 1950s, "dealt fundamentally with facts, history and analysis of events of Communist-bloc countries or of nations susceptible of a fall to Communism."[17]

The government's financial resources gave these agencies leverage to influence the types of works written and published. So far as the evidence indicates, however, they did not interfere directly with the content of the publications produced at their behest. Government officials generally contacted those individuals who shared their views and who were eager to contribute their ideas and talents to the anticommunist crusade. Still, the "stimulation of ideological materials" carried domestic propaganda implications because it endeavored to influence the types of analyses produced by American authors and published by American presses. The OCB rationalized that it was not engaging in domestic propaganda because overseas audiences were the primary targets; it was merely helping free individuals to write what they would have written anyway, and U.S. market forces would determine the influence of works published under the program. An early draft of the OCB plan reasoned, "Doctrinal warfare is not the 'propagandizing' of the American people, but it is the practical effort to develop materials *which will be acceptable to the American market* and which, in their entirety, or with minor modifications, can be distributed overseas to have a doctrinal impact." Perhaps the board protested too much. It may not have been targeting the American people per se, but its argument fails to convince. The very reason the OCB subsidized authors and publishing houses was to fund works that might not receive wide distribution. Works that might not survive the "invisible hand" of the market received an invisible nudge from U.S. psychological warfare experts.[18]

Increasing the Flow of Books Abroad

Books were considered sufficiently important to U.S. psychological warfare objectives that in February 1954 the OCB created another ad hoc working group to study ways to use books, publications, and libraries to strengthen ties between the United States and leadership groups in other countries. The working group submitted a classified sixty-four-page report to the OCB four months later. It advocated an all-out effort. "The opportunities to use books and publications abroad are of major importance to U.S. objectives," the report announced. "American world leadership, the quality of American achievements in scientific, professional, technical, and cultural fields, and the pressing need to reflect this leadership and quality of achievement throughout the world, warrant the greatest possible effort to expand the use of American books throughout the world in the present half-century." The board called for "collaborative private efforts on a major scale," stating that "a vital union of personal contacts, organization, and action, thus enhancing both distribution and follow-up, must be achieved by anti-Communist forces if the challenge is to be met."[19]

To promote the foreign sale and distribution of American books, the OCB expanded existing government publication programs and encouraged the creation of new ones. There were several mechanisms for this purpose. The Informational Media Guaranty (IMG) program, introduced in 1949, overcame currency convertibility problems by providing reimbursements to U.S. publishers that sold books to countries short on dollars. The USIA created a similar program in 1953, the Book Export Guarantee program. It

Indian students read American magazines at the U.S. Information Service library in Bombay, India. Many of the magazines that were sent overseas were donated by corporations, private groups, and ordinary Americans. Source: National Archives.

subsidized publishers that agreed to sell books making a "positive program contribution" at substantially reduced prices. A study at the end of the decade estimated that from 1953 to 1959, the USIA issued more than $53 million in export guarantee contracts to American publishers and motion-picture distributors. In the same period, the Mutual Security Agency furnished an additional $40 million. These funds provided incentives for publishers to sell American books abroad, and it gave U.S. psychological warfare experts leverage to influence which books reached readers overseas.

To make American books available to students and low-income groups in the third world, the Information Agency launched a "low-priced book program" in January 1956. It made anticommunist and pro-American books more affordable by offering special contracts to major paperback publishing companies to print and distribute books selected by the USIA. Retailers overseas sold the subsidized publications for around ten cents each. The agency set the prices low enough to be affordable to its target groups, but high enough to dispel suspicions of government manipulation of the market. "Give-aways or near give-aways," the OCB noted, "spotlight U.S. attribution which we wish to avoid." The board also did not want to interfere or compete with foreign publishing industries, which the USIA was courting to promote the sale of other American books. During the first year of the low-priced book program, over 7 million books were produced. Of these, approximately 3.3 million were published in English, the rest in eighteen different languages. By 1959, the USIA's subsidy program had assisted the publication of 1.7 billion books, most of them for sale in the developing world.[20]

A book translation program offered subsidies to contractors that printed and distributed American books in foreign languages. In 1957, the USIA published approximately eighty titles under the program. Among the most widely distributed books were *Selections from Emerson*, *The Good Earth* by Pearl S. Buck, *Abraham Lincoln* by Lord Charnwood, and a "Classics of Democracy" series containing works by Thoreau, Jefferson, Hamilton, and other American political theorists. In Japan in 1956, the USIA published translations of twenty-five different titles, totaling 625,000 books. This was in addition to the 240,000 books published under other USIA programs for the country. The same year in India, the agency subsidized the publication of nearly 2 million books in six different languages. In Thailand, agency programs translated twenty books per year. Of these, typically seven titles illustrated "U.S. strength and leadership of the Free World toward peace, freedom, and justice"; six had "anti-Communist themes"; and one recounted "the long history of Thai-American friendship." The books were sold throughout Thailand. USIA officers also donated them to libraries and gave them as gifts to Thai military and government leaders.[21]

Quasi-governmental front organizations conducted a substantial volume of USIA publishing. Franklin Publications, a purportedly private corporation funded by the agency, was one of the USIA's most important outlets for distributing American books in translation. The company was chartered in June 1952 to provide American books to the Middle East. USIA subsidies allowed it to expand and open offices around the world. By the end of 1960, Franklin Publications had printed some 10.5 million books. By 1965, it had helped to translate and to publish 43 million copies of 2,500 different titles. The books were sold commercially in countries where Franklin Books operated. Often, USIS officers worked closely with local booksellers to expedite their sale.[22]

A project in Tunisia was fairly typical. Through a complicated series of transactions involving several commercial booksellers from different countries, USIS officials helped to import American books in Arabic and in French. The process began with Franklin Publishers selling its USIA-funded books in Arabic to distributors in Cairo, Damascus, and Beirut. USIS officials in Middle Eastern posts, in turn, subsidized these distributors by buying many of the books and providing them to Tunisian booksellers. The Tunisian booksellers received a portion of the profits for their efforts. The remaining profits were deposited into a USIS account and used to purchase books from a distributor in Paris, which were imported into Tunisia. The whole process masked USIS involvement while subsidizing cooperating booksellers and increasing the number of American books available in translation. Without the USIS purchasing role, Tunisian booksellers would have had a difficult time overcoming the many fiscal and regulatory problems connected with importing books from other countries.[23]

Although Franklin Publication's identification with the government was not a classified secret, it was not common knowledge. Few consumers would have known that Franklin books were essentially USIA products. The USIA, the CIA, and the Defense Department maintained similar arrangements with Arlington Press, Overseas Editions, and the Chekhov Publishing Company. Such methods capitalized on commercial distribution networks and created the illusion to consumers that the books were privately published, rather than funded by government propagandists.[24]

Another widely used method for sending books abroad involved private organizations. Operating under OCB guidance, various government agencies worked to get professional groups, labor unions, and educational institutions to develop book-donation programs. Many donation projects were instituted by foundations as a result of suggestions made by the USIA, or through government officials' personal and professional affiliations. The USIA granted financial or shipping assistance to NGOs that donated American publications. The agency also formed book-of-the-month clubs overseas, a strategy that ensured continuous distribution of books to carefully selected targets. USIA further encouraged various American universities to establish small reading rooms of approximately 500 books in comparable institutions in foreign countries.[25]

In December 1954 the USIA developed a program, in cooperation with the nonprofit charitable foundation CARE, to distribute a collection of paperback books about American life and culture to schools and libraries in foreign countries. The USIA provided CARE with "guidance and suggestions . . . in particularly strategic areas." CARE's "American Bookshelf" assembled books of poetry, fiction, and literature, as well as books on science, technology, philosophy, history, and political science. It included nonfiction such as Alexis de Toqueville's *Democracy in America* and Dwight

D. Eisenhower's *Crusade in Europe*, and such classics of American literature as *Red Badge of Courage, The Great Gatsby, The Scarlet Letter, A Farewell to Arms, Moby-Dick*, and *The Adventures of Tom Sawyer.* Books in the collection included volumes explaining American history and the U.S. system of government, as well as a few anticommunist books such as David Shub's biography of Lenin, the *Dynamics of Soviet Society* by Walt W. Rostow, and *This is Russia Uncensored* by Edmund Stevens. Not surprisingly, it also included George Orwell's famous *1984*, a book also widely distributed by British propagandists.

To promote the program, USIA and CARE launched a countrywide PR campaign to encourage Americans and NGOs to purchase these portable libraries for distribution abroad. In a press release highlighting the importance of private cooperation in this program, USIA director Theodore Streibert explained that "the books included in the American bookshelf are potent weapons in the war for the hearts and minds of men." He called for the "cooperation and active assistance of Americans" to help "convey the truth about our country to the peoples of other lands." The executive director of CARE added, "Our own good books are the best answer to the flood of anti-American propaganda that distorts the facts about our country, our way of life and our culture."[26]

The activities of the United States Book Exchange Inc. (USBE) further illustrate the tangled web of state-private cooperation that marked these operations. The Book Exchange operated an extensive gift program to distribute books to foreign individuals and institutions. The USBE received most of its books from donations. Some came from publishing companies and libraries. Other books arrived from community book drives promoted by the People-to-People program, the Advertising Council, and civic organizations acting on their own initiative. The USIA's Office of Private Cooperation alone secured 1.5 million donations annually. Although USBE was a nonprofit, nongovernmental organization, the USIA paid operational costs for the USBE gift program. The agency also selected recipients for the donated books and determined which books were suitable for overseas distribution and which were not. The CARE foundation shipped the books overseas, receiving in turn compensation for shipping costs from the Foreign Operations Administration. A private facade camouflaged government planning and financial support.[27]

The CIA supplemented these efforts through its own extensive publishing operations. The OCB directed the CIA to "give high and continuing priority to all activities" supporting the doctrinal warfare campaign. Unfortunately, details regarding the agency's specific actions to support the doctrinal warfare campaign remain classified. Regardless, books and publications were clearly high-priority covert operations. A chief of the CIA's covert action staff wrote, "Books differ from all other propaganda media primarily because one single book can significantly change the reader's attitude and action to an extent unmatched by the impact of any other single medium [such as to] make books the most important weapon of strategic (long-range) propaganda." By 1977 the CIA had aided the publication of at least a thousand books.[28]

In addition to explicitly anticommunist publications, the agency widely supported authors and publications representing the noncommunist left. Official CIA historian Michael Warner described this tactic of supporting a liberal alternative to communism as the theoretical foundation of the agency's political operations against communism. Through its covert sponsorship of the Congress for Cultural Freedom and a web of front organizations, the CIA aided the publication of literary and political journals such as *Encounter* and the *Partisan Review*, and placed articles in such diverse publications as *Foreign Affairs, Times Literary Supplement, Encyclopedia Britannica*, and Fodor's travel guides. Many CIA publication projects were funded by large sums of money funneled through philanthropic organizations such as the Ford, Rockefeller, and Carnegie foundations. This tactic allowed the CIA to finance "a seemingly limitless range of covert action programs affecting youth groups, labor unions, universities, publishing houses, and other private institutions" without alerting recipients to the source.[29]

According to a CIA official, the agency ran its clandestine book program according to the following principles:

Get books published or distributed abroad without revealing any U.S. influence, by covertly subsidizing foreign publications or booksellers. Get books published which should not be "contaminated" by any overt tie-in with the U.S. government, especially if the position of the author is "delicate." Get books published for operational reasons, regardless of commercial viability. Initiate and subsidize indigenous national or international organizations for book publishing or distributing purposes. Stimulate the writing of politically significant books by unknown foreign authors—either by directly subsidizing the author, if covert contact is feasible, or indirectly, through literary agents or publishers.

These guidelines ostensibly directed the CIA's book publishing operations, but they also applied to a host of other publication projects initiated by agencies outside the world of secret intelligence. Other than the CIA's publishing operations, most of the government's ideological and book programs were not, strictly speaking, "covert operations," but most were not openly identified with the U.S. government either. Information about these operations was classified as secret, and some of it remains so. Despite the passage of time and this author's Freedom of Information Act requests, details pertaining to which books were published at the behest of the USIA remain unavailable. The most visible activities of the government were of course the U.S. information centers, which, as a result of the doctrinal warfare campaign, increasingly stocked books extolling the virtues of freedom and exposing the evils of communism. The angry mobs that routinely expressed their hostility to American propagandists by ransacking these centers highlighted the importance of the "camouflaged" approach. These rampages aside, U.S. efforts to increase the flow of American books abroad clearly met with some success. By 1964, U.S. book output exceeded that of Great Britain, making the United States the number one exporter of books in the world.[30]

Educational Exchange

The psychological warfare planners who devised the ideological program envisioned a total assault, an all-out effort across a wide range of intellectual endeavors. The OCB's plan specified that "ideological emphasis" should be added to all government activities involving contact with educated elites. Exchange of persons' programs received special scrutiny by the board. When properly run, exchange programs provided a means of directly reaching the most influential individuals in foreign societies, nurturing official and unofficial contacts between leaders in military, economic, political, and intellectual fields. Many propaganda experts acknowledged that exchange programs were the most effective instruments for extending American influence abroad, even though they were the least "propagandistic." One study of U.S. information efforts noted: "Exchange of persons is one of the most important elements in the whole field of international education. . . . It is a massive, concentrated and highly expensive method of modifying the competence and/or general attitude of an individual foreigner. Experience has proven that, when properly handled, it can have powerful psychological effects." Exchange programs were especially vital in winning over elites in the third world, providing as they did mechanisms for direct contact with the individuals who would determine the social and political fate of these newly independent countries.[31]

The government operated numerous such programs for bringing foreigners to the United States and sending Americans abroad. All had the underlying objective of advancing the foreign policy of the United States. The State Department administered Fulbright exchanges, which sent American scholars to teach and study abroad, and brought foreign scholars to the United States. Most Fulbright funds were spent on academic exchanges, but nearly one-quarter of these expenditures were used for "leader grants" to bring influential foreigners to the United States for training, education, or travel. The International Cooperation Administration (ICA) operated a much larger program for international education. It sponsored technical training, predominantly in the third world, designed to "contribute primarily to the balanced and integrated development of economic resources and productive capacities of underdeveloped areas." ICA programs nurtured close contacts with economic elites abroad, including business persons, engineers, and labor leaders. Foreign military officers, who often exercised power in developing countries, were reached by Defense Department military-training programs. All told, funds expended on these exchanges far exceeded any budget the USIA received in the 1950s. At the end of the decade, the United States annually spent $30 million on Fulbright exchanges, $150 million on ICA technical assistance, and $88 million on military training.[32]

The OCB's plan for doctrinal warfare called for "ideological emphasis" in all these exchange programs, as well as in forums sponsored by both governmental and private groups. Such efforts received support from the highest levels of the government. In January 1954, Vice President Nixon returned from a tour of the Far East and recommended a broadened educational exchange program there as a way of counteracting communist influence in the region. He instructed the OCB to study the problem. In March, an OCB working group submitted its report on "Expanded Educational Exchange Program in the Far East," a study that emphasized the importance of educational exchange programs as instruments of American propaganda. According to the report, the basic objective of U.S. exchange programs was "to build a receptive climate of public opinion overseas in which the actions and policies of the U.S. can be correctly interpreted, coordination and solidarity between free peoples can be enhanced, and distorted views of American life and motives can be corrected." Nineteen

This photograph of President Eisenhower seated in the White House Rose Garden with foreign exchange students was used in a USIA feature story highlighting Eisenhower's efforts to encourage more educational and cultural exchanges between the United States and other countries. Source: National Archives.

pages of proposals articulated suggestions for expanding and improving these efforts so that they better served U.S. political purposes. Of particular note, the board called for doubling U.S. expenditures for exchanges and it advocated "strenuous efforts" to enlist private sponsorship for additional educational exchanges.[33]

The working group's conclusions about the importance of educational exchanges to U.S. information work led the OCB to commission an even broader study on all government and private exchange activities. A consultant to the OCB, Harold Hoskins, prepared the final report and delivered it to the OCB in October. Hoskins stressed that the government should focus on achieving ideological and political impact: "much more could be done through . . . the overseas education programs toward the attainment of ideological objectives." The United States should go beyond "mere technical training," he advised. All international educational programs should include operations "intended to make evident the basic principles of free world ideology, to contrast the American way of life with that of the Communist-dominated world and to provide material for arguments with which to counter those of Communism." The Hoskins study particularly emphasized effective use of private overseas organizations. It called for the development of a long-range program to coordinate public and private efforts in the educational field. Toward this end, the OCB created an interdepartmental committee on overseas education to develop "suggestions with respect to the stimulation and utilization of private efforts in this field."[34]

On the basis of the recommendations of the Hoskins study and the Ideological Working Group, educational exchange programs increasingly emphasized subtle indoctrination in the workings of liberty. The State "Department added "ideological content" to the exchanges it administered. Visiting scholars were required to attend orientation programs with officials who explained to them the fundamental principles of American society. Americans traveling overseas likewise received "training, briefing or indoctrination" in support of the ideological campaign. Leaders and specialists sent abroad under State Department grants were brought to Washington for briefings on the ideological program by Foreign Service officers and USIA officials. The Fulbright program also became colored with ideological warfare overtones. The OCB indicated that Fulbright scholars should be selected according to their abilities to contribute to the ideological program and general U.S. foreign policy interests. American Fulbright grant recipients were encouraged to help disseminate American books by donating selections from their personal libraries to institutions abroad. They were also given supplemental grants to purchase books for their research and were encouraged to present these books to host institutions upon departing for home. In addition, USIS officers who managed the program from U.S. embassies used Fulbright recipients as personal contacts for disseminating publications to critical target groups. USIS Egypt reported, for example, that one American Fulbright professor distributed "innumerable" USIS publications to approximately twenty different English-language clubs in local secondary schools.[35]

The OCB especially targeted teachers for ideological indoctrination. The OCB working group on books and publications explained, "If we can get teachers to believe in our approaches, way of life, and educational methods and they enthusiastically support these ideas with their pupils, we can go far toward holding the line against Communism." The State Department managed a "teacher development" program that brought foreign teachers to the United States for a six-month stay at American universities. Upon entering and leaving the country, grant recipients spent several days in Washington receiving an "introduction to American life" and additional "orientation." Under the OCB programs, the administration sent all kinds of American textbooks overseas. It also cultivated contacts with publishing companies for the purpose of injecting prodemocracy and anticommunist themes into the texts. The Foreign Operations Administration provided grants to foreign authors who wrote textbooks that included favorable commentary on American life.

The OCB further called for using English-teaching programs as "vehicle[s] for carrying the basic concepts which we believe." The USIA conducted most of the government's English-language training, principally through the more than 100 binational centers it maintained abroad. These centers formed the core of the English teaching program, providing instruction to 160,000 students annually. The USIA also devoted portions of its media output to English instruction, including news broadcasts in "Special English," based on a vocabulary of 1,250 words. USIS posts abroad consistently reported that these English-language programs were their most popular products. To reach people with only small English vocabularies, the USIA managed a "books in simple English program." The agency translated books on American history and culture into simplified English suitable for persons with English vocabularies ranging from 1,000 to 5,000 words. Many of the translated and abridged books carried "ideological or anti-Communist themes."[36]

USIS instructors routinely incorporated material about American democracy and culture into their language lessons and course materials. In North Africa, for example, the USIS provided English instructors to the Bourguiba School of Language in Tunis, where American professors taught the language to government officials and specially selected recipients of U.S. grants. An American director supervised the school's English-language program. According to the USIS, he used his influence over courses so that "the school discreetly presents and represents American ideas and ideals so that the student absorbs—along with vocabulary and pronunciation—some important understanding of the values we live by." USIS officials in Tunisia and Morocco saw the English-teaching program there as a device for advancing American influence in North Africa. It provided a mechanism for developing a U.S. orientation among the country's elites by providing an attractive alternative to the former colonial language of French.[37]

Throughout the third world, the teaching of English emerged as one of the USIA's top priorities at the end of the decade. Many governments increasingly imposed severe limitations on the overt propaganda activities of the USIA, such as press, motion pictures, and radio. Teaching English, however, was generally tolerated. The USIA noted in a report to the NSC, "English-teaching programs are often acceptable to and sometimes even welcomed by governments which clamp down tightly on media operations. So are various cultural activities if they are not too blatant or showy."[38] The spread of the English language, moreover, contributed to the prestige and influence of the United States. A presidential commission to study U.S. information activities recommended in 1960 a "conscious and sustained effort on the part of the government to promote the acceptance of English as a universal language." Establishing English as the international language would tie third world nations culturally and intellectually to the United States. It was believed, for example, that "One of the greatest long-term opportunities Africa offers is the possibility of making English the lingua franca for Africa south. . . . The psychological impact of having English the lingua franca of the area cannot be overemphasized and every effort should be made to promote this objective." By 1960, the USIA was providing personal language instruction to 175,000 students in 46 countries and special training to an additional 6,344 local teachers of English in 37 countries.[39]

The OCB likewise promoted American Studies programs in foreign schools and universities. The State Department, operating in a "discreet fashion," worked to encourage the establishment of chairs in American studies at overseas universities. It sponsored special American Studies exchanges to build up a corps of foreign specialists who, upon returning to their respective countries, would teach American Studies courses in their own schools. Another tactic involved the creation or subsidization of research institutes abroad. Several of these were operated in Japan, where the USIS conducted a far-reaching effort to influence Japanese intellectuals. The USIS created one such research institute under the cover of a documents service. Operated by a USIS contact, the institute procured and sold reference materials and official documents. Under USIS guidance, the institute added political publications to its catalog in order to secretly distribute unattributed USIS materials. The scheme purported to reach an

The USIA believed that English teaching was an important vehicle for reaching target audiences abroad, as shown by this English-language class in Nicaragua. Teachers frequently integrated propaganda themes into their lectures. Source: National Archives.

expanded audience "under the guise of advertising" by printing abstracts of USIS-supplied documents in the catalog, which was sent to a specified target audience including pundits, educators, military experts, and media outlets.[40]

The Spiritual Roots of Freedom

The OCB's effort to enlist the support of private resources in support of the ideological warfare campaign extended to religious leaders and organizations. The board prescribed the development of programs "to enlist the cooperation of religious leaders, religious movements, etc., of all faiths . . . in promoting the objectives of the Ideological program." In promoting its ideological message, the OCB explained that the U.S. should "stress the spiritual and religious roots of freedom through all appropriate media."[41]

Buddhist monks look at a pictorial display at the entrance to a U.S. Information Center in Phnom-Penh, Cambodia. The Operations Coordinating Board developed a special program targeting Buddhists in Southeast Asia. Source: National Archives.

The OCB ideological working group created a subcommittee on the "religious factor" to coordinate these efforts. The subcommittee developed plans to increase the impact of the ideological program by appealing to religious groups and individuals of all faiths. Few of the subcommittee's actions were publicly associated with the U.S. government. The subcommittee principally worked through existing religious organizations, particularly interfaith groups like the National Conference of Christians and Jews, to present "faith as the sub-soil of freedom." The subcommittee developed "targets of opportunity for religious activities" and discreetly suggested various projects to religious leaders. The OCB stressed that "activities in this area of the religious factor must be developed so that the influence of government is insignificant or unnoticeable and so that religious organizations have the impression that activities are undertaken at their own initiative or on their own suggestion."[42]

The USIA maintained an Office of Religious Information headed by religious studies scholar Dr. Elton Trueblood. Among his many publications, *Declaration of Freedom* (published in 1955) passionately defended the "spiritual roots" of American society. It was distributed to all USIA posts. Trueblood's office functioned "to present overseas the moral and spiritual heritage which is so integral a part of American life" and acted to "counter incessant Communist efforts to represent America as interested only in materialistic matters." USIA distributed a packet of thirty-two books emphasizing the spiritual and religious foundations of freedom to all overseas posts, and it instituted a series of radio programs which emphasized "religion, human dignity, and freedom as among the motivating factors in American life."[43]

To further emphasize the spiritual dimensions of American life, the Office of Religious Information proposed that Eisenhower initiate an "International Day of Prayer for Peace." In order to avoid exposing the initiative as a "propaganda gimmick," the office developed a scheme whereby the project would "originate through the initiative of responsible church leaders." Upon receiving this "spontaneous" request, the president would then endorse the planned day of prayer and encourage religious leaders throughout the world to do so as well. This initiative developed into a plan for the president to casually mention the need for "prayers for peace" at a press conference before the Geneva Summit. Working behind the scenes, the OCB discreetly ensured that the president's casual reference received "enormous support" from religious organizations. Such efforts to impress foreign audiences with the importance of religious faith to American life were furthered by an act of Congress in June 1954, which expanded the Pledge of Allegiance to include the words "one nation under God" and stamped "In God We Trust" on U.S. currency.[44]

In January 1957 the OCB approved an operations plan to use religious themes as a means of influencing Buddhists in Southeast Asia. Because of the sensitivities accompanying government meddling with religious matters, the board directed government operators to be "as inconspicuous as possible." It explained, "Care should be taken to avoid any actions which could be construed as a political or psychological tool. Whatever is done should not appear to Buddhist leaders as a U.S. Government 'program' but rather as friendly gestures designed to be mutually helpful to Buddhist and

American lay and religious groups." The OCB sought to avoid a hostile backlash from the Buddhist "targets" and groups in the United States who might object to U.S. government activities in the religious field.[45]

The OCB's program stressed "people-to-people" activities as being the most feasible means of exerting a favorable influence on the Buddhists. Agencies implementing the OCB plan worked to stimulate a wide array of private activities directed toward Buddhist organizations and individuals in the target areas, including person-to-person contacts between American scholars and their Buddhist counterparts and the development of institutes and research centers for the study of Buddhist languages, literature, and history. The OCB arranged for a "representative group of Americans" to visit Buddhist countries to join in the celebrations for the 2,500th anniversary of Buddha. It also used government contacts to get American church and lay groups to invite Buddhists to speak to religious communities in the United States on their religion. The State Department, meanwhile, drafted a brochure for American tourists to Buddhist countries encouraging them to respect the "sensitivities, prejudices, customs and interests of Buddhists."[46]

The USIA made a concerted effort to publicize American interest in Buddhism. The agency's news services released a blitz of news and feature stories designed to illustrate American respect for the religion. Articles included a feature on a Buddhist wedding stressing the American interest in the ceremony and a story on Chinese Americans in San Francisco who raised money to build a new Buddhist temple. Other stories included "Buddhist Monk Sends Farewell to Americans," "Ambassador Bishop Thanked for Aiding in Return of Buddhist Relics," "Dulles Hails Buddhism Anniversary," and "Hawaiians of Buddhist Faith Celebrate Wesak Week." The USIA managed to get American publications to prepare special issues and articles illustrating the harmony of interests between Americans and Buddhists. The agency then purchased several hundred thousand copies of the publications for distribution as newspaper and magazine reprints, a tactic that shielded the government from criticism of the publications' content.[47]

While the Eisenhower administration recognized the potential of religious themes to counter communist ideological appeal, it was slower to put this understanding into practice when it came to Islam. Remarkably, the USIA provided no training on Muslim religious practices to its staff working in Islamic countries. Neither the Information Agency nor the Operations Coordinating Board provided guidance to U.S. personnel explaining how they should address Muslim audiences. A May 1957 OCB report

314 Chapter Nine

provided a damning admission. American personnel "have tended to rely on English-speaking, Western-educated intellectuals and to believe that these locals, and all others, reason and act much as they do. Few have any idea of the role of Islam in life and society, and they are unaware of the relationship between Islam and the present currents of nationalism and anti-foreignism."

Muslim leaders, schools, and organizations were on "mailing lists" that received USIS propaganda materials, but little of these materials expressly related to Islam. Indeed, only a few USIS activities dealt specifically with the religion. On a limited scale, USIS posts distributed pamphlets highlighting communist hostility to Islam. USIS Indonesia produced pamphlets on "Islam in the Soviet Union" and "Why a Moslem Must Reject Communism." The post disguised its role in producing these pamphlets by attributing them to the Muslim Missionary Association. USIS Iran distributed similar unattributed pamphlets, "Soviet Trained Imams are Infiltrating Islam" and "Soviet Pamphlet on Islam Denounced as Evil." In Pakistan, the USIS placed articles in newspapers on such subjects as "Genocide of Muslims in the USSR" and "Muslims in Communist China." The Voice of America, which broadcast for nine hours in Arabic, aired a sixteen-program series on "Islam and Communism," which highlighted communist repression of Muslims. The VOA also sought to attract Muslim listeners with religious programming. Each of its morning and evening shows in Arabic began with a five-minute reading from the Qur'an. In Pakistan, the Voice of America broadcast Qur'anic readings every Friday and aired special religious programs on Muslim holidays. Generally, however, U.S. efforts to appeal to Muslims *as Muslims* were restrained. The OCB acknowledged that, in the Middle East especially, this was a missed opportunity: "a powerful, regionally planned effort, to expose the incompatibility between Islam and communism is long overdue and is capable of producing far-reaching results."[48]

Militant Liberty

Psywar planners in the Defense Department joined the effort to use ideology and spirituality as foreign policy tools. They developed an unusual plan, known as "Militant Liberty," to create freedom cadres—elite evangelists who would aggressively propagate the concept of liberty in their societies. They would be missionaries for freedom. They would also be selected, organized, and trained by the American military.

Colonel John C. Broger, a consultant to the Joint Chiefs of Staff (JCS), developed the program. An evangelical Christian with powerful missionary instincts, Broger had an intriguing history. He had produced soap operas in New York before World War II. Subsequently he served as a warrant officer on an aircraft carrier in the Pacific theater. After the war, he teamed up with an evangelical pastor and a gospel singer to form the Far East Broadcasting Corporation (FEBC), a nonprofit, interdenominational radio station created to broadcast Christian messages and Bible studies to mainland China. The FEBC began transmitting in 1948 from a station in Manila with the prayer, "all hail the power of Jesus' name." The FEBC expanded rapidly. By the 1970s it had thirty-one transmitters broadcasting the Christian message in dozens of languages to Asia, Russia, Africa, the Middle East, and Latin America.[49]

A mixture of religious faith and anticommunism motivated Broger's international evangelism. In his view, religious faith could provide an antidote to communism's poisonous teachings. Communism was "not only a political party, not only a concept of a theory or economy," he declared in a speech in 1951. It was "akin to extreme religious fanaticism, for how else could Communism gain in its converts such tenacity of purpose and such complete mastery over every realm of mental, political, moral, economic and spiritual life?" The war against communism was a spiritual war, he argued. It could not be won "wholly on the basis of military manpower or production potential, gun for gun, ship for ship, propaganda for propaganda." Only "godly precepts and principles" could counter the appeal of communist ideology. Any plan to fight communism "must find its source of strength and inspiration in godly righteousness," Broger declared.[50]

Broger's evangelism and anticommunism caught the attention of the JCS chairman, Admiral Arthur W. Radford. Radford, a deeply religious Presbyterian, was impressed with Broger's "missionary zeal for spreading Christian and American principles in the Far East." Perceiving that Broger's talents might be useful in providing "spiritual stiffening" to American troops involved in the fight against communism, he hired Broger to plan troop indoctrination programs for the JCS. In this capacity, Broger developed the Militant Liberty plan, a curious project that mixed fervent anticommunism, religious zealotry, and pseudoscientific methodologies.

According to Broger, a decline in spiritual values—not nationalism or anticolonialism—explained the growing appeal of communism to people in the third world. "Most religions of these areas have lost their appeal," Broger announced in a draft of his Militant Liberty plan. "Traditional cultures have become unable to furnish an acceptable comprehension of existence in the world of today." In such a situation of "ferment and unrest," peoples in the third world were likely to turn to "a new faith, militantly propagated by articulate natives." Militant Liberty proposed to fill that void. In paternalistic language reminiscent of imperialist ideologies, Broger declared that in "the underdeveloped areas of the world the people must be furnished with a new conception of the meaning of life. This meaning will be furnished by militant Communism unless the United State takes action to propagate a stronger ideology."[51]

In Broger's analysis, two philosophic extremes concerning the individual conscience distinguished communism and freedom. According to Broger, "The least common denominator objective of Communist philosophy as it affects the individual is, first, to neutralize the conscience of the individual and then systematically annihilate it. As the conscience of the individual is annihilated, it is essential to superimpose the 'class conscience' or the 'conscience of the masses.'" In contrast, the free world rested upon a belief in the "dignity of man" and the "sensitive, conscientious actions of the individual."

On the basis of this reasoning, Broger developed a pseudoscientific method for calculating the orientation of a particular society toward freedom. Each nation would be evaluated according to six criteria: discipline, religion, economic order, civics, education, and social order. Calculated by an unknown methodology, a nation's score in each of these areas determined whether it was moving upward, toward an "evolution for freedom," or downward, toward a "revolution of aggression." The upward direction led to "individual liberty" and "responsible behavior through a sensitive conscience." The downward direction led to the "authoritarian state" that controlled the "annihilated conscience" of the individual. A half-dozen charts and graphs explained this process. All of them pointed to the fact that communism destroyed the integrity of individual choice and responsibility. One of charts evaluated the development of individual consciences on a scale of 0 to 100. It ranked the USA at around 75 on this scale of unknown units, just above England, France, and Germany. Down the scale hovered Italy, Japan, and India at between 25 and 0 units. Soviet communism ranked negatively on the scale at −100 units.[52]

Militant Liberty therefore endeavored to inculcate values of individualism and personal responsibility to weaken communal values and tendencies, which Broger perceived as leading to communism. Militant Liberty proposed to develop in target audiences "sensitive individual conscience[s]"

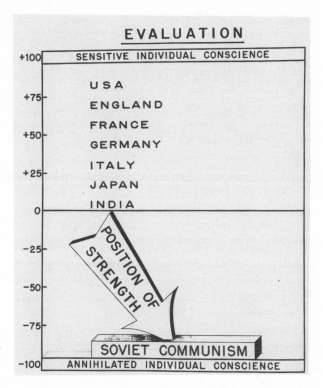

This chart accompanying the Militant Liberty plan evaluates the development
of "sensitive individual consciences" among the peoples of various nations.
According to the plan, communism annihilates the consciences of individuals.
Source: Eisenhower Library.

as foils against communist ideology that produced "annihilated individual
conscience[s]." It would do so by selecting, training, and motivating indigenous leaders to "teach and inspire their people to action in accordance
with the concept and ideals of freedom and the true worth of individual
man." Through them, the United States would "aggressively" propagate
the ideas of freedom and individual responsibility and motivate people "to
be *militant* in their belief of *liberty*."

The motivational aspect of Militant Liberty differentiated it from
other U.S. information programs, which were more concerned with shaping perceptions and attitudes than inspiring action. Broger proposed to
create cadres of activists "who have a zeal for promulgating the concepts
of liberty." They would receive "training and indoctrination" from U.S.

personnel. This would involve such things as "inspiration and education of such persons in the values and responsibilities of liberty; training them in how to use and apply these concepts in their own local country; how to take political action on the 'face to face' basis with their countrymen . . . and how to expose and defeat Communist agitation and subversion in their own countries." The ideological cadres would be brought "to places like Hawaii," shown "how freedom works there," and returned home to propagate freedom's ideology.[53]

Broger, hoping to get the Militant Liberty program implemented by several government agencies, presented a draft of the program to the OCB in December 1954. The OCB promptly created an ad hoc working group to study the plan. After just four meetings, serious divisions within the group paralyzed the program. The State Department, USIA, and CIA all opposed the "positive indoctrinating" and motivational aspects of the project. These agencies, which took great pains to hide the manipulative aspects of their information programs, apparently felt that the methods of the Militant Liberty project were too blatantly propagandistic. The OCB argued that the core concepts of Militant Liberty unnecessarily duplicated those being implemented under the ideological program. Citing "irreconcilable differences" between Defense and the other agencies, the OCB working group on Militant Liberty disbanded after nine months.

Although the OCB was cool to the plan, the JCS and Defense Department endorsed Militant Liberty. So did Nelson Rockefeller, who used his position as psychological warfare advisor to lobby for international implementation of the project. Accordingly, the Defense Department prepared an operational plan for carrying out the Militant Liberty program in Japan and Latin America. This classified version of Militant Liberty was known as "Project Action." Precise details pertaining to the project's implementation are unavailable due to security restrictions, but some information can be gleaned from a pilot project launched in Ecuador in 1955 by the Defense Department. Under the program, U.S. military missions trained members of the Ecuadorian armed forces in the Militant Liberty project "as a means of hindering the threat of communist penetration." Instructors guided "trainees" through lectures, filmstrips, audio recordings, and discussion sessions on "Civics, Social Order, Education, Religion, Economic Order, [and] Law and Order." Materials included a "Battle for Liberty Kit" containing training aids for "instruction on the free way of life versus the communist way," a text setting forth "rights and responsibilities of a free society," a Q&A sheet listing 110 questions commonly asked about freedom, and a filmstrip entitled *Liberty and Its Responsibilities*, geared to the junior high school level.[54]

The USIA, which had urged the OCB to reject Militant Liberty, nevertheless proceeded to implement a pilot project of its own that used similar concepts. It was called "Bases of Freedom: An Approach to an American Ideology." The agency developed the program partly as a response to the OCB ideological warfare initiative and the JCS's Militant Liberty project. Earl J. Wilson, a senior USIA official, oversaw the effort. He urged the agency to teach democratic concepts more assertively abroad and pushed the USIA to implement a Citizen Education Program (CEP) developed by Columbia University Teacher College. The dean of the college, William Russell, had devised the CEP idea in response to a suggestion Eisenhower had made when he was president of Columbia, urging the faculty to do more to teach democratic concepts to American youth. With Eisenhower's assistance, Russell secured a multimillion-dollar grant from the Carnegie Corporation to conduct an "Experiment in the Teaching of Freedom." Prominent political scientists, historians, and economists formulated a large catalog of reference materials on the "Premises of American Liberty." They also developed laboratory practices for inculcating democratic concepts in young people through the school system. Russell traveled about the country getting schools to implement his citizenship training program before accepting a post with the International Cooperation Agency. He began lobbying for international implementation of the program but died suddenly.

Wilson picked up the slack. At his urging, the USIA implemented three CEP projects in Guatemala, Mexico, and Korea. The agency sent out a special book collection of 100 titles called the Freedom Bookshelf, and it distributed three special books on citizen education: *The Premises of Liberty*, *When Men Are Free*, and *Laboratory Practice Technique Resources for Citizenship*. Secondary school teachers abroad were a key target audience. In Mexico, for example, teachers were reached by a magazine called *Saber*— "to know"—that Wilson developed in concert with a Mexican schoolteacher. With the magazine, the USIA distributed a pamphlet, "Senderos de Libertad" (Paths of liberty), which went out to schoolteachers across the country.[55]

Remarkably, the Militant Liberty concept, originally developed for psychological warfare abroad, was most intensively promoted within the United States. Pentagon officials enthusiastically endorsed the project for domestic implementation, remarking that "militant liberty was as much

needed in this country as it was abroad." Authorized by the OCB to re-
vise Militant Liberty for use by private domestic American organizations,
Broger proceeded with gusto. He arranged a meeting with JCS representatives and Hollywood's John Ford, Merian Cooper, John Wayne, and Ward
Bond. John Wayne reportedly agreed that themes from the Militant Liberty project "would be inserted carefully" in his pictures. John Ford, who
worked for the OSS during the war, likewise showed an interest in the project. He requested copies of the Militant Liberty materials so that he could
pass them on to his script writers. He also asked the JCS for assistance "in
putting Militant Liberty elements" in the movie *The Wings of Eagles*. The
resulting film glorified the efforts of a naval aviator who worked tirelessly
to secure funds from the parsimonious and naively pacifistic Congress in
the years preceding Pearl Harbor. The climactic scene of the film depicted
John Wayne's character struggling to regain his individual freedom after
an accident left him paralyzed from the neck down. Spurred on by sheer
willpower, he regained his movement and became a successful writer of
movies and plays about his naval exploits. Militant Liberty themes were
scarcely perceptible and overshadowed by the signature John Wayne patriotic bombast.[56]

Of greater consequence was a film prepared by a conservative Christian organization at Broger's request. Networking with his fellow evangelicals, Broger contacted the International Christian Leadership (ICL), a
highly secretive organization devoted to creating "a world organization led
by Christ." At Broger's urging, the ICL agreed to finance the production
of a film incorporating his ideas. The resulting *Militant Liberty* film emphasized religious themes to convey its anticommunist message. Broger
described it as "motion picture for training in the principles which under-
lie a Christian society in contrast to the Communist threat which challenges the free way of life." The widely used film mixed the gospel message
with a strong dose of anticommunist propaganda. It was shown to Militant
Liberty trainees abroad and to churches, schools, and clubs in the United
States.[57]

Broger also used his position as vice chairman of the Armed Service
Committee of the People-to-People program as a platform for spreading
Militant Liberty. Constantly on the move, he delivered numerous lectures
on the subject to both civilian and military audiences. He also put Militant
Liberty to work indoctrinating U.S. troops. In 1956, he was appointed
deputy director of Armed Forces Information and Education (AFIE). Four

years later, he was made AFIE's director, a position he held until the mid-1980s. In this capacity, as Anne Loveland notes, Broger "played a central role in the ideological indoctrination of armed forces personnel." The Defense Department implemented Militant Liberty to provide "unified and purposeful guiding precepts for all members of the Armed Forces." Troop indoctrination programs for active duty and reserve personnel presented communism as "the embodiment of evil." Contrasting American spiritual convictions with atheistic communism, Broger's indoctrination effort reflected his belief that patriotism alone provided an insufficient motivating factor for service personnel. A stronger set of ideological convictions was needed to insulate U.S. troops and officers from the seductive promises of communist ideology.

Underlying these programs was the assumption that U.S. security depended upon the active promotion of freedom throughout the world. Broger's Militant Liberty plan admonished, "Since World War II, it has become increasingly apparent that free men and nations, in order to remain free, cannot assume that freedom needs no explanation. The challenge of this present peril to the freedom of individuals and nations must be accepted." These operations were also shaped by a belief that the United States must duplicate communist methods in order to defeat communism. "Communism is a dynamic ideology," Broger warned, and "communist ideology can only be defeated by a stronger *dynamic* ideology."[58]

This call to arms betrayed more than a hint of insecurity that American ideas might not win the colossal ideological contest that was the Cold War. Although American officials believed that peoples in communist states were "forced to accept through regimentation a set of principles and values based on the individuals subservience to the state," they also feared that communist ideology had genuine appeal outside the borders of communist states. The Jackson Committee said as much when it cautioned that communist ideology constituted a "weapon of major importance" in the drive for world domination: "This ideology—despite all the evidence of the realities of life in the Soviet system—still has a significant appeal to many people outside the system. . . . The importance of its intellectual and national appeal, especially in Asia and Africa, should not be underestimated." The American lack of a "universally valid" doctrine was perceived as "a significant Free World weakness."[59]

The ideological programs were an attempt to meet directly these perceived weaknesses. Although the OCB never produced the universal doctrine of freedom it craved, the board did oversee a sizable effort to promote the idea of freedom as the antithesis to communism. Ideology, the OCB reported at the end of 1954, was increasingly recognized by government agencies as a useful instrument in overseas operations. Even as government officials championed the free marketplace of ideas, they took steps to make that marketplace less free than it appeared. As Frank Ninkovich noted in a different context, the United States "was not so much interested in fostering intellectual freedom as in promoting freedom as propaganda—an altogether different proposition."[60]

Chapter 10

The Power of Symbols
Psychological Strategy and the Space Race

The Soviet satellites were a genuine technological triumph, but this was exceeded by their propaganda value. To uninformed peoples in the world, Soviet success in one area led to the belief that Soviet Communism was surging ahead in all types of activity.

— Dwight D. Eisenhower, *The White House Years* (1965)

We cannot permit an image to exist that this is the end of the U.S. Golden Age . . . and the advent of a new, progressive USSR era.

— Karl G. Harr, 1959

The Soviet Union presented the United States with the biggest propaganda coup of the Eisenhower presidency when it revealed on October 4, 1957, that it had successfully launched the first artificial satellite into orbit around the earth. The United States had always been first—first to exploit the splitting of the atom and first to successfully detonate a thermonuclear device—in the Cold War's scientific arena. Sputnik, as the satellite was called, shook worldwide confidence in the one field in which American preeminence was taken for granted: science and technology. Sputnik, a powerful symbol of Soviet technological achievement, was all the more remarkable for what it suggested about the communist system. As so often happened in the battle for hearts and minds, deeds had a greater propaganda impact than words. Here was the Soviet Union—a country barely industrialized three decades earlier, supposedly shackled by totalitarian controls—overtaking the richest, strongest, and freest country on earth. Many commentators expressed doubts about the American way of life—its consumer society, its public education, and its government. Americans were lazy and complacent; they were spending too much time with their barbecues and their cars to worry about the things that really mattered. Perhaps the wave of the future lay with the communists.[1]

323

To many Americans, the military implications of Sputnik were frightening. The satellite was hurled into space atop a giant ballistic missile, theoretically capable of sending nuclear warheads to the continental United States. Sputnik presaged an era of strategic parity and created an unprecedented sense of vulnerability. Soon both superpowers would possess sufficient firepower to assure that an all-out thermonuclear war would result in mutually assured destruction. For the first time in well over a century, the American homeland was directly vulnerable to attack. In what Robert Divine has called the "Sputnik challenge," Americans demanded that the Eisenhower administration demonstrate the superiority of American technology and military power by surpassing the Soviet Union in space exploration and ballistic missile development.

Yet for Eisenhower and most of his advisors, the military ramifications of Sputnik were less worrisome than the psychological ones. Confident that the preponderant power of the United States would deter a Soviet attack, they were more concerned that Sputnik would tip the political and psychological balance of power in the Soviet Union's favor. Policy makers foresaw potentially devastating consequences of spectacular Soviet accomplishments in space. Free world morale, they concluded, would be crushed by perceptions of Soviet military superiority. American allies would be less likely to support policies that risked war; domestic audiences would press for unsafeguarded disarmament; and neutralism would spread around the world. This revolutionary scientific feat also posed an ideological challenge to the United States, with potentially grave consequences for the competition in the third world. By demonstrating the ability of the Soviet system to compete on a technological level with the West, Sputnik would lead peoples in developing countries to believe that communism, rather than capitalism, provided the surest path toward rapid modernization. American officials worried that these nations would increasingly turn to the Soviet Union to advance their national aspirations, thus tipping the world balance of power against the United States. Moreover, by substantiating Soviet claims of scientific and technological superiority over the West, Sputnik enhanced the credibility of Soviet propaganda in other fields. Henceforth it would be much more difficult to convince the world that Soviet claims were spurious. Sputnik, American officials concluded, was a first-rate propaganda victory for the communists that dealt a sharp blow to American prestige. '

Public Relations and the Historiography of Space

Sputnik sent the Eisenhower administration's public relations machinery into high gear. They could not deny that Sputnik was an impressive accomplishment—there was no getting around that. Instead, they denied the very existence of a space race. It was an incredulous claim. Both the Americans and Russians had been working to place satellites into orbit since at least 1955. They were doing so as part of the International Geophysical Year (IGY), a multinational research venture into all kinds of scientific endeavors. After the Sputnik launch, the Eisenhower administration publicly claimed that the United States had never considered itself to be in a race with the Soviet Union. It had not even tried to be first into space. American officials disparaged Sputnik as a propaganda stunt, asserting that the U.S. satellite program, unlike that of the USSR, was a peaceful, scientifically significant endeavor that would make meaningful contributions to the IGY. Sputnik had virtually no scientific instrumentation; it merely transmitted a gimmicky "beep" by radio. The American satellite—not yet launched, of course—was being designed to produce useful scientific information, not to score points before world opinion. According to this official line, the United States was not in the business of exploring outer space for military purposes, but strictly for science.

This public relations stance has exerted a powerful influence on the historiography of outer space and the Eisenhower presidency. Most historians agree that this early American space program, in contrast to the manned space expeditions that followed, was not designed with prestige or propaganda in mind: Eisenhower privileged national security and intelligence objectives over psychological considerations. Unconcerned about the prestige and psychological ramifications of spaceflight, he assigned "indubitable primacy" to ballistic missile programs over space stunts.[2] To the extent that Eisenhower was interested in outer space exploration, this view holds, it was because of the potential intelligence applications of satellite reconnaissance. The president sought to deploy spy satellites to monitor Soviet military developments, and he used the civilian satellite program as cover for these intelligence applications. According to this view, it was not until after Sputnik that psychological factors intruded into the making of U.S. outer space policy. Even then, Eisenhower resolutely resisted turning space exploration into a propaganda contest. Only when pressured by a panic-stricken press corps and opportunistic politicians did Eisenhower place a premium on programs he considered unnecessary from a national security perspective but imperative to promoting American prestige.[3]

As we have seen, however, psychological considerations shaped U.S. foreign policies from the very beginning of the Eisenhower presidency. The fledgling American space program was no exception. From the start, the U.S. scientific satellite program was predicated on maximizing American prestige at the expense of the Soviet's. From the decision to pursue the satellite program to the selection of a launch vehicle for that satellite, psychological considerations permeated U.S. outer space policy. After the Sputnik launch, these efforts were stepped up as the Eisenhower administration scrambled to regain the psychological and propaganda initiative. The space race became more than a competition to demonstrate scientific and technological superiority. At once representative of the economic, cultural, and ideological superiority of the capitalist and communist systems, space exploration became one of the most potent symbols in the struggle for hearts and minds.

Origins of U.S. Space Policy

Scientists and military personnel began investigating the feasibility of launching a "world-circling spaceship" in the aftermath of World War II. Early satellite proposals concentrated primarily on the scientific, intelligence, and military utility of satellite vehicles, but they presciently noted the political and psychological ramifications of spaceflight as well. The Air Force think tank RAND forewarned as early as 1946 that the achievement of a satellite craft "would inflame the imagination of mankind, and would probably produce repercussions in the world comparable to the explosion of the atomic bomb." A subsequent report by Manhattan Project scientist Aristid V. Grosse brought the potential propaganda consequences of a Soviet first launch directly to the top levels of the government. Grosse's farsighted report, presented to Eisenhower's assistant secretary of defense for research and development Donald Quarles in August 1953, warned that because the Soviet Union had trailed the United States in the development of atomic and hydrogen warheads, it might attempt to take the lead in the development of a satellite. Noting that the satellite "would have the enormous advantage of influencing the minds of millions of people the world over," Grosse accurately predicted that the Soviets might forego complicated instrumentation in favor of putting the satellite into orbit ahead of the United States. "If the Soviet Union should accomplish this ahead of us," he warned, "it would be a serious blow to the technical and engineering prestige of America the world over. It would be used by Soviet propaganda for all it's worth."[4]

A number of spaceflight recommendations by prominent scientists and military leaders followed Grosse's report. Of special significance was a study on surprise attack prepared by the Technological Capabilities Panel, headed by James Killian, president of the Massachusetts Institute of Technology. In addition to its important recommendations for accelerating ballistic missile programs and for improving U.S. strategic warning capabilities, the Killian committee's February 1955 report called for the use of artificial satellites for intelligence purposes. Space satellites could penetrate the veil of secrecy surrounding the USSR, providing valuable information about Soviet military capabilities.[5] This recommendation imparted a sense of immediacy to satellite and rocket R&D programs being pursued by the military services. The Air Force expedited the production of its supersecret WS-117L reconnaissance satellite, subsequently renamed Corona under CIA auspices.[6] The Navy and the Army also accelerated their satellite programs: the Navy's Project Vanguard with its Viking booster rocket, and the Army's Project Orbiter (aka Explorer) with its Jupiter ballistic missile.

At about the same time, civilian scientists urged the United States to contribute a scientific satellite to the forthcoming IGY, planned from July 1957 to December 1958, which was to be the largest international cooperative scientific venture in history. Scientists from sixty-seven countries eventually became involved in global research projects in fields such as oceanography, seismology, and solar activity. In October 1954 the U.S. National Committee for the International Geophysical Year (USNC-IGY) proposed the idea of conducting satellite experiments to the Eisenhower administration. It reacted slowly; there was little urgency. The administration did not decide to support the project until May 1955. The primary stimulus for action came when the Soviet Union announced the creation of a spaceflight commission on April 16, 1955. Now American scientists and policy makers reacted swiftly. After the Soviet announcement, Joseph Kaplan of the National Academy of Sciences wrote to the director of the National Science Foundation to urge immediate action on the IGY's satellite proposal. "I should like at this time to dwell briefly on the urgency of this matter," Kaplan wrote. If funds were not forthcoming soon, he warned, it would be virtually impossible for the satellite launch to take place before the end of the IGY in December 1958. To make perfectly clear the importance of acting fast, Kaplan enclosed a copy of the *Washington Post* article reporting the Soviet Space Commission announcement. "The critical shortage of time," he added, "cannot be over-emphasized."[7]

This sentiment was not lost on Eisenhower's advisors. Pressure for a race into space came from the highest levels of the government. On May 17, Nelson Rockefeller, then serving as Eisenhower's psychological warfare guru, sent a memorandum to the NSC calling for immediate action. Rockefeller placed considerable emphasis on beating the Soviets to the launching pad. Rockefeller noted the psychological importance of being first to launch a satellite and warned of the "costly consequences of allowing the Russian initiative to outrun ours through an achievement that will symbolize scientific and technological advancement to peoples everywhere." He implored the NSC to act expediently in order to deny the Soviets an opportunity to deprive the United States of whatever psychological and prestige awards that were to be gained. Keenly sensitive to the political ramifications of spaceflight, he continued emphatically, "The stake of prestige that is involved makes this a race that we cannot afford to lose." Things moved quickly. The next day the U.S. National Committee for the IGY approved the satellite project. Noting far more than the usual scientific, technical, and budgetary considerations, the committee drew attention to the importance of expediting the IGY satellite launch. An attachment to its IGY proposal was labeled "Factors Affecting USNC-IGY Schedule." It mentioned conspicuously: "It is of interest to note that at least one other nation has announced plans for a similar program under the direction of an extremely able physicist." The meaning was clear. The Soviet space program made American success in this field imperative.[8]

The Eisenhower administration's sensitivity to Soviet propaganda charges complicated matters. To win the psychological competition in space, officials believed, the United States needed to do more than simply get there first. Rockefeller advised the NSC that because "vigorous propaganda will be employed to exploit all possible derogatory implications of any American success that may be achieved, it is highly important that the U.S. effort be initiated under auspices that are least vulnerable to effective criticism." He identified other essential psychological parameters for the program. Even though the United States should strive to be first into space, Rockefeller cautioned, the American satellite should also possess sophisticated instrumentation. He feared that first launch by the United States of an uninstrumented satellite would be discounted by the Soviets if they followed it with a more sophisticated satellite of their own. Furthermore, the U.S. should launch its satellite under international auspices in order to ward off any Soviet propaganda billing the device as somehow militaristic or threatening. Rockefeller warned that such a spectacular scientific breakthrough would afford communist propaganda extraordinary opportunities for exploiting "superstitions" and "imputed military hazards." He advised that the United States take steps "to allay the potentially boundless fears that may be stirred up." Finally, he continued, the American project should share the information gleaned from the satellite with the international community in order to enhance its perception as a scientific—and therefore peaceful—project. From a propaganda standpoint, it would also be useful to contrast American openness with Soviet secrecy. In short, to prepare for the inevitable Soviet propaganda onslaught, the American satellite needed to appear international, scientific, and, above all, peaceful.[9]

In line with these recommendations, the NSC established three overarching principles to govern the scientific satellite program. First, the NSC mandated that the IGY satellite not interfere with military programs for the development of ballistic missiles and reconnaissance satellites. Second, it stressed that the United States needed to establish the peaceful purposes of the satellite program, both by launching it under the international auspices of the IGY and by ensuring that it provided meaningful scientific data. Third, the NSC emphasized in no uncertain terms the importance of timeliness. The United States needed to launch its satellite before the Soviets to reap the prestige benefits of being first into space and to demonstrate progress in ballistic missile development.

These parameters were spelled out in NSC 5520, dated May 20, 1955. "Considerable prestige and psychological benefits will accrue to the nation which first is successful in launching a satellite," NSC 5520 announced, noting that Russia's "top scientists" were already at work on the project. Because the satellite would be hurled into space atop a ballistic missile, the document explained, the satellite's relationship to ICBM technology would affect international perceptions of the balance of power and, correspondingly, the resolve and morale of U.S. allies. If the Soviet Union were first to establish a satellite, allies would be less likely to support strong anticommunist policies carrying with them a high risk of war. A crucial passage in NSC 5520 stated this psychological connection explicitly: "The inference of such a demonstration of advanced technology and its unmistakable relationship to intercontinental ballistic missile technology might have important repercussions on the political determination of free world countries to resist Communist threats, especially if the USSR were to be the first to establish a satellite."[10]

The importance attached to the fledgling space program reflected a broader concern of national security strategists: the United States needed to maintain its reputation as the world leader in science and technology. Such a reputation was a vital element in the psychological and political mix that established U.S. leadership in the world. The CIA elaborated on this psychological connection between the superiority of American science, the space program, and American global leadership. In a lengthy commentary on the satellite program, proposed as part of a progress report on the Technological Capabilities Panel's "surprise attack" recommendations, the agency forcefully argued that the United States needed to be first into space for psychological reasons. Noting the "psychological warfare value of launching the first earth satellite," the CIA explained that the revolutionary nature of space travel held important consequences in the realm of public opinion. The agency predicted that the first venture into outer space, like the first atomic explosion, would receive extensive publicity. International public opinion would follow news of the first satellite launch closely. The nation that first accomplished this feat would "gain incalculable prestige and recognition throughout the world."[11]

The CIA continued to explain that the U.S. reputation as the world leader in science and technology was a critical component of American leadership. The image of American superiority was a vital weapon in the battle for hearts and minds:

> The United States' reputation as the scientific and industrial leader of the world has been of immeasurable value in competing against Soviet aims in both neutral and allied states. Since the war the reputation of the United States' scientific community has been sharply challenged by Soviet progress and claims. There is little doubt the Soviet Union would like to surpass our scientific and industrial reputation in order to further her influence over neutralist states and to shake the confidence of states allied with the United States. If the Soviet Union's scientists, technicians and industrialists were apparently to surpass the United States and first explore outer space, her propaganda machine would have sensational and convincing evidence of Soviet superiority.

Because the performance in the scientific competition would have important consequences in the political arena, the United States should endeavor to beat the USSR into space.[12]

But there was an additional factor to be considered, the CIA continued. If the United States successfully launched the first satellite, "the Soviet

Union would attempt to attach hostile motivation to this development in order to cover her own inability to win the race." Thus the agency advised: "To maximize our cold war gain in prestige and to minimize the effectiveness of Soviet accusations, the satellite should be launched in an atmosphere of international good will and common scientific interest." To win the psychological competition the United States not only had to be first to orbit a satellite; it also needed to make a significant contribution to science to ensure that the journey into space was perceived as a boon, rather than a curse, to mankind. The satellite's relationship to the ICBM meant that the United States needed to demonstrate to world audiences that the American effort was peaceful. Just as the frightening implications of the leap from fission to fusion had been mitigated by the publicity surrounding the peaceful atom, so too did the achievement of a ballistic missile capacity require a peaceful facade. Rockets for space exploration would deflect attention from rockets for atomic warfare.[13]

In line with these concerns, the Department of Defense (DOD) selected the Navy's Project Vanguard to launch the scientific satellite for the IGY. Vanguard appeared the best prospect for furthering the peaceful image of the project and beating the Soviets into space, more promising than either the Air Force or Army rockets being developed. The Pentagon ruled out the Air Force's powerful Atlas rocket because it did not want the satellite to interfere with the development of this important ICBM project, and it feared the Atlas would not be operational in time for the IGY anyway. The DOD chose the Navy's Vanguard over the Army's Orbiter for reasons of image: Vanguard was a research rocket, not a weapon. It also possessed more sophisticated instrumentation. Vanguard appeared the most scientific, the least militaristic, and the most likely to beat the Russians into space.[14]

After the selection of a launch vehicle, the Eisenhower administration adopted a public relations stance stressing the scientific and peaceful objectives of the IGY satellite program. As usual, the Operations Coordinating Board (OCB) planned the PR. It prepared a public information program "to derive the maximum psychological advantage" from the U.S. decision to launch earth satellites. To secure the "mantle of benevolent world interest," the OCB stressed that the United States must announce its intention to launch an IGY satellite before the Soviet Union. "The international position of the U.S. (in terms of prestige and morality) will be somewhat damaged by the fact that the program is being implemented by the military rather than by purely scientific agencies," the OCB advised. "It will therefore be necessary to build the information carefully, giving emphasis to the international nature of the origins of the project, the scientific nature of the experiment, and the availability of the results to the international scientific community." Accordingly, the resulting announcement contained no reference to the fact that the DOD was involved in the project. When the administration revealed this information later, U.S. statements stressed that the DOD was playing an auxiliary role in the project. The Navy was merely supplying the research rockets that would propel the satellite into orbit; the project was a scientific endeavor, not a military exercise.[15]

This PR stance constrained U.S. freedom of action. Although the administration wanted to launch a satellite expeditiously, the launch needed to take place during the IGY, rather than before, in order to advance the image that the United States was working to further international scientific cooperation. Therefore, when the Army informed the Pentagon that a Jupiter missile could orbit a small satellite as early as January 1957, the proposal was rejected because it "would seriously compromise the strong moral position [of the United States] internationally." Such an early launch would occur before tracking and observation equipment were available, and before the United States possessed a follow-up program. Not only would it "flirt with failure"; it would also provide limited scientific utility and raise questions as to the peaceful intent of the American program. The United States might win the race, but it would lose the scientific competition. Even worse, a premature launch by the Army would open the U.S. to charges of warmongering.[16]

Space Exploration and the Psychological Elements of Military Power

Publicly, the Eisenhower administration distinguished the scientific satellite program from the ballistic missile programs. Privately, however, it acknowledged a very close relationship stemming from the psychological dimensions of the nuclear arms race. In the impending age of high-tech, push-button warfare, it was not enough for the United States to merely possess an effective ballistic missile capability; it needed to be demonstrated. The administration felt it had to make clear the reliability and superiority of U.S. power to deter enemies and to reassure allies. Richard Nixon commented to the NSC, "The important thing is not merely the achievement of a developed weapons capability in the ICBM field, but, from the point of view of foreign relations, that the peoples of the free world *believe* that you have achieved an ICBM." Because the ultimate weapons were ultimately unusable, the balance of power increasingly depended on perceptions of relative power. By communicating signals about the technological and military supremacy of the United States, the satellite would allow the United States to impress others with its military power without simultaneously appearing belligerent or reckless.[17]

Moreover, Eisenhower and his advisors believed that international perceptions of American military power were essential components of the political and psychological competition. The key problem was not simply deterring the Soviet Union from initiating general war. Eisenhower frequently commented that the United States had all the deterrent power it needed. Rather, because growing Soviet nuclear capabilities would affect the will of allied and neutral groups to accept strong anticommunist policies carrying with them the risk of war, the challenge for the United States was to preserve domestic and allied faith in the American ability to deter a Soviet attack so that the United States could prosecute the psychological and political war in earnest. "There was a great danger," John Foster Dulles commented, "that we should so focus our eyes on the military aspects of the struggle that we lose the cold war which is actually being waged, forgetting that an actual military conflict may never be waged." Science and technology likewise were recognized as political factors that would be viewed as evidence of military prowess and that would be regarded as keys to progress by the developing world.[18]

Such psychological concerns had a direct bearing on U.S. missile and rocket programs. In the fall of 1955, shortly after the approval of the IGY satellite, the NSC boosted the ICBM to the highest priority of all defense programs in the country and accorded the intermediate-range ballistic missile (IRBM) equally high priority. These decisions stemmed from both military considerations and from psychological calculations about perceptions of power. Similar logic determined both defense and outer space policies.

The ballistic missile programs came to the attention of the NSC on September 8, 1955 when intelligence reports suggested that the Soviets were ahead in missile development. Donald Quarles opened the NSC meeting that day. He was not especially alarmed. He explained that the Soviet lead did not likely endanger the deterrent power of the Strategic Air Command's bomber force in the immediate future, and he advised the NSC to continue pursuing the ballistic missile programs at their current priority level. The meeting devolved into a discussion concerning the precise wording of the NSC's directive. Should the ICBM programs be pursued with "all practicable speed" or "all possible speed"? Undersecretary of State Herbert Hoover Jr. interrupted to place the issue in the broader political context of American foreign relations: "If the Soviets were to demonstrate to the world that they actually had an ICBM before we had such a weapon," he warned emphatically, "the result would have the most devastating effect on the foreign relations of the United States of anything that could possibly happen." Hoover pointed out that the free world coalition was held together essentially by the knowledge that the United States could protect its allies. "If this umbrella of protection were removed," he continued, "neutralism would advance tremendously throughout the free world." President Eisenhower, who was not present at this meeting, unambiguously resolved the dispute over how best to word the NSC directive. He changed the phrase "all practicable speed" to "maximum urgency." He made the ICBM the country's highest priority R&D program.[19]

The State Department's concern for the psychological impact of a Soviet first achievement of ballistic missile capability led to an even greater effort to develop IRBMs. In response to a request from the president, the Policy Planning Staff prepared a report analyzing the political implications of ballistic missiles. The report expressed concern over the impact of a Soviet first achievement of an ICBM capability on the political will of American allies. According to the PPS report, American allies would face increased domestic pressure to adopt independent foreign policies; they would more vigorously oppose policies risking war; and they would be more likely to compromise on outstanding East-West issues. Moreover, the Soviets could exacerbate these trends toward "neutralism" by using conciliatory tactics to persuade the allies of the wisdom of accommodation. The report concluded that the IRBMs should be given priority over the intercontinental missiles, even though their military significance was about equal. The shorter-range missiles could be deployed more quickly, thus offsetting any psychological damage should the Soviets develop ICBMs first.[20]

President Eisenhower concurred that the United States needed both to possess and to demonstrate an effective ballistic missile capability as soon as possible. When Dulles presented the PPS report to the NSC in December, the president reacted in an uncharacteristically vocal and temperamental fashion. With "great warmth of feeling," the minutes recorded, Eisenhower hammered council members for apparent inaction, bureaucratic resistance, and obstructive interservice rivalries. In response to a comment by an Air Force representative that the Soviets possessed a two-year lead over the United States in ballistic missiles, Eisenhower barked that he "would like to know what had been going on since last July when he had issued his strong directive on achievement of a U.S. capability in the field of ballistic missiles." Fully subscribing to the views of the State Department as to the "profound and overriding political and psychological importance" of such weapons, Eisenhower warned that he was "absolutely determined not to tolerate any fooling with this thing." The United States had "to achieve such missiles as promptly as possible, if only because of the enormous psychological and political significance of ballistic missiles." Noting the "critical importance" of these weapons to U.S. foreign policy, Eisenhower assigned both the ICBM and the IRBM programs highest priority of all defense programs in the country. He subsequently specified, in fact, that the IRBM should have precedence over the ICBM because of the political significance of achieving some sort of ballistic missile capability before the Soviet Union. As he recalled in his memoirs, "I realized the political and psychological impact on the world of the early development of a reliable IRBM would be enormous, while its military value would, for the time being, be practically equal to that of the ICBM."[21]

In the eyes of Eisenhower and his political advisors, programs like the scientific satellite were valuable because they reinforced confidence in the capacity of the United States to resist communist blandishments through the visible display of technological prowess. The mere knowledge that the U.S. was pursuing the project, Hoover informed the NSC, "had gone a long way to help the free peoples of the world realize that we were forging ahead in our technical capabilities." Accordingly, when the NSC met in May 1956 to discuss the progress of the IGY satellite program, the NSC commanded that the satellite be given sufficient priority in relation to other R&D projects to ensure a launch early during the IGY. Time was of the essence.[22]

As the cost for the satellite program began to expand from the originally estimated $20 million to a projected $60–90 million, then to over $100 million, some of Eisenhower's advisors began to ask whether the program merited such a high price tag and urgent priority. The NSC met to discuss the IGY satellite in May 1957. "The earth satellite program was an interesting thing," the parsimonious Treasury Secretary George Humphrey told the council, "but was [it] really a pressing and urgent matter?" Should the U.S. cut an infantry division or sacrifice B-52 bombers to launch an earth satellite? Allen Dulles responded by instructing Humphrey that abandoning the program meant handing the Soviet Union a major propaganda victory. "If the Soviets succeeded in orbiting a scientific satellite and the United States did not even try to," he explained, "the USSR would have achieved a propaganda weapon which they could use to boast the superiority of Soviet scientists. In the premises, the Soviets would also emphasize the propaganda theme that our abandonment of this peaceful scientific program meant that we were devoting the resources of our scientists to warlike preparations instead of peaceful programs." Eisenhower, like Humphrey, was displeased with the escalating costs of the satellite, but he agreed with the CIA director that psychological factors required the U.S. to continue with the program. The same president who would later tell the public that he approved the project strictly on its scientific merits now complained about all the time and money scientists were wasting outfitting the satellite with fancy scientific instruments. Annoyed by the tendency to "gold plate" the satellite, Eisenhower pointedly stressed "that the element of national prestige, so strongly emphasized in NSC 5520, depended on getting a satellite into orbit, and not on the instrumentation of the scientific satellite." Eisenhower's comments indicate that not only was he fully aware of the prestige value of the satellite, but that he saw the satellite's significance primarily in psychological terms.[23]

The Sputnik Panic

Although U.S. intelligence indicated that the Soviets had been making rapid progress in ballistic missile development, and despite CIA predictions that the USSR would attempt to beat the United States in the satellite competition, the launch of Sputnik on October 4, 1957, came as a profound shock. In what has been described as the "Sputnik panic," a wave of near hysteria gripped the country. *Newsweek* wrote that Sputnik represented a "defeat in three fields: In pure science, in practical know-how, and in psychological Cold War." *Life* magazine concurred: "Let us not pretend that Sputnik is anything but a defeat for the United States." Many commentators compared Sputnik to Pearl Harbor, and many Americans concluded that U.S. defenses were inadequate. Charges of a "missile gap," an "education gap," and indeed a "propaganda gap" were levied against the Eisenhower administration, which now appeared hopelessly out of touch with the nation's security needs.[24]

Sputnik was a first-rate propaganda victory that seemed to verify the dire predictions intelligence analysts and policy makers had been making for five years. It was a serious blow to American prestige, and it was a major victory for the Soviet Union. At an NSC meeting a few days after the launch of Sputnik, Allen Dulles interpreted the event as part of a trilogy of Soviet propaganda moves: the announcement in August that the USSR had successfully tested an ICBM, the massive September hydrogen bomb test series, and now the satellite launch. Together they seemed to provide the world with convincing evidence that the Soviets possessed a substantial lead in ballistic missile technology. Furthermore, Soviet propaganda was using Sputnik to boast about the military and scientific supremacy of the USSR. Dulles warned that this "major propaganda effort" was exerting a "very wide and deep impact" in Western Europe, Africa, and Asia. Echoing this assessment, Undersecretary of State Christian Herter reported to the NSC that international reactions to the satellite were "pretty somber." Even the best allies "require assurance that we have not been surpassed scientifically and militarily by the USSR." The situation appeared even more disastrous outside the western alliance, Herter cautioned, because the Soviet feat seemed to affirm the wisdom of neutrality. The neutral countries "are chiefly engaged in patting themselves on the back and insisting that the Soviet feat proves the value and wisdom of the neutralism which these countries have adopted." Sputnik not only sapped U.S. prestige by suggesting Soviet technological superiority; it also lent added credibility to Soviet propaganda and prepared the way for a concerted communist psychological warfare campaign.[25]

The Eisenhower administration's public opinion research supported these assessments. A State Department study prepared shortly after Sputnik outlined four major effects on world public opinion:

1. Soviet claims of scientific and technological superiority over the West and especially the U.S. have won greatly widened acceptance.
2. Public opinion in friendly countries shows decided concern over the possibility that the balance of power has shifted or may soon shift in favor of the USSR.
3. The general credibility of Soviet propaganda has been greatly enhanced.
4. American prestige is viewed as having sustained a severe blow, and the American reaction, so marked by concern, discomfiture and intense interest, has itself increased the disquiet of friendly countries and increased the impact of the satellite.

The study predicted that Sputnik's repercussions would be greatest in the third world. These technologically less advanced areas of the world would be most easily "dazzled" by the feat. They were also the areas "least able to understand it" and "most vulnerable to the attractions of the Soviet system." The State Department warned that by demonstrating the ability of the Soviet system to compete on a technological level with the West, Sputnik seemed to show the developing world that communism provided the best path toward rapid modernization. This meant that developing nations would be more likely to turn to the Soviet bloc for assistance and leadership.[26]

The USIA's public opinion research pointed to similar conclusions. Polls from five countries in Western Europe indicated that Sputnik was an "attention-seizing event of the first magnitude." Nearly everyone surveyed by the agency had heard of the satellite launch and identified it with the Soviet Union. The prevailing opinion was that the USSR was ahead of the United States in scientific development. In half the countries surveyed, a majority concluded that the Soviet Union had surpassed the United States in total military strength.[27] A subsequent study by the USIA's Office of Research concluded that Sputnik tipped the perceived balance of power in the USSR's favor:

> The most significant and enduring result for world public opinion . . . was a revolutionary revision of estimates of Soviet power and standing. Before the launching of Sputnik I there was a very general belief that the Soviet Union was a long way from offering a serious challenge to the U.S. lead in science, technology, and productive power. Sputnik . . . appeared as a dramatic demonstration that the USSR was able to challenge the U.S. successfully in an endeavor where U.S. preeminence had been widely taken for granted.

Not only did it suggest Soviet military might and technological prowess, but it also possessed broader implications for the ability of Soviet propaganda to win hearts and minds. The USIA cautioned: "The USSR, by appearing to have spectacularly overtaken the U.S. in a field in which the U.S. was very generally assumed to be first by a wide margin, is now able to present itself as fully comparable to the U.S. and able to challenge it in *any field it chooses*—perhaps the most striking aspect of the propaganda impact of space developing."[28]

It all boiled down to credibility. Now that there was dramatic proof of communism's accomplishments, administration officials feared that the world would take Soviet pronouncements more seriously. This had implications for the ideological competition. If the Soviet Union could surpass the United States in science and technology, where the United States possessed a substantial lead, could not the communists surpass the capitalists in other areas as well?

Damage Control

The Eisenhower administration turned to public relations to control the damage unleashed by the Soviet satellite. In a meeting at the White House on October 8, Eisenhower set three guidelines for the administration's public posture. First, the American satellite program was a scientific endeavor, completely divorced from military projects. Second, the United States never emphasized timing, or being first, in planning its program. There was no race. Finally, the American program, unlike the Soviet one, was conducted openly; the United States would share the information gleaned from the project with the scientists of the world.[29] In other words, the administration adopted the PR line it feared the Russians would have adopted had the United States been first: "Their" program was a hasty, unscientific, militaristic propaganda stunt, whereas "ours" was peaceful and scientific.

Eisenhower parroted this line in his first press conference the next day. Reflecting public anxieties over Sputnik, a United Press reporter opened the conference with a straight shot: "Mr. President, Russia has launched an earth satellite. They also claim to have had a successful firing of an intercontinental ballistic missile, none of which this country has done. I ask you, sir, what are we going to do about it?" Eisenhower responded with a long, jumbled explanation about how the United States pursued the project merely to provide "basic research" into such things as temperature, radiation, ionization, and air pressure. He pointed out that the American satellite would provide much more useful scientific information than the Russian Sputnik, which had virtually no instrumentation. The only thing the Soviets had done was put "one small ball" in the air. "No one ever suggested to me that it was a race," the president repeated disingenuously.[30]

Eisenhower's argument that the United States did not want to be first in space provided a convenient, but ultimately unconvincing, answer for the Sputnik shock. Even less convincing was the pretended disinterest in the psychological dimensions of the satellite. From the project's inception, administration officials worked to maximize American prestige gains and minimize potential propaganda fallout by emphasizing the importance of their satellite's scientific and peaceful image as well as a timely launch. The decisions to use the Navy's Vanguard rocket, to publicly distinguish this

A USIA exhibit in Rio de Janeiro publicizes American achievements in space exploration. Source: National Archives.

"research rocket" from the other services' ballistic missiles, and to prevent the Army from launching an uninstrumented satellite before the start of the IGY, all reflected the administration's gamble that the psychological benefits of spaceflight depended on both scientific utility and speed. If anything, the shock of October 4 resulted not from the administration's insensitivity to psychological factors, but rather from its very fixation on them.[31]

The Laughing-Stock of the Whole Free World

At his October 9 press conference, Eisenhower released a statement informing reporters that Project Vanguard would launch a satellite in December. Despite the public position that the U.S. was not in a race, privately, the DOD was instructed to exert pressure on the Vanguard rocket team to ensure the launch went off on time. According to a memorandum sent to the secretary of defense on October 17, Eisenhower "made very plain" that nothing should be permitted to delay the planned launch schedule.[32]

This decision proved disastrous. Up to that point, the Navy had only tested the first stage of the Vanguard rocket. The December launch was only supposed to be the first test of all three stages of the complete rocket. It was not designed to orbit a payload, but to assess the performance of the launching vehicle itself. When the secretary of the Navy expressed concern that there was only a "probability," not a "certainty," of success, Secretary of Defense Neil McElroy ordered him to continue "with deliberate speed" anyway. "The Soviet's success with their satellite has changed the situation," McElroy wrote. "The psychological factors in this matter obviously received new emphasis."[33]

Under pressure from the president, Congress, and the international media, the Navy hastily converted its test vehicle into a "mission vehicle" equipped with a small four-pound instrumented satellite. On December 4, thousands of onlookers, including over a hundred newspaper and television reporters, descended on Cape Canaveral to witness the first American satellite launch. The countdown began at 4:30 A.M., but high-speed winds forced repeated delays. Eighteen hours later, the launch was scrubbed.

At the NSC meeting the next day, John Foster Dulles was fuming. Speaking "earnestly" (according to the minutes), he declared that in the future "we would not announce the date, the hour, and indeed the minute, that we were proposing to launch our earth satellite, until the satellite was successfully in orbit." Dulles complained that "the effect of the publicity of the last few days, culminating in the final decision to postpone the attempt to launch our first earth satellite, had had a terrible effect on the foreign relations of the United States." What happened yesterday, he continued, "had been a disaster for the United States. . . . [It] made us the laughing-stock of the whole Free World and was being most effectively exploited by the Soviets." Eisenhower was also concerned by the impact of the failed launch. He asked Donald Quarles if it were not possible to launch a satellite from a desert region rather than the densely populated Florida coast: Was there some way to conduct the test in secret, so as to avoid these kinds of embarrassments?[34]

The NSC discussion was both ironic and prophetic. As embarrassing as a delayed launch may have been, it was nothing compared to the disaster that awaited. On December 6, the Navy finally proceeded with the Vanguard launch. The rocket hovered above the launching pad for a few seconds before falling to the earth, bursting into flames, and exploding into a billowing cloud of black smoke. Broadcast on television in all its fiery glory, it was truly a national embarrassment. Newspapers pronounced Vanguard a "Flopnik," a "Dudnik," a "Stayputnik," and a "Kaputnik." When the rocket exploded, the *Nation* commented, "the whole world was watching." The *New York Times* announced in a bold headline, "Failure to launch test satellite assailed as blow to U.S. prestige."[35]

The follow-up tests for Vanguard proved no more encouraging than the first. For ten days in a row, the next Vanguard test was delayed. Ultimately the mission was aborted. The time for scientific experiments had passed. The Army took over the launch site. A Jupiter-C missile was readied with the satellite Explorer I, and on January 31, the rocket blasted into space, successfully orbiting the first American satellite. "To some extent," *Life* magazine wrote, Explorer made up "for the U.S. humiliation due to the Russian Sputniks." The nation breathed a temporary sigh of relief, but the American success was a somber one. Two months earlier the USSR had orbited a second satellite, Sputnik II, which weighed 1,121 pounds and carried a live dog. Although Explorer I contained much more sophisticated scientific instruments than either of the Soviet Sputniks, many commentators were more dazzled by the mammoth size of the Soviet satellites than the scientific gadgets sent up by the Americans. The Eisenhower administration tried to vilify Sputnik II by eliciting sympathy for Laika, the poor dog who died in orbit. In May the USSR launched Sputnik III, a huge instrumented satellite weighing 3,000 pounds. Khrushchev wasted no time using the satellite to remind the world of the Soviet lead in space technology. He told Arab diplomats visiting Moscow that the United States would need "very many satellites the size of oranges in order to catch up with the Soviet Union."[36]

Publicly, the Eisenhower administration declared that it was not impressed by the Soviet stunts. The United States would not race the Soviet Union, nor would it resort to propaganda gimmicks. It would continue to pursue an orderly and scientifically useful program. Behind the scenes, however, administration officials searched for ways to dramatize American scientific and technological prowess to the world.

Immediately after the Sputnik launch, USIA director Arthur Larson pressed the case for spectacular scientific feats to shore up American prestige. He encouraged Eisenhower to approve projects that would have a profound effect on world opinion, such as manned space vehicles or rockets that would hit the surface of the moon. The United States must plan projects to enhance American prestige abroad. "The reason for this . . . is not the value of scientific preeminence for its own sake, but the disproportionate impact that real or apparent scientific preeminence now seems

A USIA official explains a space communications satellite to men in Salisbury, the capital of the Federation of Rhodesia and Nyasaland (present-day Harare, Zimbabwe). Source: National Archives.

to have on our military position and our diplomatic bargaining power." Larson's suggestions went beyond outer space exploration to advocate any dramatic measure in the scientific field that would attract wide public interest. He wrote to the AEC suggesting some wild ideas for using "clean" hydrogen bombs to blast mountain passes, dam rivers, or create new harbors. He advised that because Sputnik had created the impression of Soviet technological and military superiority, swift action would be necessary to check this trend. Dramatic and peaceful demonstrations of American military power were imperative.[37]

The OCB also set itself to the task of devising ideas for projects that would have a "significant psychological impact." The board culled through the pages of possible projects to find the most feasible ones with the greatest psychological value. Suggestions for space spectaculars included a moon-orbiting satellite, a TV relay satellite, or a "mailbag satellite" that would record ground signals over one station and then transmit them again over another station. The OCB also considered proposals that the United States recover a live animal from orbit, circumnavigate the moon with an

American rocket, or orbit a huge inflatable satellite that would be visible to the naked eye from the earth.[38]

In general, the OCB believed that the United States could best redress the challenge posed to American scientific superiority by demonstrations of genuine technological advances on a broad front. It also believed that new and dramatic psychological warfare ideas were needed to meet the post-Sputnik propaganda challenge. A special working group generated pages upon pages of suggestions under the heading "Projects to Regain the Initiative." The OCB looked into ways to dramatically demonstrate U.S. research in such areas as solar energy and the desalination of sea-water. It proposed a program whereby American scientists would drill ten miles into the earth's crust, twice the distance of any prior drilling, to dramatize American scientific and engineering capabilities. Other ideas included spectacular development projects like a rice airlift to Indonesia, the establishment of an emergency food bank for international disasters, and programs for education in less developed areas. The board also proposed creating an "International Economic Corps for Peace," hitting on the idea for the Peace Corps in the midst of the Sputnik slump. It further recommended propaganda themes for USIA exploitation, such as publicizing failures of Soviet foreign aid, showing "how Mao has destroyed his flowers," and advertising "Soviet colonialism."[39]

The CIA submitted a list of suggestions to the working group proposing that the U.S. organize an "international medical year" or an "international biomedical year" to attack worldwide diseases and reward scientists whose research benefited humankind. Other suggestions included establishing chemical food-processing plants abroad, publicizing U.S. synthetic food research, financing hospitals in underdeveloped areas, founding a "university of the world" to benefit international students, and awarding prizes for scientific achievement. On the theme of peaceful uses of atomic energy, the CIA proposed that the United States use atomic explosions to destroy dangerous icebergs and melt ice for clearing polar passages for shipping. Amazingly, the agency also suggested that the United States fire a "clean" hydrogen bomb into a typhoon to reverse its direction or to stop it. On the space front, the CIA recommended that the United States install a transmitter on an earth satellite to broadcast a short, "catchy" musical beat that could be picked up on radios all over the world. The VOA could broadcast the tune with accompanying words such as "Freedom shall be yours." The agency reasoned, "this idea would have a tremendous propaganda effect and would induce more listeners [to the VOA] than anything which has yet been done." Similar suggestions for outer space broadcasting stunts filtered through the OCB, including proposals to broadcast "great speeches" by prominent Americans, coverage of the 1960 Olympic games, and other "high visibility" events from outer space.[40]

These suggestions indicated a siege mentality in the Eisenhower administration on the public opinion front. Documents from the administration's files evince a desperate tone, a sense of urgency that something must be done immediately to rejuvenate international confidence in American leadership. In addition to the pounding the administration was getting over the Soviet Sputniks, the Little Rock racial integration crisis had emerged as a national and international issue, the antitesting and antinuclear movements were gathering momentum, Soviet prestige was rising, Eisenhower's popularity was plummeting, and the American economy, after five years of unparalleled prosperity, was slowing. Eisenhower became eager for something—anything—that could regain the initiative. He wrote to Dulles in March 1958: "We need some vehicle to ride in order to suggest to the world, even if ever so briefly, that we are not stuck in the mud."[41]

A Call to Sacrifice

The sense of crisis was exacerbated when information from a top secret report on the nation's security became public in November 1957. The Gaither Report, named after the chair of the investigating committee H. Rowan Gaither, concluded that U.S. nuclear retaliatory forces were vulnerable to surprise attack—a hypothetical "bolt from the blue." It recommended measures to protect the Strategic Air Command, to accelerate missile production, to strengthen the U.S. ability to fight limited military operations, and to embark on a fallout shelter program to protect the civilian population. All told, the panel's recommendations would cost somewhere between $19 and $44 billion dollars above the roughly $38 billion that was already being spent on defense. Together with the Soviet Sputniks, the Gaither Report created a first-rate domestic political crisis that fueled Eisenhower's opponents and precipitated ill-founded charges of a "missile gap." The administration was pressured from all quarters to adopt Manhattan Project–style crash programs in missile development and space exploration.[42]

Eisenhower tried to calm the badly shaken nation by reassuring the American people that U.S. military power was sufficiently powerful to deter an enemy attack. In the fall of 1957, he made four televised addresses on the nation's security, each stressing the awesome power of the military.[43]

Fearful that panic would lead the country into runaway deficit spending, Eisenhower announced relatively modest steps to improve the American position. He established the President's Science Advisory Committee headed by James Killian, and he appointed a missile czar in the DOD to oversee all government missile programs. In addition, Eisenhower pursued a controversial defense reorganization plan designed to mitigate interservice rivalries and reduce waste and duplication. At the same time, Eisenhower authorized a $1.44 billion increase in the planned defense budget for the next fiscal year (1959). This was a far cry from the many more billions recommended by the Gaither Report, but it was a significant increase. In Eisenhower's view, even this relatively modest increase was largely unnecessary from a military standpoint. He estimated that "about two-thirds" of the supplementary funds were "more to stabilize public opinion" than to meet real security needs.[44]

For both domestic political and international psychological reasons, the president also accelerated the IRBM programs and worked to deploy the missiles in Europe. In December 1957, Eisenhower and Dulles attended a NATO conference in Paris to reassure European allies of American capabilities. At the conference, the allies agreed in principle to station IRBMs on their territories, a step both Eisenhower and Dulles acknowledged as unnecessary on military grounds. Eisenhower believed that the chief significance of ballistic missiles was "in the psychological area." He stressed that "the whole arms question is relative" and predicted that "Soviet ICBMs will not overmatch our bomber power in the next few years." Dulles doubted that the Soviets would risk launching a surprise attack when the price of failure was so high and the odds of success so low. But Eisenhower and Dulles realized these steps were necessary to calm domestic political opponents and to reassure U.S. allies of the credibility of the American nuclear guarantee.[45]

Concerned that domestic political obsession with missile programs would divert attention from the important nonmilitary dimensions of the Cold War, Eisenhower tried to redirect the public's focus from weapons programs, which in his view sufficiently provided for the nation's security, to other arenas of the Cold War competition. Almost immediately after Sputnik, in his radio-TV addresses, he laid the groundwork for what would become the National Defense Education Act. The act, which stressed the importance of scientific education and basic research to the nation's long-term security, was signed into law in the summer of 1958 as a stopgap measure to deal with the "educational emergency" posed by Sputnik. In subsequent moves, Eisenhower called for increased expenditures on foreign aid and mutual security programs to strengthen American alliances and win the economic war in the developing world. He also tried to secure a substantial increase in the USIA's budget for international propaganda, requesting $144 million for the agency's operations (an increase of one-third from the previous year), but was unsuccessful.[46]

These steps reflected Eisenhower's general belief that the real Cold War battles were not going to be fought with hydrogen bombs and ballistic missiles, but by nonmilitary means. The principal methods of communist aggression, he advised in one of his televised addresses, were "propaganda and subversion, economic penetration and exploitation."[47] This was the argument Eisenhower made in a remarkable State of the Union address in January 1958. Delivered in the panicky post-Sputnik atmosphere, it was one of the most revealing speeches of his presidency. The speech was a call to arms—not to the traditional weapons of war, but to the new weapons of the Cold War. The bulk of the speech argued a single thesis: the United States should not focus its full attention on military strength, but instead on the total Cold War that must be waged and won in the years ahead.[48]

On the question of the country's military posture, the address sought to reassure: Yes, steps needed to be taken, but the United States possessed the "most powerful deterrent to war in the world." In other arenas of the Cold War effort, however, urgent measures were required. Having briefly outlined the measures necessary to "*deter* a possible future war," the president turned the rest of his speech to the measures needed to "*win* a different kind of war." In alarmist language, Eisenhower explained the totality of the communist threat: "What makes the Soviet threat unique in history is its all-inclusiveness. Every human activity is pressed into service as a weapon of expansion. Trade, economic development, military power, arts, science, education, the whole world of ideas—all are harnessed to this same chariot of expansion." The Soviets, he continued, were "waging total cold war." It would be a "tragic mistake" to concentrate exclusively on military strength, the president advised. "We must never become so preoccupied with our desire for military strength that we neglect those areas of economic development, trade, diplomacy, education, ideas and principles where the foundations of real peace must be laid." Americans needed to fight fire with fire, to "wage total peace" by "bringing to bear every asset of our personal and national lives."

Eisenhower and the political advisors who helped prepare the speech—John Foster Dulles, U.N. Ambassador Henry Cabot Lodge, and USIA

Director Arthur Larson—hoped it would prevent Congress from getting so carried away with the missile and space issues that they slashed other appropriations necessary for the Cold War effort. The speech sought to capitalize on the sense of crisis generated by Sputnik to push for increases in foreign aid to meet the Soviet economic offensive in the developing world. The speech also was a call to sacrifice. It was designed to shake Americans from their complacency and rally them to the challenge ahead.[49]

There was, however, a dark irony to Eisenhower's total Cold War warning. The general who had led the final assault on Hitler's armies was perhaps unwittingly following in the footsteps of the notorious Nazi propaganda chief Joseph Goebbels, who, in the wake of the German defeat at Stalingrad, issued a plea to the German people to wage total war against the enemy. Eisenhower's call for total cold war was perhaps then a symbol of his desperation, a sign of how poorly the propaganda war was going after Sputnik; or perhaps it was less remarkable and merely an indication of how deeply ingrained the total war mentality had become to his generation. Regardless, both Goebbels and the president had similar thoughts in mind when they issued their total war pleas: they wanted to shore up the morale of their people, to steady their course, to urge them to stand tall in the face of unyielding external challenges.[50]

Eisenhower's speech also spoke to the revolution in international affairs that had been developing at least since Stalin's death. Intangible components of power—political, economic, psychological, cultural, symbolic, ideological—emerged as decisive weapons of the Cold War. Sputnik symbolized the extent to which this all-embracing competition had become a total symbolic war. In an age of push-button warfare and nuclear stalemate, perceptions, images, and symbols of power and prestige were as important as actual military force. No matter how extravagant a space race might seem, the U.S. international position depended upon the demonstration of technological, military, industrial, economic, and cultural strength that space exploration represented.

NASA and Space Strategy

Such intangible factors as image and prestige weighed heavily on the Eisenhower administration's decision to commit the United States to the long-term exploration of outer space. Shortly after Sputnik, the president instructed his new science advisor James Killian to make recommendations for how best to deal with the challenge of space exploration. After deliberating for a few months on the issue of civilian versus military direction of space projects, Killian advised Eisenhower to create a full-fledged civilian space agency, the National Aeronautics and Space Administration (NASA). The main reasons for a civilian space agency were to present the American space program as peaceful and divorced from military applications; to reap propaganda advantages from a civilian space program dedicated to international science; and to conduct basic research and development that the military might otherwise neglect. Eisenhower saw the civilian agency as a body that would appeal to scientists and to world opinion in general. NASA was officially established in August 1958.[51]

That same month, the National Security Council approved a formal statement of U.S. outer space policy, NSC 5814/1. In nine long paragraphs, NSC 5814/1 established the justification for the civilian space program. The document began with an alarming preamble: "The USSR has surpassed the United States and the Free World in scientific and technological accomplishments in outer space, which captured the imagination and admiration of the world. The USSR, if it maintains its present superiority in the exploitation of outer space, will be able to use that superiority as a means of undermining the prestige and leadership of the United States and of threatening U.S. security." The document proceeded to advise, in flowery prose, that space exploration possessed "unusual and peculiar significance" because of its powerful psychological effects on the people of the earth: "With its hint of the possibility of the discovery of fundamental truths concerning man, the earth, the solar system, and the universe, space exploration has an appeal to deep insights within man which transcend his earthbound concerns." These deep psychological implications profoundly affected U.S. foreign relations, the document advised. Echoing lines of reasoning established long before Sputnik, the NSC warned that further Soviet demonstrations of superiority in outer space technology would "dangerously impair" confidence in overall U.S. leadership. American successes in space exploration were necessary to "enhance the prestige of the United States among the people of the world and [to] create added confidence in U.S. scientific, technological, industrial and military strength."[52]

Reflecting these psychological concerns, NSC 5814/1 dictated the overall goals and parameters of the now-adolescent space program. In addition to important military and reconnaissance applications, the NSC directed NASA to "judiciously select" projects designed to achieve a "favorable world-wide psychological impact." Of the many possibilities, manned spaceflight ranked highest because, "to the layman," manned exploration represented the "true conquest" of outer space. "No unmanned experiment can substitute for manned exploration in its psychological effect on the peoples of the world," the document advised. Besides prestige-boosting space spectaculars, NSC 5814/1 called for international cooperation in space activities and advocated negotiations to ban the use of space for military purposes, both designed to establish the United States as the leader in peaceful uses of outer space. It also ordered "information and other programs" to "counter the psychological impact of Soviet outer space activities and to present U.S. outer space progress in the most favorable comparative light."[53]

Conspicuously absent from this and other Eisenhower-era documents was authorization for a lunar landing. Eisenhower refused to commit the nation to spending the millions of dollars it would cost to send a man to the moon. This decision has led most scholars to conclude that Eisenhower, unlike his successor, was not willing to stake the nation's prestige on wasteful and costly space spectaculars. This view, however, greatly exaggerates Eisenhower's aversion to using space exploration to bolster America's image abroad. According to the president, the objective of the American space effort was "to achieve a psychological advantage for ourselves" and to "break out of our psychologically and politically inferior position in the space field." Eisenhower, as we have seen, was hardly opposed to using stunts to compete in the propaganda war—he had, after all, approved of an expensive plot to send an aircraft carrier with Cinerama theaters on a world tour—but he was not sure a race to the moon was one the United States could win. Eisenhower explained to the NSC that the United States "had better not undertake space activities for psychological, political and propaganda advantage unless we are able to compete with the Russians. Otherwise we would be in a poker game with a second-best hand." The country was already trailing behind the Soviet Union; to order a moon race would flirt with failure.[54]

Eisenhower's concern that the United States would continue to suffer under the weight of Soviet "firsts" led him to adopt a restrained public posture on the space issue. He ordered his subordinates to follow suit. Successes were to be announced with a matter-of-fact tone, and new projects were to be publicized as vaguely as possible so as to avoid another Vanguard disaster. The official line of the Eisenhower administration continued to be that the United States was not in a race and that science, not propaganda, would set U.S. space priorities.[55]

Yet by the summer of 1958, as NSC 5814/1 suggested, the space race was on in earnest. Lunar probes were the focal point of the race that year. Both the United States and the Soviet Union rushed to be the first to orbit—or simply smash into—the moon. The administration released a statement to the press matter-of-factly informing reporters that the United States would try to launch "small unmanned vehicles" in the vicinity of the moon. It gave few specific details and no timetable. The statement called this effort part of an "orderly program for space exploration," but the probe had little scientific justification. Eisenhower attached a high priority to the probe in order to beat the Soviets to the moon.[56]

The United States tried to launch its first lunar probe in August, but was faced with a repeat of the Vanguard disaster when the rocket exploded seconds after takeoff. NASA took over the effort in October and created a multimillion-dollar probe outfitted with a television camera that would transmit images from the back of the moon. Despite Eisenhower's protestations that he was not interested in space stunts, the orbiting camera mission was precisely such a stunt—although Eisenhower cautiously disguised the true intent of the program "unless and until it should be a success." Pioneer I, as the probe was called, was launched atop a modified Thor missile nearly 80,000 miles into space, thirty times father than any previous attempt. It missed its goal of encircling the moon, however, because it failed to escape the earth's gravitational pull. Two subsequent attempts—Pioneer II in November and Pioneer III in December—were likewise trapped by gravity. Although these attempts were unsuccessful, they suggested the Americans were closing the space gap. In January, however, the Soviets attempted their launch of a Lunar probe, Luna I. It also failed to orbit the moon, but it did successfully escape the earth's gravitational pull—another Russian "first." Nine months later, the USSR scored another first when Luna II smashed into the surface of the moon carrying the hammer-and-sickle flag of the Soviet Union. Ridiculing the American failure to even escape earth's gravity, Khrushchev joked that capitalism made U.S. moon rockets fall into the ocean.[57]

In December 1958, NASA announced Project Discoverer, a satellite program purportedly designed to place animals and scientific instruments into space, but that actually served as cover for the clandestine Corona reconnaissance satellite project. NASA also secretly decided that Project Mercury, the effort to send a man into space, should be granted "highest priority" in order to beat the USSR in this endeavor. The same month, the United States finally secured a "first" of its own with Project Score. On December 18, Project Score placed an entire four-and-a-half-ton Atlas missile into orbit—at that time the largest object sent into space—and it orbited a small communications satellite, the first of its kind. As the satellite circled the earth, it delivered a radio Christmas message to the people of earth. Eisenhower's voice beamed from the heavens, saying: "This is the president of the United States speaking. Through the marvels of scientific advance my voice is coming to you from a satellite circling in outer space. My message is a simple one. Through this unique means I convey to you and to all mankind America's wish for peace on earth and good will toward men everywhere." The orbiting of the Atlas missile was scientifically insignificant, but Eisenhower's radio message reminded world audiences that the United States too was in space. Project Score, the *New Republic* noted, was "an extravagant propaganda stunt."[58]

An even bigger propaganda stunt came in August 1960, when the United States orbited another communications satellite called Echo. It was a balloon thirty meters in diameter that inflated upon reaching orbit. It could be seen from earth without a telescope and it was brighter than everything in the sky except the sun and the moon. Variations of the Score and Echo ideas had been earlier suggested by the OCB working group on "seizing the initiative." Both had been designed to capture public interest and dramatize in new and exciting ways U.S. accomplishments in outer space.

Other OCB ideas to dramatize U.S. scientific and technological prowess were pursued. Two of the most extravagant (and in hindsight, somewhat farcical) endeavors were Project Chariot and Project Mohole. Project Chariot was an AEC project that proposed to use nuclear explosives to create a new harbor on the north coast of Alaska. The project, as Michael Sherry notes, "was laughable in economic terms, dangerous in ecological ones, and destructive in human terms." Popular protest led to the eventual cancellation of the project, but the zeal with which the AEC pursued this atomic public works nightmare indicated the great lengths American officials would go to conquer nature for the sake of Cold War advantage. Project Mohole, conceived during the IGY, endeavored to drill a gigantic hole through the ocean floor miles into the earth's crust. The investigation had real oceanographic and geological merit, but it quickly devolved into another facet of Cold War competition. Just like the race into the heavens, this contest pitted American and Soviet scientists racing to the center of the earth to score another "first" in a major scientific undertaking.[59]

Such competitions to demonstrate scientific and technological superiority were by-products of the nuclear age and the perceived need to establish the reliability and credibility of deterrence. But to many Americans, the challenge posed by Sputnik and other Soviet outer space successes extended beyond the narrow fields of science and technology. The Sputnik challenge was also a cultural and ideological challenge. As a revision of U.S. outer space policy from January 1960 explained, the greatest challenge of the space race lay in the realm of propaganda. From the political and psychological standpoint, the most significant factor of Soviet space accomplishments was that they produced "new credibility" for Soviet statements and claims. According to the report, the Soviets were using this credibility to promote four ideas: (1) the general superiority of the Soviet system; (2) a shift in the world balance of power in favor of communism; (3) the technological equality or superiority of the Soviet Union to the United States in most respects; and (4) Soviet military superiority over the West. Soviet propagandists had been making similar claims for years, but Sputnik seemed to make them believable. "Where once the Soviet Union was not generally believed," the policy statement advised, "even its baldest propaganda claims are now apt to be accepted at face value, not only abroad but in the United States."[60]

Most remarkable about the claims identified by the document, however, is that they could not be dismissed so easily as mere propaganda; they were the core ideas that animated the ideological competition for hearts and minds. This contest was rooted in proving to the world the superiority of capitalism and communism, of the American and communist ways of life, and of cultural, economic, and scientific achievements. The hysterical response generated by Sputnik likewise reflected military, cultural, and ideological concerns. Not only did Sputnik suggest American vulnerability to strategic attack, it also called into question core ideas and assumptions taken for granted by many Americans. Perhaps communism was the way of the future; perhaps the Soviet Union would surpass the West in time; perhaps Americans were weakened by complacency and consumerism. In light of these deeply rooted insecurities, several commentators, not only the president, expressed the view that the Cold War had become a total contest requiring long-term contributions and sacrifices from the American people.[61]

Conclusion

The great struggle of our times is one of spirit. It is a struggle for the hearts and souls of
men — not merely for property, or even merely for power. It is a contest for the beliefs, the
convictions, the very innermost soul of the human being.

— Dwight D. Eisenhower, 1952

In many ways the Cold War was a different conflict at the end of the 1950s.
When the decade began, relations between the superpowers were frozen.
The world fixated its attention primarily on the military dimensions of the
Cold War as the nuclear arms race gathered steam and the war in Korea
dragged on. Propaganda from both sides was angry and strident. There
were ritualistic calls for peace, yet they appeared especially hollow in an
international environment where communication between American and
Soviet heads of state had effectively ceased.

After 1953, the climate of fear and suspicion began to clear. The rhetoric became less virulent and more focused on chances for peace. Cease-
fires were arranged in Korea and Indochina. An Austrian peace treaty was
signed. Both sides advanced new and imaginative proposals for arms control. Channels of communication opened up. At the Geneva Summit in
1955, the heads of state met in person for the first time since World War
II. Three years later, the superpowers reached the first bilateral accord of
the postwar era: a cultural agreement allowing increased person-to-person
contacts, travel, and trade between their countries. These developments
were not so much signs that an early end to the Cold War was in sight as
they were indicators that the conflict was entering a new phase, one where
the psychological and political competition assumed center stage.[1]

Counterintuitively, American anxieties became more pronounced as
Cold War tensions eased. By 1955, some American national security experts, newspaper columnists, and pollsters were asking if Russia was winning the Cold War—not because the Soviet Union was surpassing the
United States in military or economic power—far from it—but because
the tide of world opinion appeared to be shifting in its favor. The fear of
communist domination that had been so strong in the immediate aftermath of the fall of Eastern Europe and the onset of the Korean War had

helped to unify the Atlantic alliance and brought other countries into the free world camp. But Stalin's death and the new Soviet policy of "peaceful coexistence" seemed to have the opposite effect abroad. Lingering doubts about the wisdom and restraint of U.S. leadership, along with misgivings about American life and culture, rose to the surface. It was much more difficult for Americans to demonize the Soviet Union with Stalin gone. It was correspondingly difficult to maintain and expand the free world system of alliances. Many countries in the third world cast a plague on both houses, opting for a pronounced form of neutralism that Americans interpreted as virtual victories for communism. The United States, it now appeared, simply could not take it for granted that others would believe that capitalism and democracy promised a brighter future than communism.[2]

For these and other reasons, the global battle for hearts and minds intensified after 1953. The death of Stalin proved to be an especially pivotal development. The USSR's new policy of peaceful coexistence raised the stakes in the political, ideological, and cultural dimensions of the Cold War competition by challenging the very foundation of U.S. leadership: the fear of Soviet conquest that had held American alliances together. During the Stalin years of the Cold War, the Soviet Union could claim few victories in the global public relations contest. But in the years after Stalin's death, the new Soviet leadership, especially Khrushchev, actively courted world opinion. Americans saw it as an uneven playing field: they were losing ground for appearing unreasonable, while Soviet leaders were scoring points simply for being better than Stalin. U.S. officials perceived the soft line in Soviet foreign policy as representing a new phase in the Cold War where psychological, ideological, and cultural factors predominated.

Structural changes in the international system also elevated the psychological dimensions of the Cold War. The thermonuclear revolution and the approach of strategic parity made general war between the superpowers all but unthinkable, channeling the rivalry into symbolic demonstrations of national power. Decolonization and the strengthening of anticolonial nationalist sentiment in the third world further accentuated the nonmilitary dimensions of the Cold War, as both superpowers competed for the political allegiance and vital economic resources of countries in Asia, Africa, the Middle East, and Latin America. In addition, the accelerating pace of the communications revolution continued to link far-flung regions of the world to a global information network. The communications revolution provided the infrastructure that allowed the superpowers to focus their rivalry on symbolic modes of combat. The mass politics induced by the communications revolution made foreign societies more susceptible to manipulation through propaganda and covert intervention from abroad. The expanding network of mass communications also helped totalize the Cold War by filtering many social, cultural, economic, and political developments through an ideological prism.

In part because of these developments, leaders of both superpowers acknowledged, at much the same time, that the battle for world opinion was of critical importance to the Cold War struggle. In this regard, personalities mattered. Khrushchev, the same man who seemed unable to control his bombastic outbursts, appeared to understand that by engaging the West in a political and cultural contest, he was accomplishing what Stalin had been unable to do: he was breaking the Soviet Union from the international isolation that Stalin's heavy-handed tactics had created. On the American side, Eisenhower played a critical role. He had developed during World War II a deep appreciation of the importance of morale and psychological warfare as central elements of total war. His view of the Cold War was profoundly affected by his experience in World War II. In his mind, both were total conflicts demanding contributions from all aspects of American life. But because the Cold War was not primarily a shooting war, its psychological dimensions loomed larger, and President Eisenhower sought to integrate psychological warfare into all levels of U.S. national security planning.

A New Form of Diplomacy

Perhaps nothing better illustrated the increasing prominence of the propaganda war than the personal form of public diplomacy pursued by Soviet and American leaders at the end of the decade. Although the war in Korea and the possibility of a wider conflagration consumed public attention in 1950, nine years later, the world watched as the American vice president traveled to the Soviet Union on a friendly visit to open a cultural exhibit in Moscow. There, in the heart of the world's most powerful communist state, before an international audience of television viewers, Richard Nixon extolled the abundance of the capitalist system. In a pavilion stocked with American appliances, soft drinks, and Sears catalogs, he stood in a model of a typical American kitchen and had a one-on-one televised debate with Khrushchev over the merits of their rival social systems. Nixon showed the Soviet leader a color television, using it as an illustration of the wonderful commodities that capitalism provided to ordinary people. He prodded Khrushchev to open his country up to foreign ideas, advising him to "not be afraid of ideas. . . .

After all, you don't know everything." To which Khrushchev replied, "If I don't know everything, you don't know anything about communism—except fear of it." According to most American accounts at the time, Nixon "won" this so-called kitchen debate. Yet in hindsight, Khrushchev's performance was more striking. He appears jovial and approachable, even if defensive and impulsive. Although Nixon earned accolades for illustrating the miracles of consumer capitalism, Khrushchev helped his country in the global political competition simply by not appearing menacing. By opening his country up to his capitalistic adversaries, he was softening the image of the Iron Curtain, the preeminent symbol of communist totalitarianism.[3]

Perhaps it was for this reason that Khrushchev tried, almost desperately, to get Eisenhower to invite him for an informal visit to the United States at the close of the decade. Eisenhower resisted Khrushchev's entreaties for the same reason. Ignoring his many hints, the president did not want to accord the Soviet leader any additional prestige; nor did he want to give him a golden opportunity for scoring propaganda points. It was thus entirely an accident that Khrushchev got his chance to visit the United States when a State Department official mistakenly offered him an unconditional invitation. Eisenhower was furious at the official's stupidity, but the damage was done. Khrushchev arrived at Andrews Air Force Base on Tuesday, September 15. It was the first time a Soviet leader had set foot on U.S. soil.[4]

The entire visit was a bizarre propaganda play—part ideological warfare, part improvisational comedy. Khrushchev provided the opening gambit. Pointedly highlighting his country's supposed lead in the space race, he gifted the president with a model of Lunik II, a Soviet rocket that had hit the moon the day before with impeccable timing. Eisenhower, for his part, seemed almost to believe that Khrushchev would be so awed by American prosperity that he would renounce his mistaken communist beliefs. Khrushchev responded by being decidedly unimpressed by everything he saw. He viewed the rows of fine American homes stoically from Eisenhower's helicopter. The countless automobiles he saw were "jamming up the highway" and choking people with exhaust. The Hollywood can-can dancers who surrounded him and raised their skirts suggestively? Mere "pornography." The IBM computers in San Francisco? The Soviet Union had plenty, he said; it used them for rockets.

As the Soviet leader made his way across the country, Americans who encountered him seemed determined to play their parts in the propaganda play. Many made a point of appearing stone-faced. On one of Khrushchev's travel routes, an advance car reportedly carried a sign instructing people,

"No applause—No welcome to Khrushchev." The mayor of Los Angeles promised a "fight to the death." Bystanders repeatedly heckled him. Labor leaders boisterously confronted him with defenses of capitalism. Journalists were relentless. They asked him provocative questions about Stalin's terror, the repression in Hungary, and the intemperate "we will bury you" comment. Just the same, Khrushchev received a hero's welcome when he returned to Moscow. A Soviet publisher rushed to release a book-length account of the visit, *Face to Face with America*, praising the journey as a victory for the USSR. Khrushchev himself viewed the trip as his "hour of glory." He had confirmed the great power status of his country. He had met the capitalists on their own turf and fought for socialism.[5]

Eisenhower sought to steal the spotlight by embarking on two international trips of his own, one before and one after Khrushchev's arrival in the United States. A few weeks before Khrushchev's visit, Eisenhower flew to Western Europe, where cheering crowds gathered to welcome him in Britain, Germany, and France. In December, after the conclusion of Khrushchev's trip, the president embarked on an unprecedented eleven-nation goodwill tour. For three weeks he traveled by jet airplane to India, Pakistan, Afghanistan, Iran, Greece, Turkey, Italy, Spain, Tunisia, Morocco, and France. A few months later, the president took off again. This time he headed south, visiting Argentina, Brazil, Chile, Uruguay, and Puerto Rico. Khrushchev responded with more trips of his own—to India, Burma, Indonesia, Afghanistan, and France, an augmented repeat of an earlier trip he had taken in 1955. As a CBS news reporter observed, the two leaders were conducting "a new form of diplomacy by travel around the world." They were acting as the icons and chief mouthpieces of the propaganda war.[6]

Thanks to mass communications, news of their travels moved almost as quickly as they did. Every stop along the way instantly became an international media event. Heads of state had always been the principal spokespersons for their countries, of course, but such trips would have been less likely in 1950. Tensions were too high, and extensive world travel was still too cumbersome. The ability to quickly travel vast distances, however, was another feature of the communications revolution that transformed international diplomacy. Jet travel became more commonplace at the end of the decade. In August 1959, Eisenhower took his first jet flight aboard Air Force One, a large and comfortable Boeing 707 more suited for traversing the globe than its propeller-driven predecessors.[7]

Eisenhower had been urged to conduct this personal form of public diplomacy by Karl Harr, his special assistant for security operations coordination, a less glamorous version of C. D. Jackson's old post. Harr advised the president that such travel could become a major vehicle for international communication, a publicity device that could shore up American prestige in the shadow of Little Rock, Sputnik, and the fallout scare. It would provide the president a means for directly reaching the people of the world with his message. Not incidentally, it would also allow him to best Khrushchev in the global public relations contest.[8]

Eisenhower saw the journeys as chances to win hearts and minds. Asked by a reporter if he had a "particular sense of mission" as he embarked on the eleven-nation tour, Eisenhower replied, "There is a great deal of doubt remaining in the minds of many people, and including our friends, allies and other friends, as to America's real sincerity in pursuit of peace. We have tried to emphasize this point in every possible way . . . and still it doesn't seem to come through." So, Eisenhower continued, "I decided to make an effort that no President ever was called on before to make. . . . I want to prove that we are not aggressive, that we seek nobody else's territories or possessions; we do not seek to violate anybody else's rights. We are simply trying to be a good partner in this business of searching for peace." He would try, as he had many times before, to improve the image of the United States abroad.[9]

On the eve of his departure, Eisenhower delivered a radio and television address to the American people that again portrayed the trip as an effort to win world public opinion. He would personally seek to dispel the stereotypes about America that his administration had toiled to erode for years. "Our country," he announced, "has been unjustly described as one pursuing only materialistic goals; as building a culture whose hallmarks are gadgets and shallow pleasures; as prizing wealth above ideals, machines above spirit, leisure above learning, and war above peace." Eisenhower would show the world that the United States aspired only to promote human happiness, that Americans were a deeply spiritual people who believed in inalienable rights. These were themes he repeated at virtually every stop, and they were the same themes his psychological warriors had been emphasizing for years. He declared America's interest in friendship with the people of the world and their leaders. He spoke of America's religious faith, its belief in freedom, its hope for worldwide prosperity, its desire for peace. He talked of the need for greater cultural exchanges and person-to-person contacts. He stressed the defensive purposes of the country's military posture and its determination to protect freedom everywhere.[10]

Eisenhower's goodwill tours made great PR. In city after city, throngs of people—by some accounts millions of them—lined the streets to see his motorcade. Indian Prime Minister Nehru said that Eisenhower drew larger crowds than Mohandas Gandhi. The images of the cheering crowds, most from third world countries, made the trip a public relations triumph for the United States. The USIA used it for all it was worth. The agency made the journey the focal point of its global operations and publicized Eisenhower's words and movements with press articles, films, newsreels, photographs, and radio broadcasts. In Morocco, for example, USIA operatives pulled out all the stops. Information officers showed *The Life of Eisenhower* in commercial theaters and made a special film on Eisenhower's visit that they publicized before his arrival, during his stay, and for a long time afterward. They disseminated several hundred thousand pamphlets and souvenir cards featuring photographs of King Mohammed V and Eisenhower. Window displays at USIS branches and military bases hyped Eisenhower's presence. Stretching the truth, posters celebrated "Two Hundred Years of Friendship."[11]

More trips were in the offing. South America was next on the agenda, then a journey to Paris for a summit with Khrushchev, followed by a visit to Moscow. A trip to the USSR would have made a fitting end to Eisenhower's world travels—an unmatched opportunity to spread his message about the peaceful intentions of his country. Sizable crowds had greeted Nixon when he journeyed to the USSR a year earlier. Eisenhower, who had attracted multitudes on his goodwill tour, reasonably could have expected a warmer greeting. Perhaps unsettled by the prospect of being upstaged by the American president, Khrushchev used the U-2 incident as an excuse to walk out of the Paris summit in May 1960. He rescinded Eisenhower's invitation to visit Moscow. No crowds would cheer him in the Soviet Union. Nor would they in Japan, where riotous protests against the security treaty forced the cancellation of a scheduled goodwill trip to that country as well.[12]

The collapse of the Paris summit in May 1960 was a disappointment, but perhaps not much of a missed opportunity for peace. Although Eisenhower and Khrushchev had spent several years trying to convince the world of their peaceful intentions, they had taken only the tiniest steps toward détente at the end of the 1950s. The test ban treaty, so close to completion that spring, would have done nothing to slow the arms race, resolve political disputes in Europe, or restrain the competition in the third world. Détente was not on the horizon; it would not emerge until more than a decade later. Even then, the détente of the 1970s was notably conservative. Conflict persisted despite the vestiges of cooperation. Then, as in the 1950s, the pursuit of détente and the waging of the Cold War were inextricably linked. Leaders needed to appear to be seeking peace to rationalize and to facilitate the continuing struggle against the enemy. The pursuit of détente, as Jeremi Suri suggests, acted as a pressure valve in a world coming apart from revolutionary upheaval—a way to deflate the growing pressure for meaningful progress toward peace while defending the legitimacy of the leaders who pursued it.[13]

This process by which leaders employed the prospect of peace to further their own ends has a longer history. Throughout the twentieth century, world leaders used appeals for peace to bolster their legitimacy at home. They also manipulated the hope of peace to create the psychological conditions and moral space for war. They perceived, as Harold Lasswell did, that hatred and vengeance were necessary, but not sufficient, requirements of total war mobilization. Such passions needed to be softened and made morally acceptable by rhetorical bombast and propaganda framing total war as a communal sacrifice, carried by the entire nation, to bring about a more peaceful and prosperous future.

American leaders evinced a special interest in presenting their involvement in international conflicts as noble quests for higher purposes. Whether it was the senseless carnage of World War I or the global destruction of World War II, Americans were told that they were fighting not for their own prosperity or self-interest, but to preserve peace, to end all wars, to protect the basic human freedoms of themselves and others, including their enemies. The leaders of the total war generation, who commanded American foreign policy during the Cold War, learned this lesson: the American people and their allies would be unified not just by fear and hatred, but by a sense of shared positive values and aspirations for the world to come.

The Global Battle for Hearts and Minds

President Eisenhower understood, better than most, that waging war demanded the moral legitimacy of a pursuit for peace. For eight years, the general who had gained fame through war struggled to present himself as a man of peace. Whereas other leaders at other times emphasized freedom and liberty as inspirational concepts, no theme received greater attention than peace in Eisenhower's rhetoric. Although he is not generally remembered as an orator like John F. Kennedy or Franklin D. Roosevelt, few Americans spoke as eloquently for the cause of peace as Eisenhower did. His presidency both began and ended with warnings of the perils of a prolonged arms race and the dangers of a military industrial complex run amok. But for Eisenhower peace was always a distant goal. He spoke of the "path" to peace and the "road" to peace, setting this noble objective as something attainable far into the future. For the present, mutual accommodation and "coexistence" were not viable options; peace could come only through victory in the Cold War and the retraction of Soviet power. In his public rhetoric, Eisenhower framed the Cold War crusade as a quest for peace; yet his frequent references to "waging peace" suggested that peace was something to be won rather than made.

Eisenhower and many of his top advisors defined waging cold war, and in a sense they defined the Cold War itself, as a psychological contest: a battle of ideologies and a war of persuasion. The discourse of American policy makers reflected this view. Administration officials spoke often and urgently of waging "psychological warfare," "ideological warfare," and the "war of ideas." They conducted "campaigns of persuasion" to reach "target audiences" and to win the "battle for hearts and minds." Not just Eisenhower, but everyone he surrounded himself with saw public opinion as a battleground.

In this context, political warfare emerged as a critical component of the Eisenhower administration's strategy for winning the Cold War. To a remarkable extent, the president involved himself personally in the adoption of strategies intended to make U.S. propaganda more persuasive, credible, and effective. He pushed covert propaganda strategies that hid the hand of the U.S. government. Private groups and surrogate communicators carried the administration's propaganda themes directly to target audiences. A facade of objective news reporting masked propagandistic intent. Ultimately, American information campaigns charted a course between extremes, never quite developing the reputation for objectivity cherished by the BBC, but avoiding the deliberate falsehoods of Joseph Goebbels. Occasionally the anticommunist rhetoric of U.S. information agencies was intemperate and strident, but rarely on a level with the cliché-ridden and intensely ideological mouthpieces of totalitarian regimes.

American propaganda also was not the monolithic product of the state, the way it often appeared in totalitarian societies. Rather, it emerged from a network of governmental and private resources working in cooperation. In every case studied here, private individuals, NGOs, and businesses played important roles. Eisenhower worked actively to involve these private groups in the war of persuasion. He did so in part because nongovernmental voices would be more persuasive than official information programs, and in part because he understood the value of mobilizing the citizenry. Throughout, the cause of total Cold War was framed as the cause of peace. Everyday American citizens were called not so much to attack communism directly, but to prove to the world that America was a peaceful nation. They were participants in the global battle for hearts and minds, and their task was to demonstrate, through words and deeds, America's commitment to peace.

Eisenhower further shifted the emphasis of American propaganda from virulent anticommunism to the promotion of positive themes about the United States. His view of psychological warfare was more nuanced than that of many of his advisors. The general who worked tirelessly to maintain the Allied coalition during World War II and to build the North Atlantic alliance at the outset of the Cold War saw psychological warfare above all else as a means of maintaining free world unity. To Eisenhower, psychological forces were more important as tools for leading an anticommunist coalition than as magic bullets for inducing the collapse of the Soviet regime. Like George Kennan and the architects of containment, he understood that the Cold War was going to be a long struggle. He perceived that the world's communist regimes would eventually collapse by rotting from within, but that it would take time and patience. He said nothing to indicate he shared the exaggerated faith of C. D. Jackson and other psywarriors that propaganda alone could cause the spontaneous collapse of the Soviet system; in fact, he often urged his advisors to think of psychological warfare as a broader effort to boost morale at home and in allied and neutral nations.

Propaganda, narrowly defined, was only the most identifiable aspect of this effort. Eisenhower believed that the ideological competition with the Soviet Union for the allegiance of the world's peoples suffused all U.S. actions and policies with psychological significance. Propaganda was not just used as an accessory to other instruments of foreign policy. The imperative of shaping, influencing, and at times manipulating public opinion pervaded the entire U.S. approach to the Cold War. A wide-ranging study of U.S. information activities commissioned by Eisenhower in 1960 articulated what was by then familiar logic: "Diplomacy increasingly must understand and use public opinion in all countries, open and closed, old and new. It means there needs to be more emphasis on psychological factors in all aspects of our diplomatic behavior: our handling of conferences and negotiations, our representation abroad, particularly in the emerging countries, our selection and training of personnel and our treatment of foreign visitors." Diplomacy itself was becoming increasingly intertwined with efforts to affect world public opinion.[14]

This did not mean that foreign public opinion controlled all U.S. policies, that psychological considerations alone determined U.S. objectives, or that the United States pandered to public sentiment to solicit foreign approval. But psychological considerations acted as a significant influence, if not always a decisive one, on a wide range of policies relating to U.S. foreign relations. Psychological warfare shaped not only diplomatic, economic, and military policies, but also scientific exploration, cultural exchanges, tourism, the publication and production of ideas, and, indeed, everyday life.

The U.S. formula for waging the battle for hearts and minds developed from the American experience with advertising, public relations, and total war mobilization. At times, U.S. propaganda campaigns revolved around slick Madison Avenue themes like People's Capitalism, which sought to sell America through sloganeering like a bar of soap. But more often one can see the effect of public relations expertise in U.S. psychological warfare campaigns that relied on a link between action and influence. Propaganda experts did not just disseminate words and arguments. They created news through action, publicizing events and happenings that kept key themes in the global media.

The Eisenhower administration had done so most notably in the fields of peace and disarmament. The Chance for Peace, Atoms for Peace, and Open Skies initiatives began with policy statements announced by the president and were publicized by well-coordinated campaigns that kept Eisenhower's words and proposals before the world public. News accounts about the initiatives for peaceful atomic energy and international inspection were hyped to the fullest. Atoms for Peace became a multiyear international PR theme blending policy, action, and propaganda to ease world apprehensions about the nuclear arms race. In accordance with Lasswell's formulation that war needed to be presented as a fight between peace-loving people and a ruthless foe, psywarriors worked ceaselessly to portray the United States as motivated by peace and to find ways of dramatizing the irrationality of the enemy. Chance for Peace, Atoms for Peace, and Open Skies served this purpose. Their goal was not primarily to break the disarmament impasse or to pave the way for détente—quite the opposite. Their aim was to dramatize the peaceful intentions of the United States and the implacable intransigence of the USSR in order to demonstrate the necessity of continuing the Cold War struggle.

Public relations techniques also colored American cultural diplomacy. Cultural exchanges and propaganda about everyday life sought to soften the image of the United States by "humanizing" America in the eyes of the world. Like all good public relations campaigns, government-sponsored tours by cultural groups and artists generated favorable news about American life and culture to balance less pleasant representations. Propaganda about everyday life assumed special significance in the Cold War context. As an ideological competition between competing ways of life, the Cold War left few stones unturned. Soviet and American propagandists devoted enormous resources to demonstrating the superiority of their economic, political, and social systems. In this battle to allure world audiences by the promise of progress, all aspects of American life were exposed to intense international scrutiny.

The U.S. Information Agency went to great lengths to present everyday life in the United States in the best possible way. The USIA stressed progress in all of its media and in all of its campaigns. It painted a picture of the United States as a classless society characterized by spiritual vitality as well as prosperity. It was a land where free individuals struggled ceaselessly for progress, in work and labor, in family and home life, and in race relations. Employing a novel approach that differed markedly from communist propaganda tactics, the USIA openly admitted setbacks and shortcomings in race and labor relations so that it could contextualize them as bumps on a road of perpetual progress. Such efforts were not well understood by many Americans, especially segregationists who did not see racial equality as a work in progress. Their opposition to the USIA's "unfinished work" exhibit highlighted the limitations of democratic propaganda: it was difficult to push a uniform message in a society that protected freedom of expression.

But faith in the free marketplace of ideas had limits. Psychological warfare planners became anxious about the growing appeal of communist ideology. They devised a secret program of "doctrinal warfare" that used government connections and financial resources to support the production, publication, and distribution of books and articles that challenged the fundamental precepts of communism and promoted a positive vision of free world values and ideals. Some of these publications resulted from secret government payments that stimulated the authorship of officially sanctioned ideas. Others benefited from various kinds of subsidies that facilitated international distribution. The doctrinal warfare program also used English-language instruction, academic seminars, and educational exchanges as mechanisms for ideological warfare and political influence. Although the plan to promote a doctrinal text on a par with the testaments of Lenin and Stalin foundered, efforts to establish the United States as a world center of intellectual activity were more successful. By 1964 the United States had become the world's largest exporter of books. Many of these were published in translation in dozens of languages, in part because of government assistance that manipulated the global publishing market to serve U.S. interests.

Eisenhower's personal role in supporting all these programs was often pivotal. Without his support, the OCB would have languished in bureaucratic alienation as the PSB had done. The bold initiatives proposed by the psychological warfare advisors would have gone nowhere. Contrary to the once widely held image of Eisenhower kowtowing to John Foster Dulles in matters of foreign policy, Eisenhower often prodded and overruled the secretary of state when it came to the propaganda war. That being said, Eisenhower pursued bold initiatives to win world opinion more vigorously at the beginning of his presidency than at the end. He repeatedly seized the initiative during his first five years, but he struggled to conduct damage control in the remaining three. The most important and concerted information campaigns originated during the early years of the Eisenhower presidency. Two emerged from plans initially developed by Truman's PSB at the end of 1952: the Atoms for Peace and Doctrinal Warfare campaigns were in the bureaucratic pipeline before Eisenhower took office. The Emergency Fund for cultural initiatives, the Open Skies campaign, the People-to-People program, and the People's Capitalism campaign were likewise developed during the first Eisenhower administration, before 1956. One should not read too much into the chronology, however, because many of these campaigns remained in effect throughout the Eisenhower presidency, and some of them long afterward.

Nevertheless the administration had an especially difficult time controlling the message after the fall of 1957. Thereafter, the Little Rock crisis, the space race, and the nuclear testing controversy dominated international media attention. Racial conflict surrounding civil rights clashes had a calamitous impact on world opinion. This was true in Europe and even more so in the third world, where news of racial violence and segregation belied U.S. claims to be the champion of freedom and the enemy of colonialism. It was difficult to talk of progress when federal troops had to escort black children to school. The ongoing nuclear fallout controversy did not help the United States' image either. The nuclear testing debate and the global antinuclear movement revealed that all the "friendly atom" propaganda in the world could not erase anxieties about the nuclear arms race. And then there was Sputnik, a PR bombshell that shook worldwide confidence in U.S. leadership. Sputnik gave Soviet propaganda credibility. The claim that communism was the wave of the future now seemed less fantastic. The Eisenhower administration struggled to demonstrate that the United States remained the world leader in science and technology.

These developments highlighted the central truth about propaganda perceived by the Jackson Committee in 1953. In a global information age, deeds mattered more than words. The United States would be judged more by what it did than by what it said. Moreover, as the civil rights clashes made painfully clear, the American government could not control the images foreign audiences received. The actions of individual Americans could, and often did, have a greater impact on foreign perceptions of the United States.

As significant as Eisenhower's concerted psychological warfare campaigns were, perhaps of greater consequence were the less spectacular efforts to manipulate international perceptions and politics on a daily basis. American foreign relations rested on a wide array of interventionist activities that functioned as tools of empire and weapons of cold war. Through propaganda, policy initiatives, and covert operations, the American government acted directly to influence the ideas, values, beliefs, opinions, actions, politics, and cultures of other countries. It did so to extend U.S. influence abroad and to deny communist regimes economic and political gains. The technologies of the communications revolution, the influence that came with great wealth and power, and the resources for clandestine activities afforded the United States powerful mechanisms for such political warfare.

American psychological warfare experts defended their efforts by claiming that they were merely spreading the "truth" to combat communist lies. Yet although they generally avoided wild fabrications and obvious falsehoods, this simplistic description belied the manipulative elements of U.S. psychological warfare programs. Officials assumed that to win the battle against communism they would have to adopt communist methods. Intelligence estimates, national security policy papers, and diplomatic correspondence routinely declared that the Soviet Union pursued its goal of worldwide domination through subversion, propaganda, covert intervention, and support for indigenous movements. Such documents pointed to the necessity of fighting fire with fire. As a once-classified report on U.S. intelligence activities put it, "It is now clear that we are facing an implacable enemy whose avowed objective is world domination by whatever means and at whatever cost. There are no rules in such a game. Hitherto acceptable norms of human conduct do not apply. If the United States is to survive . . . we . . . must learn to subvert, sabotage and destroy our enemies by more clever, more sophisticated and more effective methods than those used against us." Eisenhower reached the same conclusion. He wrote privately "that some of our traditional ideas of international sportsmanship are scarcely applicable in the morass in which the world now founders."[15]

The experience of total war provided the prism through which Eisenhower and other officials viewed the Cold War. They perceived the Cold War as they had the world war that preceded it: as an all-embracing struggle demanding the mobilization of all the nation's resources. It was clear, the State Department announced early in the conflict, "that a 'cold war' just as much as a modern fighting war must be waged through a national mobilization of every means available for waging such a war." Admittedly the Cold War did not produce the rationing and total economic mobilization of World War II. But this was a different kind of war—an ideological and political one—that demanded a different kind of mobilization. Territory had to be won without military conquest, through persuasion and inducement as well as through the covert manipulation of perceptions and politics. Public opinion needed to be managed, populations mobilized. Just as a military division sends men and material to an exposed flank in a hot war, so too did American strategists seek to reinforce political vulnerabilities in the global battle for hearts and minds. Everything was a target, not least the psychological forces that affected the will, economic orientation, and military vitality of allies and neutrals.[16]

In line with this thinking, officials repeatedly evoked total war imagery to explain and justify their efforts to mobilize the home front for ideological warfare. Eisenhower did so time and again. In his October 1952 campaign speech on psychological strategy, Eisenhower had painted the global battle for hearts and minds as an all-encompassing effort involving such things as sporting events, cultural attractions, economic activities, education, trade, diplomacy, and scientific achievement. He reiterated these themes from the beginning of his presidency to the end. The declaration of "total cold war" in his 1958 State of the Union address was the culmination of five years of speeches instructing the American people in the duties and sacrifices of citizenship in an "age of peril." As he explained in first inaugural address, all aspects of American life were affected by the demands and the requirements of the Cold War. "No principle or treasure that we hold, from the spiritual knowledge of our free schools and churches to the creative magic of free labor and capital, nothing lies safely beyond the reach of this struggle. . . . No person, no home, no community can be beyond the reach of this call." In the anticommunist crusade, "each citizen plays an indispensable role."

Even as Eisenhower warned of the perils of the military industrial complex in his famous farewell address, he urged the American people to prepare for protracted Cold War. "We face a hostile ideology—global in scope, atheistic in character, ruthless in purpose, and insidious in method," he warned. "Unhappily the danger it poses promises to be of indefinite duration. To meet it successfully, there is called for, not so much the emotional and transitory sacrifices of crisis, but rather those which enable us to carry forward steadily, surely, and without complaint the burdens of a prolonged and complex struggle." The American people, Eisenhower might have said if he had had a different speech writer, should pay any price and bear any burden to win the Cold War.[17]

NOTES

Introduction

1. Gallup asked respondents if they thought there was "much danger of world war" or if there would be "another world war in your lifetime." The percentage of persons who answered "yes" ranged from a low of 37 percent in January 1956 to a high of 72 percent in March 1955. Most affirmative answers rested in the 60 to 70 percent range. See Gallup Poll numbers 575 (November 20, 1956), 566 (June 13, 1956), 558 (January 4, 1956), 548 (June 3, 1955), 544 (March 1, 1955), and 533 (June 30, 1954). Gallup also asked Americans on a recurring basis from 1950 to 1958 whether they expected atomic or hydrogen bombs to be used in the event of another world war. The percentage of persons who answered "yes" ranged from a low of 59 percent in 1956 to a high 75 percent in 1958. See Gallup Poll numbers 598 (April 14, 1958), 582 (April 25, 1957), 575 (November 20, 1956), 529 (April 8, 1954), and 455 (May 2, 1950). Other polls asked Americans whether they felt it was likely their communities would be attacked with nuclear weapons in the event of World War III. A total of over 50 percent replied consistently that they thought there was a "good chance" or a "fair chance" that their communities would be attacked. See Gallup Polls 529 (April 4, 1954), 517 (July 2, 1953), and 463 (October 6, 1950). Gallup Polls cited are available at http://brain.gallup.com, a database of historical polling data available through most research libraries.

2. Gallup Poll numbers 557 (December 6, 1955), 453 (February 24, 1950), and 432 (November 3, 1948); all available at http://brain.gallup.com.

3. John Lewis Gaddis, *The Long Peace: Inquiries into the History of the Cold War* (New York: Oxford University Press, 1987); William Stueck, *The Korean War: An International History* (Princeton: Princeton University Press, 1995); Marc Trachtenberg, *A Constructed Peace: The Making of the European Settlement, 1945–1963* (Princeton: Princeton University Press, 1999).

4. For a defense of the "national history" approach to U.S. foreign relations, see Robert J. McMahon, "Toward a Pluralist Vision: The Study of American Foreign Relations as International and National History," in *Explaining the History of American Foreign Relations*, ed. Michael J. Hogan and Thomas G. Paterson, 2nd ed. (New York: Cambridge University Press, 2004), 35–50.

5. I borrow the term "politically activated" from Zbigniew Brzezinski, *The Choice: Global Domination or Global Leadership* (New York: Basic Books, 2004).

6. House Committee on Foreign Affairs, *Winning the Cold War: The U.S. Ideological Offensive*, 88th Congress, 2nd Session, 1964, 6–7; William H. Jackson, "The Fourth Area of the National Effort in Foreign Affairs," November 1956, Ann Whitman File, Administration Series, box 22, William H. Jackson, Eisenhower Library, Abilene, Kansas (hereafter EL); Philip Hall Coombs, *The Fourth Dimension of Foreign Policy* (New York: Harper & Row, 1964). See also Robert S. Byfeld, *The Fifth Weapon: A Guide to Understanding What The Communists Mean* (New York: Bookmailer, 1954).

7. Nancy E. Bernhard, "Clearer than Truth: Public Affairs Television and the State Department's Domestic Information Campaign, 1947–1952," *Diplomatic History* 21 (Fall 1997): 546. The participation of private groups in Cold War propaganda campaigns has been analyzed recently by Scott Lucas, *Freedom's War: The American Crusade Against the Soviet Union* (New York: New York University Press, 1999).

8. Eisenhower to George Sloan, March 20, 1952, in *The Papers of Dwight D. Eisenhower*, ed. Alfred D. Chandler and Louis Galambos (Baltimore: Johns Hopkins University Press, 1970–1996), 13:1103. Historians who have noted Eisenhower's affinity for psychological warfare include: Blanche Wiesen Cook, *The Declassified Eisenhower: A Divided Legacy* (Garden City, N.Y.: Doubleday, 1981), 14; John Lewis Gaddis, *Strategies of Containment: A Critical Appraisal of Postwar American National Security Policy* (New York: Oxford University Press, 1982), 154–57; Walter L. Hixson, *Parting the Curtain: Propaganda, Culture, and the Cold War, 1945–1961* (New York: St. Martin's Griffin, 1997); H. W. Brands, *Cold Warriors: Eisenhower's Generation and American Foreign Policy* (New York: Columbia University Press, 1988), 117–37. These analyses have recently been supported by communication specialists who argue that Eisenhower was a skilled psychological warrior. See Martin J. Medhurst, "Eisenhower's 'Atoms for Peace' Speech: A Case Study in the Strategic Use of Language," *Communication Monographs* 54 (June 1987): 204–20; Medhurst, "Atoms for Peace and Nuclear Hegemony: The Rhetorical Structure of a Cold War Campaign," *Armed Forces and Society* 24 (Summer 1997): 571–93; Medhurst, *Dwight D. Eisenhower: Strategic Communicator* (Westport, Conn.: Greenwood Press, 1993); J. Michael Hogan, "Eisenhower and Open Skies: A Case Study in Psychological Warfare," in *Eisenhower's War of Words: Rhetoric and Leadership*, ed. Medhurst (East Lansing: Michigan State University Press, 1994), 137–56; and Shawn J. Parry-Giles, "The Eisenhower Administration's Conceptualization of the USIA: The Development of Overt and Covert Propaganda Strategies," *Presidential Studies Quarterly* 24 (Spring 1994): 263–76; Parry-Giles, "'Camouflaged' Propaganda: The Truman and Eisenhower Administrations' Covert Manipulation of News," *Western Journal of Communication* 60 (Spring 1996): 146–67.

9. Robert A. Divine, *Eisenhower and the Cold War* (New York: Oxford University Press, 1981); Robert R. Bowie and Richard H. Immerman, *Waging Peace: How*

Eisenhower Shaped an Enduring Cold War Strategy (New York: Oxford University Press, 1998); Stephen E. Ambrose, *Eisenhower: The President* (New York: Simon and Schuster, 1984). On other issues, the revisionists have been challenged by a growing body of "postrevisionist" scholars. This disparate group of historians generally accepts the revisionist view that Eisenhower presided actively over his administration but raises questions about the prudence, restraint, effectiveness, and direction of Eisenhower's leadership. Stephen G. Rabe, for example, writes that although Eisenhower was "decisive, energetic, and well-informed . . . it does not follow, as 'Eisenhower revisionists' would have it, that his stewardship served as an enduring model of presidential restraint." Or, as Chester Pach adds, "Too often, it seems, revisionists mistook Eisenhower's cognizance of policies for brilliance and his avoidance of war for the promotion of peace." Stephen G. Rabe, *Eisenhower and Latin America: The Foreign Policy of Anti-Communism* (Chapel Hill: University of North Carolina Press, 1988), 174–75; Chester J. Pach Jr. and Elmo Richardson, *The Presidency of Dwight D. Eisenhower* (Lawrence: University Press of Kansas, 1991), xiii. For historiographical discussion see John Robert Greene, "Bibliographic Essay: Eisenhower Revisionism, 1952–1992, a Reappraisal," in *Reexamining the Eisenhower Presidency*, ed. Shirley Anne Warshaw (Westport, Conn.: Greenwood Press, 1993), 209–19; Rabe, "Eisenhower Revisionism: A Decade of Scholarship," *Diplomatic History* 17 (Winter 1993): 97–115; and Robert J. McMahon, "Eisenhower and Third World Nationalism: A Critique of the Revisionists," *Political Science Quarterly* 101 (Fall 1986): 453–73.

10. The term "propaganda" has spawned as many definitions as it has euphemisms. Harold Lasswell defined it as "the management of collective attitudes by the manipulation of significant symbols." Like other social scientists in the 1930s, he emphasized the psychological elements of propaganda: propaganda was a subconscious manipulation of psychological symbols to accomplish secret objectives. Subsequent analysts stressed that propaganda could be identified by looking at the intentions of the sponsor: propaganda was defined as a *planned* and *deliberate* act of opinion management. More recently, historian Oliver Thomson defined propaganda broadly to include both deliberate and unintentional means of behavior modification—"the use of communication skills of all kinds to achieve attitudinal or behavioural changes among one group by another." A number of communication specialists have stressed that propaganda is a neutral activity, concerned only with persuasion, in order to free propaganda (and their profession) from pejorative associations. Some social scientists have abandoned the term altogether because it cannot be defined with any degree of precision. Others, like the influential French philosopher Jacques Ellul, have used the term but refused to define it because any definition would inevitably leave something out. See Oliver Thomson, *Easily Led: A History of Propaganda* (Stroud, Gloucestershire: Sutton Publishing, 1999); Jacques Ellul, *Propaganda: The Formation of Men's Attitudes* (New York: Vintage Books, 1973); William E. Daugherty and Morris Janowitz, eds., *A Psychological Warfare*

Casebook (Baltimore: Johns Hopkins University Press, 1958), 1–34; David Welch, "Power of Persuasion," *History Today* 49 (August 1999): 24–26.

11. For discussion, see Daugherty and Janowitz, *Psychological Warfare Casebook*, 1–34; Mario Del Pero, "The United States and 'Psychological Warfare' in Italy, 1948–1955," *Journal of American History* 87 (March 2001): 1304–34.

12. Welch, "Power of Persuasion." Other phrases commonly employed during the Cold War included "the war of ideas," "the struggle for the minds and wills of men," "thought war," "ideological warfare," "nerve warfare," "campaign of truth," and "war of words." Even the term "cold war" was used to refer to propaganda techniques and strategy—as in "cold war tactics" or "cold war calculations." Increasingly, psychological warfare specialists turned to less bellicose terms like "international communication," "public diplomacy," and "public communication" to make the idea of propaganda more palatable to domestic audiences. Nowadays, many of the administration's psychological warfare activities might be more readily recognized as "public relations" or simply "communication." The widespread use of these terms to refer to the manipulation of the instruments of mass communication, however, grew self-consciously out of psychological warfare planning circles to make the psychological warfare rose smell sweeter to a public that regarded propaganda as an instrument of totalitarian oppression. See Christopher Simpson, *Science of Coercion: Communication Research and Psychological Warfare, 1945–1960* (New York: Oxford University Press, 1994), 85–86; and Daugherty and Janowitz, *Psychological Warfare Casebook*, 1–34.

13. On the propaganda utility of the term "free world," see USIA Strategic Principles, n.d. [March 1954], OCB Central Files, Box 20, OCB 040 USIA (1), EL.

14. See Lucas, *Freedom's War*; Hixson, *Parting the Curtain*; Gregory Mitrovich, *Undermining the Kremlin: America's Strategy to Subvert the Soviet Bloc, 1947–1956* (Ithaca, N.Y.: Cornell University Press, 2000); and Peter Grose, *Operation Rollback: America's Secret War Behind the Iron Curtain* (Boston: Houghton Mifflin, 2000). There is also a rich body of literature on the covert broadcasting of the CIA through Radio Free Europe and Radio Liberty. The most recent is Arch Puddington, *Broadcasting Freedom: The Cold War Triumph of Radio Free Europe and Radio Liberty* (Lexington: University Press of Kentucky, 2000).

15. Although the USIA conducted polls with increasing regularity at the end of the decade, it generally operated under the assumption that media representations both reflected and determined public opinion. For the Pew survey, see http://www.pewtrusts.com/pdf/vf_pew_research_global_attitudes_0603.pdf.

Chapter 1. Regimenting the Public Mind

1. Harold Dwight Lasswell, *Propaganda Technique in the World War* (New York: Alfred A. Knopf, 1927), 34.

2. Carl von Clausewitz, *On War*, ed. and trans. Michael Howard and Peter

Paret (Princeton: Princeton University Press, 1976); Sun Tzu, *The Art of Warfare*, trans. Roger T. Ames (New York: Ballantine, 1993). On propaganda and psychological warfare in world history, see Philip M. Taylor, *Munitions of the Mind: War Propaganda from the Ancient World to the Nuclear Age* (Glasgow: Patrick Stephens, 1990); and Oliver Thomson, *Easily Led: A History of Propaganda* (Stroud, Gloucestershire: Sutton Publishing, 1999).

3. Anthony R. Pratkanis and Elliott Aronson, *Age of Propaganda: The Everyday Use and Abuse of Persuasion* (New York: W. H. Freeman, 1991); Larry Tye, *The Father of Spin: Edward L. Bernays and the Birth of Public Relations* (New York: Henry Holt, 1998), 94; David Crowley and Paul Heyers, eds., *Communication History: Technology, Culture, Society*, 2nd ed. (White Plains, N.Y.: Longman, 1995), 145–262; Edwin Emery, Phillip H. Ault, and Warren K. Agee, *Introduction to Mass Communications*, 4th ed. (New York: Dodd, Mead, 1974).

4. John Hartley, *Communications, Cultural and Media Studies: The Key Concepts* (London: Routledge, 2002), 139; Stuart Ewen, *PR! A Social History of Spin* (New York: Basic Books, 1996), 39–60; Brett Gary, *The Nervous Liberals: Propaganda Anxieties from World War I to the Cold War* (New York: Columbia University Press, 1999), 25–27. Scholars have raised serious doubts about whether such a creature as "mass society" ever existed, but it is clear that intellectuals at the turn of the century perceived it that way. For a critical assessment of the "mass society" thesis, see Richard F. Hamilton, *Mass Society, Pluralism, and Bureaucracy: Explication, Assessment, and Commentary* (Westport, Conn.: Praeger, 2001), 7–66. Gustave Le Bon, *The Crowd* (New Brunswick, N.J.: Transaction Publishers, 1995), 34–35. The original English translation is *The Crowd: A Study of the Popular Mind* (New York: Macmillan, 1896). For analysis of the meaning and impact of Le Bon's work, see Ewen, *PR!*, 60–73, 146–73.

5. Ewen, *PR!*, 64; Walter Lippmann, *Public Opinion* (New York: Harcourt, Brace, 1922), 75; Walter Lippmann, *The Phantom Public* (New York: Macmillan, 1927), 24; Tye, *Father of Spin*, 197. Other "mass society" theorists included Gabrial Tarde, *The Laws of Imitation* (New York: Henry Holt, 1903); Wilfred Trotter, *Instincts of the Herd in Peace and War* (New York: Macmillan, 1916); and Graham Wallas, *The Great Society: A Psychological Analysis* (New York: Macmillan, 1917).

6. Scott M. Cutlip, *Public Relations History: From the 17th Century to the 20th Century* (Hillsdale, N.J.: Lawrence Erlbaum Associates, 1995), 187–209; Dennis L. Wilcox, Phillip H. Ault, and Warren K. Agee, *Public Relations: Strategies and Tactics*, 3rd ed. (New York: HarperCollins, 1992), 41–43. For a good history of the public relations industry, see Roland Marchand, *Creating the Corporate Soul: The Rise of Public Relations and Corporate Imagery in American Big Business* (Berkeley: University of California Press, 2001). On Lee, see Ray Eldon Hiebert, *Courtier to the Crowd: The Story of Ivy Lee and the Development of Public Relations* (Ames: Iowa State University Press, 1966).

7. Ewen, *PR!*, 163–67. Although preceded by Lee and others, Bernays's contribution to the field was so significant that he is often identified as the "father of public relations." For discussion, see Tye, *Father of Spin*, 227–50.

8. Tye, *Father of Spin*, 55, 101, 103; Bernays, *Propaganda* (New York: Liveright Publishing, 1928), 49; Ewen, *PR!*, 160.

9. Tye, *The Father of Spin*, 28–35, 55–58, 63–69; Ewen, *PR!*, 216–17. Quote from Tye, *Father of Spin*, 40.

10. The impact of the yellow press on the Spanish-American War is still hotly debated by historians. For a succinct summary of the role of the media and historical debate, see Lyn Gorman and David McLean, *Media and Society in the Twentieth Century: A Historical Introduction* (Malden, Mass.: Blackwell, 2003), 15–17.

11. Taylor, *Munitions of the Mind*, 161; Gorman and McLean, *Media and Society*, 18–21; Brett Gary, *Nervous Liberals*, 22. Quotes from Lasswell, *Propaganda Technique*, 10.

12. Emily S. Rosenberg, *Spreading the American Dream: American Economic and Cultural Expansion, 1890–1945* (New York: Hill and Wang, 1982); Gorman and McLean, *Media and Society*, 15–20; Taylor, *Munitions of the Mind*, 163–79.

13. Quoted in Robert Jackall and Janice M. Hirota, *Image Makers: Advertising, Public Relations, and the Ethos of Advocacy* (Chicago: University of Chicago Press, 2000), 14; Rosenberg, *Spreading the American Dream*, 63–108.

14. George Creel, *Rebel at Large: Recollections of Fifty Crowded Years* (New York: G. P. Putnam's Sons, 1947), 163; Jackall and Hirota, *Image Makers*, 19, 24; Rosenberg, *Spreading the American Dream*, 79; James R. Mock and Cedric Larson, *Words that Won the War: The Story of the Committee on Public Information, 1917–1919* (Princeton: Princeton University Press, 1939); Taylor, *Munitions of the Mind*, 168–75; Lasswell, *Propaganda Technique*, 211–12.

15. Lasswell, *Propaganda Technique*, 216–17. "Such matchless skill as Wilson showed in propaganda," Harold Lasswell observed exuberantly after the war, "has never been equaled in world's history."

16. Frank Ninkovich, *The Wilsonian Century: U.S. Foreign Policy Since 1900* (Chicago: University of Chicago Press, 1999), 69; Rosenberg, *Spreading the American Dream*, 93.

17. Ninkovich, *Wilsonian Century*, 48–76. Quote on 67–68; Rosenberg, *Spreading the American Dream*, 63–95. Wilson's famous Fourteen Points speech was a notable example of psychological warfare emanating from the White House. Packaging American war aims in "short almost placard paragraphs" at the suggestion of one of Creel's agents, it was interlaced with numerous psychological goals. It sought to keep Russia in the war, encourage Germany to make peace, and at the same time exert pressure on Allied governments to pursue a nonputative peace and implement Wilson's postwar vision. See Edgar Sisson, "Sisson's Account of Wilson's Fourteen Points Speech," in *A Psychological Warfare Casebook*, ed. William E. Daugherty and Morris Janowitz (Baltimore: Johns Hopkins University Press, 1958), 89–96.

18. Gary, *Nervous Liberals*, 59, 24. On the soap-corpse atrocity story, see Lasswell, *Propaganda Technique*, 206–7.

19. Bernays, *Propaganda*, 27; Wilcox et al., *Public Relations*, 48; Gary, *Nervous Liberals*, 72. On Lippmann, see Ronald Steel, *Walter Lippmann and the American Century* (New York: Vintage Books, 1981), 172.

20. Gary, *Nervous Liberals*, 59–64. Lasswell quoted on 64. Brett Gary notes that Lasswell's classic study established him as one of the "'founding fathers' of the scientific school of mass communications" and "set the stage for four decades of scholarship on propaganda, political psychology, social science methodologies, the symbols of nationalism, and the study of power." See Gary for detailed analysis of Lasswell's ideas. For a short biographical introduction to Lasswell and his contribution to mass communication studies, see Everett M. Rogers, *A History of Communication Study: A Biographical Approach* (New York: Free Press, 1994), 203–43.

21. Lasswell, *Propaganda Technique*, 195.

22. Lasswell, *Propaganda Technique*, 47, 60, 102, 105.

23. Lasswell, *Propaganda Technique*, 11.

24. Taylor, *Munitions of the Mind*, 180–92; Thomson, *Easily Led*, 268–84.

25. Clayton D. Laurie, *The Propaganda Warriors: America's Crusade Against Nazi Germany* (Lawrence: University Press of Kansas, 1996), 9–12, 21. Laurie argues that many American observers wildly exaggerated the power of Nazi psychological warfare, accepting at face value the claims made by Hitler and his propaganda chief Joseph Goebbels that their propaganda had broken the will of their adversaries. Groups advocating a propaganda offensive included many of the leading political scientists who had developed propaganda expertise as well as others who would play a leading role in devising propaganda strategy during World War II and the Cold War, including Edward Bernays, Harold Lasswell, Hadley Cantril, Leonard Doob, Archibald MacLeish, Edmund Taylor, and C. D. Jackson. See Laurie, *Propaganda Warriors*, 29–44.

26. As John Morton Blum notes, FDR viewed Woodrow Wilson's propaganda excesses as the source of his political defeat: his crusading rhetoric raised public expectations to unrealistic levels and created a counterproductive hatred of the enemy. See John Morton Blum, *V Was for Victory: Politics and American Culture During World War II* (San Diego: Harcourt Brace Jovanovich, 1976), 8–10.

27. For general overviews, see Rosenberg, *Spreading the American Dream*, 206–9; Cary Reich, *The Life of Nelson Rockefeller: Worlds to Conquer, 1908–1958* (New York: Doubleday, 1996), 265–376; Frank A. Ninkovich, *The Diplomacy of Ideas: U.S. Foreign Policy and Cultural Relations, 1938–1950* (New York: Cambridge University Press, 1981), 35–50.

28. "Psychological Warfare," n.d., Hazeltine Papers, box 1, PWB (4), Eisenhower Library, Abilene, Kansas (hereafter EL); Alfred H. Paddock, *U.S. Army Special Warfare: Its Origins*, rev. ed. (Lawrence: University Press of Kansas, 2002), 28. For an overview of OSS morale operations, see Laurie, *Propaganda Warriors*, 192–209. On the OWI, see Allan M. Winkler, *The Politics of Propaganda: The Office*

of War Information, 1942–1945 (New Haven, Conn.: Yale University Press, 1978); Laurie, *Propaganda Warriors,* 112–27 and 210–32.

29. Paddock, *U.S. Army Special Warfare,* 6. SHAEF Operation Memorandum, "Psychological Warfare," March 11, 1944, C. D. Jackson Papers, box 10, Overlord-London (3), EL; "The Psychological Warfare Branch of A.F.H.Q. from its Inception to Nov. 1, 1943," Hazeltine Papers, box 1, PWB (3), EL; Statement by McClure, May 2, 1945, C. D. Jackson Papers, box 10, McClure-General-Paris (4), EL; C. D. Jackson, "Psychological Warfare," draft article for *New York Times Magazine,* June 16, 1944, Jackson Papers, box 10, *New York Times,* EL; and C. D. Jackson Interview, October 17, 1950, Jackson Papers, box 17, EL. On the Pacific theater, see also Allison B. Gilmore, *You Can't Fight Tanks with Bayonets: Psychological Warfare Against the Japanese Army in the Southwest Pacific* (Lincoln: University of Nebraska Press, 1998); on the European theater, see Laurie, *Propaganda Warriors.* On psywar in North Africa and the Mediterranean, see G. A. Martelli, "The Psychological Warfare Branch of A.F.H.Q.," November 5, 1943, Hazeltine Papers, box 1, PWB (3), EL; "General Outline of PWB Operations in the Light of Expansion of 'Husky,'" June 21, 1943, Hazeltine Papers, box 1, PWB (2), EL; "Report of PWB Activities, AFHQ," Hazeltine Papers, box 1, PWB (3), EL. On Normandy, see "Plan for PWD/SHAEF Psychological Warfare," C. D. Jackson Papers, box 10, Overlord-London (2), EL; "Phasing of Psychological Warfare in Connection with Overlord," C. D. Jackson Papers, box 10, Overlord-London (1), EL.

30. Blum, *V Was for Victory,* 3–52, Morganthau quotes on 16–17; William L. O'Neil, *A Democracy at War: America's Fight at Home and Abroad in World War II* (New York: Free Press, 1993), 253–55; Clayton R. Koppes and Gregory D. Black, *Hollywood Goes to War: How Politics, Profits, and Propaganda Shaped World War II Movies* (Berkeley: University of California Press, 1987); John W. Dower, *War Without Mercy: Race and Power in the Pacific War* (New York: Pantheon Books, 1986).

31. E. Lilly, "Short History of the Psychological Strategy Board," White House Office [WHO], NSC Staff Papers, OCB Secretariat Series, box 6, EL.

32. Shawn J. Parry-Giles, *The Rhetorical Presidency, Propaganda, and the Cold War, 1945–1955* (Westport, Conn.: Praeger, 2002), 5. Truman quoted in Walter Hixson, *Parting the Curtain: Propaganda, Culture, and the Cold War, 1945–1961* (New York: St. Martin's Griffin, 1997), 5. On OWI's domestic troubles, see Winkler, *Politics of Propaganda.*

33. For international overviews of Cold War propaganda battles, see Taylor, *Munitions of the Mind,* 220–32; Thomson, *Easily Led,* 268–300; and especially Gorman and McLean, *Media and Society,* 104–25. Nazi "reeducation" programs also contributed to the legitimization of U.S. propaganda activities. On these activities, see Jessica Gienow-Hecht, *Transmission Impossible: American Journalism as Cultural Diplomacy in Postwar Germany, 1945–1955* (Baton Rouge: University of Louisiana Press, 1999); Reinhold Wagnleitner, *Coca-Colonization and the Cold War: The Cul-*

tural Mission of the United States in Austria After the Second World War (Chapel Hill: University of North Carolina Press, 1994). On propaganda and the Cold War consensus, see Daniel L. Lykins, *From Total War to Total Diplomacy: The Advertising Council and the Construction of the Cold War Consensus* (Westport, Conn.: Praeger, 2003); Nancy Bernhard, *U.S. Television News and Cold War Propaganda, 1947–1960* (Cambridge: Cambridge University Press, 1999); Richard M. Fried, *The Russians Are Coming! The Russians Are Coming! Pageantry and Patriotism in Cold-War America* (New York: Oxford University Press, 1998); Michael Wala, "Selling the Marshall Plan at Home: The Committee for the Marshall Plan to Aid European Recovery," *Diplomatic History* 10 (Summer 1986): 247–65.

34. Mario Del Pero, "The United States and 'Psychological Warfare' in Italy, 1948–1955," *Journal of American History* 87 (March 2001): 1304–34; Mario Del Pero, "The Cold War as a 'Total Symbolic War': United States Psychological Warfare Plans in Italy During the 1950s," paper presented to Cold War History conference, University of California, Santa Barbara, May 2000. In classified communications, Eisenhower and American policy makers (especially those working on psychological strategy) frequently used the terms "psychological warfare" and "cold war" interchangeably, as in "cold war operations."

35. On representations of the Communist Menace in American society and culture, see Stephen J. Whitfield, *The Culture of the Cold War,* 2nd ed. (Baltimore: Johns Hopkins University Press, 1991); Michael Barson and Steven Heller, *Red Scared! The Commie Menace in Propaganda and Popular Culture* (San Francisco: Chronicle Books, 2001); and Peter J. Kuznick and James Gilbert, eds., *Rethinking Cold War Culture* (Washington, D.C.: Smithsonian Institution Press, 2001).

36. Melvyn Leffler, *A Preponderance of Power: National Security, the Truman Administration, and the Cold War* (Stanford: Stanford University Press, 1992), 210 and 359; CIA Review of the World Situation, September 26, 1947, quoted in Lucas, *Freedom's War: The American Crusade Against the Soviet Union* (New York: New York University Press, 1999), 43. According to Leffler, U.S. officials did not see the Soviet threat primarily as a military one. American national security, he writes, "was defined in terms of the potential control of resources and industrial infrastructure, which were threatened in the short run not by Soviet military capabilities but by indigenous unrest . . . Policymakers and intelligence analysts agreed that, although Soviet armies had the capability to overrun much of Europe, Stalin and his comrades wanted to avoid a military clash with the West. The men who ruled the Kremlin were not bold adventurists but domestic tyrants and prudent expansionists who knew they could not win a war against the United States." Scott Lucas adds: "The main threat of Soviet ideology was not through military deployment or diplomatic maneuver but through propaganda, subversion, and manipulation of the 'private' exchange of ideas." See Lucas, *Freedom's War,* 18; and also Gregory Mitrovich, *Undermining the Kremlin: America's Strategy to Subvert the Soviet Bloc, 1947–1956* (Ithaca, N.Y.: Cornell University Press, 2000), 82.

37. NSC 4, Report by the National Security Council on Coordination of Foreign Information Measures, December 9, 1947; and NSC 4-A, Psychological Operations, December 9, 1947; *Foreign Relations of the United States 1945–1950: Emergence of the Intelligence Establishment* (hereafter *FRUS*), Document 252 and 253, available at http://www.state.gov/about_state/history/intel/250_259.html. See also Lilly, "Short History."

38. Hixson, *Parting the Curtain*, 11; Lilly, "Short History." American misgivings about government-sponsored propaganda made the Smith-Mundt Act one of the most controversial pieces of legislation in U.S. history. By the time it was passed, it had been rewritten twice and it had acquired more than 100 amendments. It also earned more days of debate and filled more pages of the Congressional Record than the deeply controversial Taft-Hartley Act. Ultimately, the wartime experience with propaganda played a pivotal role in overcoming opposition to the legislation. The campaign to institutionalize a propaganda program received support from military personnel with psychological warfare experience, including Robert McClure, C. D. Jackson, and Eisenhower, who stressed that World War II had proven the effectiveness of propaganda and its indispensability to the achievement of national objectives. Moreover, the fact that many OWI personnel were journalists who returned to the private news industry when the war ended created a large network of opinion shapers who viewed the U.S. information program positively as a force for democratic change and progress. William Benton, who spearheaded the State Department's campaign to make the overt information program permanent, actively courted this "OWI alumni network" for editorial support, a strategy that helped give the campaign for a peacetime propaganda network a positive hearing in the press. Parry-Giles, *Rhetorical Presidency*, 6–7; John W. Henderson, *The United States Information Agency* (New York: Praeger, 1969), 41. The Voice of America remained embroiled in conflict throughout the early Cold War. See David F. Krugler, *The Voice of America and the Domestic Propaganda Battles, 1945–1953* (Columbia: University of Missouri Press, 2000).

39. Leffler, *Preponderance of Power*, 194–98; Lilly, "Short History"; Lucas, *Freedom's War*, 37–46; Hans-Jurgen Schröder, "Marshall Plan Propaganda in Austria and Western Germany," in *The Marshall Plan in Austria*, ed. Günter Bischof, Anton Pelinka, and Dieter Stiefel (New Brunswick, N.J.: Transaction Publishers, 2000), 212–46; Sallie Pisani, *The CIA and the Marshall Plan* (Lawrence: University Press of Kansas, 1991); Trevor Barnes, "The Secret Cold War: The CIA and American Foreign Policy in Europe, 1946–1956. Part I," *Historical Journal* 24 (June 1981): 399–415; and Barnes, "The Secret Cold War: The CIA and American Foreign Policy in Europe, 1946–1956. Part II," *Historical Journal* 25 (September 1982): 649–70.

40. Kennan quoted in Lucas, *Freedom's War*, 63; Policy Planning Staff Memorandum, May 4, 1948, *FRUS 1945–1950: Emergence of the Intelligence Establishment*, Document 269, available at http://www.state.gov/www/about_state/history/intel/260_269.html.

41. NSC 10/2, June 18, 1948, *FRUS, 1945–1950: Emergence of the Intelligence Establishment*, Document 292, available at http://www.state.gov/www/about_state/history/intel/290_300.html. Peter Grose, *Operation Rollback: America's Secret War Behind the Iron Curtain* (Boston: Houghton Mifflin, 2000), 8; Lilly, "Short History."

42. NSC 20/4, November 23, 1948, *FRUS 1948*, 1:662–69; George Morgan quoted in Lucas, *Freedom's War*, 145. For a review of the new literature on rollback, see Kenneth A. Osgood, "Hearts and Minds: The Unconventional Cold War," *Journal of Cold War Studies* 4 (Spring 2002): 85–107. For a provocative discussion of the containment thesis, see Sarah-Jane Corke, "History, Historians and the Naming of Foreign Policy: A Postmodern Reflection on American Strategic Thinking During the Truman Administration," *Intelligence and National Security* 16 (Fall 2001): 146–65.

43. See Grose, *Operation Rollback*.

44. Arch Puddington, *Broadcasting Freedom: The Cold War Triumph of Radio Free Europe and Radio Liberty* (Lexington: University Press of Kentucky, 2000), 14 and 43. Radio Liberty was originally named Radio Liberation from Bolshevism but changed its name in 1963.

45. Christopher Simpson, *Blowback: America's Recruitment of Nazis and Its Effects on the Cold War* (New York: Weidenfeld and Nicolson, 1988), 228; Puddington, *Broadcasting Freedom*, 21.

46. Scholars once interpreted NSC-68 as a turning point in U.S. strategy away from the original policy of containment, but it is now generally agreed, as Leffler argues, that the document's significance was simply that it called for "more, more, and more money to implement the programs and to achieve the goals already set out." See Leffler, *Preponderance of Power*, 355–60; Michael J. Hogan, *A Cross of Iron: Harry S. Truman and the Origins of the National Security State, 1945–1954* (Cambridge: Cambridge University Press, 1998), 295–312; Lucas, *Freedom's War*, 79–80; and Mitrovich, *Undermining the Kremlin*, 57–59.

47. NSC-68, April 14, 1950, *FRUS 1950*, 1:234–92. Hogan, *Cross of Iron*, 296–300; Mitrovich, *Undermining the Kremlin*, 49–59. Quotes on 49 and 58.

48. Edward Barrett quoted in Lucas, *Freedom's War*, 84, and in Hogan, *Cross of Iron*, 301. Truman in Lucas, *Freedom's War*, 84.

49. Lucas, *Freedom's War*, 128; Mitrovich, *Undermining the Kremlin*, 59. VOA quoted in Hixson, *Parting the Curtain*, 38.

50. Wallace Carroll quoted in Hixson, *Parting the Curtain*, 18. PSB D-38, Psychological Strategy Plan for Western Europe, June 30, 1953, PSB Central Files, box 17, PSB 091.4 Western Europe, EL; PSB D-22, Psychological Strategy Program for the Middle East, January 8, 1953, PSB Central Files, box 15, PSB 091.4 Middle East, EL; PSB D-21/2, February 3, 1953, Psychological Strategy with Respect to Berlin, February 3, 1953, PSB Central Files, PSB 091 Germany (2), EL;

PSB D-14, Psychological Plan for the Reduction of Communist Power in France, January 26, 1952, PSB Central Files, box 12, PSB Central Files, box 12, PSB 091 France (1), EL; Psychological Strategy Program for Japan, January 30, 1953, Freedom of Information Act. See also "One Year's Development of PSB," PSB Central Files, box 23, Background Material (1), EL; PSB Memo, n.d., WHO, NSC Staff Papers, OCB Secretariat, box 6, PSB Historical File (1), EL.

51. Gray and Cutler quoted in Lucas, "The Campaigns of Truth: The Psychological Strategy Board and American Ideology, 1951–1953," *International History Review* 18 (May 1996): 288–89. Wallace Carroll, quoted in Mitrovich, *Undermining the Kremlin*, 63.

52. Nitze quoted in Lucas, *Freedom's War*, 132. Hixson, *Parting the Curtain*, 19. For select evaluations of PSB success by board members, see WHO, NSC Series, PSB Subseries, box 22, PSB 334, President's Committee on International Activities Abroad (PCIAA) (2), EL. On the PSB's bureaucratic battles, see "One Year's Development of PSB," PSB Central Files, box 23, Background Material (1), EL. See also Sarah-Jane Corke, "Bridging the Gap: Containment, Covert Action and the Search for the Elusive Missing Link in American Cold War Policy, 1948–1953," *Journal of Strategic Studies* 20 (December 1997): 45–65; Corke, "Flexibility or Failure: Eastern Europe and the Dilemmas of Foreign Policy Coordination, 1948–1953," paper presented to the Society for Historians of American Foreign Relations, June 1997; and Corke, "Operating in a Vacuum: The American Information Program in Eastern Europe, 1948–1953," paper presented to the Canadian Association of Security and Intelligence Studies, University of Toronto, Toronto, March 1997.

Chapter 2. A New Type of Cold War

1. Eisenhower speech, October 8, 1952, C. D. Jackson Records, box 2, Robert Cutler, Eisenhower Library, Abilene, Kansas (hereafter EL).

2. On the brainwashing scare, see Susan L. Carruthers, "'The Manchurian Candidate' (1962) and the Cold War Brainwashing Scare," *Historical Journal of Film, Radio and Television* 18 (1998): 75–94. On the biological warfare campaign, see Milton Leitenberg, "Resolution of the Korean War Biological Warfare Allegations," *Critical Reviews in Microbiology* 24 (1998): 169–94; Leitenberg, "New Russian Evidence on the Korean War Biological Warfare Allegations: Background and Analysis," available at http://wwics.si.edu/index.cfm?fuseaction=library.document&topic_id=1409&id=37; Kathryn Weathersby, "Deceiving the Deceivers: Moscow, Beijing, Pyongyang, and the Allegations of Bacteriological Weapons Use in Korea," Cold War International History Project Dossier No. 1, available at http://wwics.si.edu/index.cfm?fuseaction=library.document&topic_id=1409&id=903.

3. Eisenhower speech, October 8, 1952.

4. Eisenhower to Jackson, May 8, 1952, in Eisenhower, *The Papers of Dwight D. Eisenhower*, ed. Alfred D. Chandler and Louis Galambos (Baltimore: Johns Hopkins University Press, 1970–1996), 13:1149n (hereafter *Eisenhower Papers*); Eisenhower to McClure, October 2, 1945, *Eisenhower Papers*, 8: document 367; C. D. Jackson, quoted in Blanche Wiesen Cook, *The Declassified Eisenhower: A Divided Legacy* (Garden City, N.Y.: Doubleday, 1981), 13–14; Alfred H. Paddock, *U.S. Army Special Warfare: Its Origins*, rev. ed. (Lawrence: University Press of Kansas, 2002), 14.

5. "Psychological Warfare," n.d., Hazeltine Papers, box 1, PWB (4), EL; Eisenhower to General Robert McClure, October 2, 1945, Pre-Presidential Papers, box 75, Robert A. McClure, EL; Eisenhower Speech to the National War College, October 3, 1962, Eisenhower Post-Presidential Papers, Speeches Series, box 3, October 3, 1962, Defense College (1), EL; C. D. Jackson, "Psychological Warfare," draft article for *New York Times Magazine*, June 16, 1944, Jackson Papers, box 10, "*New York Times*," EL. See also the folder, "Overlord-London (1)." Lawrence C. Soley, *Radio Warfare: OSS and CIA Subversive Propaganda* (New York: Praeger, 1989); Walter Hixson, *Parting the Curtain: Propaganda, Culture, and the Cold War, 1945–1961* (New York: St. Martin's Griffin, 1997), 3–4, 69; and Cook, *Declassified Eisenhower*, 39–62.

6. Stephen E. Ambrose, *Eisenhower: Soldier and President* (New York: Simon and Schuster, 1990), 55, 73–76; Harry C. Butcher, *My Three Years with Eisenhower* (New York: Simon and Schuster, 1946), 115, 116, 221; Eisenhower to Charles D. Herron, June 11, 1943, Pre-Presidential Papers, box 56, Charles D. Herron (2), EL. Even while Eisenhower criticized "glory hoppers" who used PR for self-promotion, he acknowledged having "learn[ed] the tricks of the trade." See Eisenhower to Charles D. Herron, June 11, 1943.

7. Butcher, *My Three Years*, 276, 388; Ambrose, *Eisenhower: Soldier and President*, 73–76; Eisenhower to Major General Russell Hartle, August 25, 1942, Pre-Presidential Papers, box 55, Russell P. Hartle, EL.

8. Eisenhower to Forestall, November 17, 1947, Pre-Presidential Papers, box 42, James Forestall (5), EL; Eisenhower to Lauris Norstad, June 19, 1947, *Eisenhower Papers*, 8:1763–64; House of Representatives, Subcommittee to Study H.R. 3342, Committee on Foreign Affairs, May 20, 1947, Pre-Presidential Papers, box 145, Hearings Vol. 4 (1), EL; U.S. Senate. Committee on Foreign Relations, Subcommittee to Study H.R. 3342, July 3, 1947, Pre-Presidential Papers, box 145, Hearings Vol. 4 (2), EL.

9. Eisenhower to Truman, December 16, 1950, Pre-Presidential Papers, box 116, H.S. Truman (1), EL. The memo was never sent directly to Truman, but it was relayed orally through Averell Harriman. On Eisenhower's involvement in the Crusade for Freedom, see Pre-Presidential Papers, box 196, Crusade for Freedom; and Pre-Presidential Papers, box 24, Clay, Lucius, EL. On Eisenhower's brief career at Columbia, see Travis Beal Jacobs, *Eisenhower at Columbia* (New Brunswick, N.J.:

Transaction Publishers, 2001). For a more comprehensive account of Eisenhower's political life between World War II and his presidency, see William B. Pickett, *Eisenhower Decides to Run: Presidential Politics and Cold War Strategy* (Chicago: Ivan R. Dee, 2000).

10. Dwight D. Eisenhower, *Mandate for Change, 1953–1956: The White House Years* (Garden City, N.Y.: Doubleday, 1963), 138–39; Eisenhower Diary, October 10, 1951, in Robert H. Ferrell, ed., *The Eisenhower Diaries* (New York: W. W. Norton, 1981), 200–201; Eisenhower to Harriman, June 30, 1951, Pre-Presidential Papers, box 55, Averell Harriman (3), EL.

11. Eisenhower to Joseph Collins, April 2, 1951; Eisenhower to Lucius Clay, April 16, 1951; and Eisenhower to Averell Harriman, June 1, 1951; all in *Eisenhower Papers*, 12:183–87, 211–12, 315–19, and 319n9. See also C. L. Sulzberger, *A Long Row of Candles: Memoirs and Diaries, 1934–1954* (New York: Macmillan, 1969), 686.

12. Eisenhower speech, September 4, 1952; Eisenhower interview, August 21, 1952; Eisenhower letter, September 9, 1952; and Eisenhower speech, October 21, 1952; all quoted in Campaign Statements of Dwight D. Eisenhower, a Reference Index, 391–93, EL.

13. Eisenhower speech, October 8, 1952; C. D. Jackson to William Jackson, March 31, 1953, White House Office [WHO], NSC Series, PSB Subseries, box 22, PSB 334, PCIIA (2), EL.

14. Abbott Washburn quoted in Hixson, *Parting the Curtain*, 21; Jackson to Rostow, December 31, 1952, C. D. Jackson Papers, box 91, Walt W. Rostow 1953–1954, EL; C. D. Jackson to William Jackson, March 31, 1953, WHO, NSC Series, PSB Subseries, box 22, PSB 334, PCIAA (2), EL; Eisenhower to Dulles, March 21, 1958, John Foster Dulles Papers, White House Memoranda Series, box 6, General Correspondence 1958 (5), EL; Eisenhower to Churchill, February 9, 1954, Ann Whitman File, Eisenhower Diary Series, box 5, February 1954 (1), EL; and Hagerty Diary, March 22, 1955, *Foreign Relations of the United States 1955–1957*, 9:521–22 (hereafter *FRUS*); Eisenhower handwritten notes, February 7, 1954, Ann Whitman File, Eisenhower Diary Series, box 4, January–November 1954 (3), EL; Eisenhower to Dulles, September 8, 1953, Ann Whitman File, Eisenhower Diary Series, box 3, August–September 1953 (2), EL; and Eisenhower to William Benton, May 1, 1953, Ann Whitman File, Eisenhower Diary Series, box 3, December 1952–July 1953 (3), EL. Author's interviews with Karl Harr, Abbott Washburn, and Andrew J. Goodpaster; Abbott Washburn Oral History and Theodore Streibert Oral History, EL. See also Eisenhower to Robert Skinner, March 27, 1953; and Eisenhower to Dulles, July 21, 1953; both in *Eisenhower Papers*, 14:130–31, 403–4.

15. Eisenhower quoted in Robert R. Bowie and Richard H. Immerman, *Waging Peace: How Eisenhower Shaped an Enduring Cold War Strategy* (New York: Oxford University Press, 1998), 44; Eisenhower to George Sloan, March 1, 1952,

in *Eisenhower Papers*, 13:1037–40; Eisenhower to Edgar Eisenhower, October 17, 1949, in *Eisenhower Papers*, 10:785–89; and Eisenhower Diary, October 10, 1951, in Robert H. Ferrell, ed., *The Eisenhower Diaries* (New York: W. W. Norton, 1981), 200–201.

16. Eisenhower Diary, October 10, 1951; and Eisenhower Diary, January 6, 1953; both in Ferrell, *Eisenhower Diaries*, 200–201, 222–24. Eisenhower Diary, January 22, 1952; Eisenhower to George Sloan, March 20, 1952; and Eisenhower to Lewis Williams Douglas, May 20, 1952; all in *Eisenhower Papers*, 13:896–902, 1097–104, 1229–31.

17. Kennan to Secretary of State, May 22, 1952, *FRUS 1952–1954*, 8:971–77; George F. Kennan, *Memoirs, 1950–1963* (Boston: Little, Brown, 1972), 145–67; Charles E. Bohlen, *Witness to History, 1929–1969* (New York: W. W. Norton, 1973), 309–36; Vojtech Mastny, *The Cold War and Soviet Insecurity: The Stalin Years* (New York: Oxford University Press, 1996), 134–70.

18. Churchill quoted in John W. Young, *Winston Churchill's Last Campaign: Britain and the Cold War 1951–5* (Oxford: Clarendon Press, 1996), 153.

19. Lloyd Gardner, "Poisoned Apples: American Responses to the Russian 'Peace Offensive,'" in *The Cold War After Stalin's Death: A New International History*, ed. Kenneth A. Osgood and Klaus Larres (Lanham, Md.: Rowman & Littlefield, forthcoming).

20. On missed opportunities, see especially Deborah Welch Larson, *Anatomy of Mistrust: U.S.-Soviet Relations During the Cold War* (Ithaca, N.Y.: Cornell University Press, 1997), 39–72; M. Steven Fish, "After Stalin's Death: The Anglo-American Debate Over a New Cold War," *Diplomatic History* 10 (Fall 1986): 333–55; James Richter, "Reexamining Soviet Policy Towards Germany During the Beria Interregnum," Cold War International History Project Working Paper No. 3 (Washington, D.C.: Woodrow Wilson International Center for Scholars, June 1992); Richter, *Khrushchev's Double Bind: International Pressures and Domestic Coalition Politics* (Baltimore: Johns Hopkins University Press, 1994), 30–50; Klaus Larres, "Eisenhower and the First Forty Days after Stalin's Death: The Incompatibility of Détente and Political Warfare," *Diplomacy and Statecraft* 6 (July 1995): 431–69; Vojtech Mastny, *Cold War and Soviet Insecurity*, 171–85; Patrick M. Morgan, "The Cold War in the 1950s: An Opportunity Missed?" in *Re-Viewing the Cold War: Domestic Factors and Foreign Policy in the East-West Confrontation*, ed. Morgan and Keith L. Nelson (Westport, Conn.: Praeger, 2000), 43–70; Jaclyn Stanke, "Danger and Opportunity: Eisenhower, Churchill, and the Soviet Union After Stalin, 1953–1955" (Ph.D. diss., Emory University, 2002); Osgood and Larres, *Cold War After Stalin's Death*.

21. Jackson to Cutler, March 4, 1953, *Declassified Documents Reference System*, 1988, microfiche 1988/2390.

22. Bohlen Memorandum, March 7, 1953, Soviet Flashpoints Collection, Doc-

ument 79639, National Security Archive, Washington, D.C. (hereafter NSA followed by document number); MemCon, March 7, 1953, NSA 76369.

23. NSC Meeting, March 11, 1953, *FRUS 1952–1954*, 8:1117–25. Other "peace" plans were developed by Walt Rostow of the Massachusetts Institute of Technology, Defense Secretary Charles Wilson, and a group of experts who met at Princeton in 1952 to discuss improvements to the U.S. propaganda effort. See Walt W. Rostow, *Europe After Stalin: Eisenhower's Three Decisions of March 11, 1953* (Austin: University of Texas Press, 1982).

24. Draft PSB plan, March 9, 1953; and Suggestions for Action to Exploit Death of Stalin, March 3, 1953; both in Jackson Records, box 1, PSB Plans for Psychological Exploitation of Stalin's Death, EL.

25. NSC Meeting, March 11, 1953, *FRUS 1952–1954*, 8:1117–25.

26. NSC Meeting, March 11, 1953; Walt W. Rostow Memorandum, May 11, 1953, *FRUS 1952–1954*, 8:1182; Hughes to Eisenhower, March 10, 1953, *FRUS 1952–1954*, 8:1113–15. Rolf Steininger, "John Foster Dulles, the European Defense Community, and the German Question," in *John Foster Dulles and the Diplomacy of the Cold War*, ed. Richard H. Immerman (Princeton: Princeton University Press, 1990), 79–108.

27. Lubell to Bernard Baruch, March 7, 1953, John Foster Dulles Papers, White House Correspondence, 1953 (5), EL; Emmet John Hughes, *The Ordeal of Power: A Political Memoir of the Eisenhower Years* (New York: Antheneum, 1963), 103; NSC Meeting, March 11, 1953.

28. *Pravda*, March 16, 1953.

29. Carlton Savage to Paul Nitze (PPS), April 1, 1953, *FRUS 1952–1954*, 8:1138; Jacob Beam to DOS, April 4, 1953, *FRUS 1952–1954*, 8:1140–43; Mastny, *Cold War and Soviet Insecurity*, 173.

30. DOS memorandum, n.d., *Declassified Documents Reference System*, microfiche 95/2722; NSC memorandum, April 8, 1953, *Declassified Documents Reference System*, microfiche 91/187.

31. PSB D-40, April 24, 1953, Jackson Records, box 1, PSB Plans for Psychological Exploitation of Stalin's Death, EL. The plan was revised after the March 11 NSC meeting and approved by the PSB on March 19. It was not formally approved by the rest of the government bureaucracies until after Eisenhower delivered the speech.

32. Eisenhower address before the American Society of Newspaper Editors, April 16, 1953, in *Public Papers of the Presidents of the United States: Dwight D. Eisenhower*, 1953 (Washington, D.C.: Government Printing Office, 1960), 179–88.

33. Revisionist historians, generally taking Eisenhower's words at face value, have portrayed the speech as a "serious bid for peace." Bowie and Immerman, for example, recently concluded that the speech amounted to "a comprehensive program for ending the Cold War" (*Waging Peace*, 225). Although the speech spoke eloquently to the perils of the arms race, it did not signal Eisenhower's quest for

Cold War détente, as revisionists have claimed. For a critique of the revisionist argument, see especially Larres, "Eisenhower and the First Forty Days," 431–69; Robert L. Ivie, "Eisenhower as Cold Warrior," in *Eisenhower's War of Words: Rhetoric and Leadership,* ed. Martin J. Medhurst (East Lansing: Michigan State University Press, 1994), 14; and Ivie, "Dwight D. Eisenhower's 'Chance for Peace': Quest or Crusade?" *Rhetoric and Public Affairs* 1 (Summer 1998): 227–44.

34. Washburn to Jackson, April 6, 1953, Jackson Records, box 7, Abbott Washburn, EL; Ivie, "Eisenhower as Cold Warrior," 15.

35. DOS quoted in Lucas, *Freedom's War: The American Crusade Against the Soviet Union* (New York: New York University Press, 1999), 174; Jackson to Morgan, April 11, 1953, Jackson Records, box 4, George Morgan, EL; Anthony Guarco to Jackson, May 11, 1953, Jackson Records, box 5, Movies, EL; Memorandum, June 4, 1953, C. D. Jackson Papers, box 104, Stalin's Death—Speech, EL; NSC 161, Status of the Psychological Program, June 30, 1953, Soviet Flashpoints Collection, Record No. 78443, NSA; DOS Telegram, April 22, 1953, *FRUS 1952–1954,* 2:1699–706.

36. Eisenhower quoted in William Stueck, *The Korean War: An International History* (Princeton: Princeton University Press, 1995), 320; PPS Memorandum, July 31, 1953, NSA 79722. On Churchill's bid for a summit see Young, *Winston Churchill's Last Campaign;* and Klaus Larres, *Churchill's Cold War: The Politics of Personal Diplomacy* (New Haven: Yale University Press, 2002).

37. Richter, *Khrushchev's Double Bind,* 45; Tsuyoshi Hasegawa, *The Northern Territories Dispute and Russo-Japanese Relations: Between War and Peace, 1697–1985* (Berkeley: University of California Press, 1998), 107; William J. Tompson, *Khrushchev: A Political Life* (New York: St. Martin's Griffin, 1995), 143–47; David Holloway, *Stalin and the Bomb: The Soviet Union and Atomic Energy, 1939–1956* (New Haven, Conn.: Yale University Press, 1994), 340–41; USIA, *Communist Propaganda: A Fact Book,* 1957–1958, 1–7.

38. Raymond L. Garthoff, *Assessing the Adversary: Estimates by the Eisenhower Administration of Soviet Intentions and Capabilities* (Washington, D.C.: Brookings Institute, 1991), 9.

39. MemCon, November 14, 1955, Ann Whitman File, Eisenhower Diary Series, box 11, DDE Diary—November 1955 (2), EL; Eisenhower to Dulles, December 5, 1955, *Eisenhower Papers,* 16:1921–23; Rockefeller to Eisenhower, August 8, 1955, Ann Whitman File, Administration Series, box 30, Nelson Rockefeller, 1952–1955 (3), EL.

40. NIE-99, October 23, 1953, *FRUS 1952–1954,* 2:551–62. Such themes remained constant in U.S. intelligence. See also NIE 11-4-54, September 14, 1954, *FRUS 1952–1954,* 8:1248–53; NIE 100-7-55, November 1, 1955, *FRUS 1955–1957,* 19:131–45; NIE 11-4-57, November 12, 1957, *Declassified Documents Reference System,* microfiche 1979/128A.

41. NIE-99, October 23, 1953; NIE 100-7-55, November 1, 1955; NIE 11-4-

Notes to Pages 70–76

60, n.d., in Donald P. Steury, ed., *Intentions and Capabilities: Estimates on Soviet Strategic Forces, 1950–1983* (Washington, D.C.: Center for the Study of Intelligence, 1996), 141–46. On the Eisenhower administration's reinterpretation of the Soviet threat, see Robert J. McMahon, "The Illusion of Vulnerability: American Reassessments of the Soviet Threat, 1955–1956," *International History Review* 18 (August 1996): 591–619.

42. JCS to Wilson, June 23, 1954, *FRUS 1952–1954*, 2:680–81; Dulles to the NSC, November 15, 1954, *FRUS 1952–1954*, 2:772–76; Streibert to the NSC, November 19, 1954, *FRUS 1952–1954*, 2:784; and Allen Dulles to the NSC, November 18, 1954, *Declassified Documents Reference System*, microfiche 1981/415B; Eisenhower to Dulles, March 21, 1958; Eisenhower to Sid Richardson, August 21, 1957, Ann Whitman File, Eisenhower Diary, box 26, August 1957—DDE Dictation, EL; Telephone Conversation, Eisenhower and Senator Lyndon Johnson, May 18, 1956, Eisenhower Diary, box 15, May 1956 Phone Calls, EL. McMahon, "Illusion of Vulnerability"; Melvyn P. Leffler, *A Preponderance of Power: National Security, the Truman Administration, and the Cold War* (Stanford: Stanford University Press, 1992).

43. Bowie and Immerman, *Waging Peace*; Andreas Wenger, *Living with Peril: Eisenhower, Kennedy, and Nuclear Weapons* (Lanham, Md.: Rowman & Littlefield, 1997); Saki Dockrill, *Eisenhower's New-Look National Security Policy, 1953–61* (New York: St. Martin's Press, 1996).

44. John Lewis Gaddis, *Strategies of Containment: A Critical Appraisal of Postwar American National Security Policy* (New York: Oxford University Press, 1982), 127–97.

45. Eisenhower quoted in Gaddis, *Strategies of Containment*, 128; Dulles, "A Policy of Boldness," *Life*, May 19, 1952, 146.

46. NSC 162/2, October 30, 1953, *FRUS 1952–1954*, 2:577–97.

47. NSC 5501, January 7, 1955, WHO, Office of the Special Assistant for National Security Affairs [OSANSA], NSC Series, Policy Papers Subseries, box 14, NCS 5501, EL.

48. NSC 5611, Status of the USIA Program, June 30, 1956, WHO, OSANSA, NSC Staff Papers, Status of Projects Series, box 7, NSC 5611, Part II (3), EL; NSC 5525, Status of the USIA Program, August 31, 1955, *FRUS 1955–1957*, 9:529–48.

49. NSC 5509, Status of the USIA Program, March 2, 1955, *FRUS 1955–1957*, 9:504–21.

Chapter 3. Camouflaged Propaganda

1. For a history of investigations into the information and cultural programs, see Lois Roth, "Public Diplomacy and the Past: The Studies of U.S. Information and Cultural Programs (1952–1975)," *United States Department of State: Foreign*

Service Institute Executive Seminar in National and International Affairs, 23rd Session, 1980–1981.

2. Quoted in John W. Henderson, *The United States Information Agency* (New York: Praeger, 1969), 46–47.

3. C. D. Jackson to William H. Jackson, March 31, 1953, NSC Staff Papers, PSB Central Files, box 22, PSB 334, PCIAA (2), Eisenhower Library, Abilene, Kansas (hereafter EL).

4. Eisenhower to William Benton, May 1, 1953, Ann Whitman File, Eisenhower Diary Series, box 3, December 1952–July 1953 (3), EL.

5. The heavy reliance of the United States on covert and unattributed propaganda materials has gone largely unnoticed by historians. Although it is now well known that the CIA engaged in extensive covert propaganda operations, the fact most USIA propaganda was unattributed to the United States government has not been explored. Parry-Giles also discusses the concept of "camouflaged propaganda" but defines it differently than I do here. Shawn J. Parry-Giles, "The Eisenhower Administration's Conceptualization of the USIA: The Development of Overt and Covert Propaganda Strategies," *Presidential Studies Quarterly* 24 (Spring 1994): 263–76; and Parry-Giles, "Camouflaged Propaganda: The Truman and Eisenhower Administrations' Covert Manipulation of News," *Western Journal of Communication* 60 (Spring 1996): 146–67.

6. On "words and deeds," see, for example, Eisenhower Diary, January 6, 1953, in Robert H. Ferrell, ed., *The Eisenhower Diaries* (New York: W. W. Norton, 1981), 223; Eisenhower to Emmet Hughes, December 10, 1953, *The Papers of Dwight D. Eisenhower*, ed. Alfred D. Chandler and Louis Galambos (Baltimore: Johns Hopkins University Press, 1970–1996), 15:748–53 (hereafter *Eisenhower Papers*); Eisenhower speech, October 8, 1952, C. D. Jackson Records, box 2, Robert Cutler, EL.

7. Roth, "Public Diplomacy and the Past." Statement by the President, January 26, 1953, Eisenhower, *Public Papers of the Presidents of the United States: Dwight D. Eisenhower, 1953* (Washington, D.C.: Government Printing Office, 1960), 8; William H. Jackson, "The Fourth Area of the National Effort in Foreign Affairs," n.d. [November 1956], Ann Whitman File, box 22, William H. Jackson, EL.

8. Report of the President's Committee on International Information Activities (Jackson Committee Report), *Foreign Relations of the United States 1952–1954: Emergence of the Intelligence Establishment* (hereafter *FRUS*), 2:1795–1896. Quotes on 1797, 1800, and 1811.

9. Most of the committee's recommendations for anti-Soviet psychological warfare concerned increasing the credibility, cost-effectiveness, and efficiency of the various overt and covert radio broadcasts behind the Iron Curtain. Jackson Committee Report, *FRUS 1952–1954*, 2:1817, 1824.

10. Jackson Committee Report, *FRUS 1952–1954*, 2:1836; Jackson Committee to Eisenhower, May 2, 1953, *FRUS 1952–1954*, 2:1869.

11. NSC Meeting, July 3, 1953, Ann Whitman File, NSC Series, box 4, EL; Remarks of Abbott Washburn, conference on Public Discourse in Cold War America, Texas A&M University, March 5–8, 1998; Eisenhower to Rockefeller, August 5, 1955, Ann Whitman File, Eisenhower Diary Series, box 11, August 1955 (1), EL.

12. Jackson, "Fourth Area."

13. Jackson to Eisenhower, December 17, 1952, C. D. Jackson Papers, box 50, Eisenhower Correspondence 1952, EL; Informal Note by Robert Cutler, March 28, 1953, White House Office [WHO], NSC Staff Papers, PSB Central Files, box 22, PSB 334, PCIAA (2), EL; Washburn to C. D. Jackson, April 11, 1953, C. D. Jackson Records, box 7, Abbott Washburn, EL; author's interviews with Abbott Washburn, General Andrew J. Goodpaster, and Karl Harr. In an interview with the author, Abbott Washburn explained that C. D. Jackson left the White House under pressure from Henry Luce, who told Jackson he would lose his position at Time-Life if he did not return soon. On C. D. Jackson, see also Scott Lucas, *Freedom's War: The American Crusade Against the Soviet Union* (New York: New York University Press, 1999), 166; Blanche Wiesen Cook, "First Comes the Lie: C. D. Jackson and Political Warfare," *Radical History Review* 31 (1984): 42–70; H. W. Brands, *Cold Warriors: Eisenhower's Generation and American Foreign Policy* (New York: Columbia University Press, 1988), 117–37; Valur Ingimundarson, "Containing the Offensive: The 'Chief of the Cold War' and the Eisenhower Administration's German Policy," *Presidential Studies Quarterly* 27 (Summer 1997): 480–96. On the Council for Democracy, see Clayton D. Laurie, *The Propaganda Warriors: America's Crusade Against Nazi Germany* (Lawrence: University Press of Kansas, 1996), 32–33.

14. Eisenhower to Jackson, October 21, 1957, Ann Whitman File, Administration Series, Jackson, 1956–1957, EL; Eisenhower to Jackson, May 17, 1954, Jackson Papers, box 50, Eisenhower Correspondence, 1954 (1), EL; Eisenhower to Jackson, June 29, 1955, Jackson Papers, box 69, Jackson Log, EL.

15. Cary Reich, *The Life of Nelson Rockefeller: Worlds to Conquer, 1908–1958* (New York: Doubleday, 1996), 549–634; Karl Harr Oral History; Jackson, "Fourth Area." On the Office of Non-Military Warfare, see Nelson A. Rockefeller Papers, Record Group [RG] 4, Washington, D.C., Files, Subseries 7, box 3, Rockefeller Archive Center, Sleepy Hollow, New York (hereafter RAC). On the Planning Coordination Group, see Reich, *Life of Nelson Rockefeller*, 560–61, 617–18; and James D. Marchio, "The Planning Coordination Group: Bureaucratic Casualty in the Cold War Campaign to Exploit Soviet-Bloc Vulnerabilities," *Journal of Cold War Studies* 4 (Fall 2002): 3–28. Marchio sees the failure of the PCG as evidence that Eisenhower did not support psychological warfare, underestimating the bureaucratic opposition to Rockefeller's empire building and conflating psychological warfare too narrowly with anti-Soviet rollback operations. The PCG intruded not only on State Department turf, but also on CIA and Defense operations. The heads of these agencies did their best to ignore and marginalize the PCG until eventually Rockefeller

disbanded it. The PCG also duplicated the activities of the OCB and too closely resembled the Psychological Strategy Board, an experiment most administration officials agreed had failed.

16. Eisenhower to Dulles, October 24, 1953, Ann Whitman File, Eisenhower Diary Series, box 3, October 1953 (2), EL; Telephone Conversation, Eisenhower and Dulles, May 24, 1955, *FRUS 1955–1957*, 9:524–25; Eisenhower to Dulles, September 8, 1953, Ann Whitman File, Eisenhower Diary Series, box 3, August–September 1953 (2), EL. See also Eisenhower to Dulles, March 21, 1958, John Foster Dulles Papers, White House Memo Series, box 6, White House, General Correspondence 1958 (5), EL; Eisenhower to Dulles, July 21, 1959, WHO, Office of the Special Assistant for National Security Affairs [OSANSA], OCB Series, Subject Subseries, box 1, Coordination of Information and Public Opinion Aspects of National Security Policy, EL; Eisenhower to Dulles, October 24, 1953, Ann Whitman File, Eisenhower Diary Series, box 3, October 1953 (2), EL; Eisenhower Diary, January 24, 1958, in Robert H. Ferrell, ed., *The Eisenhower Diaries* (New York: W. W. Norton, 1981), 350. On Dulles's view of the role of Eisenhower psychological warfare advisors, see Reich, *Life of Nelson Rockefeller*, 610–11. In 1958, Eisenhower tried to talk Dulles into appointing C. D. Jackson as undersecretary of state to oversee Cold War and psychological warfare planning. (Eisenhower noted, however, that Jackson would be given a "euphemistic" title.) The proposal illustrates the importance Eisenhower attached to both C. D. Jackson and psychological warfare, but Dulles, not wanting to share the spotlight with Jackson, opposed the idea.

17. Seventh Semiannual Report of the United States Advisory Commission on Information, House Document No. 94, 83rd Congress, 1st Session; Eisenhower to Dulles, July 21, 1959, WHO, OSANSA, OCB Series, Subject Subseries, box 1, Coordination of Information and Public Opinion Aspects of National Security Policy, EL.

18. Lucas, *Freedom's War*, 147.

19. Quotes from Karl Harr interview; Eisenhower Speech to the National War College, October 3, 1962, DDE Post-Presidential Papers, Speeches Series, box 3, October 3, 1962 Defense College (1), EL; Roy M. Melbourne, *Conflict and Crisis: A Foreign Service Story* (Lanham, Md.: University Press of America, 1993), 175–91; Karl G. Harr Jr., "Eisenhower's Approach to National Security Decisionmaking," in *The Eisenhower Presidency*, ed. Kenneth W. Thompson (Lanham, Md.: University Press of America, 1984), 89–111; Robert Bowie and Richard Immerman, *Waging Peace: How Eisenhower Shaped an Enduring Cold War Strategy* (Oxford: Oxford University Press, 1998), 93–95; and Lucas, *Freedom's War*, 177, 209–10. Lucas mistakenly reports that the OCB had a budget of $450 million. Because the OCB was a coordinating body rather than an operational agency, the OCB's budget was considerably smaller. Presumably the $450 million figure Lucas refers to is the overall expenditure of the Eisenhower administration on activities to influence foreign opinion.

20. Robert Blum quoted in Bowie and Immerman, *Waging Peace*, 94; Senate Committee on Government Operations, Organizational History of the National Security Council, 86th Congress, 2nd Session, 1960, 25; U.S. Resources for Foreign Communication and Political Warfare, May 9, 1960, Records of the President's Committee on Information Activities Abroad (Sprague Committee), box 20, PCIAA #9, EL.

21. U.S. Resources for Foreign Communication and Political Warfare, May 9, 1960; Jackson, "Fourth Area"; Harr, "Eisenhower's Approach to National Security."

22. Rockefeller to OCB, April 20, 1955, Nelson A. Rockefeller Papers, Washington, D.C., Files, Subseries 7, box 2, OCB Effectiveness, RAC; OCB Check List, n.d., Karl Harr Papers, author's collection; "A descriptive statement of the organization, functions, and procedures of the OCB," September 1955, EL.

23. In this respect, the OCB was a sort of Cold War precursor to the White House Office of Communications, which has managed the news and public relations strategies for every president since Nixon. But where the Office of Communications has concerned itself primarily with domestic politics and the managing of the media on a day-to-day basis, the OCB was more concerned with orchestrating a few select themes and managing public debate over particularly contentious issues relating to the Cold War. On the Office of Communications, see John Anthony Maltese, *Spin Control: The White House Office of Communications and the Management of Presidential News*, 2nd ed. (Chapel Hill: University of North Carolina Press, 1994).

24. Author's interview with Elmer B. Staats; Karl Harr and Elmer B. Staats Oral Histories, EL.

25. President's Advisory Committee on Government Organization to Eisenhower, April 7, 1953, *FRUS 1952–1954*, 2:1691–97; Seventh Semiannual Report of the United States Advisory Commission on Information, House Document No. 94, 83rd Congress, 1st Session; Jackson Committee Report, *FRUS 1952–1954*, 2:1861–62.

26. Stephen E. Ambrose, *Eisenhower: The President* (New York: Simon and Schuster, 1984), 61; Sigurd Larmon and Theodore Streibert oral histories, E. L. Hixson notes that Eisenhower favored a strong information agency within the State Department but reluctantly bowed to Dulles's wishes and acquiesced in a separate USIA. I have uncovered no evidence definitively indicating that Eisenhower had a strong preference either way. Some evidence suggests that Eisenhower thought the information program would be marginalized and timid if it remained in the State Department. It is worth noting that C. D. Jackson and Robert Cutler supported removing the information program from the State Department. See Walter L. Hixson, *Parting the Curtain: Propaganda, Culture, and the Cold War, 1945–1961* (New York: St. Martin's Griffin, 1997), 25–27; C. D. Jackson to William H. Jackson, March 31, 1953; and Informal Note, March 28, 1953; both in WHO, NSC Staff Papers, PSB Central Files, box 22, PSB 334, PCIAA (2), EL.

27. Henderson, *United States Information Agency,* 53, 71; Barnes to Barclay, Foreign Office, Information Policy Department, April 27, 1953, FO 953/1419, Public Record Office, Kew, United Kingdom (hereafter PRO). Hixson concluded that the decision to remove the information program from the State Department "ensured a certain marginalization of the overseas effort" because "psychological considerations" were placed outside the department where policy was made. Hixson, *Parting the Curtain,* 27. Eisenhower, however, never intended for the USIA to be a policy-making agency. He wanted the USIA to focus primarily on operations, looking instead to his special assistants, the NSC, and the OCB for guidance on "psychological strategy."

28. U.S. Resources for Foreign Communication and Political Warfare, May 9, 1960; Guy Oakes, *The Imaginary War: Civil Defense and American Cold War Culture* (New York: Oxford University Press, 1994). See also Laura McEnaney, *Civil Defense Begins at Home: Militarization Meets Everyday Life in the Fifties* (Princeton: Princeton University Press, 2000); and Rachel Mihalovich, "Fighting 'Don't Know, Don't Care': The FCDA's Public Education Quest" (M.A. thesis, Old Dominion University, 2001); Martin J. Medhurst, ed. *Eisenhower's War of Words: Rhetoric and Leadership* (East Lansing: Michigan State University Press, 1994); Robert Griffith, "The Selling of America: The Advertising Council and American Politics, 1942–1960," *Business Historical Review* 57 (Autumn 1983): 388–412; Daniel L. Lykins, *From Total War to Total Diplomacy: The Advertising Council and the Construction of the Cold War Consensus* (Westport, Conn.: Praeger, 2003). Reliable figures for the CIA are not available. Christopher Simpson estimates that total expenditures may have reached $1 billion annually during the 1950s. Christopher Simpson, *Science of Coercion: Communication Research and Psychological Warfare, 1945–1960* (New York: Oxford University Press, 1994), 176n.

29. Directive Approved by the President for the Guidance of the United States Information Agency, October 28, 1953, Eisenhower, *Public Papers,* 1953:728; Streibert to Eisenhower, October 27, 1953, *FRUS 1952–1954,* 2:1755.

30. USIA Strategic Principles, n.d. [March 1954], OCB Central Files, box 20, OCB 040 USIA (1), EL.

31. USIA Strategic Principles; Foreign Office, Information Policy Department Minutes, April 12, 1954, FO 953/1528, PRO. The figures cited for the number of posts are taken from the agency's December 1960 report to Congress, the only year during the decade this information was reported. USIA 15th Report to Congress, December 1960.

32. Available USIA policy documents do not provide clear statements of the agency's priorities (possibly because pertinent documents remain unavailable to researchers). It is difficult to arrive at precise statistics for the allocation of USIA resources because of vagueness, lack of precision, and inconsistencies in the USIA's semiannual reports to Congress. My deduction of the relative size of the posts is based on the numbers of Americans and locals employed at each post as listed in

the December 1956 report to Congress (which I have compared against other reports). Figures for expenditures are an unreliable estimate of priority because they need to be weighed against currency conversion and balanced according to the ratio of American to "local" personnel (because Americans cost more to employ). Personnel figures provide a better estimation of priority, but this too provides an imperfect estimate because it does not take into consideration the relative size of the countries. If ranked by funding, a similar order of priorities is produced, except that the programs in Mexico and Great Britain move up the scale because they were funded more generously than they were staffed. After 1956, the agency no longer provided this information in its congressional reports. Budget figures listed in the text are rounded. USIA 3rd Report to Congress, December 1954; USIA 6th Report to Congress, June 1956.

33. OCB, Principles to Assure the Coordination of Gray Activities, May 18, 1954, *Declassified Documents Reference System*, CD-ROM, document 1992090102989.

34. Arthur Larson Memorandum Book, March 1957, Arthur Larson Papers, box 22, EL; Eisenhower to William Benton, May 1, 1953, Ann Whitman File, Eisenhower Diary Series, box 3, December 1952–July 1953 (3), EL; Eisenhower to Averell Harriman, June 1, 1951, *Eisenhower Papers*, 12:315–19; MemCon, Eisenhower and George Allen, March 23, 1959, Ann Whitman File, Eisenhower Diary Series, box 40, Staff Notes, March 15–31, 1959, EL. Fred I. Greenstein, *The Hidden-Hand Presidency: Eisenhower as Leader*, 2nd ed. (Baltimore: Johns Hopkins University Press, 1994).

35. Jackson Committee Report, *FRUS 1952–1954*, 2:1849, 1841, 1872; Progress Report on Jackson Committee Report, September 30, 1953, WHO, NSC Staff Papers, PSB Central Files, box 22, PSB 334, PCIAA, EL. This document lists some of the Jackson Committee's recommendations not declassified in the report published in *FRUS*. Although radio was deemphasized in propaganda to free world areas, it continued to be almost the only medium for reaching areas behind the Iron Curtain. Most of VOA's budget went to broadcasting in communist countries. In 1953, only $9 million of VOA's $22 million budget went to "free world" broadcasting: 63 percent of the total budget went to broadcasting in the Soviet Union, Eastern Europe, and the People's Republic of China. Jackson Committee Report, *FRUS 1952–1954*, 2:1827, 1824, 1845; Oren Stephens, *Facts to a Candid World: America's Overseas Information Program* (Stanford: Stanford University Press, 1955), 81.

36. Jackson Committee Report, *FRUS 1952–1954*, 2:1842; Progress Report on Jackson Committee Report, September 30, 1953, WHO, NSC Staff Papers, PSB Central Files, box 22, PSB 334, PCIAA, EL; USIA Strategic Principles, n.d. [March 1954], OCB Central Files, box 20, OCB 040 USIA (1), EL.

37. Jackson Committee Report, chap. 6, Freedom of Information Act (FOIA). Two and a half pages of this chapter were released in response to this author's FOIA

request, but unfortunately the released portions were of a decidedly general nature. Every other chapter of the report has been virtually declassified in full.

38. Christopher Andrew, *For the President's Eyes Only: Secret Intelligence and the American Presidency from Washington to Bush* (New York: Harper Collins, 1996), 202, 212; NSC 5412/2, December 28, 1955, WHO, OSANSA, Special Assistant Series, President's Papers Subseries, box 2, 1955 (1), EL.

39. NSC 5412/2; Allen Dulles speech to the Council on Foreign Relations, June 5, 1953, quoted in Washburn to William H. Jackson, June 11, 1953, White House Central Files, Official File, box 674, 133-M-1 PCIAA (3), EL. On Allen Dulles, see Peter Grose, *Gentleman Spy: The Life of Allen Dulles* (Boston: Houghton Mifflin, 1994).

40. House Select Committee on Intelligence, *The CIA and the Media*, 95th Congress, 1st and 2nd Sessions, 1978, 21. In his testimony, Colby refers specifically to "political and paramilitary work." The context of his statement, plus the fact that agency parlance generally described propaganda as "political action," indicates that Colby was including propaganda activities in that figure. Jackson Committee Report, chap. 6, FOIA.

41. The U.S. Information Program Since July 1953, Records of the President's Committee on Information Activities Abroad (Sprague Committee), box 19, USIA (2), EL; U.S. Resources for Foreign Communication and Political Warfare, May 9, 1960, Records of the President's Committee on Information Activities Abroad (Sprague Committee), box 20, PCIAA #9, EL; Killian to Eisenhower, December 20, 1956, WHO, OSANSA, Special Assistant Series, Chronological Subseries, box 7, President's Board of Consultants on Foreign Intelligence Activities First Report to the President (1), EL; Allen Dulles to Arthur Larson, Action on Recommendation No. 6 of the Report to the President by the President's Board of Consultants on Foreign Intelligence Activities, April 30, 1957, Mandatory Declassification Review, EL. The "gray treaty" was: Principles to Assure the Coordination of Gray Activities, May 18, 1954, *Declassified Documents Reference System*, CD-ROM, document 1992090102989. This top secret report referred to the CIA as the "OCB designee."

42. Memorandum Book, March 1957, Arthur Larson Papers, box 22 (emphasis in original), EL; Eisenhower to William Benton, May 1, 1953, Ann Whitman File, Eisenhower Diary Series, box 3, December 1952–July 1953 (3), EL; MemCon, Eisenhower and George Allen, March 23, 1959, Ann Whitman File, Eisenhower Diary Series, box 40, Staff Notes, March 15–31, 1959, EL; Jackson Committee to Eisenhower, May 2, 1953, *FRUS 1952–1954*, 2:1868–69; USIA Strategic Principles; Some Suggestions on How the Label of "Propaganda" Can Be Avoided Without Sacrificing Effectiveness, n.d. [1953], RG 306, Special "S" Reports of the Office of Research, 1953–63, box 2, National Archives, College Park, Maryland (hereafter NA). For the VOA, the transition from hard-hitting anticommunist ma-

terials to objective-sounding news and entertainment materials was slow and pain-
ful. Many of the VOA broadcasters were émigrés who tended to offer stridently
anticommunist commentary. As late as 1957, Eisenhower was lecturing the USIA
director Arthur Larson that the VOA must eliminate overtly propagandistic materials and concentrate on factual information. By the end of the decade, however,
the Information Agency reported that "objective news reporting now provides the
primary basis of VOA programs. Exhortation and abuse are avoided altogether."
See remarks of Barry Zorthian, October 11, 1990, in Hans N. Tuch and G. Lewis
Schmidt, eds., *Ike and USIA: A Commemorative Symposium* (Washington, D.C.:
Public Diplomacy Foundation, 1991), 16; The U.S. Information Program Since
July 1953.

43. Jackson Committee Report, *FRUS 1952–1954*, 2:1871; USIA Strategic
Principles.

44. Parry-Giles, "'Camouflaged' Propaganda," 161; Stephens, *Facts to a Candid
World*, 110.

45. Foreign Office, Information Policy Department Minutes, June 10, 1954,
FO 953/1528, PRO; Watson to Marett, Foreign Office, Information Policy De-
partment, October 22, 1954, FO 953/1529, PRO; Jackson Committee Report,
FRUS 1952–1954, 2:1838; Some Suggestions on How the Label of "Propaganda"
Can Be Avoided Without Sacrificing Effectiveness.

46. NSC 5509, Status of USIA Program, March 2, 1955, *FRUS 1955–1957*,
9:506–7.

47. The Operational Importance of Goals that Promise Hope; and Harr to
Eisenhower, May 22, 1959; both in WHO, OSANSA, OCB Series, Subject Subseries, box 5, Miscellaneous (7), EL.

48. NSC 162/2, October 30, 1953, *FRUS 1952–1954*, 2:577–97.

49. Foreign Office, Information Policy Department Minutes, April 12, 1954,
FO 953/1528.

50. Quoted in White to IIA Plans Board, n.d. [May 25, 1953], RG 306, Special
"S" Reports of the Office of Research, 1953–63, box 2, NA; The U.S. Information
Program Since July 1953.

Chapter 4. Secret Empire

1. For a critical review of recent literature on U.S. cultural diplomacy in Europe, see Jessica Gienow-Hecht, "Shame on U.S.? Academics, Cultural Transfer, and the Cold War: A Critical Review," *Diplomatic History* 24 (Summer 2000):
465–94. Notable works include Gienow-Hecht, *Transmission Impossible: American
Journalism as Cultural Diplomacy in Postwar Germany, 1945–1955* (Baton Rouge:
Louisiana State University Press, 1999); Gienow-Hecht and Frank Schumacher, eds., *Culture and International History* (New York: Berghahn Books, 2003);

Richard Pells, *Not Like U.S.: How Europeans Have Loved, Hated, and Transformed American Culture* (New York: Basic Books, 1997); Richard F. Kuisel, *Seducing the French: The Dilemma of Americanization* (Berkeley: University of California Press, 1993); Michael Ermarth, ed., *America and the Shaping of German Society, 1945–1955* (Providence, R.I.: Berg, 1993). CIA funding of the Congress for Cultural Freedom has elicited especially lively debate. See Frances Stonor Saunders, *The Cultural Cold War: The CIA and the World of Arts and Letters* (New York: New Press, 1999); Peter Coleman, *The Liberal Conspiracy: The Congress for Cultural Freedom and the Struggle for the Mind of Postwar Europe* (New York: Free Press, 1989); Hugh Wilford, *The CIA, the British Left, and the Cold War: Calling the Tune?* (London: Frank Cass, 2003); Volker R. Berghahn, *America and the Intellectual Cold Wars in Europe* (Princeton: Princeton University Press, 2001).

2. Robert W. Murphy to Nelson Rockefeller, August 19, 1955, Planning Coordinating Group files, box 2, Bandung, Eisenhower Library, Abilene, Kansas (hereafter EL); PSB Draft of Western European Plan, March 17, 1953, Mandatory Declassification Review, EL. Frank Schumacher explores the American reaction to "emotional neutralism" in his work. See Schumacher, "Cold War Propaganda and Alliance Management: The United States and West Germany in the 1950s," paper presented to the Society for Historians of American Foreign Relations, June 2000; Schumacher, "Democratization and Hegemonic Control: American Propaganda and the West German Public's Foreign Policy Orientation, 1949–1955," in *The American Nation, National Identity, Nationalism*, ed. Knud Krakau, 285–316 (Münster: Transaction Publishers, 1997). See also Schumacher, *Kalter Krieg und Propaganda: Die USA, der Kampf um die Weltmeinung und die ideelle Westbindung der Bundesrepublik Deutschland, 1945–1955* (Trier: Wissenschaftlicher Verlag, 2000).

3. On Eisenhower and European integration, see Pascaline Winand, *Eisenhower, Kennedy, and the United States of Europe* (New York: St. Martin's Press, 1993); Geir Lundestad, *Empire by Integration: The United States and European Integration, 1945–1997* (Oxford: Oxford University Press, 1998); Robert R. Bowie and Richard H. Immerman, *Waging Peace: How Eisenhower Shaped an Enduring Cold War Strategy* (New York: Oxford University Press, 1998), 205–9; John Lewis Gaddis, *Strategies of Containment: A Critical Appraisal of Postwar American National Security Policy* (Oxford: Oxford University Press, 1982), 152–54.

4. PSB Draft of Western European Plan, March 17, 1953; USIA, Third Report to Congress, December 1954.

5. USIS Bonn to DOS, July 14, 1960, Record Group [RG] 306, Foreign Service Despatches, box 1, Europe, National Archives, College Park, Maryland (hereafter NA); NSC 5525, Status of the USIA Program, August 31, 1955, August 31, 1955, *Foreign Relations of the United States 1955–1957*, 9:529–48 (hereafter *FRUS*); PSB Draft of Western European Plan, March 17, 1953; NSC 5509, Status of the USIA Program, March 2, 1955, *FRUS 1955–1957*, 9:504–21; NSC 5611, Status of

the USIA Program, June 30, 1956, White House Office [WHO], Office of the Special Assistant for National Security Affairs [OSANSA], NSC Staff Papers, Status of Projects Series, box 7, NSC 5611, Part II (3), EL.

6. PSB Draft of Western European Plan, March 17, 1953; NSC 5509, Status of the USIA Program, March 2, 1955; OCB Operations Plan for France, February 28, 1958, RG 59, Records Relating to DOS participation in NSC and OCB, 1947–63, Lot 62D 430, box 18, France, NA. On psychological warfare in Italy, see Mario Del Pero, "The United States and 'Psychological Warfare' in Italy, 1948–1955," *Journal of American History* 87 (March 2001): 1304–34.

7. USIA, Third Report to Congress, December 1954; USIA, Sixth Report to Congress, June 1956; USIS Bonn to DOS, July 14, 1960, RG 306, Foreign Service Despatches, box 1, Europe, NA. The German information program was so large in part because the maintenance of strong, democratic Germany closely aligned with the West and free to participate in military alliances was a cardinal goal of American foreign policy in the 1950s, but also because most USIS operations against the Soviet bloc and Eastern Europe originated in Germany. Staff for the program in Germany progressively declined over the decade, as resources were increasingly allocated elsewhere, but the program remained disproportionately large nevertheless.

8. USIS Reykjavik to USIA, January 6 and 13, 1955, RG 306, Office of Research, Country Project Correspondence 1952–63, box 10, Iceland 1955, NA.

9. USIS Reykjavik to USIA, September 26, 1955, RG 306, Office of Research, Country Project Correspondence 1952–63, box 10, Iceland 1955, NA; OCB Progress Report on Iceland, October 24, 1956, RG 59, Records Relating to DOS Participation in NSC and OCB, 1947–63, Lot 62D 430, box 19, Iceland 1956–1957, NA; NSC 5611, Status of the USIA Program, June 30, 1956; USIS Rejkavik to USIA, July 29, 1957, Freedom of Information Act (FOIA); U.S. Embassy Paris to DOS, March 27, 1957, FOIA.

10. The term "third world" was coined in the middle of the 1950s to refer to countries that were not aligned with either of the superpowers. It later came to denote developing countries in Asia, Africa, Latin America, and the Middle East.

11. Alvin Z. Rubinstein, *Soviet Foreign Policy Since World War II: Imperial and Global*, 4th ed. (New York: HarperCollins, 1992), 175–82; James G. Richter, *Khrushchev's Double Bind: International Pressures and Domestic Coalition Politics* (Baltimore: Johns Hopkins University Press, 1994), 71.

12. Marc Frey, "Tools of Empire: Persuasion and the United States's Modernizing Mission in Southeast Asia," *Diplomatic History* 27 (September 2003): 543–68. I would like to thank Marc Frey for providing me with an advance copy of this article. USIA reactions to Congressional budget cuts provide one indication of the agency's priorities. When Congress slashed the 1953 appropriation by 37 percent, the USIA cut its European programming in half. When Congress cut the agency's budget again in 1956 (by 16 percent), European operations again bore the lion's share of the burden, taking a 27 percent hit. See NSC 5407, Status of

the USIA Program, March 1, 1954, *Declassified Documents Reference System*, microfiche, 1993/488; NSC 5720, Status of the USIA Program, June 30, 1957, *FRUS 1955–1957*, 9:594–612.

13. Asia, Report to the President's Committee on Information Activities Abroad, July 11, 1960, Records of the President's Committee on Information Activities Abroad, box 23, PCIAA #30 (1), EL (hereafter PCIAA Report on Asia).

14. Michael Schaller, *Altered States: The United States and Japan Since the Occupation* (New York: Oxford University Press, 1997); Walter LaFeber, *The Clash: A History of U.S.-Japan Relations* (New York: W. W. Norton, 1997).

15. USIA, Sixth Report to Congress, June 1956; John W. Allison to Dulles, January 24, 1958, RG 59, Central Decimal File, FW 103.USIA/1-2456, NA; PSB D-27, Psychological Strategy Program for Japan, Mandatory Declassification Review, EL; OCB Report on Japanese Intellectuals, December 13, 1955, OCB Central Files, box 48, OCB 091 Japan (File #4) (2), EL. Relative size of information posts is deduced according to the number of personnel employed. The U.S. military also developed an extensive public information program for Japan. See Far East Command Headquarters, Policy Direction No. 10-4, November 29, 1956, RG 59, Central Decimal File, 511.94/12-656, NA. Fumiko Fujita, "U.S. Cultural Diplomacy Toward Japan in the 1950s," paper presented to New York University, March 2003. On censorship during the occupation, see John Dower, *Embracing Defeat: Japan in the Wake of World War II* (New York: W. W. Norton, 1999), 405–43. On the occupation, see also Michael Schaller, *The American Occupation of Japan: The Origins of the Cold War in Asia* (New York: Oxford University Press, 1985).

16. Walter Nichols Oral History, quoted in Fumiko Fujita, "U.S. Cultural Diplomacy"; Max W. Bishop to Robertson, November 10, 1955, RG 59, Central Decimal File, 511.942/11-1055, NA; USIS Tokyo to USIA, December 14, 1954, RG 306, Office of Research, Country Project Correspondence 1952–63, box 13, Japan 1954, NA; NSC 5509, Status of the USIA Program, March 2, 1955; Mark A. May, Report on USIS-Japan, June–July 1959, USIA Historical Collection, Washington, D.C. The May report assessed the USIS program in Japan during the 1950s and provides revealing insights into the "sources and methods" of USIS operations. The document includes an extraordinary level of information on "unattributed" operations, surpassing any other document I have found in the archives for its comprehensive, illuminating detail on USIS gray propaganda.

17. May, Report on USIS-Japan.

18. Henry (Hank) Gosho, Oral History Interview, USIA Alumni Association; May, Report on USIS-Japan.

19. NSC 5405, U.S. Objectives and Courses of Action with Respect to Southeast Asia, January 16, 1954, *FRUS 1952–1954*, 13:971–76; Report of President's Special Committee on Indo-China, March 2, 1954, *FRUS 1952–1954*, 13:1109–16; NSC 5407, Status of the USIA Program, March 1, 1954; USIS Saigon to USIA, October 12, 1957, RG 306, Foreign Service Despatches 1954–1963, box 1, Asia,

NA. David Schmidtz, *Thank God They're On Our Side: The United States and Right-Wing Dictatorships, 1921–1965* (Chapel Hill: University of North Carolina Press, 1999), 198–208; David Anderson, *Trapped By Success: The Eisenhower Administration and Vietnam, 1953–1961* (New York: Columbia University Press, 1991). For an overview of U.S. policy toward the region, see Robert J. McMahon, *The Limits of Empire: The United States and Southeast Asia Since World War II* (New York: Columbia University Press, 1999). For analysis of U.S. psychological warfare in Southeast Asia, see Frey, "Tools of Empire." Matthew B. Masur explores in detail U.S. programs to boost the legitimacy of Diem's government in "Hearts and Minds: Cultural Nation-Building in South Vietnam, 1954–1964" (Ph.D. diss., Ohio State University, 2004). See also Masur's "People of Plenty: American Cultural and Economic Programs in South Vietnam, 1954–1963," paper presented to the Ohio Academy of History, 2002.

20. Frey, "Tools of Empire"; OCB, Detailed Development of Major Actions Relating to Southeast Asia, June 22, 1956, RG 59, Records Relating to Department of State Participation in NSC and OCB, Lot 62 D430, box, 30, Southeast Asia (5), NA; NSC 6013, Status of the USIA Program, June 30, 1960, WHO, OSANSA, NSC Series, Status of Projects Subseries, box 9, NSC 6013 (4), EL; Memorandum on Civic Action Program in South Vietnam, December 30, 1955, *Declassified Documents Reference System* online, document no. CK3100288657.

21. OCB, Detailed Development of Major Actions Relating to Southeast Asia, June 22, 1956; "Edward Lansdale Report on CIA Operations in Vietnam, 1954–1955," in *The Pentagon Papers: Abridged Edition*, ed. George C. Herring (New York: McGraw-Hill, 1993); Cecil B. Currey, *Edward Lansdale: The Unquiet American* (Washington, D.C.: Brassey's, 1998), 134–85; Nicholas J. Cull, *Selling America: U.S. International Propaganda and Public Diplomacy Since 1945* (New York: Cambridge University Press, forthcoming), chap. 3; NSC 5525, Status of the USIA Program, August 31, 1955.

22. Statler notes that the American program of "nation building" in South Vietnam was not all that different from the "civilizing mission" of the French colonial regime. Kathryn Statler, "Building a Colony: The Eisenhower Administration and South Vietnam, 1953–1961," in *Managing an Earthquake: The Eisenhower Administration, the Third World, and the Globalization of the Cold War,* ed. Andrew L. Johns and Kathryn C. Stattler (Lanham, MD: Rowman and Littlefield, forthcoming); Frey, "Tools of Empire." For a detailed analysis of the establishment of French colonialism in Indochina, see Nicola Cooper, *France in Indochina: Colonial Encounters* (Oxford: Berg, 2001).

23. NSC 5611, Status of the USIA Program, June 30, 1956; OCB, Detailed Development of Major Actions Relating to Southeast Asia, June 22, 1956; Communist Threat to American Interests in Singapore and Malaya and Possible Countermeasures, December 14, 1955, RG 59, Lot 62 D430, Records Relating to Department of State Participation in NSC and OCB, box 18, Far East, NA; NSC 5525,

Status of the USIA Program, August 31, 1955; PCIAA Report on Asia; Frey, "Tools of Empire."

24. OCB Report on Southeast Asia, June 4, 1958, Ann Whitman File, NSC Series, box 10, filed under 368th NSC Meeting, EL; OCB, Detailed Development of Major Actions Relating to Southeast Asia, June 22, 1956; PCIAA Report on Asia. On the CIA's intervention in Indonesia, see Audrey R. Kahin and George McT. Kahin, *Subversion as Foreign Policy: The Secret Eisenhower and Dulles Debacle in Indonesia* (Seattle: University of Washington Press, 1995); McMahon, *Limits of Empire*, 84–90; Robert J. McMahon, "'The Point of No Return': The Eisenhower Administration and Indonesia, 1953–1960," in Johns and Statler, *Managing an Earthquake*. McMahon called it "one of the most misguided, ill-conceived, and ultimately counterproductive covert operations of the entire Cold War era" (*Limits of Empire*, 88).

25. OCB, Detailed Development of Major Actions Relating to Southeast Asia, June 22, 1956; American Embassy Bangkok to DOS, April 25, 1953, RG 59, Central Decimal File, 511.92/4-2553, NA.

26. "An Action Program for Asia," OCB Central Files, box 64, OCB 091.4 Asia (File #1) (1) EL; OCB, Detailed Development of Major Actions Relating to Southeast Asia, June 22, 1956; American Embassy Bangkok to DOS, April 25, 1953, RG 59, Central Decimal File, 511.92/4-2553, NA. On U.S. policy toward Thailand, see Daniel Fineman, *A Special Relationship: The United States and Military Government in Thailand, 1947–1958* (Honolulu: University of Hawaii Press, 1997).

27. NSC 5430, Status of the USIA Program, August 18, 1954, *FRUS 1952–1954*, 2:1778–90; NSC 5525, Status of the USIA Program, August 31, 1955; NSC 5611, Status of the USIA Program, June 30, 1956; USIS Bangkok to USIA, August 2, 1957, and July 29, 1958, both in RG 306, Foreign Service Despatches, box 1, Asia, NA; American Embassy Bangkok to DOS, April 25, 1953, RG 59, Central Decimal File, 511.92/4-2553, NA; OCB, Detailed Development of Major Actions Relating to Southeast Asia, June 22, 1956.

28. NSC 5720, Status of the USIA Program, June 30, 1957; PCIAA Report on Asia.

29. Quoted in Jason Parker, "Small Victory, Missed Chance: The Eisenhower Administration, The Bandung Conference, and the Turning of the Cold War," in Johns and Statler, *Managing an Earthquake*. For a superb analysis of decolonization and Eisenhower's foreign policy, see Thomas Borstelmann, *The Cold War and the Color Line: American Race Relations in the Global Arena* (Cambridge: Harvard University Press, 2001), 85–134.

30. NSC 161, Status of the Psychological Program, June 30, 1953, Soviet Flashpoints Collection, Record No. 78443, National Security Archive, Washington, D.C.; NSC 5407, Status of the USIA Program, March 1, 1954; NSC 5430, Status of the USIA Program, August 18, 1954; NSC 5509, Status of the USIA Program, March 2, 1955; NSC 5525, Status of the USIA Program, August 31, 1955; NSC 5611, Status of the USIA Program, June 30, 1956; NSC 5720, Status of the USIA

Program, June 30, 1957. NSC 5819, Status of the USIA Program, June 30, 1958; NSC 5912, Status of the USIA Program, June 30, 1959; and NSC 6013, Status of the USIA Program, June 30, 1960; all in WHO, OSANSA, NSC Series, Status of Projects Subseries, boxes 8–9, EL.

31. Working Paper, A US Position on North African Problems, August 31, 1955, Nelson A. Rockefeller Papers, RG 4, Washington, D.C., Files, Subseries 7, box 8, "Quantico II: Background Papers," Rockefeller Archive Center, Sleepy Hollow, New York; NSC 5720, Status of the USIA Program, June 30, 1957; Borstelmann, *The Cold War and the Color Line*, 113 and 117; Caroline Pruden, *Conditional Partners: Eisenhower, the United Nations, and the Search for a Permanent Peace* (Baton Rouge: Louisiana State University Press, 1998), 174–75.

32. On U.S. policy toward North Africa and France, see Martin Thomas, "Defending a Lost Cause? France and the United States Vision of French North Africa, 1945–1956," *Diplomatic History* 26 (Spring 2002): 215–47; Egya N. Sangmuah, "Eisenhower and Containment in North Africa, 1956–1960," *Middle East Journal* 44 (Winter 1990): 76–91; Paul J. Zingg, "The Cold War in North Africa: American Foreign Policy and Postwar Muslim Nationalism, 1945–1962," *Historian* 39 (November 1976): 439–66; Yahia H. Zoubir, "U.S. and Soviet Policies Toward France's Struggle with Anticolonial Nationalism in North Africa," *Canadian Journal of History* 30 (December 1995): 439–66; Matthew Connelly, *A Diplomatic Revolution: Algeria's Fight for Independence and the Origins of the Post–Cold War Era* (Oxford: Oxford University Press, 2002); Irwin M. Wall, *France, the United States, and the Algerian War* (Berkeley: University of California Press, 2001).

33. OCB Progress Report on United States Policy in North Africa, April 4, 1956, OCB Central Files, box 61, OCB 091.4 Africa (file #1) (8), EL; OCB Operational Guidance with Respect to Tunisia, Morocco and Algeria in Implementation of NSC 5614/1, February 27, 1957, OCB Central Files, box 61, OCB 091.4 Africa (file #2) (12), EL.

34. Working Paper, A U.S. Position on North African Problems, August 31, 1955; NSC 5611, Status of the USIA Program, June 30, 1956; OCB Progress Report on United States Policy in North Africa, April 4, 1956; Thomas, "Defending a Lost Cause," 218; Pruden, *Conditional Partners*, 173–97.

35. OCB Progress Report on United States Policy in North Africa, April 4, 1956; Lodge to Eisenhower, in cable from Dillon to Dulles, March 7, 1956, Ann Whitman File, Administration Series, box 24, Henry Cabot Lodge 1956 (3), EL. Thomas, "Defending a Lost Cause," 242.

36. NSC 5611, Status of the USIA Program, June 30, 1956; OCB Progress Report on United States Policy in North Africa, April 4, 1956.

37. OCB Outline Plan of Operations with Respect to Tunisia, Morocco, Algeria, October 22, 1956, Mandatory Declassification Review, EL; OCB Progress Report on United States Policy in North Africa, April 4, 1956; OCB Operational

Guidance with Respect to Tunisia, Morocco and Algeria in Implementation of NSC 5614/1, February 27, 1957.

38. Ted Morgan, *A Covert Life: Jay Lovestone: Communist, Anti-Communist, and Spymaster* (New York: Random House, 1999), 285–94, quote on 286; Thomas, "Defending a Lost Cause," quote on 231. OCB Outline Plan of Operations with Respect to Tunisia, Morocco, Algeria, October 22, 1956; OCB Progress Report on United States Policy in North Africa, April 4, 1956.

39. OCB Operational Guidance with Respect to Tunisia, Morocco and Algeria, February 27, 1957; OCB Outline Plan of Operations with Respect to Tunisia, Morocco, Algeria, October 22, 1956; Joseph S. Nye, "Soft Power," *Foreign Policy* 80 (Fall 1990): 153–71.

40. NSC 5720, Status of the USIA Program, June 30, 1957. On U.S. policy toward African decolonization, see Steven Metz, "American Attitudes toward Decolonization in Africa," *Political Science Quarterly* 99 (Fall 1984): 515–33; Thomas J. Noer, *Cold War and Black Liberation: The United States and White Rule in Africa, 1948–1968* (Columbia: University of Missouri Press, 1985); James Hunter Meriwether, *Proudly We Can Be Africans: Black Americans and Africa, 1935–1961* (Chapel Hill: University of North Carolina Press, 2002); Borstelmann, *The Cold War and the Color Line.*

41. Africa, Report to the President's Committee on Information Activities Abroad, July 11, 1960, Records of the President's Committee on Information Activities Abroad, box 23, PCIAA #31 (1), EL (hereafter PCIAA Report on Africa).

42. PCIAA Report on Africa, July 11, 1960; NSC 6013, Status of the USIA Program, June 30, 1960; NSC 5912, Status of the USIA Program, June 30, 1959.

43. During the early Cold War, Americans often referred to this region as the Near East. For the purposes of this discussion, the Middle East refers to Egypt, Iran, Iraq, Israel, Jordan, Lebanon, Libya, Syria, Turkey, Saudi Arabia, and the United Arab Republic (union of Egypt and Syria formed in 1958). Algeria, Morocco, Sudan, and Tunisia, occasionally considered part of the region, are discussed earlier in this chapter. British protectorates that achieved independence in subsequent decades are not included in this discussion.

44. NSC 5611, Status of the USIA Program, June 30, 1956. NSC 5820, the Middle East policy paper approved in November 1958, articulated this goal of dual containment. It urged the United States to "establish an effective working relationship with Arab nationalism while at the same time seeking . . . to contain its outward thrust." See NSC 5820, November 4, 1958, *FRUS 1958–1960*, 12:187–99. Recent scholarship on Eisenhower's policy for the Middle East stresses the objective of containing Arab nationalism. See especially Ray Takeyh, *The Origins of the Eisenhower Doctrine: The U.S., Britain and Nasser's Egypt, 1953–57* (London: Macmillan, 2000); and Salim Yaqub, *Containing Arab Nationalism: The Eisenhower Doctrine and the Middle East* (Chapel Hill: University of North Carolina Press, 2004).

45. The region's oil reserves, as well as critical waterways and proximity to the

Soviet Union, gave it supreme economic and strategic value. "As far as sheer value of territory is concerned there is no place more strategically important than the Middle East," Eisenhower remarked. "The oil of the Arab world has grown increasingly important to all of Europe. The economy of the European states would collapse if these supplies were cut off." Takeyh, *Origins of the Eisenhower Doctrine*, 8. On oil and international politics, see Daniel Yergin, *The Prize: The Epic Quest for Oil, Money, and Power* (New York: Simon and Schuster, 1991).

46. NSC 5428, U.S. Objectives and Policies with Respect to the Near East, July 23, 1954, National Security Archive Electronic Briefing Book No. 78, Early Cold War U.S. Propaganda Activities in the Middle East, available at http://www.gwu.edu/~nsarchiv/NSAEBB/NSAEBB78/docs.htm, document 127; NSC 5810/1, Long-Range U.S. policy toward the Middle East, January 24, 1958, *Declassified Documents Reference System* online, Document Number: CK3100084048; The Middle East, Report to the President's Committee on Information Activities Abroad, July 25, 1960, Records of the President's Committee on Information Activities Abroad, box 24, PCIAA #36 (1), EL (hereafter PCIAA Report on the Middle East); OCB Report, Inventory of U.S. Government and Private Organization Activity Regarding Islamic Organizations as an Aspect of Overseas Operations, May 3, 1957, *Declassified Documents Reference System* online, document no. CK3100496987. USIS Country Plans for the region consistently identified nationalism as a greater concern, in relation to public opinion, than communism. See Foreign Service Despatches, box 3, Africa, NA. H. W. Brands, *The Specter of Neutralism: The United States and the Emergence of the Third World, 1947–1960* (New York: Columbia University Press, 1989), 230–31, 245.

47. NSC 5428, U.S. Objectives and Policies with Respect to the Near East, July 23, 1954; PCIAA Report on the Middle East, July 25, 1960, EL; Brands, *Specter of Neutralism*, 229–41. The Truman administration had tried unsuccessfully to create a Middle Eastern Defense Organization (MEDO) centering around Egypt. Egypt's opposition, stemming from the belief that Israel and Britain posed greater threats to Arabs than communism, led Eisenhower and Dulles to scrap the MEDO concept. They turned instead to the "northern tier" strategy. Although the United States initially developed the concept of the Baghdad Pact, it opposed Britain's joining the pact in April 1955 and refused to join thereafter because it had become tainted by British imperialism. U.S. relations with Israel also made it difficult on the domestic political side for the administration to create a formal alliance with Baghdad. The United States joined the military committee of the Baghdad Pact after the Suez crisis but did not become a member. On the Baghdad Pact, see Persson Magnus, *Great Britain, the United States, and the Security of the Middle East: The Formation of the Baghdad Pact* (Lund, Sweden: Lund University Press, 1998). For an overview of the MEDO concept and U.S. policy toward Egypt, see Peter Hahn, *The United States, Great Britain, and Egypt, 1945–1956* (Chapel Hill, University of North Carolina Press, 1991), 147–65.

48. PSB D-22, January 8, 1953, PSB Central Files, box 15, PSB 091.4, Middle East, EL; NSC 161, Status of the Psychological Program, June 30, 1953; NSC 5428, U.S. Objectives and Policies with Respect to the Near East, July 23, 1954.

49. Country Plan for USIS Libya, June 20, 1956, RG 306, Foreign Service Despatches, box 3, Africa, NA; Andrew Rathmell, *Secret War in the Middle East: The Covert Struggle for Syria, 1949–1961* (London: Tauris Academic Studies, 1995), 96; NSC 5428, U.S. Objectives and Policies with Respect to the Near East, July 23, 1954; NSC 5525, Status of the USIA Program, August 31, 1955; NSC 5611, Status of the USIA Program, June 30, 1956. Peter L. Hahn, *Caught in the Middle East: U.S. Policy Toward the Arab-Israeli Conflict, 1945–1961* (Chapel Hill: University of North Carolina Press, 2004); Isaac Alteras, *Eisenhower and Israel: U.S.-Israeli Relations, 1953–1960* (Gainsville: University Press of Florida, 1993).

50. NSC 5428, U.S. Objectives and Policies with Respect to the Near East, July 23, 1954; James Vaughan, "Propaganda by Proxy? Britain, America, and Arab Radio Broadcasting, 1953–1957," *Historical Journal of Film, Radio and Television* 22 (2002): 157–72. For a superb account of Anglo-American propaganda activities in the Middle East during this period, see Vaughan, "The Anglo-American Relationship and Propaganda Strategies in the Middle East, 1953–1957" (Ph.D. diss., University of London, 2001). Judged by personnel, not one Middle Eastern nation was included in the top ten countries targeted by the USIA. The largest information program in the region was in Iran, with 132 employees (23 Americans and 109 locals). It ranked fifteenth out of the 76 countries targeted in 1956, the year of highest agency funding during the decade. Egypt fell just below this, with 111 employees (20 Americans and 91 locals). Other area programs ranked in the middle or bottom tiers: Turkey and Iraq employed around 70 persons each; Syria and Lebanon had staffs of under 50 each; and Israel, Libya, and Jordan lay at the bottom of the bracket, with 30 employees or fewer. USIA Sixth Report to Congress, June 1956.

51. Vaughan, "Propaganda by Proxy."

52. Embassy Baghdad to USIA/DOS, January 13, 1954; and January 19, 1954; both in National Security Archive Briefing Book No. 78, documents 118 and 119; Vaughan, "Propaganda by Proxy."

53. Embassy Baghdad to USIA/DOS, March 16, 1954, National Security Archive Briefing Book No. 78, document 120.

54. Country Plan for USIS Libya, June 20, 1956, RG 306, Foreign Service Despatches, box 3, Africa, NA; NSC 5430, Status of the USIA Program, August 18, 1954; NSC 5611, Status of the USIA Program, June 30, 1956; OCB Report, Inventory of U.S. Government and Private Organization Activity Regarding Islamic Organizations as an Aspect of Overseas Operations, May 3, 1957. Joyce Battle, December 13, 2002, "U.S. Propaganda in the Middle East—The Early Cold War Version," available at http://www.gwu.edu/~nsarchiv/NSAEBB/NSAEBB78/essay.htm. Takeyh, *Origins of the Eisenhower Doctrine*, 113. On U.S. relations with Saudi Arabia and Eisenhower's efforts to boost the prestige of King Saud, see Nathan

Citino, *From Arab Nationalism to OPEC: Eisenhower, King Saud, and the Making of U.S.-Saudi Relations* (Bloomington: Indiana University Press, 2002), esp. chap. 4.

55. Robert J. McMahon, *The Cold War on the Periphery: The United States, India, and Pakistan* (Columbia: Columbia University Press, 1994), 198–213; USIA, Third Report to Congress, December 1954; USIA, Sixth Report to Congress, June 1956. Together, the USIS programs in India and Pakistan employed more persons than the regions of the Middle East, Latin America, or Africa. On USIS operations in India, see the country plans and correspondence in RG 306, Office of Research, Country Project Correspondence, 1952–63, box 10, India, NA.

56. An internal CIA history of the operation and planning documents were disclosed by the *New York Times* on April 16 and June 18, 2000. See Dr. Donald N. Wilber, Overthrow of Premier Mossadeq, March 1954; and Initial Operation Plan for TPAJAX, June 1, 1953, available at http://www.nytimes.com/library/world/mideast/041600iran-cia-index.html; Loy Henderson to DOS, November 21, 1951, National Security Archive Briefing Book No. 78, document 41; Mark J. Gasiorowski, "The 1953 Coup d'Etat in Iran," *International Journal of Middle East Studies* 19 (August 1987): 261–86; Stephen Kinzer, *All the Shah's Men: An American Coup and the Roots of Middle East Terror* (Hoboken, N.J.: John Wiley & Sons, 2003); Moyara de Moraes Ruehsen, "Operation 'Ajax' Revisited: Iran, 1953," *Middle Eastern Studies* 29 (July 1993): 467–86.

57. Dulles to U.S. Embassy Tehran, June 26, 1953; DOS to Embassy Tehran, October 13, 1953; and American Embassy Tehran to DOS, November 7, 1953, National Security Archive Briefing Book No. 78, documents 97, 111, and 114.

58. USIS Tehran to USIA, October 2, 1953; American Embassy Tehran to USIA, September 15, 1953, National Security Archive Briefing Book No. 78, documents 110 and 108; PCIAA Report on the Middle East.

59. Christopher Simpson, *Blowback: America's Recruitment of Nazis and Its Effects on the Cold War* (New York: Weidenfeld and Nicolson, 1988), 249–52; Wilbur Crane Eveland, *Ropes of Sand: America's Failure in the Middle East* (London: W. W. Norton, 1980), 103; Miles Copeland, *The Game of Nations* (New York: Simon and Schuster, 1969), 100, 113; Semi-Annual USIS Report for Egypt, August 4, 1955, Foreign Service Despatches, box 3, Africa, NA; Vaughan, "Propaganda by Proxy." The extent of CIA involvement in recruiting former Nazi officials to assist Nasser's government is still unknown. According to Timothy Naftali, recently released documents under the Nazi War Crimes Disclosure Act suggest that the agency was more of an "onlooker" than a "sponsor" of these activities. See Timothy Naftali, "Reinhard Gehlen and the United States," in *U.S. Intelligence and the Nazis*, ed. Richard Breitman, Norman J. W. Goda, Timothy Naftali, and Robert Wolfe, 375–418 (New York: Cambridge University Press, 2005).

60. NSC 5407, Status of the USIA Program, March 1, 1954; Semi-Annual USIS Report for Egypt, August 4, 1955, Foreign Service Despatches, box 3, Africa, NA; Vaughan, "Propaganda by Proxy."

61. Gary D. Rawnsley, *Radio Diplomacy and Propaganda: The BBC and VOA in International Politics, 1956–64* (New York: St. Martin's Press, 1996), 20–35. Nasser quoted in Takeyh, *Origins of the Eisenhower Doctrine*, 61 and 64. The Suez crisis has received voluminous treatment by scholars. For a summary of the literature, see Tore Tingvold Peterson, *The Middle East Between the Great Powers: Anglo-American Conflict and Cooperation, 1952–7* (New York: St. Martin's Press, 2000), 65–75.

62. Quotes from Takeyh, *Origins of the Eisenhower Doctrine*, 126; Chester J. Pach Jr. and Elmo Richardson, *The Presidency of Dwight D. Eisenhower*, rev. ed. (Lawrence: University Press of Kansas, 1991), 132–33; and Pruden, *Conditional Partners*, 238.

63. NSC 5819, Status of the USIA Program, June 30, 1958.

64. Dulles to Eisenhower, March 28, 1956, John Foster Dulles Papers, White House Memoranda Series, box 4, White House Correspondence-General 1956 (6), EL; Malcolm H. Kerr, *The Arab Cold War: Gamal 'abd Al-Nasir and His Rivals* (London: Oxford University Press, 1971). On "divide and rule," see Nathan Citino, "Middle East Cold War(s): Oil and Arab Nationalism in U.S.-Iraqi Relations, 1958–1961," in Johns and Statler, *Managing an Earthquake*. On Omega, see Ray Takeyh, *Origins of the Eisenhower Doctrine*, 105–23; Yaqub, *Containing Arab Nationalism*, 42–46, 59–601; Hahn, *United States*, 200–204, 209–10.

65. Arthur Larson to Eisenhower, September 28, 1957, Ann Whitman File, box 37, USIA (1), EL. On Syria, see Anthony Gorst and W. Scott Lucas, "The Other Collusion: Operation Straggle and Anglo-American Intervention in Syria, 1955–56," *Intelligence and National Security* 4 (July 1988): 576–95; Douglas Little, "Cold War and Covert Action: The United States and Syria 1945–1958," *Middle East Journal* 44 (1990): 51–75; Philip Anderson, "'Summer Madness': The Crisis in Syria, August–October 1957," *British Journal of Middle Eastern Studies* 22 (1995): 21–42; Rathmell, *Secret War*. On Lebanon and the Eisenhower Doctrine, see Douglas Little, "His Finest Hour? Eisenhower, Lebanon, and the 1958 Middle East Crisis," *Diplomatic History* 20 (Winter 1996): 27–54; Ray Takeyh, *Origins of the Eisenhower Doctrine*; Brands, *Specter of Neutralism*, 72–79; Robert J. McMahon, "Credibility and World Power: Exploring the Psychological Dimension in Postwar American Diplomacy," *Diplomatic History* 15 (Fall 1991): 455–71, esp. 464–65.

66. USIA to Douglas MacArthur II, April 10, 1956, RG 59, Decimal File, 103-USIA/4-1056, NA; Ad Hoc Committee on Middle Eastern Informational Activities, November 13, 1956, NSC Staff Papers, OCB Central Files, OCB 091.4 Middle East (12-17-56), EL; Rawnsley, *Radio Diplomacy*, 30. On Iraq, see Nathan Citino, "Middle East Cold War(s)."

67. Takeyh, *Origins of the Eisenhower Doctrine*, 52, 113, 119, 148, 151–52.

68. On the Eisenhower administration's perception of the nationalist challenge in Latin America, see James F. Siekmeier, *Aid, Nationalism, and Inter-American Relations: Guatemala, Bolivia, and the United States, 1945–1961* (Lewiston, N.Y.: Edwin Mellon Press, 1999). For detailed analysis of the Eisenhower administration's evolving national security policies toward the region, see Stephen G. Rabe, *Eisenhower*

and Latin America: The Foreign Policy of Anti-Communism (Chapel Hill: University of North Carolina Press, 1988).

69. NSC 5613/1, U.S. Policy Toward Latin America, September 25, 1956, *FRUS 1955–1957*, 6:119–28; NSC 5432/1, September 3, 1954, *FRUS 1952–1954*, 4:81–86; OCB Outline Plan of Operations Against Communism in Latin America, April 18, 1956, *FRUS 1955–1957*, 6:61–82.

70. OCB Progress Report on NSC 5432/1, January 19, 1955, *FRUS 1952–1954*, 4, no. 89–115, esp. 106–9; NSC 5525, Status of the USIA Program, August 31, 1955. Milton Eisenhower's report is in *U.S. Department of State Bulletin*, November 22, 1953, 695–717. On implementation of the report, see Deputy Director of the Office of South American Affairs (Bennett) to Assistant Secretary of State for Inter-American Affairs (Holland), August 20, 1954, *FRUS 1952–1954*, 4:229–50.

71. NSC 5525, Status of the USIA Program, August 31, 1955; OCB Progress Report on NSC 5432/1, March 28, 1956, *FRUS 1955–1957*, 6:46–45; OCB Progress Report on NSC 5432/1, January 19, 1955, *FRUS 1952–1954*, 4:89–115, esp. 106–9; USIS Santiago to USIA, June 13, 1956, FOIA.

72. Nick Cullather, *Secret History: The CIA's Classified Account of Its Operations in Guatemala, 1952–1954* (Stanford: Stanford University Press, 1999), esp. 39–41, 52, 56–58, 61, 63–81, 94. Also on the psychological elements of the operation, see E. Howard Hunt, *Undercover: Memoirs of an American Secret Agent* (New York: Berkeley Publishing, 1974), 96–101; David Atlee Phillips, *The Night Watch* (New York: Atheneum, 1977), 37–68. Other important works include Stephen Schlesinger and Stephen Kinzer, *Bitter Fruit: The Untold Story of the American Coup in Guatemala* (Garden City, N.Y.: Doubleday, 1982); Richard H. Immerman, *The CIA in Guatemala: The Foreign Policy of Intervention* (Austin: University of Texas Press, 1982); Piero Gleijeses, *Shattered Hope: The Guatemalan Revolution and the United States, 1944–1954* (Princeton: Princeton University Press, 1991). For a historiographical review, see Stephen M. Streeter, "Interpreting the 1954 U.S. Intervention in Guatemala: Realist, Revisionist, and Postrevisionist Perspectives," *History Teacher* 34 (2000), May 5, 2002, available at http://www.historycooperative.org/journals/ht/34.1/streeter.html. On Bernays' role, see Edward L. Bernays, *Biography of an Idea: Memoirs of a Public Relations Counsel* (New York: Simon and Schuster, 1965), 762–66.

73. Report on Actions Taken by the United States Information Agency in the Guatemalan Situation, July 27, 1954, *FRUS 1952–1954, Guatemala*, 432–37; USIS-Habana to USIA, August 11, 1954, RG 306, Foreign Service Despatches, box 1, Latin America, NA.

74. Report on Actions Taken by the United States Information Agency in the Guatemalan Situation, July 27, 1954; Airgram from Operation PBSUCCESS Headquarters in Florida to CIA Station Chief in Guatemala, May 19, 1954; and CIA Memorandum, May 12, 1975, in *FRUS 1952–1954, Guatemala*, 292–93, 448–50. The weapons deception, code-named WASHTUB, was undertaken in coop-

eration with the government of Nicaragua, which paid Nicaraguan fishermen to discover the arms cache. See also Gleijeses, *Shattered Hope*, 294, 299; Cullather, *Secret History*, 57, 79.

75. Report on Actions Taken by the United States Information Agency in the Guatemalan Situation, July 27, 1954; NSC 5525, Status of the USIA Program, August 31, 1955; Cullather, *Secret History*, 116; Telegram from CIA to CIA Station in Guatemala, August 9, 1954, *FRUS 1952–1954, Guatemala*, 440; Editorial Note, *FRUS 1952–1954, Guatemala*, 447; Cullather, *Secret History*, 106–7. The CIA reported that the Guatemalan press was "unusually cooperative" in exploiting the PBHISTORY documents, but that other Latin American media were not. A few years after the coup, Ronald M. Schneider analyzed the PBHISTORY documents and found little evidence of Soviet control over the Guatemalan communists in *Communism in Guatemala, 1944–1954* (New York: Praeger, 1958). For a more comprehensive and complex analysis of the role of communism in Arbenz's government, see Gleijeses, *Shattered Hope*.

76. USIS Santiago to USIA, June 13, 1956, FOIA; OCB Outline Plan of Operations Against Communism in Latin America, November 29, 1955; and OCB Operations Plan for Latin America, May 28, 1958; both in RG 59, Records Relating to DOS Participation in NSC and OCB, 1947–63, Lot 62 D430, boxes 23 and 24, NA; OCB Progress Report on NSC 5432/1, January 19, 1955, *FRUS 1952–1954*, 4:89–115, esp. 106–9; NSC 5613/1, U.S. Policy Toward Latin America, September 25, 1956, *FRUS 1955–1957*, 6:119–28; A Program of Covert Action Against the Castro Regime, March 16, 1960, in *Psywar on Cuba: The Declassified History of U.S. Anti-Castro Propaganda*, ed. Jon Elliston (Melbourne: Ocean Press, 1999), 16–19. Elliston's volume includes documents pertaining to American psychological warfare operations against Cuba from the 1950s through the 1990s. On anti-Castro operations, see also Fabian Escalante, *The Secret War: CIA Covert Operations Against Cuba, 1959–1962*, trans. Maxine Shaw (New York: Talman Company for Ocean, 1995). For an international account of U.S.-Cuba-Soviet relations in the early Cold War, see Aleksandr Fursenko and Timothy Naftali, *"One Hell of a Gamble": Khrushchev, Castro, and Kennedy, 1958–1964* (New York: W. W. Norton, 1997).

77. Niall Ferguson, *Empire: The Rise and Demise of the British World Order and the Lessons for Global Power* (New York: Basic Books, 2004). Arguments that the American empire was based on consensus-building and according to an "invitation" by Europeans have been made by John Lewis Gaddis in *We Now Know: Rethinking Cold War History* (Oxford: Clarendon Press, 1997), esp. 26–53, 284–86; and Geir Lundestad, *The United States and Western Europe Since 1945: From "Empire" by Invitation to Transatlantic Drift* (Oxford: Oxford University Press, 2003); Lundestad, "Empire by Invitation? The United States and Western Europe, 1945–1952," *Journal of Peace Research* 23 (September 1986): 263–77. Economic expansion as a formula

for empire has been widely explored. The classic treatment is William Appleman Williams, *The Tragedy of American Diplomacy* (New York : W. W. Norton, 1972).

Chapter 5. Spinning the Friendly Atom

1. Winkler, *Life Under a Cloud: American Anxiety About the Atom* (Oxford: Oxford University Press, 1993), 75–77; Richard Smoke, *National Security and the Nuclear Dilemma: An Introduction to the American Experience in the Cold War,* 3rd ed. (New York: McGraw-Hill, 1993), 57–58.

2. Such sentiments pervade American national security documents during the 1950s. See, for example, the global studies on neutralism attached to Robert Murphy to Nelson Rockefeller, August 19, 1955, White House Office [WHO], NSC Staff Papers, Planning Coordination Group series, box 2, Bandung, Eisenhower Library, Abilene, Kansas (hereafter EL).

3. Paul Boyer, *By the Bomb's Early Light: American Thought and Culture at the Dawn of the Atomic Age* (Chapel Hill: University of North Carolina Press, 1985), 291–333; Eisenhower address, December 8, 1953, *Public Papers of the Presidents of the United States: Dwight D. Eisenhower, 1953* (Washington, D.C.: Government Printing Office, 1961), 813–22 (hereafter *Public Papers*).

4. Historians have tended to take Eisenhower's words at face value, describing Atoms for Peace as a sincere, if flawed, attempt to reduce Cold War tensions. Most recently, Robert R. Bowie and Richard H. Immerman defended the proposal as a novel approach to arms control, premised on a "modest, long-term strategy for making gradual progress by small steps," in *Waging Peace: How Eisenhower Shaped an Enduring Cold War Strategy* (New York: Oxford University Press, 1998), 225. Others who support the "revisionist" view of Atoms for Peace include Stephen Ambrose, *Eisenhower: The President* (New York: Simon and Schuster, 1984); Robert A. Divine, *Eisenhower and the Cold War* (Oxford: Oxford University Press, 1981); Thomas F. Soapes, "A Cold Warrior Seeks Peace: Eisenhower's Strategy for Nuclear Disarmament," *Diplomatic History* 4 (Winter 1980): 57–71; Robert A. Strong, "Eisenhower and Arms Control," in *Reevaluating Eisenhower: American Foreign Policy in the 1950s,* ed. Richard A. Melanson and David Mayers (Urbana: University of Illinois Press, 1987), 241–66; Charles Albert Appleby Jr., "Eisenhower and Arms Control, 1953–1961: A Balance of Risks" (Ph.D. diss., Johns Hopkins University, 1987); Richard G. Hewlett and Jack M. Holl, *Atoms for Peace and War, 1953–1961: Eisenhower and the Atomic Energy Commission* (Berkeley: University of California Press, 1989); McGeorge Bundy, *Danger and Survival: Choices About the Bomb in the First Fifty Years* (New York: Random House, 1988), 236–357. For contrary views, see Walter L. Hixson, *Parting the Curtain: Propaganda, Culture, and the Cold War, 1945–1961* (New York: St. Martin's Griffin, 1997); Andreas Wenger, *Living with Peril: Eisenhower, Kennedy, and Nuclear Weapons* (Lantham, Md.: Rowman

& Littlefield, 1997), 94–99; Kenneth A. Osgood, "Form Before Substance: Eisenhower's Commitment to Psychological Warfare and Negotiations with the Enemy," *Diplomatic History* 24 (Summer 2000): 405–33; Martin J. Medhurst, "Eisenhower's 'Atoms for Peace' Speech: A Case Study in the Strategic Use of Language," *Communication Monographs* 54 (June 1987): 204–20; Medhurst, "Atoms for Peace and Nuclear Hegemony: The Rhetorical Structure of a Cold War Campaign," *Armed Forces and Society* 24 (Summer 1997): 571–93; and Ira Chernus, *Eisenhower's Atoms for Peace* (College Station: Texas A&M Press, 2002).

5. Winkler, *Life Under a Cloud;* Boyer, *By the Bomb's Early Light;* Boyer, *Fallout: A Historian Reflects on America's Half-Century Encounter with Nuclear Weapons* (Columbus: Ohio State University Press, 1998); Boyer, "Exotic Resonances: Hiroshima in American Memory," *Diplomatic History* 19 (Spring 1995): 297–318; Margot A. Henriksen, *Dr. Strangelove's America: Society and Culture in the Atomic Age* (Berkeley: University of California Press, 1997). Lawrence Wittner explores government attempts to restrain antinuclear sentiment in *One World or None: A History of the World Nuclear Disarmament Movement Through 1953* (Stanford: Stanford University Press, 1993); and Wittner, *Resisting the Bomb: A History of the World Nuclear Disarmament Movement, 1954–1970* (Stanford: Stanford University Press, 1997). The use of civil defense to ease public fears has been expertly treated by Guy Oakes, *The Imaginary War: Civil Defense and American Cold War Culture* (Oxford: Oxford University Press, 1994).

6. Eisenhower offered this view of an ideal effort in his San Francisco speech on psychological warfare: "we must bring the dozens of agencies and bureaus into concerted action under an overall scheme of strategy" and by so doing "bear all our forces at the same time under the same plan on the same target. . . . Every significant act of government should be so timed and so directed at a principal target, and so related to other governmental actions, that it will produce maximum effect." Eisenhower Speech, October 8, 1952. As Cutler described it, an ideal psychological warfare initiative involved "the marshaling and integration of the powers of our national government so that when our national government takes a significant action, in any field, all those powers can be brought to bear at once and a single coordinated massive blow or series of blows." Informal Note by Robert Cutler, February 28, 1953, *Declassified Documents Reference System,* microfiche 1980/331B.

7. Stefan Possony, An Outline of American Atomic Strategy in the Non-Military Fields, October 6, 1952, WHO, NSC Staff Papers, OCB Secretariat series, box 3, Ideological Documents (4), EL.

8. Panel of Consultants report, Armaments and American Policy, January 1953; NSC meeting, February 18, 1953; NSC meeting, February 25, 1953; NSC meeting, May 27, 1953; and NSC Action No. 717b; all in *Foreign Relations of the United States 1952–1954,* 2:1056–91, 1106–14, 1169–74 (hereafter *FRUS*). Bowie and Immerman, *Waging Peace,* 223–24.

9. Eisenhower quoted in Wittner, *Resisting the Bomb*, 148. Eisenhower press conferences, June 17, 1953, July 8, 1953, and September 30, 1953, all in *Public Papers*, 1953: 432–33, 475–76, 617–18. Murray S. Levine, "Civil Defense vs. Public Apathy," *Bulletin of Atomic Scientists*, February 1953, 27; Walter Lapp, "Operation Candor Versus Atomic Secrecy," *Reporter*, September 1, 1953; G. Samuels, "Plea for Candor About the Atom," *New York Times Magazine*, June 21, 1953; David Lilienthal, "Case for Candor on National Security," *Bulletin of Atomic Scientists*, December 1953; and the *New York Times*, October 9, 1953, and October 10, 1953.

10. The film released on the Operation Ivy test series was carefully choreographed to both frighten and calm. Nearly an hour of deliberate (indeed, boring) narration preceded the footage of the hydrogen weapon's detonation to make the test seem routine and noneventful. The bomb's explosion was accompanied by a dramatic cacophony and followed by the narrator, walking on a peaceful South Pacific beach, commenting that one day the world will enjoy the peaceful applications of this tremendous force. The film and the press surrounding it were coordinated to the last detail by the OCB. See the records in WHO, NSC Staff Papers, OCB Central Files, box 122, OCB 388.3; and 185th NSC Meeting, February 18, 1954, Ann Whitman File, box 5, EL.

11. NSC meeting, February 25, 1953; and NSC meeting, May 27, 1953; Lambie to Cutler, July 29, 1953, White House Central Files [WHCF], Confidential Series, Subject Subseries, box 12, Candor and U.N. Speech, 12/8/53 (1), EL; Lambie to Adams, August 5, 1953, WHCF, Confidential Series, Subject Subseries, box 12, Candor and U.N. Speech, 12/8/53 (13), EL.

12. "Project 'Candor'—to inform the public of the realities of the 'Age of Peril,'" July 22, 1953; Lambie to Adams, August 5, 1953, and July 9, 1953; Report on Project Candor, July 22, 1953; Lambie to Cutler, July 29, 1953; Memorandum on A Public Information Program in Support of Operation Candor, July 8, 1953; all in WHCF, Confidential Series, Subject Subseries, box 12, Candor and U.N. Speech (1) and (13), EL. See also WHO, NSC Staff Papers, Psychological Strategy Board (PSB) Central Files, box 28, PSB 338.3, EL.

13. C. D. Jackson Memorandum, September 20, 1954, C. D. Jackson Papers, box 29, Atoms for Peace, Evolution (1), EL; John W. Young, *Winston Churchill's Last Campaign: Britain and the Cold War, 1951–5* (Oxford: Clarendon Press, 1996); Vojtech Mastny, *The Cold War and Soviet Insecurity: The Stalin Years* (New York: Oxford University Press, 1996), 171–98; and Vladislav Zubok and Constantine Pleshakov, *Inside the Kremlin's Cold War: From Stalin to Khrushchev* (Cambridge: Harvard University Press, 1996), 138–235.

14. Jackson Committee Report; 134th NSC meeting, February 25, 1953; and 165th NSC meeting, October 7, 1953; all in *FRUS 1952–1954*, 2:514–34, 1110–14, 1665–66, 1800; Lambie to Adams, July 9, 1953, WHCF, Subject Series, box 12, Candor and U.N. Speech (1), EL; Hagerty Diary, March 22, 1955, *FRUS 1955–*

1957, 9:521. NSC meeting July 2, 1953; NSC meeting, July 16, 1953; and NSC meeting, July 30, 1953; all in *FRUS 1952–1954*, 2:1877–79, 394–98, 435–40.

15. Eisenhower to Cutler, September 10, 1953, *FRUS 1952–1954*, 2:1213; Strauss to Eisenhower, September 17, 1953, Ann Whitman File, box 5, Atoms for Peace.

16. Cutler comments on State Department draft, October 19, 1953, Jackson Papers, box 30, Atoms for Peace, Evolution (5), EL; NSC planning board discussion, October 19, 1953; and Dulles to Eisenhower, October 23, 1953; both in *FRUS 1952–1954*, 2:1227–29, 1234–35. Wenger, *Living with Peril*, 95.

17. Eisenhower address, December 8, 1953. For in-depth analysis of the speech itself, see Medhurst, "Eisenhower's 'Atoms for Peace' Speech"; and Chernus, *Eisenhower's Atoms for Peace*.

18. Bundy, *Danger and Survival*, 295.

19. The plan went through a number of revisions but the working drafts guided U.S. policy until the final paper was formally approved in March 1954. The various drafts of the OCB plan can be found in OCB Central Files, box 121, OCB 388.3, EL.

20. Jackson to OCB, WHO, NSC Series, PSB Subseries, box 28, PSB 388.3, EL; National Operations Plan to Exploit the President's U.N. Speech, December 24, 1953, OCB Central Files, box 121, OCB 388.3 (File #1) (2), EL. The Atoms for Peace campaign also served the domestic political objective of persuading the U.S. Congress to amend the Atomic Energy Act of 1946, a necessary prelude to the nuclearization of NATO forces. It also advanced U.S. economic interests by encouraging the growth of a domestic atomic power industry with a strong export market. See Medhurst, "Atoms for Peace and Nuclear Hegemony," 581–86.

21. Streibert to Eisenhower, February 27, 1954, OCB Central Files, box 122, OCB 388.3 (File #1) (5), EL; OCB Progress Report on Implementation of U.N. Speech, April 23, 1954, OCB Central Files, OCB 388.3 (File #2) (2), EL.

22. OCB Progress Report on Implementation of U.N. Speech, April 23, 1954; Marie McCrum Oral History Interview, May 15, 1975, EL.

23. National Operations Plan to Exploit the President's U.N. Speech, December 24, 1953, OCB Central Files, box 121, OCB 388.3 (File #1) (2), EL.

24. National Operations Plan to Exploit the President's U.N. Speech, December 24, 1953; and OCB Memo, U.S. Postage Stamp 'Atoms for Peace,' November 2, 1954, Record Group [RG] 59, Records Relating to DOS Participation in the OCB and NSC, box 35, Atomic and Nuclear Energy, 1954, National Archives, College Park, Maryland (hereafter NA); "Suggested Domestic Implementation Check List," n.d., WHO, NSC Series, PSB Subseries, box 28, PSB 388.3, EL; OCB Minutes, March 8, 1954, OCB Central Files, box 122, OCB 388.3 (File #1) (7), EL; "Stamp Boosts Atoms for Peace," *Chicago Tribune*, July 7, 1955; "Atoms Stamp Artist to Visit Eisenhower," *New York Long Island Press*, July 25, 1955.

25. National Operations Plan to Exploit the President's U.N. Speech, December 24, 1953; OCB Progress Report on Implementation of U.N. Speech, April 23, 1954.

26. USIA Summary of Events Planned for Exploitation of U.N. Speech, January 20, 1954, OCB Central Files, box 122, OCB 388.3 (File #1) (3), EL; USIA Biweekly Summary of Events Planned for Exploitation of U.N. Speech, April 28, 1954, OCB Central Files, OCB 388.3 (File #2) (3) EL; USIA Summary of Events Planned for Exploitation of U.N. Speech, March 17, 1954, RG 306, Special Papers of the Coordinator for Psychological Intelligence, 1952–1954, box 1, Entry 1044, NA.

27. *New York Times*, December 22, 1953; Hewlett and Holl, *Atoms for Peace and War*, 221. On Soviet foreign policy and Atoms for Peace, see David Holloway, *Stalin and the Bomb: The Soviet Union and Atomic Energy, 1939–1956* (New Haven, Conn.: Yale University Press, 1994), 346–63.

28. National Operations Plan to Exploit the President's U.N. Speech, December 24, 1953; Jackson to Eisenhower, December 29, 1953; and Jackson Memorandum, December 28, 1953, Jackson Papers, box 29, Atoms for Peace, Evolution (2), EL.

29. National Operations Plan to Exploit the President's U.N. Speech, December 24, 1953; Jackson to Eisenhower, December 29, 1953; Jackson Memorandum, December 28, 1953; and Memorandum for the Record, December 29, 1953, WHCF, Subject Series, Confidential Subseries, box 13, Operation Candor (19); OCB Memorandum, January 6, 1953, OCB Central Files, box 121, OCB 388.3 (File #1) (2), EL.

30. Eisenhower to Jackson, December 31, 1953, Ann Whitman File, Eisenhower Diary Series, box 4; Jackson Log, January 16, 1954, Jackson Papers, box 68, Log 1954 (1), EL.

31. OCB Minutes, January 27, 1954, OCB Central Files, box 122, OCB 388.3 (File #1) (3), EL; Jackson to OCB, n.d., WHO, NSC Series, PSB Subseries, box 28, PSB 388.3, EL; Jackson to Department and Agency Information Officers, January 19, 1954, OCB Central Files, box 121, OCB 388.3 (File #1) (2), EL.

32. Medhurst, "Atoms for Peace and Nuclear Hegemony." U.N. Ambassador Lodge pledged 100 kilograms of material during a speech before the United Nations in November 1954. In a commencement address at Pennsylvania State University, Eisenhower expanded the amount to 200 kilograms and made his "50–50" offer. In 1956, the pool of fissionable materials included 20,000 kilograms nationally and 20,000 for international use. For documentation on these efforts, see Nelson A. Rockefeller Papers, RG 4, Washington, D.C., Files, Subseries 7, boxes 13 and 14, Rockefeller Archive Center, Sleepy Hollow, New York.

33. OCB Minutes, February 17, 1954; and DOS Memorandum, February 4, 1954; both in OCB Central Files, box 122, OCB 388.3 (File #1) (5), EL; National Operations Plan to Exploit the President's U.N. Speech, December 24, 1953. NSC 5431/1, August 13, 1954; and NSC Meeting, August 12, 1954; both in *FRUS 1952–1954*, 2:1482–500.

34. Top Secret extract from OCB Minutes, January 6, 1954; and A Program to Exploit the A-Bank Proposals, March 10, 1954; both in OCB Central Files, box 121, OCB 388.3 (File #1) (2), EL. Current Information Report, USIS Bonn to USIA, November 2, 1954, USIA Historical Collection, Washington, D.C.

35. OCB Minutes, January 6, 1954, OCB Central Files, box 121, OCB 388.3 (File #1) (2), EL; AEC Memorandum, Salisbury to Hirsch, February 3, 1954, OCB Central Files, box 122, OCB 388.3 (File #1) (3), EL; National Operations Plan to Exploit the President's U.N. Speech, December 24, 1953; OCB Progress Report on Implementation of U.N. Speech, April 23, 1954.

36. Global Theme IV, April 29, 1955, RG 59, Bureau of Public Affairs, International Education and Exchange Service, Correspondence, Memoranda, Reports, and Other Records of the Program Development Staff, box 6, Atomic Energy Program, NA; Streibert to Eisenhower, February 27, 1954; USIA Biweekly Summary of Events Planned for Exploitation of U.N. Speech, December 17, 1953; USIA Biweekly Summary of Events Planned for Exploitation of U.N. Speech, January 20, 1954; USIA Biweekly Summary of Events Planned for Exploitation of U.N. Speech, March 17, 1954. "Atoms for Peace Activities of the U.S. Information Agency," Report to the Joint Committee on Atomic Energy, June 1956, Freedom of Information Act (FOIA).

37. "Atomic Power for Peace: International Press Service Background and Action Kit 24," RG 306, Feature Packets: Non-Recurring Subjects, 1953–1958, Entry 1002, box 2, NA. The kit was sent to all USIA posts and included forty-five separate items for publicity and speech use, picture stories and displays, pamphlets and reprints, background information.

38. "Atomic Bomb Blast Waves," *Scientific American*, April 1953; "When an Atomic Blast Hits Your Home or Auto," *U.S. News and World Report*, March 27, 1953; "Have Atom Bomb Tests Fouled Up the Weather?" *Look Magazine*, August 11, 1953; "Soon the World Will Need the Atom for Energy," *Business Week*, December 19, 1953; "Those Wonderful Atoms," *Catholic World*, May 1954; "Atoms for Industry," *Newsweek*, May 31, 1954; "The Taming of the Atom," *Newsweek*, November 29, 1954; "Atom Power for Homes in Five Years," *U.S. News and World Report*, June 25, 1954. Reprints of the *Today's Health* and *New York Times Magazine* articles are in "Atomic Power for Peace: International Press Service Background and Action Kit 24," RG 306, Feature Packets: Non-Recurring Subjects, 1953–1958, box 2. On the link between gendered images and the bomb, see Elaine Tyler May, *Homeward Bound: American Families in the Cold War Era* (New York: Basic Books, 1988), 92–113.

39. F. Barrows Colton, "Man's New Servant, the Friendly Atom," *National Geographic*, January 1954; USIA Biweekly Summary of Events Planned for Exploitation of U.N. Speech, April 28, 1954; "Atomic Power for Peace: International Press Service Background and Action Kit," RG 306, Feature Packets: Non-Recurring Subjects, 1953–1958, box 2, NA.

40. Current Information Report, USIS Bonn to USIA, November 2, 1954, and February 10, 1955; both in USIA Historical Collection.

41. Outline for USIA Exhibit at San Paulo Quadricentennial Exposition and USIS Briefing Manual for Demonstrators at the Atoms for Peace Exhibit, both in USIA Historical Collection; "Atom Model Enroute to Germany Soon," *Marlboro Enterprise*, August 29, 1956; "Atoms for Peace Activities of the U.S. Information Agency," Report to the Joint Committee on Atomic Energy, June 1956, USIA Files, box 455, Accession No. W306-63Z0190EB, Atomic Energy, National Records Center, Suitland, Maryland.

42. USIS Briefing Manual for Demonstrators at the Atoms for Peace Exhibit, USIA Historical Collection.

43. USIS Bonn to USIA, February 10, 1955; and USIS Buenos Aires to USIA, November 9, 1955; both in USIA Historical Collection; USIS Accra, Ghana to USIA, March 15, 1957 and USIS Tokyo to USIA, March 5, 1956, both in RG 306, World Project Files of the Office of Research, box 2, NA; Atoms for Peace Exhibit in Zagreb and Belgrade, USIS Belgrade, December 15, 1955, RG 306, World Project Files of the Office of Research, box 2, NA.

44. Preliminary Report on the Mobile Exhibit, USIS Rome, June 18, 1954, USIA Historical Collection.

45. "Worth Seeing," *Evening Star*, June 7, 1955; "Atoms Show Stresses Good Tidings for All," *Washington Post*, June 3, 1955; "Atoms in Action," *Washington Post*, June 4, 1955; Current Information Report, USIS Bonn to USIA, February 10, 1955, USIA Historical Collection. Associated Press quoted in Preliminary Report on the Mobile Exhibit, USIS Rome, June 18, 1954, USIA Historical Collection. See also "Japanese Hail Avco Exhibit at Big 'Atoms for Peace' Show," *Bridgeport Post*, January 29, 1956; Robert Trumball, "Japan Welcomes Peace Atom Show," *New York Times*, November 1, 1955; "Big Atom Exhibit Presented to Cuba," *Times of Havana*, April 13, 1957; "Atomic Exhibit in Iraq Proves Popular," *Bee Hive*, March 1955, "Atoms into Plowshares," *Beaver Falls News-Tribune*, August 12, 1955; "Quiet, But Important," *San Bernardino Telegram*, August 12, 1955; "Mammoth Conference on Atoms," *Asheville Citizen*, August 8, 1955.

46. Shawn J. Parry-Giles, "'Camouflaged' Propaganda: The Truman and Eisenhower Administrations' Covert Manipulation of News," *Western Journal of Communication* 60 (Spring 1996): 146–67.

47. U.S. House Permanent Select Committee on Intelligence, *The CIA and the Media*, 95th Congress, 1st and 2nd Sessions, 1978.

48. Hewlett and Holl, *Atoms for Peace and War*, 253; Nancy E. Bernhard, "Clearer than Truth: Public Affairs Television and the State Department's Domestic Information Campaigns, 1947–1952," *Diplomatic History* 21 (Fall 1997): 552–53; Bernhard, *U.S. Television News and Cold War Propaganda, 1947–1960* (Cambridge: Cambridge University Press, 1999); Christopher Simpson, *Science of*

Coercion: Communication Research and Psychological Warfare, 1945–1960 (Oxford: Oxford University Press, 1994); and Parry-Giles, "'Camouflaged' Propaganda."

49. USIS India Annual Assessment Report, November 4, 1958, RG 306, Foreign Service Despatches, box 1, Asia, NA; NSC 5611, WHO, Office of the Special Assistant for National Security Affairs, NSC Series, Status of Projects Subseries, box 7, EL.

50. Winkler, *Life Under a Cloud*, 142; Elizabeth Walker Mechling and Jay Mechling, "The Atom According to Disney," *Quarterly Journal of Speech* 81 (1995): 436–53. The song "Atomic Power" was recorded in 1946 by The Buchanons. This song and others about "the bomb" can be heard at http://www.authentichistory. com/audio/1950s/atomic_music_01.html.

51. Robert J. Lifton, *Death in Life: Survivors in Hiroshima* (New York: Random House, 1967); Lifton and Greg Mitchell, "The Age of Numbing," *Technology Review* 98 (August–September 1995): 58–60; Lifton and Richard Falk, *Indefensible Weapons: The Political and Psychological Case Against Nuclearism* (New York: Basic Books, 1982), 82; Lifton and Greg Mitchell, *Hiroshima in America: Fifty Years of Denial* (New York: G. P. Putnam's Sons, 1995), 337–441.

Chapter 6. *The Illusory Spirit of Geneva*

1. Henry Kissinger, "Psychological and Pressure Aspects of Negotiations with the USSR," November 1955, in Psychological Aspects of United States Strategy: Source Book of Individual Papers, White House Office [WHO], Office of the Special Assistant for National Security Affairs [OSANSA], NSC Series, Subject Subseries, box 10, Eisenhower Library, Abilene, Kansas (hereafter EL); Kissinger, "Reflections on American Diplomacy," *Foreign Affairs* 35 (October 1956): 37–56.

2. According to the PPS, "the very process of repeating carefully selected words or phrases in U.S. Government media and official statements will tend to bring these words and phrases into current usage in the commercial media," thereby affecting popular perceptions of the negotiations. Policy Planning Staff [PPS], "Semantics in Foreign Policy," February 24, 1958, Record Group [RG] 59, PPS Records 1957–1961, Lot 67 D548, box 120, Information Policy, National Archives, College Park, Maryland (hereafter NA). Other studies that addressed the "new diplomacy" include: William H. Jackson, "The Fourth Area of the National Effort in Foreign Affairs," n.d. [November 1956], Ann Whitman File, box 22, William H. Jackson, EL; Committee on Foreign Affairs Personnel, *Personnel for the New Diplomacy: Report of the Committee on Foreign Affairs Personnel* (Washington, D.C.: Carnegie Endowment for International Peace, 1962), 1–9; House Committee on Foreign Affairs, *Winning the Cold War: The U.S. Ideological Offensive*, 88th Congress, 2nd Session, 1964, 6–7; Sir Roger Makins, "The World Since the War: The Third Phase," *Foreign Affairs* 33 (October 1954): 1–16; Chester Bowles, "Toward

a New Diplomacy," *Foreign Affairs* 40 (January 1962): 244–51; Harold Nicolson, "Diplomacy Then and Now," *Foreign Affairs* 40 (October 1961): 39–49; John W. Henderson, *The United States Information Agency* (New York: Praeger, 1969), vii–x, 3–20.

3. Scholars have explored other psychological connections with foreign relations, including the psychological aspects of deterrence, the psychological dimensions of the "credibility" imperative, and the psychology of foreign policy decision makers themselves. See, for example, Alexander L. George, "Psychological Dimensions of the U.S.-Soviet Conflict," in *Reexamining the Soviet Experience: Essays in Honor of Alexander Dallin*, ed. David Holloway and Norman Naimark (Boulder, Colo.: Westview Press, 1996), 101–18; Robert J. McMahon, "Credibility and World Power: Exploring the Psychological Dimension in Postwar American Diplomacy," *Diplomatic History* 15 (Fall 1991): 455–71; Deborah Welch Larson, *Anatomy of Mistrust: U.S.-Soviet Relations During the Cold War* (Ithaca, N.Y.: Cornell University Press, 1997); Larson, *Origins of Containment: A Psychological Explanation* (Princeton: Princeton University Press, 1985); Robert Jervis, *Perception and Misperception in International Politics* (Princeton: Princeton University Press, 1976).

4. David Holloway, *Stalin and the Bomb: The Soviet Union and Atomic Energy, 1939–1956* (New Haven, Conn.: Yale University Press, 1994), 340; Caroline Pruden, *Conditional Partners: Eisenhower, the United Nations, and the Search for a Permanent Peace* (Baton Rouge: Louisiana State University Press, 1998), 144–71; Lawrence S. Wittner, *One World or None: A History of the World Nuclear Disarmament Movement Through 1953* (Stanford: Stanford University Press, 1993), 251–57, 273–74.

5. Revisionists generally accept Robert Divine's conclusion that the "overriding aim" of Eisenhower's foreign policy was "reducing Cold War tensions and achieving détente with the Soviet Union." Robert A. Divine, *Eisenhower and the Cold War* (Oxford: Oxford University Press, 1981), 105. See Robert R. Bowie and Richard H. Immerman, *Waging Peace: How Eisenhower Shaped an Enduring Cold War Strategy* (New York: Oxford University Press, 1998); Stephen E. Ambrose, *Eisenhower: The President* (New York: Simon and Schuster, 1984); Robert A. Strong, "Eisenhower and Arms Control," in *Reevaluating Eisenhower: American Foreign Policy in the 1950s*, ed. Richard A. Melanson and David Mayers (Urbana: University of Illinois Press, 1987), 241–66; Charles Albert Appleby Jr., "Eisenhower and Arms Control, 1953–1961: A Balance of Risks" (Ph.D. diss., Johns Hopkins University, 1987); Larson, *Anatomy of Mistrust*; Richard Immerman, "Confessions of an Eisenhower Revisionist: An Agonizing Reappraisal," *Diplomatic History* 14 (Summer 1990): 319–42; Richard G. Hewlett and Jack M. Holl, *Atoms for Peace and War, 1953–1961: Eisenhower and the Atomic Energy Commission* (Berkeley: University of California Press, 1989); Walt W. Rostow, *Open Skies: Eisenhower's Proposal of July 21, 1955* (Austin: University of Texas Press, 1982); Thomas F. Soapes, "A Cold Warrior Seeks Peace: Eisenhower's Strategy for Nuclear Disarmament," *Diplomatic History* 4 (Winter 1980): 57–71. Some studies agree

with the revisionist claim that Eisenhower earnestly sought disarmament but find fault with his pursuit of that goal. See Piers Brendon, *Ike: His Life and Times* (New York: Harper & Row, 1986); Andreas Wenger, *Living with Peril: Eisenhower, Kennedy, and Nuclear Weapons* (Lantham, Md.: Rowman & Littlefield, 1997); McGeorge Bundy, *Danger and Survival: Choices About the Bomb in the First Fifty Years* (New York: Random House, 1988), 236–319; and Pruden, *Conditional Partners*. More skeptical "postrevisionist" interpretations include: Martha Smith-Norris, "The Eisenhower Administration and the Nuclear Test Ban Talks, 1958–1960: Another Challenge to 'Revisionism,'" *Diplomatic History* 27 (September 2003): 503–41; J. Michael Hogan, "Eisenhower and Open Skies: A Case Study in Psychological Warfare," in *Eisenhower's War of Words: Rhetoric and Leadership*, ed. Martin J. Medhurst (East Lansing: Michigan State University Press, 1994), 137–56; Jeremi Suri, "America's Search for a Technological Solution to the Arms Race: The Surprise Attack Conference of 1958 and a Challenge for 'Eisenhower Revisionists,'" *Diplomatic History* 21 (Summer 1997): 417–52; and Walter L. Hixson, *Parting the Curtain: Propaganda, Culture, and the Cold War* (New York: St. Martin's Griffin, 1997).

6. Panel of Consultants report, Armaments and American Policy, January 1953; NSC meeting, February 18, 1953; NSC meeting, February 25, 1953; and NSC Action No. 717b; all in *Foreign Relations of the United States 1952–1954*, 2:1056–91, 1106–14 (hereafter *FRUS*).

7. NSC-112/1, September 1, 1953, *FRUS 1952–1954*, 2:1190–206.

8. Quotations from Pruden, *Conditional Partners*, 154–55. Vishinsky's proposals are in U.S. Department of State, *Documents on Disarmament* (Washington, D.C.: Government Printing Office, 1960), 1945–1959, 1:431–33.

9. James Richter, *Khrushchev's Double Bind: International Pressures and Domestic Coalition Politics* (Baltimore: Johns Hopkins University Press, 1994), 74. Recent scholarship agrees that the May 10, proposal represented a potential breakthrough in disarmament discussions. See Pruden, *Conditional Partners*, 156; Larson, *Anatomy of Mistrust*, 64–65; Holloway, *Stalin and the Bomb*, 340–41. For Malik's proposal, see *Documents on Disarmament*, 1945–1959, 1:456–67.

10. New scholarship suggests that the Soviet leadership was divided on the question of negotiating with the United States and its allies—divisions that accounted for the uneven character of the Soviet peace offensive in the two years after Stalin's death. When Georgii Malenkov assumed the premiership in March 1953, he stressed the importance of bringing about a relaxation of tensions in order to divert the country's resources from heavy to light industry. Such a move would lesson the danger of war and allow the regime to devote more attention to the production of consumer goods. Another faction of the leadership, led by the old Stalinist Vyasheslav Molotov, advocated negotiations only to bring about a breathing spell while the regime weathered the succession. From Stalin's death until the beginning of 1955, Nikita Khrushchev, head of the Communist Party, sided with the hardliners. He used his opposition to the soft approach as a tool to defeat his rivals and

force Malenkov to retreat. By April 1954, Malenkov had publicly recanted his conciliatory strategy, and in February 1955, he resigned. After Malenkov's resignation, Khrushchev adopted many of the positions of his fallen rival. After Khrushchev solidified his hold on power, he advocated a conciliatory line in favor of peaceful coexistence with the West and authorized the generous May 10 proposal. The relationship between Soviet domestic political disputes and foreign policy is analyzed in Richter, *Khrushchev's Double Bind*. Also see David Holloway, *Stalin and the Bomb*, 320–63; Vladislav Zubok and Constantin Pleshakov, *Inside the Kremlin's Cold War: From Stalin to Khrushchev* (Cambridge: Harvard University Press, 1996), 138–209. Although these and other recent works on Soviet foreign policy during the Cold War incorporate many new findings from Russian archives, most scholars have not had access to the most important high-level documents in the Russian Presidential Archive. Evaluations of Soviet intentions remain primarily based on memoirs, fragmentary documentation, and public record sources. Unfortunately, the historical evidence remains too fragmentary to reach firm conclusions about the true intentions of the Soviet leadership.

11. NSC meeting, June 30, 1955, *FRUS 1955–1957*, 20:144–45. Before this meeting, Dulles wrote, "The Soviet bloc economy cannot indefinitely sustain the effort to match our military output, particularly in high-priced modern weapons. Already there is evidence that the Soviet economy is feeling the strain of their present effort and that their rulers are seeking relief. They have been conducting a vast propaganda effort to bring about the abolition of atomic weapons and they now offer to reduce their land armies if they can thereby get relief in terms of new weapons." See Memorandum by the Secretary of State, June 29, 1955, *FRUS 1955–1957*, 20:140–42. On another occasion, Dulles wrote to the president: "Undoubtedly, one of the Soviet desires is to relieve itself of the economic burdens of the present arms race . . . [The Soviets] believe that if world opinion can be aroused and focused upon us, we may accept disarmament under hastily devised and perhaps imprudent conditions." Dulles to Eisenhower, June 18, 1955, John Foster Dulles Papers, White House Memoranda Series, box 3, EL.

12. Wadsworth to Lodge, May 11, 1955, *FRUS 1955–1957*, 20:78–81; Holloway, *Stalin and the Bomb*, 341; Larson, *Anatomy of Mistrust*, 64–65; Pruden, *Conditional Partners*, 156. For U.S. analysis of the proposals, see Special Staff Study for the President—NSC Action No. 1328, May 26, 1555, *FRUS 1955–1957*, 20:93–109.

13. Eisenhower to Churchill, December 1, 1954, in *The Papers of Dwight D. Eisenhower*, ed. Alfred D. Chandler and Louis Galambos (Baltimore: Johns Hopkins University Press, 1970–1996) (hereafter *Eisenhower Papers*), 15:1444–47. For detailed treatment see John W. Young, *Winston Churchill's Last Campaign: Britain and the Cold War, 1951–5* (Oxford: Clarendon Press, 1996); and Klaus Larres, *Churchill's Cold War: The Politics of Personal Diplomacy* (New Haven: Yale University Press, 2002). On

the Geneva conference, see Günter Bishof and Saki Dockrill, eds., *Cold War Respite: The Geneva Summit of 1955* (Baton Rouge: Louisiana University Press, 2000).

14. Dwight D. Eisenhower, *Mandate for Change, 1953–1956: The White House Years* (Garden City, N.Y.: Doubleday, 1963), 506; Dulles quoted in Hixson, *Parting the Curtain*, 100; Lawrence Wittner, *Resisting the Bomb: A History of the World Nuclear Disarmament Movement, 1954–1970* (Stanford: Stanford University Press, 1997), esp. 125–60. Select Western European Public Reaction of United States Efforts for Peace in 1955, February 20, 1956, USIA Files, box 216, Legislation—1956 Disarmament Hearings, National Records Center, Suitland, Maryland (hereafter NRC); British Views About Nuclear Weapons: An Interpretation, December 18, 1957, RG 306, Office of Research, Production Division Reports, 1956–1959, box 4, NA.

15. From Quantico to Geneva, C. D. Jackson Papers, box 68, Log 1955 (1), EL. C. D. Jackson to Rockefeller, May 28, 1955, Nelson A. Rockefeller Papers, Washington, D.C., Files, Subseries 7, box 4, Four-Power Conference—Pre-Conference (3), Rockefeller Archive Center, Sleepy Hollow, New York (hereafter RAC); Cary Reich, *The Life of Nelson Rockefeller: Worlds to Conquer, 1908–1958* (New York: Doubleday, 1996), 579.

16. Report of the Quantico Vulnerabilities Panel; and Annex B to the Report of the Quantico Vulnerabilities Panel; both in Nelson A. Rockefeller Papers, box 4, Quantico Vulnerabilities Panel, RAC; Rockefeller to Eisenhower, June 13, 1955, Ann Whitman File, Administration Series, box 30, Nelson Rockefeller, 1952–1955 (4), EL; C. D. Jackson to Dulles, June 13, 1955, Jackson Papers, box 49, John Foster Dulles (3), EL; Rostow to Rockefeller, June 17, 1955, Jackson Papers, box 91, Walt W. Rostow 1955, EL; Jackson to Luce, June 21, 1955, Ann Whitman File, Administration Series, box 22, Jackson, C. D. 1955, EL; From Quantico to Geneva, Jackson Papers, box 68, Log 1955 (1), EL.

17. Dulles quoted in Reich, *Life of Nelson Rockefeller*, 592–93; Dulles to Eisenhower, June 18, 1955, Ann Whitman File, International Meetings Series, box 1, Geneva Conference July 18–23, 1955 (#1), EL; From Quantico to Geneva, Jackson Papers, box 68, Log 1955 (1), EL.

18. Rockefeller explained that since U.S. alliances depended on "the attitudes, aspirations and allegiances of the people in the . . . countries concerned," international public opinion, particularly in allied countries, must be the principal target of the summit's diplomacy. The United States, therefore, "should do everything possible" in the negotiations to convey three impressions to world audiences: (1) "we are being 'reasonable' and 'open-minded'"; (2) "we share their aspirations for 'peace'"; and (3) "we are sincerely working with them, as a partner, to accomplish these purposes." Psychological Aspects of U.S. Position at Conference, Rockefeller to Eisenhower, July 6, 1955; and Chronology of Mutual Inspection Proposal; both in Nelson A. Rockefeller Papers, box 4, Four-Power Conference—Pre-Conference (3), RAC.

19. Disarmament Proposal for Four Power Conference, Rockefeller to Eisenhower, July 6, 1955, Nelson A. Rockefeller Papers, box 4, Four-Power Conference—Pre-Conference (3), RAC; Bundy, *Danger and Survival,* 298–99.

20. Reich, *Life of Nelson Rockefeller,* 590. As deputy director of the USIA Abbott Washburn explained it, "The emotional yearning for peace is so great that whether we like it or not people will deeply hope and secretly expect results from the conference. If nothing concrete materializes, they will tend to believe the enemy's propaganda that the U.S. came with the attitude of demanding concessions rather than a genuine desire to relax tensions. This is tough to buck on the world opinion front." Washburn to Jackson, June 20, 1955, Nelson A. Rockefeller Papers, box 4, Four-Power Conference—Pre-Conference (3), RAC.

21. Don Irwin to Rockefeller, July 12, 1955; Irwin to Rockefeller, May 20, 1955, Hirsch to Craig, May 19, 1955; Policy Information Statement for USIA, May 25, 1955; Policy Information Statement for USIA, July 8, 1955; and Memorandum of Meeting of the Ad Hoc Working Group on Four-Power Conference, May 31, 1955; all in Nelson A. Rockefeller Papers, box 4, Four-Power Conference—Pre-Conference (4), RAC.

22. Statement on Disarmament Presented at Geneva Conference, July 21, 1955, *Public Papers of the Presidents of the United States: Dwight D. Eisenhower,* 1955 (Washington, D.C.: Government Printing Office, 1961), 713–16. On the last-minute behind-the-scenes maneuvering that led to Eisenhower's use of the Open Skies proposal at the conference, see From Quantico to Geneva, Jackson Papers, box 68, Log 1955 (1), EL; Chronology of Mutual Inspection Proposal, Nelson A. Rockefeller Papers, box 4, Four-Power Conference—Pre-Conference (3), RAC; and Reich, *Life of Nelson Rockefeller,* 596–605.

23. Eisenhower later admitted that he doubted the USSR would accept the proposal. See NSC Meeting, February 18, 1960, *FRUS 1958–1960,* 3:836–45.

24. Opinion Trends in the Aftermath of Geneva, September 23, 1955, RG 306, Special "S" Reports of the Office of Research, 1953–63, box 10, NA.

25. MemCon, August 11, 1955; and United States Post-Geneva Policy, August 15, 1955; both in Nelson A. Rockefeller Papers, box 6, Four-Power—Post Geneva, RAC.

26. Summary of discussion at Quantico II meeting, August 26, 1955, Nelson A. Nelson A. Rockefeller Papers, RG 4, Washington, D.C., Files, Subseries 7, box 8, Quantico II: Establishment, Membership, Meetings, and Correspondence, RAC; Policy Information Statement for USIA, August 5, 1955; United States Post-Geneva Policy, August 15, 1955; Irwin to Rockefeller, August 1, 1955; and Irwin to Livermore, October 18, 1955; all in Nelson A. Rockefeller Papers, box 6, Four-Power—Post Geneva, RAC.

27. Bundy, *Danger and Survival,* 328. "Revisionist" accounts explain Eisenhower's pursuit of a test ban as part of his overall strategy for limiting or ending the strategic arms race. The most comprehensive such treatment is Appleby, "Eisen-

hower and Arms Control." Allen G. Greb and Gerald W. Johnson have written a thorough, but unfortunately unpublished, history, "The Test Ban Treaty Reconsidered: Arms Control Policy in the Eisenhower Administration" (paper presented to the Center for Cold War Studies, University of California, Santa Barbara, 1998). Deborah Larson supports the revisionist interpretation but portrays Eisenhower as paralyzed by his distrust of the Soviet Union in *Anatomy of Distrust*, 72–106.

28. Pruden, *Conditional Partners*, 158–59.

29. USIA Effectiveness Report, January 27, 1956, RG 59, Lot 58 D133, box 214, Public Relations, NA; USIS Bern to USIA, January 5, 1956, RG 59, Lot 58D133, box 206, Open Skies for Peace; Mutual Inspection for Peace, RG 59, Lot 58 D133, box 206, Exhibits, NA; Disarmament and Inspection Theme and Projects, RG 306, USIA Files, box 216, Accession No. W306-68T4933AA, National Records Center, Suitland, Maryland (hereafter NRC).

30. Rope to Springer, November 2, 1955, RG 59, Records Relating to Public Affairs Activities 1944–1965, Lot 66 D257, box 12, Disarmament 1954 and 1955, NA; Springer to Stassen, September 7, 1955, RG 59, Lot 58 D133, box 206, Exhibits, NA; Aerial Inspection Exhibit Committee Summary Record, October 13, 1955, RG 59, Lot 58 D133, box 204, NA; Brief Outline for 10-minute documentary, RG 306, USIA Files, box 216, Accession No. W306-68T4933AA, NRC.

31. Quoted in Pruden, *Conditional Partners*, 160–61. Eisenhower's instructions to Stassen included the instructions that he should "spearhead efforts to inform and instruct the American people in . . . this vital subject." Eisenhower to Stassen, March 1, 1955, Ann Whitman File, Administration Series, box 34, Stassen, 1954–55 (2), EL.

32. Quotes from Wittner, *Resisting the Bomb*, 177; Pruden, *Conditional Partners*, 159; and Eisenhower to Dulles, January 3, 1958, Ann Whitman File, Dulles-Herter Series, box 7, Dulles—January 1958 (2), EL; John P. Meagher to Harry W. Seamans, October 4, 1957; Wilkinson to Meagher, February 7, 1956, RG 59, Records Relating to Public Affairs Activities 1944–1965, Lot 66 D257, box 12, Disarmament 1954 and 1955, NA; John P. Meagher to Harry W. Seamans, October 4, 1957; Wilkinson to Meagher, February 7, 1956, RG 59, Records Relating to Public Affairs Activities 1944–1965, Lot 66 D257, box 12, Disarmament 1954 and 1955, NA.

33. Eisenhower to Whitney, June 11, 1957; and Stassen to Zorin, May 31, 1957; both in *FRUS 1955–1957*, 20:616, 574–83; Pruden, *Conditional Partners*, 161–63.

34. Robert A. Divine, *Blowing on the Wind: The Nuclear Test Ban Debate, 1954–1960* (New York: Oxford University Press, 1978), 139; Hewlett and Holl, *Atoms for Peace and War*, 472–76, 454–55; Allan M. Winkler, *Life Under a Cloud: American Anxiety About the Atom* (New York: Oxford University Press, 1993), 84–108; Bundy, *Danger and Survival*, 329–331. For USIA public opinion surveys on atomic energy and disarmament, see the USIA's Office of Research Files in RG 306, NA.

35. OCB Memorandum, August 14, 1956; OCB Progress Report on Nuclear Energy Projects, August 15, 1956; Report of the OCB Nuclear Energy Working Group, September 28, 1956; Terms of Reference for Working Group on Nuclear Energy, June 6, 1956; all in RG 59, Records Relating to Department of State Participation in the OCB and NSC, box 35, Atomic and Nuclear Energy 1956, NA; Greb and Johnson, "Test Ban Treaty." On the information activities of Stassen's disarmament staff, see RG 59, Lot 58 D133, box 214, Public Relations-N60 Liaison, NA.

36. Larson, *Anatomy of Mistrust*, 74.

37. Bulganin-Eisenhower Letter Exchange: Propaganda Aims of the Bulganin Letters and Chronology, March 3, 1958, RG 306, Research Notes, 1958–1961, box 1, NA; Larson, *Anatomy of Mistrust*, 76.

38. NSC Meeting, January 6, 1958, *FRUS 1958–1960*, 3:533–45.

39. Abbott Washburn Memorandum, September 24, 1957, RG 59, Records Relating to Public Affairs Activities 1944–1965, Lot 66 D257, box 12, Disarmament, NA; Washburn to Eisenhower, November 11, 1957, John Foster Dulles Papers, White House Memoranda Series, box 5, Meetings with the President, 1957 (1), EL; Williams to Melbourne, August 5, 1948, WHO, NSC Staff Papers, OCB Secretariat, box 1, Disarmament Information Program—Ad Hoc Committee On (1), EL; Dearborn to Cutler, October 1, 1957, WHO, OSANSA, NSC Staff Papers, Briefing Notes, box 18, USIA (2), EL.

40. Increased Understanding and Support for Established U.S. Position on Disarmament, January 31, 1958; OCB Memorandum, January 22, 1958; and OCB Meeting, February 12, 1958; all in WHO, NSC Staff Papers, OCB Secretariat, box 1, Disarmament Information Program—Ad Hoc Committee On (1), EL; Stassen to Eisenhower, January 14, 1958, *FRUS 1958–1960*, 3:547–51.

41. Eisenhower to Dulles, March 21, 1958, John Foster Dulles Papers, White House Memoranda Series, box 6, General Correspondence 1958 (5), EL.

42. MemCon, March 24, 1958, *FRUS 1958–1960*, 3:567–72.

43. Quotes from Larson, *Anatomy of Mistrust*, 82. Bundy, *Danger and Survival*, 331.

44. Dulles to Eisenhower, April 30, 1958; MemCon, August 8, 1958; MemCon, April 8, 1958; all in *FRUS 1958–1960*, 3:604–6, 624–30; 590–97; MemCon, April 26, 1958, Ann Whitman File, Dulles-Herter Series, box 8, Dulles—April 1958 (1), EL. In a revealing conversation with NATO commander General Alfred M. Gruenther, Dulles compared the American situation to that of the German government before World War I. Referring to the "disastrous consequences of being regarded throughout the world as militaristic," Dulles stressed that it was vital to continue to seek arms limitations, even if they were undesirable or unattainable. MemCon, Dulles and Gruenther, February 19, 1958, *FRUS 1958–1960*, 3:553–54. See also the editorial note on pages 554–55.

45. NSC Meeting, January 6, 1958, *FRUS 1958–1960*, 3:533–45. See also John A. McCone, Notes for the Files, March 10, 1960; MemCon, 350th NSC Meeting,

January 6, 1958; and MemCon, May 5, 1959; all in *FRUS 1958–1960*, 3:533–45, 737–41, 846–47.

46. Greb and Johnson, "Test Ban Treaty."

47. JCS to Secretary of Defense, March 13, 1958, *FRUS 1958–1960*, 3:555–57; Greb and Johnson, "Test Ban Treaty." On PSAC, see Zuoyue Wang, "American Science and the Cold War: The Rise of the U.S. President's Science Advisory Committee" (Ph.D. diss., University of California, Santa Barbara, 1994).

48. MemCon, August 8, 1958, *FRUS 1958–1960*, 3:624–30; Eisenhower quoted in Wittner, *Resisting the Bomb*, 182; and MemCon, August 18, 1958, *FRUS 1958–1960*, 3:647–48; MemCon, April 26, 1958, Ann Whitman File, Dulles-Herter Series, box 8, Dulles—April 1958 (1), EL.

49. Divine, *Blowing on the Wind*, 231–34, 262; Winkler, *Life Under a Cloud*, 103; Wittner, *Resisting the Bomb*, 183. On the fallout scare, see especially Divine, chap. 10.

50. MemCon, March 17, 1959, *FRUS 1958–1960*, 3:714–15.

51. MemCon, May 5, 1959, *FRUS 1958–1960*, 3:737–40.

52. MemCon, March 19, 1959; and Editorial Note recounting meeting on March 22, 1959; both in *FRUS 1958–1960*, 3:716–18, 728.

53. Larson, *Anatomy of Mistrust*, 91–99.

54. On the U2 crisis, see Richter, *Khrushchev's Double Bind*, 129–32; and Michael R. Beschloss, *Mayday: Eisenhower, Khrushchev and the U-2 Affair* (New York: Harper & Row, 1986).

55. Dwight D. Eisenhower, *Waging Peace, 1956–1961: The White House Years* (Garden City, N.Y.: Doubleday, 1965), 467–68, 653. Revisionists supporting Eisenhower's claim of pursuing a "small steps" approach to disarmament generally point to his memoirs as well as to a small quantity of personal correspondence to friends and relatives. Such sources usually convey a laundry list of motives, including both psychological warfare and disarmament reasons. Nowhere in official meetings, memoranda, or policy documents did Eisenhower advocate, develop, or implement such a strategy.

56. Evidence of Eisenhower's distrust of the Soviet Union is plentiful. See, for example: MemCon, November 14, 1955, Ann Whitman File, Eisenhower Diary Series, box 11, November 1955 (2), EL; Jackson memo, January 16, 1954, Jackson Papers, box 68, Log 54 (1), EL; and Eisenhower to Jackson, December 31, 1953, Eisenhower Diary, December 1953 (1), EL; Eisenhower to Churchill, July 8, 1954, John Foster Dulles Papers, Subject Series, box 11, Churchill-Eden Correspondence 1954 (3), EL; Eisenhower to Gruenther, February 1, 1955, Alfred M. Gruenther Papers, Eisenhower Correspondence Series, box 1, D.E. Eisenhower (1955) (2), EL; and Eisenhower to Churchill, July 22, 1954, and December 14, 1954, both in *Eisenhower Papers*, 15:1207–11, 1444–47; See also *FRUS 1952–1954*, 8:1116, 1166; *FRUS 1952–1954*, 4:1170; and Larson, *Anatomy of Mistrust*, 73–74.

57. Kissinger, "Reflections on American Diplomacy."

Chapter 7. Every Man an Ambassador

1. Eisenhower to President of the Senate, July 27, 1954, OCB Central Files, box 14, OCB 007 (File #1) (1), Eisenhower Library, Abilene, Kansas (hereafter EL); OCB Memorandum, December 2, 1955, OCB Central Files, box 15, OCB 007 (File #3) (1), EL; General Policy and Operating Guidance for the President's Special International Program, April 26, 1957, OCB Central Files, box 16, OCB 007 (File #5) (2), EL.

2. USIA Basic Guidance and Planning Paper No. 2, "The Cultural Program of USIA—A Basic Paper," September 17, 1958, USIA Historical Collection, Washington, D.C.

3. William Harlan Hale, "Every Man an Ambassador," reprinted by USIA from the *Reporter*, Record Group [RG] 59, Lot 61 D127, box 20, V-20, Original Amateur Hour, National Archives, College Park, Maryland (hereafter NA).

4. Chairman for the State Committee for Cultural Relations with Foreign Countries quoted in USIA, *Communist Propaganda: A Fact Book, 1957–1958* (1958); Progress Report on the Activities of the OCB Trade Fair Committee, February 9, 1956, OCB Central Files, box 15, OCB 007 (file #3) (3), EL. Charles A. Thomson and Walter H. C. Laves, *Cultural Relations and U.S. Foreign Policy* (Bloomington: Indiana University Press, 1963), 127; Walter L. Hixson, *Parting the Curtain: Propaganda, Culture, and the Cold War, 1945–1961* (New York: St. Martin's Griffin, 1997), 102; Ronald Grigor Suny, *The Soviet Experiment: Russia, the USSR, and the Successor States* (Oxford: Oxford University Press, 1998), 404–7.

5. USIA Position Paper, January 4, 1955, OCB Central Files, box 14, OCB 007 (File #1) (3), EL.

6. House of Representatives, Subcommittee to Study H.R. 3342, Committee on Foreign Affairs, May 20, 1947, Pre-Presidential Papers, box 145, Hearings Vol. 4 (1), EL; Cabinet Meeting, November 19, 1954, OCB Central Files, box 14, OCB 007 (File #1) (2), EL; Abbott Washburn and Theodore Streibert Oral Histories, EL; Eisenhower to Edgar Eisenhower, November 22, 1955, Ann Whitman File, Eisenhower Diary Series, box 11, Eisenhower Diary—November 1955 (1), EL; Elmer Staats to Ralph Becker, n.d., OCB Central Files, box 16, OCB 007 (File #4) (5), EL; Author's interview with General Andrew J. Goodpaster.

7. Until the early 1960s, legislation dealing with cultural activities was a patchwork of unrelated acts. The 1961 Fulbright-Hays Act codified and amplified the existing mass of legislation. Thomson and Laves, *Cultural Relations*, 27–141; Charles Frankel, *The Neglected Aspect of Foreign Affairs: American Educational and Cultural Policy Abroad* (Washington, D.C.: Brookings Institution, 1966), 25–28.

8. Eisenhower to Streibert, August 18, 1954; and Taquey to Executive Officer, September 2, 1954; President's Emergency Fund General Policy and Operating Agreement—Fiscal Year 1956, June 10, 1955; Terms of Reference for Working Groups, Oc-

tober 8, 1954; all in OCB Central Files, box 14, OCB 007 (File #1) (1), (File #1) (6), and (File #2) (2), EL. General Policy and Operating Guidance for the President's Special International Program, April 26, 1957; Frank Ninkovich, *U.S. Information Policy and Cultural Diplomacy* (New York: Foreign Policy Association, 1996), 21.

9. Progress Report on Activities of the OCB Cultural Presentation Committee, March 15, 1956, OCB Central Files, box 15, OCB 007 (file #3) (4), EL; President's Emergency Fund General Policy and Operating Agreement—Fiscal Year 1956, June 10, 1955; General Policy and Operating Guidance for the President's Special International Program, April 26, 1957. "Europe happens to be the area of lowest priority in the program," an OCB memo noted: Ralph R. Busick to Leslie S. Brady, January 15, 1957, OCB Central Files, box 16, OCB 007 (File #4) (5), EL.

10. Quoted in Robert H. Haddow, *Pavilions of Plenty: Exhibiting American Culture Abroad in the 1950s* (Washington, D.C.: Smithsonian Institution Press, 1997), 41. On U.S. participation in trade fairs, see Robert W. Rydell, *All the World's a Fair: Visions of Empire at American International Expositions, 1876–1916* (Chicago: University of Chicago Press, 1984); and Rydell, *World of Fairs: The Century-of-Progress Expositions* (Chicago: University of Chicago Press, 1993).

11. Taquey to Executive Officer, September 2, 1954; Eisenhower to President of the Senate, July 27, 1954; NSC 5430, Status of USIA Program, August 18, 1954, *Foreign Relations of the United States 1952–1954:* 2:1780. (hereafter *FRUS*).

12. Hixson, *Parting the Curtain,* 138; Sixth Quarterly Report, President's Emergency Fund for Participation in International Affairs, October 1–December 31, 1955, OCB Central Files, box 15, OCB 007 (File #3) (3), EL; Third Quarterly Report, President's Emergency Fund for Participation in International Affairs, January 1, 1955–March 31, 1955, OCB Central Files, box 15, OCB 007 (File #1) (5), EL; Fifth Quarterly Report, President's Emergency Fund for Participation in International Affairs, July 1, 1955–September 30, 1955, OCB Central Files, box 15, OCB 007 (File #2) (8), EL.

13. Sixth Quarterly Report, President's Emergency Fund for Participation in International Affairs; Progress Report on Activities of the OCB Trade Fair Committee, February 9, 1956; President's Emergency Fund General Policy and Operating Agreement—Fiscal Year 1956, June 10, 1955; Fourth Quarterly Report, President's Emergency Fund for Participation in International Affairs, April 1, 1955–June 30, 1955, 1955, OCB Central Files, box 15, OCB 007 (File #2) (2), EL.

14. Earl J. Wilson Oral History, Georgetown University Library, Special Collections.

15. Report by G. H, Greenhalgh, January 6, 1958, British Atomic Energy Publicity Committee, BT 213/83, Public Record Office, Kew, United Kingdom; Progress Report on Activities of OCB Cultural Presentation Committee, July 20, 1955, OCB Central Files, box 15, OCB 007 (file #2) (2), EL; Sixth Quarterly Report, President's Emergency Fund for Participation in International Affairs; USIA, *Vol-*

unteers for International Communication: Report of USIA Private Sector Committees (1984).

16. Quoted in Fourth Quarterly Report, President's Emergency Fund for Participation in International Affairs; Thomson and Laves, *Cultural Relations*, 123.

17. President's Emergency Fund General Policy and Operating Agreement—Fiscal Year 1956, June 10, 1955; General Policy and Operating Guidance for the President's Special International Program, April 26, 1957.

18. Busick to Brady, January 15, 1057, OCB Central Files, box 16, OCB 007 (File #4) (5), EL; General Policy and Operating Guidance for the President's Special International Program, April 26, 1957.

19. Fourth Quarterly Report, President's Emergency Fund for Participation in International Affairs; Third Quarterly Report, President's Emergency Fund for Participation in International Affairs; Sixth Quarterly Report, President's Emergency Fund for Participation in International Affairs; Progress Report on Activities of the OCB Cultural Presentation Committee, July 13, 1956, OCB Central Files, box 16, OCB 007 (File #3) (9), EL.

20. General Policy and Operating Guidance for the President's Special International Program, April 26, 1957; The U.S. Information Program Since July 1953, Records of the President's Committee on Information Activities Abroad (Sprague Committee), box 19, USIA (2), EL.

21. *New York Times* quoted in Progress Report on Activities of the OCB Cultural Presentation Committee, February 9, 1956, OCB Central Files, box 16, OCB 007 (File #3) (3), EL; Progress Report on Activities of the OCB Cultural Presentation Committee, July 13, 1956, OCB Central Files, box 16, OCB 007 (File #3) (9), EL. See also Penny M. Von Eschen, *Satchmo Blows Up the World: Jazz Ambassadors Play the Cold War* (Cambridge: Harvard University Press, 2004).

22. *New York Times* quoted in Third Quarterly Report, President's Emergency Fund for Participation in International Affairs; Seventh Quarterly Report, President's Emergency Fund for Participation in International Affairs, January 1, 1956–March 31, 1956, OCB Central Files, box 16, OCB 007 (File #3) (8), EL; NSC 5525, Status of USIA Program, August 31, 1955, *FRUS 1955–1957*, 9:533.

23. USIA Staff Study, September 20, 1955, Nelson A. Rockefeller Papers, RG 4, Subseries 4, box 20, "Cinerama; Porgy & Bess; and Soviet Union," Rockefeller Archive Center, Sleepy Hollow, New York. *Sunday Standard* quoted in Fourth Quarterly Report, President's Emergency Fund for Participation in International Affairs.

24. Third Quarterly Report, President's Emergency Fund for Participation in International Affairs; Progress Report on Activities of the OCB Cultural Presentation Committee, April 16, 1957.

25. USIA Position Paper, January 4, 1955; President's Emergency Fund General Policy and Operating Agreement—Fiscal Year 1956, June 10, 1955; NSC 5525, Status of USIA Program, August 31, 1955. NSC 5819, Status of USIA Program,

June 30, 1958; and NSC 5912, Status of USIA Program, June 30, 1959; both in White House Office [WHO], Office of the Special Assistant for National Security Affairs [OSANSA], NSC Series, Status of Projects Subseries, box 8, EL.

26. Naima Prevots, *Dance for Export: Cultural Diplomacy and the Cold War* (Hanover, N.H.: Wesleyan University Press, 1998). Prevots has written the fullest account of the President's Emergency Fund. This useful resource on "cultural diplomacy" unfortunately did not use the records of the Operations Coordinating Board, which cast serious doubt on Prevots's argument that the exchange programs were not influenced by propaganda considerations.

27. Allan M. Winkler, *The Politics of Propaganda: The Office of War Information, 1942–1945* (New Haven, Conn.: Yale University Press, 1978); Robert Jackall and Janice M. Hirota, *Image Makers: Advertising, Public Relations, and the Ethos of Advocacy* (Chicago: University of Chicago Press, 2000), chap. 1; Emily S. Rosenberg, *Spreading the American Dream: American Economic and Cultural Expansion, 1890–1945* (New York: Hill and Wang, 1982), 79; Frank A. Ninkovich, *The Diplomacy of Ideas: U.S. Foreign Policy and Cultural Relations, 1938–1950* (Cambridge: Cambridge University Press, 1981); Clayton R. Koppes and Gregory D. Black, *Hollywood Goes to War: How Politics, Profits and Propaganda Shaped World War II Movies* (Berkeley: University of California Press, 1987); Daniel L. Lykins, *From Total War to Total Diplomacy: The Advertising Council and the Construction of the Cold War Consensus* (Westport, Conn.: Praeger, 2003).

28. DOS Memorandum, Urgent Need to Mobilize United States Private Industry for Pro-Democratic Information Program Throughout the Free World, undated [n.d.], Freedom of Information Act (FOIA); and DOS Memorandum, Private Enterprise Cooperation, FOIA; Hixson, *Parting the Curtain*, 11. Quote from John E. Juergensmeyer, *The President, the Foundations, and the People-to-People Program*, Intra-University Case Program No. 84 (Indianapolis: Bobbs-Merrill, 1965), 9. Juergensmeyer wrote this study on the People-to-People program for the Eisenhower administration in 1959. It can be found at the Eisenhower Library.

29. Activities of the Private Enterprise Cooperation Staff, International Information Administration, n.d. [1952], FOIA; Private Enterprise Cooperation, draft of pamphlet, n.d. [1952], FOIA; Monthly Information Sheet, November–December 1951, FOIA; Bi-weekly Report on Principal Activities of Cooperation with Private Enterprise Unit, July 14, 1951, FOIA; Helen Laville, "The Committee of Correspondence: CIA Funding of Women's Groups, 1952–1967," *Intelligence and National Security* 12 (January 1997): 104–21; Scott Lucas, "Mobilising Culture: The State-Private Network and the CIA in the Early Cold War," in *War and Cold War in American Foreign Policy, 1942–1962*, ed. Dale Carter and Robin Clifton (New York: Palgrave, 2002), 83–107.

30. Hendershot to Block, Program Suggestion, April 15, 1952, FOIA; Private Enterprise Cooperation, draft of pamphlet, n.d. [1952], FOIA.

31. Raish to Weightman, Briefing on IE Projects, January 2, 1952, FOIA.

32. Jackson Committee Report, *FRUS 1952–1954*, 2:1849, 1847, 1872; USIA Strategic Principles, n.d. [March 1954], OCB Central Files, box 20, OCB 040 USIA (1), EL; Juergensmeyer, *The President, the Foundations.*

33. USIA Statement of Basic Objectives and Procedures For the Business Council for International Understanding, n.d., FOIA. Quotes from Juergensmeyer, *The President, the Foundations.*

34. Streibert to Eisenhower, September 4, 1955, John Foster Dulles Papers, General Correspondence and Memoranda Series, box 3, Strictly Confidential—Q-S (4), EL. Streibert's proposal was drafted by Abbott Washburn, who originally devised the People-to-People program.

35. Quotes from Juergensmeyer, *The President, the Foundations.*

36. Streibert to Dulles, September 15, 1955, John Foster Dulles Papers, General Correspondence and Memoranda Series, box 3, Strictly Confidential—Q-S (4), EL; McCardle to Murphy, October 5, 1957, RG 59, Lot 66 D257, box 7, People-to-People Partnership Program, NA; *United Press* quoted in June Activities Report, July 7, 1958, WHO, OSANSA, OCB Series, Subject Subseries, box 5, People-to-People (2), EL. Memorandum of Conference with the President, February 8, 1956, Ann Whitman File, Eisenhower Diary Series, box 12, February 1956, Goodpaster, EL; Eisenhower to Adams, August 12, 1957, Ann Whitman File, Eisenhower Diary Series, box 26, August 1957—Eisenhower Dictation, EL; Eisenhower to Dulles, September 21, 1957, John Foster Dulles Papers, White House Memoranda Series, box 5, White House Correspondence—General 1957 (4), EL. Author's interview with Abbott Washburn; Juergensmeyer, *The President, the Foundations.*

37. Statement of Alexander Wiley, July 24, 1958, Congressional Record, 85th Congress, 2nd Session, 13639–13642; Hale, "Every Man an Ambassador."

38. Eisenhower letter, RG 59, Lot 66 D257, box 7, People-to-People Partnership Program, NA.

39. Eisenhower remarks at the People-to-People conference, September 11, 1956, *Public Papers of the Presidents of the United States: Dwight D. Eisenhower,* 1956 (Washington, D.C.: Government Printing Office, 1958): 749–52; Summary Report on the White House Conference on a Program for People-to-People Partnership, n.d., FOIA. The USIA had intended the conference to take place several months earlier, but it was delayed by Eisenhower's heart attack.

40. Senate Appropriations Committee, Survey of United States Information Service Operations: Western Europe, February 1957, 12; USIA, Brief History of the Office of Private Cooperation, USIA Historical Collection; USIA, *Volunteers for International Communication.*

41. Terms of Reference for the OCB Committee on the People-to-People Program, June 25, 1957, OCB Central Files, box 61, OCB 091.4, EL; Terms of Reference for the OCB Committee on the People-to-People Program, July 3, 1957, RG 59, Lot 66 D257, box 8, People-to-People Conference, Message, General Information, NA; Juergensmeyer, *The President, the Foundations,* 27.

42. Arthur Larson meeting with the president, October 10, 1957; and Larson meeting with the president, April 4, 1957; both in Larson Papers, box 22, Memorandum Book, October 1952 #1, March 1957, EL; Washburn to Adams, August 21, 1957, FOIA; Ford Foundation Turndown of People-to-People Request for Five Million Dollar Grant, n.d., FOIA; Summary Report on the White House Conference on A Program for People-to-People Partnership, n.d., FOIA; Juergensmeyer, *The President, the Foundations*, 33. Considering the ties of many of these foundations to the CIA, their reservations about participating in a government propaganda project seem incredulous. More than likely, it was the overt connection to the government that was worrisome. On the CIA's connection with nonprofit foundations, see Frances Stonor Saunders, *The Cultural Cold War: The CIA and the World of Arts and Letters* (New York: New Press, 1999).

43. Except where noted otherwise, the ensuing discussion of People-to-People activities is based on the lengthy and detailed quarterly reviews of activities located in the National Archives, RG 59, Lot 66 D257, box 7, and the Eisenhower Library, WHO, OSANSA, OCB Series, Subject Subseries, box 5.

44. Review of Activities, November 1957, RG 59, Lot 66 D257, box 7, People-to-People Annual Surveys and Review of Activities, NA; "Books on America Asked for World," *New York Times*, May 3, 1957.

45. "Friendship Abroad Through Gardening," *New York Times*, January 25, 1959.

46. "Truth Ambassadors," *Bayonne Times*, January 3, 1957; "People are the Best Propaganda," *Current Events*, January 12–16, 1959, "Here's Your Chance to be a Diplomat," *Baltimore News-Post*, March 22, 1957; "People-to-People Plan Seen as Waging Peace," Star, May 1, 1957; "True Story of America," *Memphis Press-Scimitar*, January 22, 1957; "Doctors as Diplomats," *New York Times*, January 19, 1958.

47. *Richmond Times-Dispatch* quoted in June Activities Report, July 7, 1958, WHO, OSANSA, OCB Series, Subject Subseries, box 5, People-to-People (2), EL; USIA, *People-to-People: A Program of International Friendship* (1957); Review of Activities, November 1957.

48. C. D. Jackson to George Allen, October 6, 1959, WHO, OSANSA, NSC Briefing Notes, box 18, USIA (3), EL. Washburn to OCB, December 15, 1960; and Washburn to Gray, December 20, 1960; both in WHO, OSANSA, NSC Briefing Notes, box 18, USIA (3), EL. DOS Telegrams, RG 59, Central Decimal File, 511.00/2-1259 and 511.006/12-458, NA. *New York Times*, September 15, 1961; September 11, 1960; February 8, 1959; February 11, 1959; April 29, 1961; and August 27, 1961. Author's interview with Abbott Washburn.

49. Allen to Herter, November 21, 1958, WHO, OSANSA, OCB Series, Subject Subseries, box 5, People-to-People (2), EL.

50. Statement of Senator Alexander Wiley, July 24, 1958; "Your Community in World Affairs," RG 59, Lot 66 D257, box 7, People-to-People—Civic Committee,

NA; "How to Join the People-to-People Program," RG 59, Lot 66 D257, box 7, People-to-People—Civic Committee, NA.

51. Statement of Senator Alexander Wiley, August 12, 1958, Congressional Record, 85th Congress, 2nd Session, 15700–15704; "Make a Friend this Trip: For Yourself, For Your Business, For Your Community," WHO, OSANSA, OCB Series, Subject Subseries, box 5, Passport and Military Personnel Overseas Letters, EL; Review of Activities, November 1957. JCS Cold War Activities Report, June 30, 1959; and Memorandum for the Secretary of Defense, Military Activities During the Cold War, November 30, 1959; both in RG 218, JCS Central Files 1959, box 33, 3310 Psychological Warfare, NA; JCS Memorandum, Cold War Activities, June 23, 1958, RG 218, JCS Decimal Files 1958, box 77, Sec. 48, NA; JCS Report of Military Activities During the Cold War, October 31, 1959, JCS Central Files 1960, box 18, 3310 Psychological Warfare, NA.

52. Koons to Dearborn, May 27, 1957; and Loomis to Larson, June 19, 1957; both in WHO, OSANSA, OCB Series, Subject Subseries, box 5, People-to-People (1), EL. MacArthur quoted in Michael Schaller, *Altered States: The United States and Japan Since the Occupation* (New York: Oxford University Press, 1997), 127–28.

53. Carter to Wright, Foreign Office, Information Policy Department Minutes, February 24, 1059, FO 953/1724, Public Record Office, Kew, United Kingdom; Robert Schulzinger, *A Time For War: The United States and Vietnam, 1941–1975* (New York: Oxford University Press, 1997): 98–99; William Lederer and Eugene Burdick, *The Ugly American* (New York: W. W. Norton, 1958).

54. Koons to Dearborn, May 27, 1957; Robert Cutler Memorandum, July 7, 1957, WHO, OSANSA, OCB Series, Subject Subseries, box 5, People-to-People (1), EL; NSC Meeting, July 19, 1957, Ann Whitman File, NSC Series, box 9, EL; NSC Meeting, April 25, 1958, Ann Whitman File, NSC Series, box 10, EL; Harr to Eisenhower, July 28, 1959, WHO, OSANSA, OCB Series, Subject Subseries, box 5, People-to-People, EL. "Ike Will Ask Travelers to Act as U.S. 'Envoys,'" *Washington Post*, July 26, 1957; Press Release, July 25, 1957, WHO, OSANSA, OCB Series, Subject Subseries, box 5, Passport and Military Personnel Overseas Letters, EL; Sprague Memorandum, July 17, 1957, WHO, OSANSA, OCB Series, Subject Subseries, box 5, Passport and Military Personnel Overseas Letters, EL.

55. George Allen, The Image of America, WHO, OSANSA, OCB Series, Subject Subseries, box 3, Image of America, EL; "Americans Abroad: Spokesmen for the U.S.A.," RG 59, Lot 66 D257, box 8, People-to-People Conference, Message, General Information, NA; OCB Memorandum, September 6, 1957, RG 59, Lot 59 D193, Miscellaneous Records of the Bureau of Public Affairs, box 22, Committees-OCB-People-to-People Program, NA; OCB Memorandum, September 24, 1957, RG 59, Lot 66 D257, box 8, People-to-People Conference, Message, General Information, NA. See also Common Council for American Unity, "What Should I Know When I Travel Abroad?" James Lambie Records, box 22, Letters from America 1955, EL.

56. Letter to the Secretary of State, n.d. [1961], RG 59, Records Relating to Public Affairs Activities 1944–1965, Lot 66 D257, box 8, People-to-People Conference, Messages, General Information, NA.

57. U.S. Resources for Foreign Communication and Political Warfare, May 9, 1960; "People-to-People Activities," June 20, 1960, Records of the President's Committee on Information Activities Abroad (Sprague Committee), box 23, PCIAA #29, EL.

58. Jackson Committee Report, *FRUS 1952–1954*, 2:1847; U.S. Resources for Foreign Communication and Political Warfare, May 9, 1960.

59. NSC 5611, Status of USIA Program, June 30, 1956, WHO, OSANSA, NSC Series, Status of Projects Subseries, box 7, EL; NSC 5819, Status of USIA Program, June 30, 1958.

60. U.S. Resources for Foreign Communication and Political Warfare, May 9, 1960.

61. "People-to-People Activities," June 20, 1960.

62. Hale, "Every Man an Ambassador."

Chapter 8. Facts About the United States

1. USIA, *Facts About the United States* (1956).

2. USIA, *Telling America's Story Abroad Through Press and Publications*, USIA Historical Collection, Washington, D.C.; [NSC 165/1] Mission of the USIA, October 24, 1953, *Foreign Relations of the United States 1952–1954*, 2:1753–54 (hereafter *FRUS*); Basic Guidance and Planning Paper No. 2, "The Cultural Program of USIA—A Basic Paper," September 17, 1958, USIA Historical Collection. Author's interviews with General Andrew J. Goodpaster and Abbott Washburn.

3. Images of everyday life in USIA propaganda are analyzed in Laura A. Belmonte, "Defending a Way of Life: American Propaganda and the Cold War, 1945–1959" (Ph.D. diss., University of Virginia, 1996); and Belmonte, "Almost Everyone Is a Capitalist: The USIA Presents the American Economy, 1953–1959," paper presented to the Society for Historians of American Foreign Relations, June 1994.

4. USIA Basic Guidance and Planning Paper No. 2, "The Cultural Program of USIA—A Basic Paper."

5. USIA Basic Guidance and Planning Paper No. 10, "Themes on American Life and Culture," July 14, 1959, USIA Historical Collection.

6. USIA Basic Guidance and Planning Paper No. 10, "Themes on American Life and Culture," July 14, 1959, USIA Historical Collection.

7. USIA Basic Guidance and Planning Paper No. 10, "Themes on American Life and Culture," July 14, 1959, USIA Historical Collection.

8. "A Visit with Mrs. Forster," USIS Picture Story, Record Group [RG] 306, Feature Packets, Recurring Subjects, box 16, USIS Special Women's Packet #1, National Archives, College Park, Maryland (hereafter NA). Details about Mrs.

Forster not listed in the USIA story are from http://www.dalyfamilyhistory.com/story4.htm.

9. USIA, "Themes on American Life and Culture"; USIA, Basic Guidance and Planning Paper No. 12, "Women's Activities (Part II)," August 13, 1959, USIA Historical Collection; Belmonte, "Almost Everyone Is a Capitalist."

10. USIA, "Women's Activities (Part II)"; Belmonte, "Defending a Way of Life," 249.

11. "Ester Williams' Pet Project," RG 306, Feature Packets, Recurring Subjects, box 22, USIS Women's Packet #43, NA; "Helping the Blind—Vocation and Avocation for Alice Rohrback," RG 306, Feature Packets, Recurring Subjects, box 16, USIS Special Women's Packet # 1, NA; "Home with a Heart," USIS Picture Story, RG 306, Feature Packets, Recurring Subjects, box 18, USIS Special Women's Packet #10, NA.

12. USIA, "Women's Activities (Part II)." "Gail Borden: Pioneer Preserver of Milk"; "American Women Help Scholars of Many Lands"; "U.S. Women Serve at All Levels of Government"; and "As Others See U.S. Women"; all in RG 306, Feature Packets, Recurring Subjects, box 18, USIS Women's Packet #10, NA; "U.S. Women on the Political Scene," RG 306, Feature Packets, Recurring Subjects, box 22, USIS Women's Packet #43, NA. These articles are representative of others included in USIS feature packets on American women.

13. USIA, "Women's Activities (Part II)."

14. USIA, "Women's Activities (Part II)."

15. Carlton Savage to Charles Yost, July 2, 1958, RG 59, Policy Planning Staff Records 1957–1961, Lot 67 D548, box 120, Information Policy, NA; USIA, "Women's Activities (Part II)"; USIA Basic Guidance and Planning Paper No. 6, "Women's Activities."

16. USIS Mexico to USIA, November 4, 1957, RG 306, Office of Research, Country Project Correspondence 1952–63, box 15, Mexico 1957, NA. *Young People*, June 1953; "Thousands of U.S. 'Little Merchant' Newsboys Conduct Own Business"; and "Teen-Age Employment Agency Gets Jobs for Youngsters"; all in RG 306, Feature Packets, Recurring Subjects, boxes 1 and 6, USIS Special Youth Packets 4 and 41, NA.

17. "Rosemary Jones: A Leading U.S. Woman Basketball Player"; "Stella Walsh Inspires U.S. Women Athletes"; "Judy Devlin: Outstanding World Badminton Player"; "Thirteen-Year-Old Girl Enters U.S. Golfing Scene"; "Jim Thorpe, U.S.A."; and "Willie Mays—Wonder Boy of Baseball"; all in RG 306, Feature Packets, Recurring Subjects, boxes 25 and 34, USIS Sports Packets #3 and 34, NA; "Soccer Gaining Popularity Among U.S. Servicemen in Europe," RG 306, Feature Packets, Recurring Subjects, box 25, USIS Sports Packet #17, NA. A covering note to the public affairs officer explained, "Soccer is probably the most popular international sport. This article tells how the activity is being adopted by U.S. servicemen in Europe and how it serves to promote international goodwill." "Polish Refugee

Track Performer Draws Attention of U.S. Sports Fans," RG 306, Feature Packets, Recurring Subjects, box 22, USIS Sports Packet #36, NA.

18. USIA Basic Guidance and Planning Paper No. 3, "Labor Information Program," October 20, 1958, USIA Historical Collection.

19. USIA Basic Guidance and Planning Paper No. 3, "Labor Information Program," October 20, 1958, USIA Historical Collection; U.S. Labor Background and Suggestions, RG 306, Feature Packets, Non-Recurring Subjects, box 2, Background Information Suggestions, NA; "Living and Working in a Free Enterprise System," RG 306, Feature Packets, Recurring Subjects, box 14, USIS Labor Packet #51, NA. OCB Handbook of Policy Statements on Overseas Labor Activities, May 28, 1956; Secretary of Commerce to Elmer Staats, May 2, 1955; Commerce Department Position Paper on Draft Statement of U.S. International Labor Policy; all in OCB Central Files, box 13, OCB 004.06 [Overseas Labor] (File #1) (7), (File #1) (2), and (File #1) (4), Eisenhower Library, Abilene, Kansas (hereafter EL).

20. "Living and Working in a Free Enterprise System," RG 306, Feature Packets, Recurring Subjects, box 14, USIS Labor Packet #51, NA; American Labor Review, February 1960, PCIAA, Sprague Committee, box 3, American Labor in International Affairs #15 (1), EL; USIA, *Facts About the United States*.

21. Edmund Nash, "Purchasing Power of Soviet Workers," magazine reprint from the *Monthly Labor Review*, RG 306, Feature Packets, Recurring Subjects, box 7, USIS Labor Packet #5, NA. See also RG 306, Feature Packets, Recurring Subjects, box 8, USIS Labor Packet #7; and box 7, USIS Labor Packet #2, NA.

22. Ralph Wright, "Guilty as Charged," magazine reprint from the American Federationist, RG 306, Feature Packets, Recurring Subjects, box 8, USIS Labor Packet #7, NA; "Workers Win 10-Year Fight Against Forced Labor," RG 306, Feature Packets, Recurring Subjects, box 14, USIS Labor Packet #51, NA; "Food Cost in Working Minutes," RG 306, Feature Packets, Non-Recurring Subjects, box 2, Labor Photographs, Graphics, NA; USIA, "Labor Information Program."

23. USIA, "Themes on American Life and Culture"; USIA, "Labor Information Program"; Commerce Department Position Paper on Draft Statement of U.S. International Labor Policy. "AFL Show Reflects Labor-Management Teamwork"; "UAW Expands Area of Collective Bargaining"; "Union's Training Program Aids Workers and Industry"; "American Unions Winning Shorter Workweek"; "Union Wins 'Conversion' Pay for Displaced Workers"; "U.S. Workers Get Cost of Living Increases"; "Labor Union Finds New Way to Serve Community"; and "Labor Briefs and Short Features"; all in RG 306, Feature Packets, Recurring Subjects, box 11, USIS Labor Packets #24 and 28, NA.

24. USIA Basic Guidance and Planning Paper No. 11, "The American Economy," July 16, 1959, USIA Historical Collection; USIA, "Themes on American Life and Culture"; USIA, "Labor Information Program"; "U.S. Labor's Struggle Against Communism," RG 306, Feature Packets, Non-Recurring Subjects, box 2,

U.S. Labor, Articles, Publicity and Speech Materials, NA; and RG 306, Feature Packets, Recurring Subjects, box 14, USIS Labor Packet #51, NA.

25. Information Policy on the American Economy, October 3, 1955; and USIA, "American Economy"; both in USIA Historical Collection. USIA, "Labor Information Program"; USIA, "Themes on American Life and Culture"; Belmonte, "Defending a Way of Life," 327–28.

26. Statement by Theodore Repplier, in USIA to all USIS posts, February 29, 1956, White House Office [WHO], NSC Staff Series, OCB Secretariat Subseries, box 3, Ideological Documents File (7), EL; MemCon Eisenhower and Repplier, August 3, 1955, James Lambie Records, box 23, People's Capitalism 1955, EL.

27. "People's Capitalism in the U.S.A.," WHO, NSC Staff Series, OCB Secretariat Subseries, box 3, Ideological Documents File (7), EL; Washburn Memorandum, October 26, 1955, OCB Central Files, box 15, OCB 007 (file#2) (8), EL; Statement by Theodore Repplier, in USIA to all USIS posts, February 29, 1956. Emphasis in original.

28. Description of People's Capitalism exhibit; and Streibert to Country Public Affairs Officers, January 10, 1956; both in Lambie Records, box 23, People's Capitalism 1955, EL. My description of the exhibit is also based on photographs in the USIA Historical Collection files.

29. "People's Capitalism Is Theme of General Electric Advertisement," *Public Service Advertising*, October 1956, in Lambie Records, box 31, People's Capitalism 1956 (1), EL; "Steel Union Chief Proud of 'People's Capitalism,'" September 18, 1956, *Los Angeles Times;* "People's Capitalism," *New York World-Telegram and Sun*, September 25, 1956; David S. Goodman, "What Kind of Capitalism?" Torch, September 1956; Henry Hazlitt, "People's Capitalism," *Newsweek*, September 17, 1956.

30. Monthly Report on People's Capitalism for September 1957, Lambie Records, box 38, People's Capitalism 1957, EL. Conference Report, August 14, 1956; Meeting of the People's Capitalism Committee, May 15, 1956; and Meeting of the People's Capitalism Committee, July 12, 1956; all in Lambie Records, box 31, *People's Capitalism* 1956 (1), EL. Henry C. Fleisher, "The Basic Concepts of America's 'People's Capitalism," USIS Feature, in RG 306, Feature Packets, Recurring Subjects, box 14, USIS Labor Packet #51, NA; Marcus Nadler, *People's Capitalism* (New York: Hanover Bank, 1956); Jacob M. Budish, *People's Capitalism: Stock Ownership and Production* (New York: International Publishers, 1958); David M. Potter, *The American Round Table Discussions on People's Capitalism at Yale University, New Haven, Connecticut, November 16 and 17, 1956* (New York: Advertising Council, 1957); David M. Potter, *An Inquiry into the Social and Cultural Trends in America Under Our System of Widely Shared Materials Benefits: [Discussions] at the Yale Club, New York, New York, May 22, 1957* (New York: Advertising Council, 1957).

31. Washburn Memorandum, October 26, 1955; USIA to all USIS posts, June 12, 1956, Lambie Records, box 31, People's Capitalism 1956 (1), EL; Monthly reports on People's Capitalism, Lambie Records, box 38, People's Capitalism 1957, EL; USIA circular, March 16, 1959, USIA Historical Collection.

32. USIA Basic Guidance and Planning Paper No. 5, "Basic Planning Paper on Minorities," December 4, 1958, USIA Historical Collection; "Treatment of Minorities in the United States—Impact on Our Foreign Relations," quoted in Michael L. Krenn, "'Unfinished Business': Segregation and U.S. Diplomacy at the 1958 World's Fair," *Diplomatic History* 20 (Fall 1996): 591–92; Mary L. Dudziak, "Desegregation as a Cold War Imperative," *Stanford Law Review* 41 (November 1988): 61–120; Dudziak, *Cold War Civil Rights: Race and the Image of American Democracy* (Princeton: Princeton University Press, 2001). The international dimensions of the fight against racial discrimination and colonialism have received extensive treatment in recent scholarship. See especially Thomas Borstelmann, *The Cold War and the Color Line: American Race Relations in the Global Arena* (Cambridge: Harvard University Press, 2001); Brenda Gayle Plummer, *Rising Wind: Black Americans and U.S. Foreign Affairs, 1935–1960* (Chapel Hill: University of North Carolina Press, 1996); Penny M. Von Eschen, *Black Americans and Anticolonialism, 1937–1957* (Ithaca, N.Y.: Cornell University Press, 1997); Carol Anderson, *Eyes Off the Prize: The United Nations and the African American Struggle for Human Rights, 1944–1955* (Cambridge: Cambridge University Press, 2003); Michael L. Krenn, *Black Diplomacy: African Americans and the State Department, 1945–1969* (Armonk, N.Y.: M. E. Sharpe, 1999); James Hunter Meriwether, *Proudly We Can Be Africans: Black Americans and Africa, 1935–1961* (Chapel Hill: University of North Carolina Press, 2002).

33. Quotes in Dudziak, *Cold War Civil Rights*, 37–38. U.S. Department of State, Policy Planning Staff, Recent Moscow Comment, April 4, 1961, RG 59, PPS Records 1957–1961, Lot 67 D548, box 120, "Information Policy," NA. American diplomats and information officers frequently complained about communist propaganda on the race issue. USIS Italy's complaint was typical. It reported that communists in Italy "inspire systematic propaganda against U.S. race practices, attempting to give a deliberately distorted and generalized impression of American race discrimination and hatred." USIS Italy Country Assessment Report for 1960, February 8, 1961, RG 306, Foreign Service Despatches, box 1, Europe, NA. Other such complaints are well documented in Dudziak, *Cold War Civil Rights*.

34. Haitian incident in Dudziak, *Cold War Civil Rights*, 41; Donald M. Wilson to Frederick G. Dutton, April 27, 1961, RG 59, PPS Records 1957–1961, Lot 67 D548, box 120, "Information Policy," NA.

35. NSC 5720, Status of USIA Program, September 11, 1957, *FRUS 1955–1957*, 9:607; USIS India Annual Assessment Report, November 4, 1958, RG 306, Foreign Service Despatches, box 1, Asia, NA; NSC 5611, Status of USIA Program,

June 30, 1956, WHO, Office of the Special Assistant for National Security Affairs [OSANSA], NSC Series, Status of Projects Subseries, box 7, EL; U.S. Embassy Brussels to DOS, March 20, 1956, RG 59, Central Decimal File 811.411/3-2056, NA.

36. Information Policy on the American Economy, October 3, 1955; USIA, "Basic Planning Paper on Minorities."

37. NSC 5819, Status of USIA Program, June 30, 1958, WHO, OSANSA, NSC Series, Status of Projects Subseries, box 8; NSC 5720, Status of USIA Program, September 11, 1957; USIS Italy Country Assessment Report for 1960, February 8, 1961, RG 306, Foreign Service Despatches, box 1, Europe, NA. USIS Addis Ababa to USIA, October 3, 1956; USIS Country Plan for Nigeria, June 26, 1959; and USIS Lagos to USIA, March 15, 1960; all in RG 306, Foreign Service Despatches, box 3, Africa, NA. VOA quoted in Walter L. Hixson, *Parting the Curtain: Propaganda, Culture, and the Cold War, 1945–1961* (New York: St. Martin's Griffin, 1997), 130.

38. USIS Bangkok to USIA, Revision of Country Plan for Thailand, July 29, 1958, RG 306, Foreign Service Despatches, box 1, Asia, NA; USIS Lagos to USIA, March 15, 1960, RG 306, Foreign Service Despatches, box 3, Africa, NA.

39. USIS Addis Ababa to USIA, October 3, 1956, RG 306, Foreign Service Despatches, box 3, Africa, NA. Author's interview with James Halsema.

40. NSC 5819, Status of USIA Program, June 30, 1958; Cary Fraser, "Crossing the Color Line in Little Rock: The Eisenhower Administration and the Dilemma of Race for U.S. Foreign Policy," *Diplomatic History* 24 (Spring 2000): 233–64; Chester J. Pach Jr. and Elmo Richardson, *The Presidency of Dwight D. Eisenhower*, rev. ed. (Lawrence: University Press of Kansas, 1991), 137–58; and Robert Fredrick Burk, *The Eisenhower Administration and Black Civil Rights* (Knoxville: University of Tennessee Press, 1984).

41. Fraser, "Crossing the Color Line," 234; Status of USIA Program, NSC 5430, August 18, 1954, *FRUS 1952–1954*, 2:1785; USIS Bombay to USIA, August 20, 1954, RG 306, Foreign Service Despatches, box 1, Asia, NA. For a description of the pamphlet "The Negro in American Life," see Dudziak, *Cold War Civil Rights*, 49–54.

42. Status of USIA Program, NSC 5819, June 30, 1958; and Status of USIA Program, NSC 5611, June 30, 1956; Status of USIA Program, NSC 5720, September 11, 1957.

43. Dudziak, *Cold War Civil Rights*, 115–18; Fraser, "Crossing the Color Line"; Telephone Conversation, Dulles to Brownell, September 24, 1957, *FRUS 1955–1957*, 9:612–13.

44. Fraser, "Crossing the Color Line"; Dwight D. Eisenhower, *Waging Peace, 1956–1961: The White House Years* (Garden City, N.Y.: Doubleday, 1965), 168; Radio and television address, September 24, 1957, *Public Papers of the Presidents of the United States: Dwight D. Eisenhower, 1957* (Washington, D.C.: Government Printing Office, 1961): 689–94.

45. Dudziak, *Cold War Civil Rights*, 133; Stephens to Toner, September 24, 1957, WHO, Staff Research Group Series, USIA 1-350, EL; USIS Mexico to USIA Washington, November 1, 1957, RG 306, Office of Research, Country Project Correspondence 1952–63, box 15, Mexico 1957, NA.

46. DOS Circular Telegram, October 10, 1957, RG 59, Central Decimal File, 811.411/10-1057, NA; USIS Geneva to USIA, November 26, 1957, RG 306, Foreign Service Despatches, box 1, Europe, NA.

47. USIS India Annual Assessment Report, November 4, 1958, RG 306, Foreign Service Despatches, box 1, Asia, NA; DOS Circular Telegram, October 10, 1957, RG 59, Central Decimal File, 811.411/10-1057, NA; USIS Paris quoted in Dudziak, *Cold War Civil Rights*, 135.

48. "International Prestige is Secondary," *Charleston Post*, September 14, 1956.

49. Quoted in Krenn, "Unfinished Business," 597.

50. Quotes from Hixson, *Parting the Curtain*, 141–50; and Krenn, "Unfinished Business," 598.

51. Quotes from Krenn, "Unfinished Business," 600–601, 606.

52. Eisenhower to Frank Alexander Willard, March 7, 1951, in *The Papers of Dwight D. Eisenhower*, ed. Alfred D. Chandler and Louis Galambos (Baltimore: Johns Hopkins University Press, 1970–1996), 12:102–4.

53. Hixson, *Parting the Curtain*, 146.

54. USIA, "American Economy."

Chapter 9. A New "Magna Carta" of Freedom

1. Although some observers have interpreted the film as a rant against conformity, the anticommunist message seems much stronger, particularly in the context in which it was made.

2. NSC report on internal security, quoted in Scott Lucas, *Freedom's War: America's Crusade Against the Soviet Union* (New York: New York University Press, 1999), 59. Stephen J. Whitfield, *The Culture of the Cold War*, 2nd ed. (Baltimore: Johns Hopkins University Press, 1991); Michael Barson and Steven Heller, *Red Scared! The Commie Menace in Propaganda and Popular Culture* (San Francisco: Chronicle Books, 2001).

3. Allen Dulles speech to the Council on Foreign Relations, June 5, 1953, quoted in Washburn to William H. Jackson, June 11, 1953, White House Central Files, Official File, box 674, 133-M-1 PCIAA (3), Eisenhower Library, Abilene, Kansas (hereafter EL); Conyers Read quoted in Whitfield, *Culture of the Cold War*, 58. The call for an American "ideology" was not universally accepted. Many intellectuals recoiled at the notion that the United States should craft a formalized doctrine. For discussion, see Whitfield, *Culture of the Cold War*, 53–76.

4. PSB D-33, U.S. Doctrinal Program, June 29, 1953, Freedom of Information Act (FOIA).

5. Statement on Doctrinal Warfare Targets, February 6, 1953, *Declassified Documents Reference System* 1992/2945; USIA, United States Doctrinal Program, n.d. [June 1954], OCB Central Files, box 70, OCB 091.4 Ideological Program (File #1) (1), EL.

6. Lilly to PSB, March 31, 1953; and Draft National Psychological Strategy Plan for the Use of Doctrinal Warfare, March 31, 1953; both in *Declassified Documents Reference System* microfiche 1991/2952. Memorandum for the Record, December 18, 1953, OCB Central Files, box 71, OCB 091.4 Ideological Program (File #1) (1), EL; Ideological Program, n.d. [December 1954], OCB Central Files, box 70, OCB 091.4 Ideological Program (File #2) (3), EL; Outline Plan of Operations for the U.S. Ideological Program (D-33), December 15, 1954, OCB Central Files, box 70, OCB 091.4 Ideological Program (File #2) (3), EL; Study on Ideological Strategy, June 9, 1954, OCB Central Files, box 70, OCB 091.4 Ideological Program (File #1) (3), EL. Emphasis in original.

7. Outline Plan of Operations for the U.S. Ideological Program (D-33), December 15, 1954, OCB Central Files, box 70, OCB 091.4 Ideological Program (File #2) (3), EL; Ideological Program, n.d. [December 1954]. Outline Plan of Operations for the U.S. Ideological Program (D-33), February 16, 1955, Record Group [RG] 59, Central Decimal File, 511.00/4-155, National Archives, College Park, Maryland (hereafter NA).

8. OCB Field Reports on Ideological Outline of Operations, n.d. [1955], OCB Central Files, box 37, OCB 091 India (File #1) (9), EL; Summary of Field Comment on the Ideological Plan, July 28, 1955, RG 59, Bureau of Public Affairs, International Exchange Service Correspondence, Memoranda, Reports, and Other Records of the Program Development Staff, box 4, Reports and Minutes of OCB Working Group on the Ideological Plan, NA.

9. U.S. Embassy Saigon to DOS, June 23, 1955, RG 59, Central Decimal Files 511.00/6-2355, NA; OCB, Field Reports on Ideological Outline of Operations, n.d. [1955]. U.S. Embassy Brussels to DOS, June 28, 1955, RG 59, Central Decimal Files 511.00/6-2855, NA; U.S. Embassy Saigon to DOS, June 23, 1955, RG 59, Central Decimal Files 511.00/6-2355, NA. On U.S. support for right-wing dictatorships, see David Schmidtz, *Thank God They're On Our Side: The United States and Right-Wing Dictatorships, 1921–1965* (Chapel Hill: The University of North Carolina Press, 1999).

10. Outline Plan of Operations for the U.S. Ideological Program (D-33), December 15, 1954; Ideological Program, n.d. [December 1954]; OCB Ideological Planning Guide, February 24, 1956, OCB Central Files, box 70, OCB 091.4 Ideological Program (File #3) (8), EL.

11. U.S. Embassy Brussels to DOS, June 28, 1955, RG 59; Ideological Program, n.d. [December 1954]; McCardle to Hoover, February 15, 1955, RG 59, Bureau of Public Affairs, International Exchange Service, Correspondence, Memoranda, Reports and Other Records of the Program Development Staff, box 4, OCB Working

Group on the Ideological Program (Correspondence), NA. Draft Outline Plan of Operations, n.d., OCB Central Files, box 70, OCB 091.4 Ideological Program (File #1) (6), EL.

12. McCardle to Hoover, February 15, 1955; Summary Statement, June 28, 1954, OCB Central Files, box 70, OCB 091.4 Ideological Program (File #1) (3), EL; USIA, United States Doctrinal Program, n.d. [June 1954], OCB Central Files, box 70, OCB 091.4 Ideological Program (File #1) (1), EL.

13. PSB D-33, June 29, 1953.

14. Eighth Semiannual Report on Educational Exchange Activities, House Doc. 35, 83d Congress, 1st session (1953), quoted in Charles A. Thomson and Walter H. C. Laves, *Cultural Relations and U.S. Foreign Policy* (Bloomington: Indiana University Press, 1963), 85; Wilson P. Dizard, *The Strategy of Truth: The Story of the U.S. Information Service* (Washington, D.C.: Public Affairs Press, 1961), 137; Thomson and Laves, *Cultural Relations*, 52–53; Reinhold Wagnleitner, *Coca-Colonization and the Cold War: The Cultural Mission of the United States in Austria After the Second World War* (Chapel Hill: University of North Carolina Press, 1994), 129.

15. Wagnleitner, *Coca-Colonization*, 136–39; Frances Stonor Saunders, *The Cultural Cold War: The CIA and the World of Arts and Letters* (New York: New Press, 1999), 191–93; Dizard, *Strategy of Truth*, 139–42; Ellen Schrecker, *Many Are the Crimes: McCarthyism in America* (Boston: Little, Brown, 1998), 256–57.

16. USIA, United States Doctrinal Program, n.d. [June 1954]; PSB, Doctrinal Warfare, March 31, 1953, *Declassified Documents Reference System* microfiche 1991/2952; Outline Plan of Operations for the U.S. Ideological Program (D-33), February 16, 1955, RG 59, 511.00/4-155, NA; Mark A. May, Report on USIS-Japan, June–July 1959, USIA Historical Collection, Washington, D.C.

17. Ideological Program, n.d. [December 1954]; OCB Progress Report on the U.S. Ideological Program, August 24, 1955, OCB Central Files, box 7, OCB 091.4 Ideological Program (File #3) (5), EL. It is difficult to determine with precision the steps taken to "stimulate the production of ideological materials." The available documentation offers a general picture of some of the ways the OCB's objectives were achieved, but such details as who was "stimulated" to produce what probably reside in the substantial volume of files on this program that remain classified. This author's FOIA requests, repeatedly filed over a six-year period, failed to yield specific evidence as to which manuscripts the USIA commissioned in the 1950s. In 1966, the USIA's covert commissioning of manuscripts was uncovered by a Congressional inquiry and caused quite a stir in Washington. This inquiry remains the only source of concrete details pertaining to the USIA's manuscript commission efforts. The *New York Times* reported that works published in 1965 included *President Kennedy and Africa, The Truth About the Dominican Republic, American Negro Reference Book, The Communist Front as a Weapon of Political Warfare,* and *From Colonialism to Communism*. *New York Times*, October 3, 1966, 34:1; *New York Times*, September

29, 1966, 27:1; *New York Times*, September 28, 1966, 13:1. *New York Times*, February 24, 1967; Report of OCB Working Group on Books, Publications and Libraries, June 10, 1954, White House Office [WHO], NSC Staff Papers, OCB Secretariat Series, box 1, Book Program; RG 306, Foreign Service Despatches, boxes 1–3, NA.

18. Detailed Annex, March 31, 1953, *Declassified Documents Reference System* microfiche 1992/2949; Summary Statement, June 28, 1954, OCB Central Files, box 70, OCB 091.4 Ideological Program (File #1) (3), EL.

19. Report of OCB Working Group on Books, Publications and Libraries, June 10, 1954, WHO, NSC Staff Papers, OCB Secretariat Series, box 1, Book Program, EL.

20. The Low-Priced Book Program, December 27, 1956, RG 59, Records Relating to DOS participation in the OCB and NSC, box 36, Books and Publications, NA.

21. The Low-Priced Book Program, December 27, 1956; USIA, *Volunteers for International Communication: Report of USIA Private Sector Committees* (1984); Report of OCB Working Group on Books, Publications and Libraries, June 10, 1954; The U.S. Information Program Since July 1953, Records of the President's Committee on Information Activities Abroad (Sprague Committee), box 19, USIA (2), EL; American Embassy Bangkok to DOS, April 25, 1953, RG 59, Central Decimal File, 511.92/4-2553, NA.

22. "Franklin Publications, Inc.," January 25, 1956; and "Memorandum on the Rationale and Methods of Promotion and Distribution"; both in USIA Historical Collection. *New York Times*, May 28, 1965, and October 7, 1964; Report of OCB Working Group on Books, Publications and Libraries, June 10, 1954; Annual Survey of People-to-People Activities, January 1959, RG 59, Lot 66 D257, box 7, People to People—Annual Surveys and Reviews of Activities, NA; Lilly to Leonard, August 15, 1955, OCB Central Files, box 71, OCB 091.4 Ideological Programs (File #3) (4), EL; Outline Plan of Operations for the U.S. Ideological Program (D-33), February 16, 1955; Dizard, *Strategy of Truth*, 149–51.

23. USIS Tunis to USIA Washington, June 29, 1960, Country Plan for USIS Tunisia, July 29, 1960, RG 306, Foreign Service Despatches, 1954–65, box 3, Africa, NA.

24. Saunders, *Cultural Cold War*, 246. According to Saunders, the Chekhov Publishing Company was subsidized by the CIA. The company also was funded by the USIA and the Defense Department.

25. Report of OCB Working Group on Books, Publications and Libraries, June 10, 1954.

26. Report of OCB Working Group on Books, Publications and Libraries, June 10, 1954; USIA Press Release, December 7, 1954, RG 59, International Exchange Service, Correspondence, Memoranda, Reports, and Other Records of the Program Development Staff, box 4, Reports and Minutes of OCB Working Group

on the Ideological Program, NA. On the use of *1984* by British propagandists, see Paul Lashmar and James Oliver, *Britain's Secret Propaganda War, 1948–1977* (Stroud, Gloucestershire: Sutton Publishing, 1998).

27. Report of OCB Working Group on Books, Publications and Libraries, June 10, 1954; USIA, *Volunteers for International Communication*; Annual Survey of People-to-People Activities, January 1959.

28. Detailed Annex, March 31, 1953; quote from Saunders, *Cultural Cold War,* 245.

29. Michael Warner, "Origins of the Congress for Cultural Freedom," *Studies in Intelligence* 38 (Summer 1995), available at http://www.odci.gov/csi/studies/95unclas/war.html, accessed June 10, 2003. An indication of the scale of these operations was revealed by the church committee's finding that during one three-year period in the 1960s, 108 of 700 foundation grants of over $10,000 involved the CIA. Moreover, many foundations that sponsored CIA publication projects were CIA fronts that "*didn't exist* except on paper." Final Report of the Church Committee and Tom Braden interview, in Saunders, *Cultural Cold War,* 127, 134–35. See also Hugh Wilford, *The CIA, the British Left, and the Cold War: Calling the Tune?* (London: Frank Cass, 2003); Wilford, "Playing the CIA's Tune? The *New Leader* and the Cultural Cold War," *Diplomatic History* 27 (Winter 2003): 15–34; Giles Scott-Smith, *The Politics of Apolitical Culture: The Congress for Cultural Freedom, the CIA, and Post-War American Hegemony* (London: Routledge, 2002); and Peter Coleman, *The Liberal Conspiracy: The Congress for Cultural Freedom and the Struggle for the Mind of Postwar Europe* (New York: Free Press, 1989).

30. Quote from Saunders, *Cultural Cold War,* 245. In 1965, the Advisory Commission on Information reported that in the preceding twenty years there had been sixty-eight separate incidents in which USIS libraries around the world had been attached by mobs. The *New York Times* indicated that the number was far greater, citing ninety-seven such incidents. John W. Henderson, *The United States Information Agency* (New York: Praeger, 1969), 130; Memorandum of a Meeting Between the President and the Republican Leadership, March 13, 1956, in *Foreign Relations of the United States 1955–1957,* 2:565 (hereafter *FRUS*); *New York Times*, May 1, 1965.

31. Outline Plan of Operations for the U.S. Ideological Program (D-33), February 16, 1955, RG 59, 511.00/4-155, NA; PCIAA Study No. 40, The Exchange of Persons Program of the U.S. Government, August 8, 1960, Mandatory Declassification Review, EL.

32. PCIAA Study No. 40, The Exchange of Persons Program of the U.S. Government, August 8, 1960, Mandatory Declassification Review, EL.

33. Outline Plan of Operations for the U.S. Ideological Program (D-33), February 16, 1955; Expanded Educational Exchange Program in the Far East, March 15, 1954, RG 59, Bureau of Public Affairs, International Exchange Service, Staff Studies and Reports and OCB Working Papers, box 1, OCB Minutes, NA.

34. Implementation of OCB Study of American Overseas Education Programs, December 13, 1955, RG 59, Records Relating to DOS Participation in the OCB and NSC 1953–1960, box 34, American Overseas Education 1955, NA; Brief on OCB Study, June 8, 1955; OCB Study on Overseas Education, December 1, 1954; and Report on the Implementation of the OCB Study on American Overseas Education, April 26, 1955, all in Nelson A. Rockefeller Papers, RG 4, Washington, D.C., Files, Subseries 7, box 12, OCB—Overseas Education—Hoskins Study, Rockefeller Archive Center, Sleepy Hollow, New York (hereafter RAC).

35. DOS Progress Report on Ideological Program, January through June 1955, OCB Central Files, box 71, OCB 091.4 Ideological Program (File #3) (3), EL; OCB Ideological Planning Guide, February 24, 1956, OCB Central Files, box 70, OCB 091.4 Ideological Program (File #3) (8), EL; DOS Progress Report on Ideological Program, January through June 1955, OCB Central Files, box 71, OCB 091.4 Ideological Program (File #3) (3), EL; Semi-Annual USIS Report for Egypt, August 4, 1955, RG 306, Foreign Service Despatches, box 3, Africa, NA.

36. DOS Progress Report on Ideological Program, January through June 1955; Report of OCB Working Group on Books, Publications and Libraries, June 10, 1954; USIS Bangkok to USIA, Revision of Country Plan for Thailand, July 29, 1958, RG 306, Foreign Service Despatches, box 1, Asia, NA; PCIAA Study No. 7, The English Teaching Program, May 23, 1960, Mandatory Declassification Review, EL; NSC 6013, Status of the USIA Program, June 30, 1960, White House Office [WHO], Office of the Special Assistant for National Security Affairs [OSANSA], NSC Series, Status of Projects Subseries, box 9, NSC 6013 (4), EL.

37. USIS Tunis to USIA Washington, June 29, 1960, Country Plan for USIS Tunisia, July 29, 1960, RG 306, Foreign Service Despatches, 1954–65, box 3, Africa, NA; USUS-Rabat to USIA, May 2, 1960, RG 306, Foreign Service Despatches, box 3, Africa, NA.

38. "The Agency increased its cultural operations in a number of places where there were no bothersome restrictions, but relied most heavily upon them in those countries where restrictions were severe." NSC 6013, Status of the USIA Program, 30 June 1960, WHO, OSANSA, NSC Series, Status of Projects Subseries, box 9, NSC 6013 (4), EL.

39. English-Teaching Program, Report to the President's Committee on Information Activities Abroad, July 11, 1960, Records of the President's Committee on Information Activities Abroad, box 23, PCIAA #7 (1), EL; Africa, Report to the President's Committee on Information Activities Abroad, July 11, 1960, Records of the President's Committee on Information Activities Abroad, box 23, PCIAA #31 (1), EL; NSC 6013, Status of the USIA Program, 30 June 1960, WHO, OSANSA, NSC Series, Status of Projects Subseries, box 9, NSC 6013 (4), EL.

40. Report on the Implementation of the OCB Study on American Overseas Education, April 26, 1955; DOS Progress Report on Ideological Program, January through June 1955; Expanded Educational Exchange Program in the Far East,

March 15, 1954; USIS Tokyo to USIA, Japan Country Plan—Annex No. 1, February 26, 1959, RG 306, Foreign Service Despatches, box 1, Asia; Mark A. May, Report on USIS-Japan, June–July 1959.

41. Outline Plan of Operations for the U.S. Ideological Program (D-33), February 16, 1955.

42. Memoranda of Meetings of the Ideological Subcommittee on the Religious Factor, May 19, 1955, June 8, 1955, and June 20, 1955, both in OCB Central Files, box 2, OCB 000.3 [Religion] (File #1) (2), EL.

43. Harlow to O'Hara, June 6, 1955, White House Central Files, Official File, box 910, OF 247 1955 (2), EL.

44. MacKnight to McIlvaine, June 3, 1955; and Sanger to MacKnight May 31, 1955; both in RG 59, Lot 61 D53, box 69, P-MacKnight, USIA-State-Friends 1954–1955, NA; Memoranda of Meetings of the Ideological Subcommittee on the Religious Factor, May 19, 1955, June 8, 1955, and June 20, 1955, in OCB Central Files, box 2, OCB 000.3 [Religion] (File #1) (2), EL; Rockefeller to Eisenhower, April 7, 1955, Nelson A. Rockefeller Papers, box 16, "Misc. Projects," RAC; *New York Times*, June 8, June 18, and July 12, 1955. Saunders, *Cultural Cold War*, 279.

45. Outline Plan Regarding Buddhist Organizations in Ceylon, Burma, Thailand, Laos, and Cambodia, January 16, 1957, OCB Central Files, box 2, OCB 000.3 [Religion] (File #2) (1), EL.

46. Outline Plan Regarding Buddhist Organizations in Ceylon, Burma, Thailand, Laos, and Cambodia, January 16, 1957.

47. Outline Plan Regarding Buddhist Organizations in Ceylon, Burma, Thailand, Laos, and Cambodia, January 16, 1957; Keller to Sias, July 11, 1956, OCB Central Files, box 2, OCB 000.3 [Religion] (File #1) (4), EL.

48. OCB Report, Inventory of U.S. Government and Private Organization Activity Regarding Islamic Organizations as an Aspect of Overseas Operations, May 3, 1957, *Declassified Documents Reference System* online, document no. CK3100496987.

49. Gleason H. Ledyard, *Sky Waves: The Incredible Far East Broadcasting Company Story* (Chicago: Moody Press, 1968); Anne C. Loveland, *American Evangelicals and the U.S. Military, 1942–1993* (Baton Rouge: Louisiana State University Press, 1996), 56–66; Mark Ellis, "Far East Broadcasting Founder Watched God Do the Impossible with Radio," ASSIST News Service, December 29, 2003, available at http://www.assistnews.net/stories/s03120113.htm; "June 4, 1948: The Triumph of the Far East Broadcasting Co.," available at http://www.gospelcom.net/chi/DAI-LYF/2003/06/daily-06-04-2003.shtml. Archival holdings pertaining to the FEBC are in the Billy Graham Center in Wheaton, Illinois, collection 59.

50. Quoted in Loveland, *American Evangelicals*, 59.

51. Militant Liberty Outline Plan, November 5, 1954, OCB Central Files, box 70, OCB 091.4 Ideological Programs (File #1) (8), EL; Proposed Implementation Plan for Project Action, July 15, 1955, OCB Central Files, box 71, OCB 091.4

Ideological Programs (File #3) (4), EL. The basic concept of Militant Liberty was unclassified but the Defense Department also developed a classified version of this program designated Project Action.

52. U.S. Department of Defense, *Militant Liberty: A Program of Evaluation and Assessment of Freedom* (Washington, D.C.: Government Printing Office, 1955).

53. Bonesteel to Cutler, November 26, 1954, *Declassified Documents Reference System*, CD-ROM, document no. 1990030100665; Militant Liberty Outline Plan, November 5, 1954. Emphasis in original.

54. Memorandum for the Secretary of Defense, July 26, 1957, WHO, OSAN-SA, OCB Series, Subject Subseries, box 3, Militant Liberty, EL; OCB Concern with Militant Liberty, NA. It is difficult to determine precisely where the Militant Liberty project was implemented because documents are not declassified. It may have been implemented in France, Iran, and Norway. See USIS Paris to USIA, January 31, 1956, RG 218, JCS Central Files 1954–1956, box 124, sec. 18, NA; and Memorandum for JCS Chairman, December 9, 1954, RG 218, JCS Central Files 1954–1956, box 122, sec. 95, NA. Documents do not specify whether the program was actually implemented in Japan. The Embassy in Japan was not receptive to the idea. In a cable to Washington, the pubic affairs officer wrote that the program was "suspiciously facile and shallow. . . . [A]n indoctrination program of the nature suggested would be interpreted as subverting the social and governmental structures of a number of . . . governments in the Far East." U.S. Embassy Japan to DOS, April 29, 1955, *Declassified Documents Reference System*, CD-ROM, document no. 1994010100252. Lucas, *Freedom's War,* 269; Cary Reich, *The Life of Nelson Rockefeller: Worlds to Conquer, 1908–1958* (New York: Doubleday, 1996), 563; Nancy Bernhard, *U.S. Television News and Cold War Propaganda, 1947–1960* (Cambridge: Cambridge University Press, 1999), 149.

55. Earl J. Wilson Oral History, Georgetown Library, Special Collections, Georgetown University, Washington, D.C. On Eisenhower's efforts to promote citizenship education at Columbia, see Travel Beal Jacobs, *Eisenhower at Columbia* (New Brunswick, N.J.: Transaction Publishers, 2001).

56. Smith to Staats, February 21, 1955, OCB Central Files, box 71, OCB 091.4 Ideological Programs (File #2) (7), EL; OCB Concern with Militant Liberty, WHO, OSANSA, OCB Series, Subject Subseries, box 3, Militant Liberty, EL; Memorandum for JCS Chairman, July 5, 1956, RG 218, JCS Central Files 1954–1956, box 125, sec 24, NA; Memorandum to Chief of Naval Operations, December 16, 1955, RG 218, JCS Central Files 1954–1956, box 124, sec. 16, NA; Militant Liberty Outline Plan, November 5, 1954; Saunders, *Cultural Cold War,* 285–86.

57. Nelson A. Rockefeller Papers, box 18, "Militant Liberty (Action) Book 3," RAC; Loveland, *American Evangelicals,* 60–63. The ICL—today called the Fellowship and known for its sponsorship of the National Prayer Breakfast—has main-

tained access to the highest levels of the American government since its founding in the 1930s. In 1958, Broger was elected vice president of ICL. On the ICL and the Fellowship, see Lisa Getter, "Showing Faith in Discretion," *Los Angeles Times*, September 27, 2002; Jeffrey Sharlet, "Jesus Plus Nothing," *Harper's Magazine*, March 2003, available at http://www.harpers.org/JesusPlusNothing.html?pg=1; Anthony Lappé, "Meet 'The Family,'" *Guerrilla News Network*, June 13, 2003, available at http://www.alternet.org/story/16167. The fellowship's archives are located at the Billy Graham Center in Wheaton, Illinois. See collection 459.

58. JCS, Militant Liberty Outline Plan, November 5, 1954, OCB Central Files, box 70, OCB 091.4 Ideological Programs (File #1) (8), EL. Emphasis in original.

59. OCB Ideological Planning Guide, April 4, 1956, OCB Central Files, box 71, OCB 091.4 Ideological Program (File #4) (1), EL; Jackson Committee Report, *FRUS 1952–1954*, 2:1806; JCS, Militant Liberty Outline Plan, November 5, 1954.

60. OCB Study on Overseas Education, December 1, 1954; Frank Ninkovich, *The Diplomacy of Ideas: U.S. Foreign Policy and Cultural Relations, 1938–1950* (New York: Cambridge University Press, 1981), 163.

Chapter 10. The Power of Symbols

1. Andreas Wenger, *Living with Peril: Eisenhower, Kennedy, and Nuclear Weapons* (Lanham, Md.: Rowman & Littlefield, 1997), 154–55; Piers Brendon, *Ike: His Life and Times* (New York: Harper & Row, 1986), 347–49; Walter A. McDougall, *The Heavens and the Earth: A Political History of the Space Age* (Baltimore: Johns Hopkins University Press, 1985), 7.

2. McDougall, *The Heavens and the Earth*, 119–21.

3. This view has been widely accepted by revisionist historians of the Eisenhower presidency and by most historians of U.S. space policy. For a concise summary of the revisionist argument, see Giles Alston, "Eisenhower: Leadership in Space Policy," in *Reexamining the Eisenhower Presidency*, ed. Shirley Anne Warshaw (Westport, Conn.: Greenwood Press, 1993). See also McDougall, *The Heavens and the Earth*; Robert A. Divine, *The Sputnik Challenge: Eisenhower's Response to the Soviet Satellite* (New York: Oxford University Press, 1993); Stephen Ambrose, *Eisenhower: The President* (New York: Simon and Schuster, 1984), 423–61; R. Cargill Hall, "Origins of U.S. Space Policy: Eisenhower, Open Skies, and Freedom of Space," in *Exploring the Unknown: Selected Documents in the History of the U.S. Civil Space Program*, ed. John M. Logsdon (Washington, D.C.: National Aeronautics and Space Administration, 1995), 213–29; Dwayne A. Day, "Cover Stories and Hidden Agendas: Early American Space and National Security Policy," in *Reconsidering Sputnik: Forty Years Since the Soviet Satellite*, ed. Roger P. Launius, John M. Logsdon, and Robert W. Smith (Australia: Harwood Academic Publishers, 2000), 161–95. For critical analyses of the Sputnik historiography, see Rip Bulkeley, *The Sputniks Crisis*

and Early United States Space Policy: A Critique of the Historiography of Space (Bloomington: Indiana University Press, 1991); and Kenneth A. Osgood, "Before Sputnik: National Security and the Formation of U.S. Outer Space Policy," in Launius et al., *Reconsidering Sputnik*, 197–229. Bulkeley demonstrates the influence of U.S. public relations on the historiography of space. He writes: "Unfortunately, the historical perception . . . of Eisenhower's early space policy has been coloured by the fact that after Sputnik 1 the administration staged a damage-limiting public relations exercise . . . which drew a bogus but meretricious distinction between its own peaceful, scientific and allegedly disinterested satellite project and the somehow more sinister and less noble, if more effective, Soviet one." Bulkeley, *Sputniks Crisis*, 160.

4. Douglas Aircraft Company, Inc., Preliminary Design of an Experimental World-Circling Spaceship, May 2, 1946; and Grosse to Quarles, August 25, 1953; both in Logsdon, *Exploring the Unknown*, 236–44, 267–69. Hall, "Origins of U.S. Space Policy," 213–29.

5. Portions of the report remain classified, although a general summary is in *Foreign Relations of the United States 1955–1957*, 2:41–56 (hereafter *FRUS*). The declassified sections of the TCP report and related documents may be found in White House Office [WHO], Office of the Special Assistant for National Security Affairs [OSANSA], NSC Series, Subject Subseries, box 11, Eisenhower Library, Abilene, Kansas (hereafter EL). The report expressed some concern that orbiting satellites might be perceived as violations of national airspace so it also urged the United States to reexamine international law to verify the principle of "freedom of space." The United States did not want the first earth satellite to be a military or intelligence satellite because it wanted to establish a precedent of space overflight in international law. For detailed discussion of the "freedom of space" issue, see Hall, "Origins of U.S. Space Policy"; and Day, "Cover Stories." Although these authors are correct in noting that freedom of space was an important objective of U.S. outer space policy, they exaggerate its importance and underestimate the administration's concerns for prestige and psychological benefits from the scientific satellite program. I develop this argument in detail in "Before Sputnik."

6. On Corona and the spy satellite program, see Dwayne A. Day, ed., *Eye in the Sky: The Story of the Corona Spy Satellites* (Washington, D.C.: Smithsonian Institution Press, 1999); Kevin C. Ruffner, ed., *Corona: America's First Satellite Program* (Washington, D.C.: Center for the Study of Intelligence, 1995); Jeffrey T. Richelson, *America's Secret Eyes in Space: The U.S. Keyhole Spy Satellite Program* (New York: Harper & Row, 1990).

7. *Washington Post*, April 17, 1955; Kaplan to Alan T. Waterman, May 6, 1955 both in Logsdon, *Exploring the Unknown*, 295–308.

8. USNC-IGY meeting, May 18, 1955, in Logsdon, *Exploring the Unknown*, 295–308; Rockefeller to Lay, May 17, 1955, Ann Whitman File, Administration Series, box 31, Nelson Rockefeller 1952–1955, EL.

9. Rockefeller to Lay, May 17, 1955, Ann Whitman File, Administration Series, box 31, Nelson Rockefeller 1952–1955, EL.

10. NSC 5520, May 20, 1955, WHO, OSANSA, NSC Series, Policy Papers Subseries, box 16, EL; 283d NSC Meeting, May 3, 1956, Ann Whitman File, box 7, EL. The copy of NSC 5520 at the Eisenhower Library is more thoroughly declassified than the one published in *FRUS*.

11. NSC 5522, June 8, 1955, WHO, OSANSA, NSC Series, Policy Papers Subseries, box 16, EL. In addition to demonstrating the importance of prestige and psychological factors to the American satellite program, this document also illustrates that the "freedom of space" issue was not as important as other historians have suggested. NSC 5522 reported that the Departments of State, Treasury, Defense, and Justice all concurred that the launching of an artificial satellite was permissible under international law. The departments reported: "By customary law every State has exclusive sovereignty 'over the space above its territory.' However, *air space ends with the atmosphere*. There has been no recognition that sovereignty extends into the airless space beyond the atmosphere." The departments expressed no concern for U.S. vulnerability to criticism on that front. Significantly, the CIA, which had a material interest in reconnaissance satellites, did not even mention the freedom of space issue in its comments. The agency was primarily concerned with the prestige and psychological aspects of the scientific satellite. According to the Pentagon, which was also involved in intelligence projects, the IGY satellite was intended "for propaganda and scientific purposes."

12. NSC 5522, June 8, 1955.

13. NSC 5522, June 8, 1955.

14. The job of selecting a launch vehicle fell to Assistant Secretary of Defense Donald Quarles, to whom the 1953 Grosse report had been addressed. Quarles appointed a subcommittee directed by Homer J. Stewart to recommend a launch vehicle for the satellite. The Stewart Committee selected Project Vanguard because it promised the best combination of all three parameters established by the National Security Council: noninterference with ICBM development, peaceful image, and timeliness. Despite the claims of McDougall and others that speed was not a consideration in the selection of a launch vehicle, it is clear that timeliness weighed heavily on the Stewart Committee's decision. The first test launches of the Air Force Atlas-B were scheduled for January, February, and March 1958, at best an uncomfortably narrow margin for the IGY. Contrary to McDougall's claim that there was "little doubt" that the Army proposal promised a satellite soonest, half of the Stewart Committee members thought the Army's Jupiter-C missile was less likely to succeed than the Navy's Viking rocket. See Constance McLaughlin Green and Milton Lomask, *Vanguard: A History* (Washington, D.C.: National Aeronautics and Space Administration, 1970), 41–56. McDougall, *The Heavens and the Earth*, 123.

15. Public Information Program with Respect to the Implementation of NSC 5520, July 18, 1955; Stacy May to Rockefeller, July 25, 1955; MemCon, August 3, 1955; Don Irwin to Rockefeller, October 12, 1955; all in Nelson A. Rockefeller Papers, Record Group [RG] 4, Washington, D.C., Files, Subseries 7, box 11, NSC—Scientific Satellite Program, Rockefeller Archive Center, Sleepy Hollow, New York.

16. Stewart to DOD, June 22, 1956, and E.V. Murphree to DOD, July 5, 1956, in WHO, Staff Secretary Records, box 6, Missiles and Satellites, EL; Day, "Cover Stories." Most historians have viewed the tension between speed and science in "either-or" terms—either the United States wanted to be first, or it wanted scientific utility. But such a dichotomy misses the point. In order to secure maximum psychological benefit from the satellite, it needed to be both.

17. NSC meeting, September 8, 1955, Ann Whitman File, NSC Series, box 7, EL; The Impact of Achievements in Science and Technology Upon the Image Abroad of the United States, June 6, 1960, Records of the President's Committee on Information Activities Abroad (Sprague Committee), box 22, PCIAA #23 (1), EL. John Lewis Gaddis, *Strategies of Containment: A Critical Appraisal of Postwar American National Security Policy* (Oxford: Oxford University Press, 1982), 92 and 144. Emphasis added.

18. NSC Meeting, November 7, 1957; and NIE 100-5-55, June 14, 1955; both in *FRUS 1955–1957*, 19:85–87, 630–35; NSC meeting, May 31, 1956, Ann Whitman File, NSC Series, box 7, EL. Wenger, *Living with Peril*, 158; McDougall, *The Heavens and the Earth*, 112.

19. NSC meeting, September 8, 1955, Ann Whitman File, NSC Series, box 7, EL. For official chronologies of important developments in the ICBM and satellite programs see the *Declassified Documents Reference System*, microfiches 1994/2000 and 1981/221A.

20. State Department memorandum, n.d., *FRUS 1955–1957*, 19:154–61.

21. NSC meeting, October 1, 1955, *FRUS 1955–1957*, 19:166–70; see also n. 9, 19:170. See Divine, *Sputnik Challenge*, 24; and Dwight D. Eisenhower, *Mandate for Change, 1953–1956: The White House Years* (Garden City, NY: Doubleday, 1963), 457.

22. NSC meeting, September 8, 1955; NSC meeting, May 3, 1956, Ann Whitman File, NSC Series, box 7, EL.

23. NSC meeting, May 11, 1957, Ann Whitman File, NSC Series, box 8, EL.

24. *Newsweek*, October 14, 1957; *Life*, October 21, 1957; CIA Consultants to Allen Dulles, October 23, 1957, *Declassified Documents Reference System*, microfiche 95/3241; Michael S. Sherry, *In the Shadow of War: The United States Since the 1930s* (New Haven, Conn.: Yale University Press, 1995), 216. Nuclear physicist Edward Teller and Senator Lyndon Johnson were the two most notable figures to compare Sputnik to Pearl Harbor.

25. NSC meeting, October 10, 1957, Ann Whitman File, NSC Series, box 9,

EL; Hagerty to Eisenhower, October 7, 1957, *Declassified Documents Reference System,* microfiche 87/126.

26. State Department memorandum, n.d [October 16, 1957], *Declassified Documents Reference System,* microfiche 81/373A. For similar expressions see also: CIA Consultants to Allen Dulles, October 23, 1957, *Declassified Documents Reference System,* microfiche 95/3241; USIA survey, December 1957, *Declassified Documents Reference System,* microfiche 86/2225; USIA Circular Telegram, October 16, 1957, *FRUS 1955–1957,* 24:167–68; NIE 11-4-57, November 23, 1957, *FRUS 1955–1957,* 19:665–72; Arneson to Dulles, November 14, 1957, *FRUS 1955–1957,* 11:768–69; Soviet Embassy to DOS, November 16, 1957, *FRUS 1955–1957,* 24:185–87; USIA memo, December 17, 1957, *FRUS 1955–1957,* 11:779–82; and USIA Report, WHO, OSANSA, OCB Series, Subject Subseries, Space Satellites, Rockets, Etc. (1), EL.

27. USIA, Post-Sputnik Opinion in Western Europe, *Declassified Documents Reference System* microfiche 86/2225; George V. Allen to Eisenhower, December 13, 1957, Ann Whitman File, Administration Series, box 37, USIA (1), EL. The USIA surveyed public opinion in Britain, West Germany, Italy, France, and Norway.

28. USIA Office of Research and Analysis, Impact of U.S. and Soviet Space Programs on World Opinion, July 7, 1959, WHO, OSANSA, OCB Series, Subject Subseries, box 7, Space Council (1), EL. Emphasis added.

29. MemCon, October 8, 1957, *Declassified Documents Reference System* microfiche 88/1695; Secretary of Defense to Eisenhower, October 7, 1957, *Declassified Documents Reference System* microfiche 87/126; Proposed White House Press Release, n.d. [October 1957], *Declassified Documents Reference System* microfiche 86/2232.

30. Press Conference, October 9, 1957, *Public Papers of the Presidents of the United States: Dwight D. Eisenhower, 1957* (Washington, D.C.: Government Printing Office, 1961): 719–32 (hereafter *Public Papers*).

31. In looking for an explanation for why the United States did not beat the USSR into space, one must also consider the impact of bureaucratic momentum. In order for the NSC to reverse the decision to use the Navy's Vanguard rocket, convincing evidence of its certain failure needed to present itself. Such evidence did not materialize until after Sputnik. Before this, the administration received repeated assurances that the launch would occur on time, with the first attempt scheduled for October 1957. Furthermore, for all the warnings about Soviet scientific advances, many Americans (both in and out of the administration) found it difficult to believe that the Soviet Union might outpace the United States in a technological competition. Coming from such a perspective—as opposed to the perspective of the underdog—it would have been difficult to rationalize exorbitant expenditures to win a race that most simply assumed the United States would win.

32. To be safe, Eisenhower also instructed the Secretary of Defense to have the Army prepare a rocket as backup for the Navy Vanguard. Statement by the president, October 9, 1957, *Public Papers,* 1957, 733–35; Robert Cutler to Secretary

of Defense, October 17, 1957, *Declassified Documents Reference System* microfiche 1986/3442; Green and Lomask, *Vanguard*, 197–98. The statement released to the press on October 9 stressed that the American program was "designed from its inception for maximum results in scientific research." It downplayed the military's role in the project so as to reassure domestic audiences that the American missile programs were unaffected by Sputnik and to remind the world that the American program, unlike that of the USSR, was a peaceful and scientific endeavor. Furthermore, the statement asserted that the American satellite program was never conducted as a race with other nations.

33. Thomas S. Gates to Neil McElroy, October 22, 1957; and McElroy to Gates, October 29, 1957; in *FRUS 1955–1957*, 11:766–67.

34. NSC Meeting, December 5, 1957, *FRUS 1955–1957*, 11:771–73.

35. *The Nation*, December 21, 1957; *New York Times*, December 7, 1957; Quarles to Eisenhower, December 6, 1957, *FRUS 1955–1957*, 11:775–76. Divine, *Sputnik Challenge*, 70.

36. *Life*, February 10, 1958. Divine, *Sputnik Challenge*, 144.

37. NSC Meeting, October 10, 1957, Ann Whitman File, NSC Series, box 9, EL; Larson to Eisenhower, October 15, 1957, Ann Whitman File, Administration Series, box 37, USIA (1), EL; Larson to Strauss, October 15, 1957, Ann Whitman File, Administration Series, box 37, U.S. Satellites, EL.

38. R. V. Mrozinski to Harr, August 3, 1958, WHO, OSANSA, OCB Series, Subject Subseries, box 8, Space, Satellites, Rockets, Etc. (3), EL; OCB Memorandum, October 29, 1957, WHO, OSANSA, OCB Series, Subject Subseries, box 8, Weapons and Technological Field (2), EL; W. H. Godel to J. I. Coffey, July 27, 1959, WHO, OSANSA, OCB Series, Subject Subseries, box 8, Space, Satellites, Rockets, Etc. (6), EL.

39. OCB Memorandum, October 29, 1957, WHO, OSANSA, OCB Series, Subject Subseries, box 8, Weapons and Technological Field (2), EL. Projects to Regain the Initiative, January 24, 1958; Elmer Staats to OCB, January 23, 1958; Charles E. Johnson to OCB, January 23, 1958; Suggestions for Improving the Position of the United States in the Face of the Communist Challenge, January 2, 1958; all in WHO, OSANSA, OCB Series, Administration Subseries, Special OCB Committee (Chronological File) (4) (6), EL.

40. CIA Suggestions Relating to the International Posture of the United States, n.d. [January 1958?], OSANA, OCB Series, Administration Subseries, box 5, Special OCB Committee (Chronological File) (7), EL; CIA Suggestions; and OCB Memorandum, February 14, 1958; both in OSANA, OCB Series, Administration Subseries, box 5, Special OCB Committee (FM Dearborn) (2), EL; W. H. Godel to J. I. Coffey, July 27, 1959, WHO, OSANSA, OCB Series, Subject Subseries, box 8, Space, Satellites, Rockets, Etc. (6), EL.

41. Eisenhower to Dulles, March 21, 1958, John Foster Dulles Papers, White House Memo Series, box 6, White House, General Correspondence 1958 (5), EL.

42. The Gaither report is in *FRUS 1955–1957*, 19:638–61. On the Gaither report and Eisenhower's response to the missile gap, see Peter J. Roman, *Eisenhower and the Missile Gap* (Ithaca, N.Y.: Cornell University Press, 1995); David L. Snead, *The Gaither Committee, Eisenhower, and the Cold War* (Columbus: Ohio State University Press, 1999); McGeorge Bundy, *Danger and Survival: Choices About the Bomb in the First Fifty Years* (New York: Random House, 1988), 236–358; Wenger, *Living with Peril*, 154–66.

43. Craig Allen, *Eisenhower and the Mass Media: Peace, Prosperity, and Prime-Time TV* (Chapel Hill: University of North Carolina Press, 1993), 156–58. Radio and Television Address, November 7, 1957, and November 13, 1957, both in *Public Papers*, 1957, 789–99, 809–16.

44. MemCons, November 27, 1957, and December 5, 1957, both in *FRUS 1955–1957*, 19:697–98, 702–4. Wenger, *Living with Peril*, 160–61; Divine, *Sputnik Challenge*, 73–74.

45. MemCons, November 7, 1957, and November 22, 1957, *FRUS 1955–1957*, 19:638, 687–88; NSC Meeting, November 22, 1957, *FRUS 1955–1957*, 19:689–95.

46. Allen, *Eisenhower and the Mass Media*, 154–55; Sherry, *In the Shadow of War*, 227–28. Eisenhower's appointment of the partisan Republican Arthur Larson to the position of director of the USIA doomed the president's requested appropriations for the agency. Larson had earlier written a controversial book attacking the Democratic Party and singling out such powerful figures as Senator Lyndon Johnson. As payback, the Senate slashed the USIA budget to pre-1956 levels, a move that prompted Eisenhower to replace Larson a year later with the less controversial George Allen.

47. Radio and Television Address, November 13, 1957, *Public Papers*, 1957, 809–16.

48. Eisenhower address, January 9, 1958, *Public Papers*, 1958, 2–15. Emphasis added.

49. Lodge to Eisenhower, November 18, 1957, Ann Whitman File, Administration Series, box 24, Henry Cabot Lodge 1957–58 (2), EL; Dulles to Eisenhower, November 25, 1957, *FRUS 1955–1957*, 19:696–97.

50. On Goebbels's "total war" campaign, see Ralf Georg Reuth, *Goebbels*, trans. Krishna Winston (New York: Harcourt Brace, 1993), 293–330; and Philip M. Taylor, *Munitions of the Mind: War Propaganda from the Ancient World to the Nuclear Age* (Glasgow: Patrick Stephens, 1990), 218–19. For the text of Goebbels's speech, see http://www.calvin.edu/academic/cas/gpa/goeb36.htm.

51. McDougall, *The Heavens and the Earth*, 195; Divine, *Sputnik Challenge*, 104–5, 111–12, 145–54. NASA replaced the National Aeronautics and Space Council.

52. NSC 5814/1, August 18, 1958, WHO, OSANSA, NSC Series, Policy Papers Subseries, box 25, EL; OCB, Report on Preliminary U.S. Policy on Outer Space (NSC 5814), September 16, 1959, Freedom of Information Act.

53. In line with NSC 5814/1, the USIA diverted $3.5 million previously allocated for a disarmament information campaign to a new effort to promote American scientific prowess. The agency developed special programs for promoting American scientific and technological accomplishments, and it worked closely with NASA on the propaganda aspects of space exploration. See Abbott Washburn to Eisenhower, November 11, 1957, John Foster Dulles Papers, White House Memoranda Series, box 5, Meetings with the President, 1957 (1), EL; Washburn to Keith Glennan, May 10, 1960, WHO, OSANSA, OCB Series, Subject Subseries, box 7, Space Council (16), EL. USIA, Basic Guidance and Planning Paper No. 4: Science and Technology, November 18, 1958; and USIA Program in Science and Technology, March 18, 1958, both in USIA Historical Collection, Washington, D.C.

54. NSC Meeting, January 12, 1960, AWF, NSC Series, box 12, EL.

55. Public Information on U.S. Space Activities, November 19, 1958, WHO, OSANSA, OCB Series, Subject Subseries, box 7, Space Council (2), EL; OCB, Public Information Policy on U.S. Space Activities, December 16, 1958, RG 59, Records Relating to State Department Participation in the OCB and NSC, box 43, Public Information Guidelines Re Nuclear and Outer Space Activities, National Archives, College Park, Maryland. Allen, *Eisenhower and the Mass Media*, 162–63; Divine, *Sputnik Challenge*, 186, 191, 204.

56. Divine, *Sputnik Challenge*, 109–110.

57. Divine, *Sputnik Challenge*, 155, 188–89; McDougall, *The Heavens and the Earth*, 202.

58. Message Relayed from the Atlas Satellite, December 19, 1958, *Public Papers*, 1958, 865. Divine, *Sputnik Challenge*, 191, 205; McDougall, *The Heavens and the Earth*, 190.

59. Sherry, *In the Shadow of War*, 221; Willard Bascom, *A Hole in the Bottom of the Sea: The Story of the Mohole Project* (Garden City, N.Y.: Doubleday, 1961); Jacob Darwin Hamblin, "Oceanography and International Cooperation During the Early Cold War" (Ph.D. diss., University of California, Santa Barbara, 2001), 259–65.

60. Report by the National Aeronautics and Space Council, January 26, 1960, *FRUS 1958–1960*, 2:921–35.

61. McDougall, *The Heavens and the Earth*, 155, 199; Sherry, *In the Shadow of War*, 226–27.

Conclusion

1. On the cultural agreement, see Walter L. Hixson, *Parting the Curtain: Propaganda, Culture, and the Cold War, 1945–1961* (New York: St. Martin's Griffin, 1997), 151–61.

2. Don Irwin to Nelson Rockefeller, April 12, 1955, Nelson A. Rockefeller Papers, Record Group [RG] 4, Washington, D.C., Files, Subseries 7, box 4, "Four-

Power Conference—Pre-Conference (2), Rockefeller Archive Center, Sleepy Hollow, New York; David Sarnoff, "Program for a Political Offensive Against World Communism," White House Office [WHO], Office of the Special Assistant for National Security Affairs [OSANSA], Special Assistant Series, Name Subseries, box 4, S-General, Eisenhower Library, Abilene, Kansas (hereafter EL); "Right Now the Reds Are Winning," *Washington Daily News*, December 28, 1956; "Public Evenly Divided on Who's Ahead in Cold War," *Public Opinion News Service*, January 20, 1956. A Gallup Poll in December 1956 revealed that 28 percent of Americans believed Russia was winning the Cold War, 37 percent believed the United States was winning, and 33 percent had no opinion. See Gallup Poll 557, available at http://brain.gallup.com.

3. Hixson, *Parting the Curtain*, 161–83.

4. Khrushchev, in the last volume of "memoirs," repeatedly returns to the theme of how Stalin's foreign policy had isolated the USSR in international affairs. See Nikita S. Khrushchev, *Khrushchev Remembers: The Glasnost Tapes*, ed. and trans. Jerrold L. Schecter with Vyacheslav V. Luchkov (Boston: Little, Brown, 1990), 68–160. On Khrushchev's relentless efforts to secure an invitation to visit the United States, see William Taubman, *Khrushchev: The Man and His Era* (New York: W. W. Norton, 2003), 396–424. According to Taubman, Eisenhower instructed Undersecretary of State Robert Murphy to issue a qualified invitation to the United States, contingent upon satisfactory progress of ongoing negotiations in Geneva. Murphy mistakenly issued the invitation without qualification. Campbell Craig argues that this snafu was deliberately engineered by Eisenhower because he, like Khrushchev, secretly desired a summit meeting. See Craig, *Destroying the Village: Eisenhower and Thermonuclear War* (New York: Columbia University Press, 1998), 105.

5. Taubman, *Khrushchev*, 424–41; Stephen Ambrose, *Eisenhower: The President* (New York: Simon and Schuster, 1984), 541–45; M. Kharlamov, ed., *Face to Face with America: The Story of N. S. Khrushchev's Visit to the USA, September 15–27, 1959* (Moscow: Foreign Language Publishing House, 1960). For Khrushchev's recollections of the trip, see Nikita S. Khrushchev, *Khrushchev Remembers: The Last Testament*, ed. and trans. Strobe Talbott (Boston: Little, Brown, 1974), 368–416. For Eisenhower's, see Dwight D. Eisenhower, *Waging Peace, 1956–1961: The White House Years* (Garden City, N.Y.: Doubleday, 1965), 432–49.

6. Press conference, August 12, 1959, *Public Papers of the Presidents of the United States: Dwight D. Eisenhower, 1959* (Washington, D.C.: Government Printing Office, 1961), 574 (hereafter *Public Papers*); Ambrose, *Eisenhower*, 538–41, 551–53; Eisenhower, *Waging Peace*, 485–513.

7. Ambrose, *Eisenhower*, 534–37. "Air Force One" is a radio call sign rather than the name of a specific aircraft. It came into use early in the Eisenhower administration to indicate that the president was on board a particular flight. The need arose in 1953 when a mix-up confused Eisenhower's plane with a commercial one using

the same call sign. For most of Eisenhower's presidency, he flew a four-engine pro-peller plane known popularly as the *Columbine*, so named by Mamie Eisenhower after the state flower of Colorado. In late 1958, the Air Force added two jet aircraft to the presidential fleet.

8. Author's interview with Karl Harr. The president related in his memoirs the PR objectives that lay behind his mission. He "wanted to try to raise the morale or struggling and underprivileged peoples, to enhance confidence in the value of friendship with the United States, and to give them assurance of their own security and chances for progress." See Eisenhower, *Waging Peace*, 486.

9. Press conference, December 2, 1959, *Public Papers*, 1959, 786.

10. Radio and Television address, December 3, 1959, *Public Papers*, 1959, 795–99. Eisenhower's public statements on his travels are documented in *Public Papers*, 1959, 799–883, and *Public Papers*, 1960–61, 202–82.

11. Ambrose, *Eisenhower*, 552–53; USIS-Rabat to USIA, May 2, 1960, RG 306, Foreign Service Despatches, box 3, Africa, National Archives, College Park, Mary-land; NSC 6013, Status of the USIA Program, June 30, 1960, WHO, OSANSA, NSC Series, Status of Projects Subseries, box 9, NSC 6013 (4), EL.

12. Michael Schaller, *Altered States: The United States and Japan Since the Occupa-tion* (New York: Oxford University Press, 1997), 143–62; Taubman, *Khrushchev*, 442–68.

13. Jeremi Suri, *Power and Protest: Global Revolution and the Rise of Détente* (Cam-bridge: Harvard University Press, 2003).

14. Sprague Committee Report, "New Dimensions of Diplomacy," Ann Whit-man File, Administration Series, box 33, Sprague Committee (2), EL.

15. James H. Doolittle, Report by the Special Study Group on Covert Activi-ties, September 30, 1954, quoted in Christopher Andrew, *For the President's Eyes Only: Secret Intelligence and the American Presidency from Washington to Bush* (New York: Harper Collins, 1996), 211. Eisenhower quoted on page 202.

16. DOS Memorandum on Urgent Need to Mobilize United States Private Industry for Pro-Democratic Information Program Throughout the Free World, n.d., Freedom of Information Act. Although this document is undated, it was prob-ably authored in 1948 or 1949.

17. Radio Address, May 19, 1953, *Public Papers*, 1961, 306–16; Inaugural ad-dress, January 20, 1953, *Public Papers*, 1953, 1–8; Farewell Address, January 17, 1961, *Public Papers*, 1961, 1035–40.

BIBLIOGRAPHY

Archival Sources

Dwight D. Eisenhower Library. Abilene, Kansas.

Couch, Virgil. Papers.
Dulles, John Foster. Papers.
Eisenhower, Dwight D. Papers as President.
Eisenhower, Dwight D. Pre-Presidential Papers.
Flemming, Arthur. Papers.
Gruenther, Alfred M. Papers.
Hazeltine, Charles B. Papers.
Herter, Christian. Papers.
Howard, Katherine. Papers.
Jackson, C. D. Papers.
Jackson, C. D. Records.
Lambie, James M., Jr. Papers.
Larson, Arthur. Papers.
McCardle, Carl. Papers.
National Security Council Staff. Papers.
Office of the Special Assistant for National Security Affairs. Records.
Operations Coordinating Board Central Files.
Planning Coordination Group. Records.
President's Committee on Information Activities Abroad (Jackson Committee). Records.
President's Committee on Information Activities Abroad (Sprague Committee). Records.
Psychological Strategy Board Central Files.
Schooley, Herschel. Papers.
White House Central Files.
Whitman, Ann. File.

National Archives. College Park, Maryland.

Record Group 59—Department of State.
Record Group 218—Joint Chiefs of Staff.
Record Group 263—Central Intelligence Agency.
Record Group 273—National Security Council.
Record Group 306—United States Information Agency.

Record Group 319—U.S. Army.
Record Group 326—Atomic Energy Commission.
Record Group 330—Department of Defense.

National Records Center. Suitland, Maryland.
Record Group 306—United States Information Agency.

National Security Archive. Washington, D.C.
Soviet Flashpoints Collection.

Public Record Office. Kew, United Kingdom.
Cabinet Office. Records.
Foreign Office. Records.
Prime Minister's Office. Records.
United Kingdom Atomic Energy Authority. Records.

Rockefeller Archive Center. Sleepy Hollow, New York.
Rockefeller, Nelson A. Papers.

Seeley G. Mudd Library. Princeton University, Princeton, New Jersey.
Dulles, Allen. Papers.
Dulles, John Foster. Papers.

USIA Archive. Washington, D.C.
Subject Files.
Pamphlet Files.

Author's Collection.
Halsema, James. Diary.
Harr, Karl. Papers.

Billy Graham Center. Wheaton, Illinois.
Records of the Fellowship Foundation.

Periodicals

Asheville Citizen
Baltimore News-Post
Bayonne Times
Beaver Falls News-Tribune
Bee Hive
Bridgeport Post
Bulletin of Atomic Scientists
Charleston Post

Chicago Tribune
Current Events
Evening Star
Foreign Affairs
Guerilla News Network
Life
Look Magazine
Los Angeles Times
Marlboro Enterprise
Memphis Press-Scimitar
The Nation
National Geographic
New Republic
New Times
New York Herald Tribune
New York Long Island Press
New York Times
New York Times Magazine
New York World-Telegram and Sun
Newsweek
Pravda
Propaganda Analysis
Reporter
San Bernadino Telegram
Scientific American
Star
Time
Times of Havana
Torch
U.S. Department of State Bulletin
U.S. News and World Report
Washington Post

Oral Histories and Interviews

Allen, George. Oral History, Dwight D. Eisenhower Library.
Beecham, Charles Robert. USIA Oral History Project.
Berding, Andrew. John Foster Dulles Oral History Project, Princeton University.
Burleson, Hugh. USIA Oral History Project.
Goodpaster, Andrew J. Author's Interview.
Gosho, Henry. USIA Oral History Project.
Halsema, Alice. Author's Interview.

Halsema, James. Author's Interview.

Harr, Karl G., Jr. Author's Interview.

Harr, Karl G., Jr. Oral History, Dwight D. Eisenhower Library.

Hummel, Arthur W., Jr. USIA Oral History Project.

Hutchison, John N. USIA Oral History Project.

Lambie, James. Oral History, Dwight D. Eisenhower Library.

Larmon, Sigurd. Oral History, Dwight D. Eisenhower Library.

Larson, Arthur. John Foster Dulles Oral History Project, Princeton University.

Loomis, Henry. USIA Oral History Project.

McCardle, Carl W. Oral History, Dwight D. Eisenhower Library.

McCardle, Carl W. John Foster Dulles Oral History Project, Princeton University.

Nichols, Robert L. USIA Oral History Project.

O'Brien, John R. USIA Oral History Project.

Picon, Leon. USIA Oral History Project.

Schmidt, G. Lewis. Author's Interview.

Schmidt, Kyoko Edayoshi. Author's Interview.

Staats, Elmer B. Oral History, Dwight D. Eisenhower Library.

Staats, Elmer B. Author's Interview.

Stassen, Harold E. John Foster Dulles Oral History Project, Princeton University.

Streibert, Theodore. Oral History, Dwight D. Eisenhower Library.

Streibert, Theodore. John Foster Dulles Oral History Project, Princeton University.

Walsh, Joe. USIA Oral History Project.

Washburn, Abbott. Author's Interview.

Wilson, Earl J. Foreign Affairs Oral History Program, Georgetown University.

Published Sources

Adams, Sherman. *First-Hand Report: The Story of the Eisenhower Administration.* New York: Harper & Brothers, 1961.

Aguilar, Manuela. *Cultural Diplomacy and Foreign Policy: German-American Relations, 1955–1968.* New York: Peter Lang, 1996.

Aldrich, Richard J., Gary D. Rawnsley, and Ming-Yeh T. Rawnsley, eds. *The Clandestine Cold War in Asia: Western Intelligence, Propaganda and Special Operations.* London: Frank Cass, 2000.

Allen, Craig. *Eisenhower and the Mass Media: Peace, Prosperity, and Prime-Time TV.* Chapel Hill: University of North Carolina Press, 1993.

Alston, Giles "Eisenhower: Leadership in Space Policy." In *Reexamining the Eisenhower Presidency,* edited by Shirley Anne Warshaw, 103–119. Westport, Conn.: Greenwood Press, 1993.

Alteras, Isaac. *Eisenhower and Israel: U.S.-Israeli Relations, 1953–1960.* Gainesville: University Press of Florida, 1993.

Ambrose, Stephen E. *Eisenhower: The President*. New York: Simon and Schuster, 1984.
———. *Eisenhower: Soldier and President*. New York: Simon and Schuster, 1990.
Anderson, Carol. *Eyes Off the Prize: The United Nations and the African American Struggle for Human Rights, 1944–1955*. Cambridge: Cambridge University Press, 2003.
Anderson, David L. *Trapped By Success: The Eisenhower Administration and Vietnam, 1953–1961*. New York: Columbia University Press, 1991.
Anderson, Philip. "'Summer Madness': The Crisis in Syria, August–October 1957." *British Journal of Middle Eastern Studies* 22 (1995): 21–42.
Andrew, Christopher. *For the President's Eyes Only: Secret Intelligence and the American Presidency from Washington to Bush*. New York: Harper Collins, 1996.
Appleby, Charles Albert, Jr. "Eisenhower and Arms Control, 1953–1961: A Balance of Risks." Ph.D. diss., Johns Hopkins University, 1987.
Barnes, Trevor. "The Secret Cold War: The CIA and American Foreign Policy in Europe, 1946–1956. Part I." *Historical Journal* 24 (June 1981): 399–415.
———. "The Secret Cold War: The CIA and American Foreign Policy in Europe, 1946–1956. Part II." *Historical Journal* 25 (September 1982): 649–70.
Barrett, Edward W. *Truth Is Our Weapon*. New York: Funk and Wagnalls, 1953.
Barson, Michael, and Steven Heller. *Red Scared! The Commie Menace in Propaganda and Popular Culture*. San Francisco: Chronicle Books, 2001.
Bascom, Willard. *A Hole in the Bottom of the Sea: The Story of the Mohole Project*. Garden City, N.Y.: Doubleday, 1961.
Belmonte, Laura A. "Almost Everyone Is a Capitalist: The USIA Presents the American Economy, 1953–1959." Paper presented to the Society for Historians of American Foreign Relations, June 1994.
———. "Defending a Way of Life: American Propaganda and the Cold War, 1945–1959." Ph.D. diss., University of Virginia, 1996.
Berghahn, Volker R. *America and the Intellectual Cold Wars in Europe*. Princeton: Princeton University Press, 2001.
Bernays, Edward L. *Biography of an Idea: Memoirs of a Public Relations Counsel*. New York: Simon and Schuster, 1965.
———. *Crystallizing Public Opinion*. New York: Liveright Publishing, 1923.
———. *Propaganda*. New York: Liveright Publishing, 1928.
Bernhard, Nancy E. "Clearer than Truth: Public Affairs Television and the State Department's Domestic Information Campaigns, 1947–1952." *Diplomatic History* 21 (Fall 1997): 545–68.
———. *U.S. Television News and Cold War Propaganda, 1947–1960*. Cambridge: Cambridge University Press, 1999.
Beschloss, Michael R. *Mayday: Eisenhower, Khrushchev and the U-2 Affair*. New York: Harper & Row, 1986.
Bischof, Günter, and Stephen E. Ambrose. *Eisenhower: A Centenary Assessment*. Baton Rouge: Louisiana University Press, 1995.

Bischof, Günter, and Saki Dockrill, eds. *Cold War Respite: The Geneva Summit of 1955*. Baton Rouge: Louisiana University Press, 2000.

Bischof, Günter, Anton Pelinka, and Dieter Stiefel, eds. *The Marshall Plan in Austria*. New Brunswick, N.J.: Transaction Publishers, 2000.

Bissell, Richard M., Jr. *Reflections of a Cold Warrior: From Yalta to the Bay of Pigs*. New Haven, Conn.: Yale University Press, 1996.

Blum, John Morton. *V Was for Victory: Politics and American Culture During World War II*. San Diego: Harcourt Brace Jovanovich, 1976.

Bogart, Leo. *Cool Words, Cold War: A New Look at USIA's Premises for Propaganda*. Rev. ed. Washington, D.C.: American University Press, 1995.

———. "A Study of the Operating Assumptions of the U.S. Information Agency." *Public Opinion Quarterly* 19 (Winter 1955–1956): 369–79.

Bohlen, Charles E. *Witness to History, 1929–1969*. New York: W. W. Norton, 1973.

Borhi, László. "Rollback, Containment, or Inaction? U.S. Policy and Eastern Europe in the 1950s." *Journal of Cold War Studies* 1 (Fall 1999): 67–110.

Borstelmann, Thomas. *The Cold War and the Color Line: American Race Relations in the Global Arena*. Cambridge: Harvard University Press, 2001.

Bowie, Robert R., and Richard H. Immerman. *Waging Peace: How Eisenhower Shaped an Enduring Cold War Strategy*. New York: Oxford University Press, 1998.

Boyer, Paul. *By the Bomb's Early Light: American Thought and Culture at the Dawn of the Atomic Age*. Chapel Hill: University of North Carolina Press, 1985.

———. "Exotic Resonances: Hiroshima in American Memory." *Diplomatic History* 19 (Spring 1995): 297–318.

———. *Fallout: A Historian Reflects on America's Half-Century Encounter with Nuclear Weapons*. Columbus: Ohio State University Press, 1998.

Brands, H. W. "The End of Vulnerability: Eisenhower and the National Security State." *American Historical Review* 94 (October 1989): 963–89.

———. *Cold Warriors: Eisenhower's Generation and American Foreign Policy*. New York: Columbia University Press, 1988.

———. *The Specter of Neutralism: The United States and the Emergence of the Third World, 1947–1960*. New York: Columbia University Press, 1989.

———. "Testing Massive Retaliation: Credibility and Crisis Management in the Taiwan Strait." *International Security* 12 (Spring 1988): 124–51.

Breitman, Richard, Norman J. W. Goda, Timothy Naftali, and Robert Wolfe. *U.S. Intelligence and the Nazis*. New York: Cambridge University Press, 2005.

Brendon, Piers. *Ike: His Life and Times*. New York: Harper & Row, 1986.

Brockriede, Wayne, and Robert L. Scott. *Moments in the Rhetoric of the Cold War*. New York: Random House, 1970.

Brzezinski, Zbigniew. *The Choice: Global Domination or Global Leadership*. New York: Basic Books, 2004.

Budish, Jacob M. *People's Capitalism: Stock Ownership and Production*. New York: International Publishers, 1958.

Bulkeley, Rip. *The Sputniks Crisis and Early United States Space Policy: A Critique of the Historiography of Space.* Bloomington: Indiana University Press, 1991.

Bundy, McGeorge. *Danger and Survival: Choices About the Bomb in the First Fifty Years.* New York: Random House, 1988.

Burk, Robert Frederick. *The Eisenhower Administration and Black Civil Rights.* Knoxville: University of Tennessee Press, 1984.

Butcher, Harry C. *My Three Years with Eisenhower.* New York: Simon and Schuster, 1946.

Byfeld, Robert S. *The Fifth Weapon: A Guide to Understanding What the Communists Mean.* New York: Bookmailer, 1954.

Carew, Anthony. "The American Labor Movement in Fizzland: The Free Trade Union Committee and the CIA." *Labor History* 39 (1998): 25–42.

Carruthers, Susan L. "'The Manchurian Candidate' (1962) and the Cold War Brainwashing Scare." *Historical Journal of Film, Radio and Television* 18 (1998): 75–94.

Carter, Dale, and Robin Clifton, eds. *War and Cold War in American Foreign Policy, 1942–1962.* New York: Palgrave Macmillan, 2002.

Castle, Eugene. *Billions, Blunders, and Baloney: The Fantastic Story of How Uncle Sam Is Squandering Your Money Overseas.* New York: Devin-Adair, 1955.

Chernus, Ira. *Eisenhower's Atoms for Peace.* College Station: Texas A&M Press, 2002.

———. "Eisenhower's Ideology in World War II." *Armed Forces and Society* 23 (Summer 1997): 595–613.

Childs, Marquis. *Eisenhower: Captive Hero.* New York: Harcourt, Brace, 1958.

Chilton, Paul A. *Security Metaphors: Cold War Discourse from Containment to the Common House.* New York: Peter Lang, 1996.

Citino, Nathan. *From Arab Nationalism to OPEC: Eisenhower, King Saud, and the Making of U.S.-Saudi Relations.* Bloomington: Indiana University Press, 2002.

———. "Middle East Cold War(s): Oil and Arab Nationalism in U.S.-Iraqi Relations, 1958–1961." In Johns and Statler, *Managing an Earthquake.*

Clausewitz, Carl von. *On War.* Edited and translated by Michael Howard and Peter Paret. Princeton: Princeton University Press, 1976.

Coleman, Peter. *The Liberal Conspiracy: The Congress for Cultural Freedom and the Struggle for the Mind of Postwar Europe.* New York: Free Press, 1989.

Committee on Foreign Affairs Personnel. *Personnel for the New Diplomacy: Report of the Committee on Foreign Affairs Personnel.* Washington, D.C.: Carnegie Endowment for International Peace, 1962.

Connelly, Matthew. *A Diplomatic Revolution: Algeria's Fight for Independence and the Origins of the Post–Cold War Era.* Oxford: Oxford University Press, 2002.

Cook, Blanche Wiesen. *The Declassified Eisenhower: A Divided Legacy.* Garden City, N.Y.: Doubleday, 1981.

———. "First Comes the Lie: C. D. Jackson and Political Warfare." *Radical History Review* 31 (1984): 42–70.

Coombs, Philip Hall. *The Fourth Dimension of Foreign Policy*. New York: Harper & Row, 1964.

Cooper, Nicola. *France in Indochina: Colonial Encounters*. Oxford: Berg, 2001.

Copeland, Miles. *The Game of Nations*. New York: Simon and Schuster, 1969.

Corke, Sarah-Jane. "Bridging the Gap: Containment, Covert Action and the Search for the Elusive Missing Link in American Cold War Policy, 1948–1953." *Journal of Strategic Studies* 20 (December 1997): 45–65.

———. "Bridging the Gap: Containment, Psychological Warfare and the Search for the Missing Link in American Cold War Policy, 1948–1953." Paper presented to the Society for Historians of American Foreign Relations, June 1998.

———. "Flexibility or Failure: Eastern Europe and the Dilemmas of Foreign Policy Coordination, 1948–1953." Paper presented to the Society for Historians of American Foreign Relations, June 1997.

———. "History, Historians and the Naming of Foreign Policy: A Postmodern Reflection on American Strategic Thinking During the Truman Administration." *Intelligence and National Security* 16 (Fall 2001): 146–65.

———. "Operating in a Vacuum: The American Information Program in Eastern Europe, 1948–1953." Paper presented to the Canadian Association of Security and Intelligence Studies, University of Toronto, Toronto, March 1997.

Craig, Campbell. *Destroying the Village: Eisenhower and Thermonuclear War*. New York: Columbia University Press, 1998.

Creel, George. *How We Advertised America*. New York: Harper & Brothers, 1920.

———. *Rebel at Large: Recollections of Fifty Crowded Years*. New York: G. P. Putnam's Sons, 1947.

Crowley, David, and Paul Heyers, eds. *Communication History: Technology, Culture, Society*. 2nd ed. White Plains, N.Y.: Longman, 1995.

Cull, Nicholas J. *Selling America: U.S. International Propaganda and Public Diplomacy Since 1945*. New York: Cambridge University Press, forthcoming.

Cullather, Nick. *Secret History: The CIA's Classified Account of Its Operations in Guatemala, 1952–1954*. Stanford: Stanford University Press, 1999.

Currey, Cecil B. *Edward Lansdale: The Unquiet American*. Washington, D.C.: Brassey's, 1998.

Cutler, Robert. *No Time for Rest*. Boston: Little, Brown, 1963.

Cutlip, Scott M. *Public Relations History: From the 17th Century to the 20th Century*. Hillsdale, N.J.: Lawrence Erlbaum Associates, 1995.

Daugherty, William E., and Morris Janowitz, eds. *A Psychological Warfare Casebook*. Baltimore: Johns Hopkins University Press, 1958.

Davidson, Philip. *Propaganda and the American Revolution, 1763–1783*. Chapel Hill: University of North Carolina Press, 1941.

Day, Dwayne A. "Cover Stories and Hidden Agendas: Early American Space and National Security Policy." Launius et al., *Reconsidering Sputnik*, 161–95.

Day, Dwayne A., ed. *Eye in the Sky: The Story of the Corona Spy Satellites.* Washington, D.C.: Smithsonian Institution Press, 1999.

de Moraes Ruehsen, Moyara. "Operation 'Ajax' Revisited: Iran, 1953." *Middle Eastern Studies* 29 (July 1993): 467–86.

Declassified Documents Reference System. Woodbridge, Conn.: Primary Source, 1976–.

Defty, Andrew. *Britain, American and Anti-Communist Propaganda, 1945–1958:* The Information Research Department. London: Frank Cass, 2003.

Deibel, Terry L. *Culture and Information: Two Foreign Policy Functions.* Beverly Hills: Sage Publications, 1976.

Del Pero, Mario. "The Cold War as a 'Total Symbolic War': United States Psychological Warfare Plans in Italy During the 1950s." Paper presented to the Center for Cold War Studies, University of California, Santa Barbara, May 2000.

———. "The United States and 'Psychological Warfare' in Italy, 1948–1955." *Journal of American History* 87 (March 2001): 1304–34.

Dennis, Everette E., George Gerbner, and Yassen N. Zassoursky, eds. *Beyond the Cold War: Soviet and American Media Images.* Newbury Park, Calif.: Sage, 1991.

Divine, Robert A. *Blowing on the Wind: The Nuclear Test Ban Debate, 1954–1960.* New York: Oxford University Press, 1978.

———. *Eisenhower and the Cold War.* New York: Oxford University Press, 1981.

———. *The Sputnik Challenge: Eisenhower's Response to the Soviet Satellite.* New York: Oxford University Press, 1993.

Dizard, Wilson P. *The Strategy of Truth: The Story of the U.S. Information Service.* Washington, D.C.: Public Affairs Press, 1961.

Dockrill, Saki. *Eisenhower's New-Look National Security Policy, 1953–61.* New York: St. Martin's Press, 1996.

Dower, John W. *Embracing Defeat: Japan in the Wake of World War II.* New York: W. W. Norton, 1999.

———. *War Without Mercy: Race and Power in the Pacific War.* New York: Pantheon Books, 1986.

Dudziak, Mary L. *Cold War Civil Rights: Race and the Image of American Democracy.* Princeton: Princeton University Press, 2000.

———. "Desegregation as a Cold War Imperative." *Stanford Law Review* 41 (November 1988): 61–120.

Eisenhower, Dwight D. *The Papers of Dwight D. Eisenhower,* edited by Alfred D. Chandler and Louis Galambos. 17 vols. Baltimore: Johns Hopkins University Press, 1970–1996.

———. *Public Papers of the Presidents of the United States: Dwight D. Eisenhower, 1953–1961.* 8 vols. Washington, D.C.: Government Printing Office, 1960–1961.

———. *Mandate for Change, 1953–1956: The White House Years.* Garden City, N.Y.: Doubleday, 1963.

————. *Waging Peace, 1956–1961: The White House Years.* Garden City, N.Y.: Doubleday, 1965.

Elder, Robert Ellsworth. *The Information Machine: The United States Information Agency and American Foreign Policy.* Syracuse: Syracuse University Press, 1967.

Eldridge, David N. "'Dear Owen': The CIA, Luigi Luraschi and Hollywood, 1953." *Historical Journal of Film, Radio and Television* 20 (2000): 149–96.

Elliston, Jon, ed. *Psywar on Cuba: The Declassified History of U.S. Anti-Castro Propaganda.* Melbourne: Ocean Press, 1999.

Ellul, Jacques. *Propaganda: The Formation of Men's Attitudes.* New York: Vintage Books, 1973.

Emery, Edwin, Phillip H. Ault, and Warren K. Agee. *Introduction to Mass Communications.* 4th ed. New York: Dodd, Mead, 1974.

Ermarth, Michael, ed. *America and the Shaping of German Society, 1945–1955.* Providence, R.I.: Berg, 1993.

Escalante, Fabian. *The Secret War: CIA Covert Operations Against Cuba, 1959–1962.* Translated by Maxine Shaw. New York: Talman Company for Ocean, 1995.

Eveland, Wilbur Crane. *Ropes of Sand: America's Failure in the Middle East.* London: W. W. Norton, 1980.

Ewen, Stuart. *PR! A Social History of Spin.* New York: Basic Books, 1996.

Ferguson, Niall. *Empire: The Rise and Demise of the British World Order and the Lessons for Global Power.* New York: Basic Books, 2004.

Ferrell, Robert H., ed. *The Diary of James C. Hagerty: Eisenhower in Mid-Course, 1954–1955.* Bloomington: Indiana University Press, 1983.

————. *The Eisenhower Diaries.* New York: W. W. Norton, 1981.

Fineman, Daniel. *A Special Relationship: The United States and Military Government in Thailand, 1947–1958.* Honolulu: University of Hawaii Press, 1997.

Fish, M. Steven. "After Stalin's Death: The Ango-American Debate Over a New Cold War." *Diplomatic History* 10 (Fall 1986): 333–55.

Frankel, Charles. *The Neglected Aspect of Foreign Affairs: American Educational and Cultural Policy Abroad.* Washington, D.C.: Brookings Institution, 1966.

Fraser, Cary. "Crossing the Color Line in Little Rock: The Eisenhower Administration and the Dilemma of Race for U.S. Foreign Policy." *Diplomatic History* 24 (Spring 2000): 233–64.

Freiberger, Steven Z. *Dawn Over Suez: The Rise of American Power in the Middle East, 1953–1957.* Chicago: Ivan R. Dee, 1992.

Frey, Marc. "Tools of Empire: Persuasion and the United States's Modernizing Mission in Southeast Asia." *Diplomatic History* 27 (September 2003): 543–68.

Frezza, Daria. "Psychological Warfare and the Building of National Morale During World War II: The Role of Non-Government Agencies." In *Aspects of War in American History,* edited by David Adams and Cornelius V. van Minnen, 173–96. Keele, England: Keele University Press, 1997.

Fried, Richard M. *The Russians Are Coming! The Russians Are Coming! Pageantry and Patriotism in Cold-War America.* New York: Oxford University Press, 1998.

Fujita, Fumiko. "U.S. Cultural Diplomacy Toward Japan in the 1950s." Paper presented to New York University, March 2003.

Fursenko, Aleksandr, and Timothy Naftali. *"One Hell of a Gamble": Khrushchev, Castro, and Kennedy, 1958–1964.* New York: W. W. Norton, 1997.

Gaddis, John Lewis. *The Long Peace: Inquiries into the History of the Cold War.* New York: Oxford University Press, 1987.

———. *Strategies of Containment: A Critical Appraisal of Postwar American National Security Policy.* Oxford: Oxford University Press, 1982.

———. *The United States and the End of the Cold War: Implications, Reconsiderations, Provocations.* New York: Oxford University Press, 1992.

———. *We Now Know: Rethinking Cold War History.* Oxford: Clarendon Press, 1997.

Gardner, Lloyd. "Poisoned Apples: American Responses to the Russian 'Peace Offensive.'" In *The Cold War After Stalin's Death: A New International History,* edited by Kenneth A. Osgood and Klaus Larres. Lanham, Md.: Rowman & Littlefield, forthcoming.

Garthoff, Raymond L. *Assessing the Adversary: Estimates by the Eisenhower Administration of Soviet Intentions and Capabilities.* Washington, D.C.: Brookings Institution, 1991.

Gary, Brett. *The Nervous Liberals: Propaganda Anxieties from World War I to the Cold War.* New York: Columbia University Press, 1999.

Gasiorowski, Mark J. "The 1953 Coup d'Etat in Iran." *International Journal of Middle East Studies* 19 (August 1987): 261–86.

George, Alexander L. "Psychological Dimensions of the U.S.-Soviet Conflict." In *Reexamining the Soviet Experience: Essays in Honor of Alexander Dallin,* edited by David Holloway and Norman Naimark, 101–18. Boulder, Colo.: Westview Press, 1996.

Gienow-Hecht, Jessica C. E. *Transmission Impossible: American Journalism as Cultural Diplomacy in Postwar Germany, 1945–1955.* Baton Rouge: Louisiana State University Press, 1999.

———. "Shame on U.S.? Academics, Cultural Transfer, and the Cold War: A Critical Review." *Diplomatic History* 24 (Summer 2000): 465–94.

Gienow-Hecht, Jessica C. E., and Frank Schumacher, eds. *Culture and International History.* New York: Berghahn Books, 2003.

Gilmore, Allison B. *You Can't Fight Tanks with Bayonets: Psychological Warfare Against the Japanese Army in the Southwest Pacific.* Lincoln: University of Nebraska Press, 1998.

Gleijeses, Piero. *Shattered Hope: The Guatemalan Revolution and the United States, 1944–1954.* Princeton: Princeton University Press, 1991.

Gorman, Lyn, and David McLean. *Media and Society in the Twentieth Century: A Historical Introduction.* Malden, Mass.: Blackwell, 2003.

Gorst, Anthony, and W. Scott Lucas. "The Other Collusion: Operation Straggle and Anglo-American Intervention in Syria, 1955–56." *Intelligence and National Security* 4 (July 1988): 576–95.

Gould-Davies, Nigel. "Rethinking the Role of Ideology in International Politics During the Cold War." *Journal of Cold War Studies* 1 (Winter 1999): 90–109.

Greb, G. Allen, and Gerald W. Johnson. "The Test Ban Treaty Reconsidered: Arms Control Policy in the Eisenhower Administration." Paper presented to the Center for Cold War Studies, University of California, Santa Barbara, 1998.

Green, Constance McLaughlin, and Milton Lomask. *Vanguard: A History.* Washington, D.C.: National Aeronautics and Space Administration, 1970.

Green, Fitzhugh. *American Propaganda Abroad.* New York: Hippocrene Books, 1988.

Greene, John Robert. "Bibliographic Essay: Eisenhower Revisionism, 1952–1992, a Reappraisal." In *Reexamining the Eisenhower Presidency,* edited by Shirley Anne Warshaw, 209–19. Westport, Conn.: Greenwood Press, 1993.

Greenstein, Fred I. *The Hidden-Hand Presidency: Eisenhower as Leader.* 2nd ed. Baltimore: Johns Hopkins University Press, 1994.

Griffith, Robert. "The Selling of America: The Advertising Council and American Politics, 1942–1960." *Business History Review* 57 (Autumn 1983): 388–412.

Grose, Peter. *Gentleman Spy: The Life of Allen Dulles.* Boston: Houghton Mifflin, 1994.

———. *Operation Rollback: America's Secret War Behind the Iron Curtain.* Boston: Houghton Mifflin, 2000.

Haefele, Mark. "John F. Kennedy, USIA, and World Public Opinion." *Diplomatic History* 25 (Winter 2001): 63–84.

Haddow, Robert H. *Pavilions of Plenty: Exhibiting American Culture Abroad in the 1950s.* Washington, D.C.: Smithsonian Institution Press, 1997.

Hahn, Peter L. *Caught in the Middle East: U.S. Policy toward the Arab-Israeli Conflict, 1945–1961.* Chapel Hill: University of North Carolina Press, 2004.

———. *The United States, Great Britain, and Egypt, 1945–1956.* Chapel Hill, University of North Carolina Press, 1991.

Halberstam, David. *The Fifties.* New York: Fawcett Columbine, 1993.

Hall, R. Cargill. "Origins of U.S. Space Policy: Eisenhower, Open Skies, and Freedom of Space." In *Exploring the Unknown: Selected Documents in the History of the U.S. Civil Space Program,* edited by John M. Logsdon, 213–29. Washington, D.C.: National Aeronautics and Space Administration, 1995.

Hamblin, Jacob Darwin. "Oceanography and International Cooperation During the Early Cold War." Ph.D. diss., University of California, Santa Barbara, 2001.

Hamilton, Richard F. *Mass Society, Pluralism, and Bureaucracy: Explication, Assessment, and Commentary.* Westport, Conn.: Praeger, 2001.

Harr, Karl G., Jr., "Eisenhower's Approach to National Security Decisionmaking." In *The Eisenhower Presidency,* edited by Kenneth W. Thompson, 89–111. Lanham, Md.: University Press of America, 1984.

Hartley, John. *Communications, Cultural and Media Studies: The Key Concepts.* London: Routledge, 2002.

Hasegawa, Tsuyoshi. *The Northern Territories Dispute and Russo-Japanese Relations: Between War and Peace, 1697–1985.* Berkeley: University of California Press, 1998.

Henderson, John W. *The United States Information Agency.* New York: Praeger, 1969.

Henriksen, Margot A. *Dr. Strangelove's America: Society and Culture in the Atomic Age.* Berkeley: University of California Press, 1997.

Herring, George C., ed. *The Pentagon Papers: Abridged Edition.* New York: McGraw-Hill, 1993.

Hewlett, Richard G., and Jack M. Holl. *Atoms for Peace and War, 1953–1961: Eisenhower and the Atomic Energy Commission.* Berkeley: University of California Press, 1989.

Hiebert, Ray Eldon. *Courtier to the Crowd: The Story of Ivy Lee and the Development of Public Relations.* Ames: Iowa State University Press, 1966.

Hitchcock, David I. *U.S. Public Diplomacy.* Washington, D.C.: Center for Strategic and International Studies, 1988.

Hixson, Walter L. *Parting the Curtain: Propaganda, Culture, and the Cold War, 1945–1961.* New York: St. Martin's Griffin, 1997.

Hoffman, Elizabeth Cobbs. *All You Need Is Love: The Peace Corps and the Spirit of the 1960s.* Cambridge: Harvard University Press, 1998.

Hogan, J. Michael. "Eisenhower and Open Skies: A Case Study in Psychological Warfare." In *Eisenhower's War of Words: Rhetoric and Leadership,* edited by Martin J. Medhurst, 137–56. East Lansing: Michigan State University Press, 1994.

Hogan, Michael J. *A Cross of Iron: Harry S. Truman and the Origins of the National Security State, 1945–1954.* Cambridge: Cambridge University Press, 1998.

Hogan, Michael J., and Thomas G. Paterson, eds. *Explaining the History of American Foreign Relations.* 2nd ed. New York: Cambridge University Press, 2004.

Holloway, David. *Stalin and the Bomb: The Soviet Union and Atomic Energy, 1939–1956.* New Haven, Conn.: Yale University Press, 1994.

Holt, Robert T., and Robert W. van de Velde. *Strategic Psychological Operations and American Foreign Policy.* Chicago: University of Chicago Press, 1960.

Hughes, Emmet John. *The Ordeal of Power: A Political Memoir of the Eisenhower Years.* New York: Antheneum, 1963.

Hunt, E. Howard. *Undercover: Memoirs of an American Secret Agent.* New York: Berkeley Publishing, 1974.

Hunt, Michael H. *Ideology and U.S. Foreign Policy.* New Haven, Conn.: Yale University Press, 1987.

Immerman, Richard H. *The CIA in Guatemala: The Foreign Policy of Intervention.* Austin: University of Texas Press, 1982.

———. "Confessions of an Eisenhower Revisionist: An Agonizing Reappraisal." *Diplomatic History* 14 (Summer 1990): 319–42.

Immerman, Richard H., ed. *John Foster Dulles and the Diplomacy of the Cold War.* Princeton: Princeton University Press, 1990.

Ingimundarson, Valur. "Containing the Offensive: The 'Chief of the Cold War' and the Eisenhower Administration's German Policy." *Presidential Studies Quarterly* 27 (Summer 1997): 480–96.

———. "The Eisenhower Administration, the Adenauer Government, and the Political Uses of the East German Uprising in 1953." *Diplomatic History* 20 (Summer 1996): 381–410.

Ivie, Robert L. "Dwight D. Eisenhower's 'Chance for Peace': Quest or Crusade?" *Rhetoric and Public Affairs* 1 (Summer 1998): 227–44.

———. "Eisenhower as Cold Warrior." In *Eisenhower's War of Words: Rhetoric and Leadership*, edited by Martin J. Medhurst, 7–25. East Lansing: Michigan State University Press, 1994.

Jackall, Robert, and Janice M. Hirota. *Image Makers: Advertising, Public Relations, and the Ethos of Advocacy.* Chicago: University of Chicago Press, 2000.

Jacobs, Travis Beal. *Eisenhower at Columbia.* New Brunswick, N.J.: Transaction Publishers, 2001.

Jacobson, Mark R. "'Minds Then Hearts': U.S. Political and Psychological Warfare During the Korean War." Ph.D. diss., Ohio State University, 2005.

Jervis, Robert. *The Logic of Images in International Relations.* Princeton: Princeton University Press, 1970.

———. *Perception and Misperception in International Politics.* Princeton: Princeton University Press, 1976.

Jervis, Robert, Richard Ned Lebow, and Janice Gross Stein. *Psychology and Deterrence.* Baltimore: Johns Hopkins University Press, 1985.

Johns, Andrew L., and Kathryn C. Statler, eds. *Managing an Earthquake: The Eisenhower Administration, the Third World, and the Globalization of the Cold War.* Lanham, MD: Rowman and Littlefield, forthcoming.

Joseph, Franz M. *As Others See Us: The United States Through Foreign Eyes.* Princeton: Princeton University Press, 1959.

Joyce, Walter. *The Propaganda Gap.* New York: Harper & Row, 1963.

Juergensmeyer, John E. *The President, the Foundations, and the People-to-People Program.* Intra-University Case Program No. 84. Indianapolis: Bobbs-Merrill, 1965.

———. "A Short History of the People to People Program." Ph.D. diss., Princeton University, 1958.

Kahin, Audrey R., and George McT. Kahin. *Subversion as Foreign Policy: The Secret Eisenhower and Dulles Debacle in Indonesia.* Seattle: University of Washington Press, 1995.

Kennan, George F. *Memoirs: 1925–1950.* Boston: Little, Brown, 1967.

———. *Memoirs: 1950–1963.* Boston: Little, Brown, 1972.

Kerr, Malcolm H. *The Arab Cold War: Gamal 'abd Al-Nasir and His Rivals.* London: Oxford University Press, 1971.

Kharlamov, M., ed. *Face to Face with America: The Story of N. S. Khrushchev's Visit to the USA, September 15–27, 1959.* Moscow: Foreign Language Publishing House, 1960.

Khrushchev, Nikita S. *Khrushchev Remembers: The Glasnost Tapes.* Edited and translated by Jerrold L. Schecter with Vyacheslav V. Luchkov. Boston: Little, Brown, 1990.

———. *Khrushchev Remembers: The Last Testament.* Edited and translated by Strobe Talbott. Boston: Little, Brown, 1974.

Killian, James. *Sputnik, Scientists, and Eisenhower.* Cambridge: MIT Press, 1977.

Kinzer, Stephen. *All the Shah's Men: An American Coup and the Roots of Middle East Terror.* Hoboken, N.J.: John Wiley & Sons, 2003.

Kistiakowsky, George B. *A Scientist at the White House: The Private Diary of President Eisenhower's Special Assistant for Science and Technology.* Cambridge: Harvard University Press, 1971.

Koppes, Clayton R., and Gregory D. Black. *Hollywood Goes to War: How Politics, Profits, and Propaganda Shaped World War II Movies.* Berkeley: University of California Press, 1987.

Kornher, Kenneth L. "The Psychological Instrument of U.S. Foreign Policy: USIA 1953–1963." Ph.D. diss., Georgetown University, 1969.

Kramer, Mark. "The Early Post-Stalin Succession Struggle and Upheavals in East-Central Europe: Internal-External Linkages in Soviet Policy Making (Part 1)." *Journal of Cold War Studies* 1 (Winter 1999): 3–55.

———. "The Early Post-Stalin Succession Struggle and Upheavals in East-Central Europe: Internal-External Linkages in Soviet Policy Making (Part 2)." *Journal of Cold War Studies* 1 (Spring 1999): 3–38.

———. "The Early Post-Stalin Succession Struggle and Upheavals in East-Central Europe: Internal-External Linkages in Soviet Policy Making (Part 3)." *Journal of Cold War Studies* 1 (Fall 1999): 3–66.

Kraske, Gary E. *Missionaries of the Book: The American Library Profession and the Origins of American Cultural Diplomacy.* Westport, Conn.: Greenwood Press, 1985.

Krenn, Michael L. *Fall-out Shelters for the Human Spirit: American Art and the Cold War.* Chapel Hill: University of North Carolina Press, 2005.

———. *Black Diplomacy: African Americans and the State Department, 1945–1969.* Armonk, N.Y.: M. E. Sharpe, 1999.

———. "'Unfinished Business': Segregation and U.S. Diplomacy at the 1958 World's Fair." *Diplomatic History* 20 (Fall 1996): 591–92.

Krugler, David F. *The Voice of America and the Domestic Propaganda Battles, 1945–1953.* Columbia: University of Missouri Press, 2000.

Kuisel, Richard F. *Seducing the French: The Dilemma of Americanization.* Berkeley: University of California Press, 1993.

Kuznick, Peter J., and James Gilbert, eds. *Rethinking Cold War Culture.* Washington, D.C.: Smithsonian Institution Press, 2001.

Kyle, Keith. *Suez*. New York: St. Martin's Press, 1991.

L'Etang, Jacquie. "State Propaganda and Bureaucratic Intelligence: The Creation of Public Relations in 20 Century Britain." *Public Relations Review* 24 (Winter 1998): 413–441.

LaFeber, Walter. *The Clash: A History of U.S.-Japan Relations*. New York: W. W. Norton, 1997.

Larres, Klaus. *Churchill's Cold War: The Politics of Personal Diplomacy*. New Haven: Yale University Press, 2002.

———. "Eisenhower and the First Forty Days After Stalin's Death: The Incompatibility of Détente and Political Warfare." *Diplomacy and Statecraft* 6 (July 1995): 431–69.

Larson, Arthur. *Eisenhower: The President Nobody Knew*. New York: Scribner, 1968.

Larson, Deborah Welch. *Anatomy of Mistrust: U.S.-Soviet Relations During the Cold War*. Ithaca, N.Y.: Cornell University Press, 1997.

———. *Origins of Containment: A Psychological Explanation*. Princeton: Princeton University Press, 1985.

Lashmar, Paul, and James Oliver. *Britain's Secret Propaganda War, 1948–1977*. Stroud, Gloucestershire: Sutton Publishing, 1998.

Lasswell, Harold Dwight. *Propaganda Technique in the World War*. New York: Alfred A. Knopf, 1927.

Lasswell, Harold D., Daniel Lerner, and Hans Speier, eds. *Propaganda and Communication in World History*. Honolulu: University Press of Hawaii, 1979.

Launius, Roger D. *NASA: A History of the U.S. Civil Space Program*. Malabar, Fla.: Krieger Publishing, 1994.

Launius, Roger D., John M. Logsdon, and Robert W. Smith, eds. *Reconsidering Sputnik: Forty Years Since the Soviet Satellite*. Australia: Harwood Academic Publishers, 2000.

Laurie, Clayton D. *The Propaganda Warriors: America's Crusade Against Nazi Germany*. Lawrence: University Press of Kansas, 1996.

Laville, Helen. "The Committee of Correspondence: CIA Funding of Women's Groups, 1952–1967." *Intelligence and National Security* 12 (January 1997): 104–21.

Lavine, Harold. *War Propaganda and the United States*. Rev. ed. New York: Garland Publishers, 1972.

Le Bon, Gustave. *The Crowd: A Study of the Popular Mind*. New York: Macmillan, 1896.

———. *The Crowd*. Introduction by Robert A. Nye. New Brunswick, N.J.: Transaction Publishers, 1995.

Lederer, William, and Eugene Burdick. *The Ugly American*. New York: W. W. Norton, 1958.

Ledyard, Gleason H. *Sky Waves: The Incredible Far East Broadcasting Company Story*. Chicago: Moody Press, 1968.

Leffler, Melvyn P. *A Preponderance of Power: National Security, the Truman Adminis-tration, and the Cold War*. Stanford: Stanford University Press, 1992.

Lehman, Kenneth. "Resolution of the Korean War Biological Warfare Allega-tions." *Critical Reviews in Microbiology* 24 (1998): 169–94.

Lerner, Daniel, ed. *Propaganda in War and Crisis*. New York: Arno Press, 1972.

———. *Sykewar: Psychological Warfare against Germany, D-Day to VE-Day*. New York: G. W. Stewart, 1949.

Lifton, Robert Jay. *Death in Life: Survivors in Hiroshima*. New York: Random House, 1967.

Lifton, Robert Jay, and Richard Falk. *Indefensible Weapons: The Political and Psycho-logical Case Against Nuclearism*. New York: Basic Books, 1982.

Lifton, Robert Jay, and Greg Mitchell. "The Age of Numbing." *Technology Review* 98 (August–September 1995): 58–60.

———. *Hiroshima in America: Fifty Years of Denial*. New York: G. P. Putnam's Sons, 1995.

Lilly, Edward P. "The Psychological Strategy Board and Its Predecessors: For-eign Policy Coordination, 1938–1953." In *Studies in Modern History*, edited by Gaetano L. Vincitorio, 337–53. New York: St. Johns University Press, 1968.

Lippmann, Walter. *Public Opinion*. New York: Harcourt, Brace, 1922.

———. *The Phantom Public*. New York: Macmillan, 1927.

Lisle, Leslie. *United States Information Agency, 1953–1983*. Washington, D.C.: United States Information Agency, 1984.

Little, Douglas. "Cold War and Covert Action: The United States and Syria 1945–1958." *Middle East Journal* 44 (1990): 51–75.

———. "His Finest Hour? Eisenhower, Lebanon, and the 1958 Middle East Cri-sis." *Diplomatic History* 20 (Winter 1996): 27–54.

Logsdon, John M., ed. *Exploring the Unknown: Selected Documents in the History of the U.S. Civil Space Program*. Vol. 1. Washington, D.C.: National Aeronautics and Space Administration, 1995.

Loth, Wilfried. *Europe, Cold War and Co-existence, 1953–1965*. London: Frank Cass, 2003.

Louis, William Roger, and Roger Owen, eds. *Suez 1956: The Crisis and Its Conse-quences*. Oxford: Clarendon Press, 1989.

Loveland, Anne C. *American Evangelicals and the U.S. Military, 1942–1993*. Baton Rouge: Louisiana State University Press, 1996.

Lucas, Scott. "The Campaigns of Truth: The Psychological Strategy Board and Ameri-can Ideology, 1951–1953." *International History Review* 18 (May 1996): 279–302.

———. *Freedom's War: The American Crusade Against the Soviet Union*. New York: New York University Press, 1999.

———. "Mobilising Culture: The State-Private Network and the CIA in the Early Cold War." In *War and Cold War in American Foreign Policy, 1942–62*, edited by Dale Carter and Robin Clifton, 83–107. New York: Palgrave, 2002.

Lundestad, Geir. "Empire by Invitation? The United States and Western Europe, 1945–1952." *Journal of Peace Research* 23 (September 1986): 263–77.

———. *Empire by Integration: The United States and European Integration, 1945–1997.* Oxford: Oxford University Press, 1998.

———. *The United States and Western Europe Since 1945: From "Empire" by Invitation to Transatlantic Drift.* Oxford: Oxford University Press, 2003.

Lykins, Daniel L. *From Total War to Total Diplomacy: The Advertising Council and the Construction of the Cold War Consensus.* Westport, Conn.: Praeger, 2003.

MacCann, Richard Dyer. *The People's Films: A Political History of U.S. Government Motion Pictures.* New York: Hastings House, 1973.

MacDonald, J. Fred. *Television and the Red Menace: The Video Road to Vietnam.* New York: Praeger, 1985.

Magnus, Persson. *Great Britain, the United States, and the Security of the Middle East: The Formation of the Baghdad Pact.* Lund, Sweden: Lund University Press, 1998.

Maltese, John Anthony. *Spin Control: The White House Office of Communications and the Management of Presidential News.* 2nd ed. Chapel Hill: University of North Carolina Press, 1994.

Marchand, Roland. *Creating the Corporate Soul: The Rise of Public Relations and Corporate Imagery in American Big Business.* Berkeley: University of California Press, 2001.

Marchio, James D. "The Planning Coordination Group: Bureaucratic Casualty in the Cold War Campaign to Exploit Soviet-Bloc Vulnerabilities." *Journal of Cold War Studies* 4 (Fall 2002): 3–28.

———. "Resistance Potential and Rollback: U.S. Intelligence and the Eisenhower Administration's Policies Toward Eastern Europe, 1953–56." *Intelligence and National Security* 10 (April 1995): 219–41.

———. "Rhetoric and Reality: The Eisenhower Administration and Unrest in Eastern Europe, 1953–1959." Ph.D. diss., American University, 1990.

Marks, Frederick W., III. *Power and Peace: The Diplomacy of John Foster Dulles.* Westport, Conn.: Praeger, 1993.

Mastny, Vojtech. *The Cold War and Soviet Insecurity: The Stalin Years.* New York: Oxford University Press, 1996.

Masur, Matthew B. "Hearts and Minds: Cultural Nation-Building in South Vietnam, 1954–1964." Ph.D. diss., Ohio State University, 2004.

———. "People of Plenty: American Cultural and Economic Programs in South Vietnam, 1954–1963." Paper presented to the Ohio Academy of History, 2002.

Matthews, Jane DeHart. "Art and Politics in Cold War America." *American Historical Review* 81 (October 1976): 762–87.

May, Elaine Tyler. *Homeward Bound: American Families in the Cold War Era.* New York: Basic Books, 1988.

May, Lary, ed. *Recasting America: Culture and Politics in the Age of Cold War.* Chicago: University of Chicago Press, 1989.

McDougall, Walter A. *The Heavens and the Earth: A Political History of the Space Age.* Baltimore: Johns Hopkins University Press, 1985.

McEnaney, Laura. *Civil Defense Begins at Home: Militarization Meets Everyday Life in the Fifties.* Princeton: Princeton University Press, 2000.

McLaurin, Ronald De, et al., eds. *The Art and Science of Psychological Operations: Case Studies of Military Application.* Washington, D.C.: Department of the Army, 1976.

McMahon, Robert J. *The Cold War on the Periphery: The United States, India, and Pakistan.* New York: Columbia University Press, 1994.

———. "Credibility and World Power: Exploring the Psychological Dimension in Postwar American Diplomacy." *Diplomatic History* 15 (Fall 1991): 455–71.

———. "Eisenhower and Third World Nationalism: A Critique of the Revisionists." *Political Science Quarterly* 101 (Fall 1986): 453–73.

———. "The Illusion of Vulnerability: American Reassessments of the Soviet Threat, 1955–1956." *International History Review* 18 (August 1996): 591–619.

———. *The Limits of Empire: The United States and Southeast Asia Since World War II.* New York: Columbia University Press, 1999.

———. "'The Point of No Return': The Eisenhower Administration and Indonesia, 1953–1960." In Johns and Statler, *Managing an Earthquake.*

———. "Toward a Pluralist Vision: The Study of American Foreign Relations as International and National History." In *Explaining the History of American Foreign Relations,* edited by Michael J. Hogan and Thomas G. Paterson, 35–50. 2nd ed. Cambridge: Cambridge University Press, 2004.

Mechling, Elizabeth Walker, and Jay Mechling. "The Atom According to Disney." *Quarterly Journal of Speech* 81 (1995): 436–53.

———. "The Campaign for Civil Defense and the Struggle to Naturalize the Bomb." *Western Journal of Speech Communication* 55 (Spring 1991): 105–33.

Medhurst, Martin J. "Atoms for Peace and Nuclear Hegemony: The Rhetorical Structure of a Cold War Campaign." *Armed Forces and Society* 24 (Summer 1997): 571–93.

———. *Dwight D. Eisenhower: Strategic Communicator.* Westport, Conn.: Greenwood Press, 1993.

———. "Eisenhower's 'Atoms for Peace' Speech: A Case Study in the Strategic Use of Language." *Communication Monographs* 54 (June 1987): 204–20.

Medhurst, Martin J., ed. *Eisenhower's War of Words: Rhetoric and Leadership.* East Lansing: Michigan State University Press, 1994.

Medhurst, Martin J., Robert L. Ivie, Philip Wander, and Robert L. Scott. *Cold War Rhetoric: Strategy, Metaphor, and Ideology.* East Lansing: Michigan State University Press, 1990.

Melanson, Richard A., and David Mayers, eds. *Reevaluating Eisenhower: American Foreign Policy in the 1950s.* Urbana: University of Illinois Press, 1987.

Melbourne, Roy M. *Conflict and Crisis: A Foreign Service Story.* Lanham, Md.: University Press of America, 1993.

Meriwether, James Hunter. *Proudly We Can Be Africans: Black Americans and Africa, 1935–1961.* Chapel Hill: University of North Carolina Press, 2002.

Metz, Steven. "American Attitudes Toward Decolonization in Africa." *Political Science Quarterly* 99 (Fall 1984): 515–33.

Meyerhoff, Arthur E. *The Strategy of Persuasion: The Use of Advertising Skills in Fighting the Cold War.* New York: Coward-McCann, 1965.

Meyerowitz, Joanne. *Not June Cleaver: Women and Gender in Postwar America, 1945–1960.* Philadelphia: Temple University Press, 1994.

Mickelson, Sig. *America's Other Voices: Radio Free Europe and Radio Liberty.* New York: Praeger, 1983.

Mihalovich, Rachel M. "Fighting 'Don't Know, Don't Care': The FCDA's Public Education Quest." M.A. thesis, Old Dominion University, 2001.

Mitrovich, Gregory. *Undermining the Kremlin: America's Strategy to Subvert the Soviet Bloc, 1947–1956.* Ithaca, N.Y.: Cornell University Press, 2000.

Mock, James R., and Cedric Larson. *Words that Won the War: The Story of the Committee on Public Information, 1917–1919.* Princeton: Princeton University Press, 1939.

Morgan, Patrick M. "The Cold War in the 1950s: An Opportunity Missed?" In *Re-Viewing the Cold War: Domestic Factors and Foreign Policy in the East-West Confrontation,* edited by Morgan and Keith L. Nelson, 43–70. Westport, Conn.: Praeger, 2000.

Morgan, Patrick M., and Keith L. Nelson, eds. *Re-Viewing the Cold War: Domestic Factors and Foreign Policy in the East-West Confrontation.* Westport, Conn.: Praeger, 2000.

Morgan, Ted. *A Covert Life: Jay Lovestone: Communist, Anti-Communist, and Spymaster.* New York: Random House, 1999.

Nadler, Marcus. *People's Capitalism.* New York: Hanover Bank, 1956.

Naftali, Timothy. "Reinhard Gehlen and the United States." In *U.S. Intelligence and the Nazis,* edited by Richard Breitman, Norman J. W. Goda, Timothy Naftali, Robert Wolfe, Richard Breitman, and Norman J. W. Goda, 375–418. New York: Cambridge University Press, 2005.

Needell, Allan A. "'Truth Is Our Weapon': Project TROY, Political Warfare, and Government-Academic Relations in the National Security State." *Diplomatic History* 17 (Summer 1993): 399–420.

Nelson, Michael. *War of the Black Heavens: The Battles of Western Broadcasting in the Cold War.* London: Brassey's, 1997.

Ninkovich, Frank A. *The Diplomacy of Ideas: U.S. Foreign Policy and Cultural Relations, 1938–1950.* New York: Cambridge University Press, 1981.

———. *U.S. Information Policy and Cultural Diplomacy.* New York: Foreign Policy Association, 1996.

————. *The Wilsonian Century: U.S. Foreign Policy Since 1900.* Chicago: University of Chicago Press, 1999.

Noer, Thomas J. *Cold War and Black Liberation: The United States and White Rule in Africa, 1948–1968.* Columbia: University of Missouri Press, 1985.

Nye, Joseph S. "Soft Power." *Foreign Policy* 80 (Fall 1990): 153–71.

O'Neil, William L. *A Democracy at War: America's Fight at Home and Abroad in World War II.* New York: Free Press, 1993.

Oakes, Guy. *The Imaginary War: Civil Defense and American Cold War Culture.* New York: Oxford University Press, 1994.

Osgood, Kenneth A. "Before Sputnik: National Security and the Formation of U.S. Outer Space Policy." In Launius et al., *Reconsidering Sputnik*, 197–229.

————. "Form Before Substance: Eisenhower's Commitment to Psychological Warfare and Negotiations with the Enemy." *Diplomatic History* 24 (Summer 2000): 405–33.

————. "Hearts and Minds: The Unconventional Cold War." *Journal of Cold War Studies* 4 (Spring 2002): 85–107.

Osgood, Kenneth A., and Klaus Larres, eds. *The Cold War After Stalin's Death: A New International History.* Lanham, Md.: Rowman & Littlefield, forthcoming.

Ostermann, Christian F. "The United States, The East German Uprising of 1953, and the Limits of Rollback." Cold War International History Project Working Paper No. 11. Washington, D.C.: Woodrow Wilson International Center for Scholars, December 1994.

Pach, Chester J., Jr., and Elmo Richardson. *The Presidency of Dwight D. Eisenhower.* Rev. ed. Lawrence: University Press of Kansas, 1991.

Paddock, Alfred H. *U.S. Army Special Warfare: Its Origins.* Rev. ed. Lawrence: University Press of Kansas, 2002.

Page, Caroline. *U.S. Official Propaganda During the Vietnam War, 1965–1973: The Limits of Persuasion.* London: Leicester University Press, 1996.

Parks, J. D. *Culture, Conflict and Coexistence: American-Soviet Cultural Relations, 1917–1958.* Jefferson, N.C.: McFarland, 1983.

Parker, Jason. "Small Victory, Missed Chance: The Eisenhower Administration, The Bandung Conference, and the Turning of the Cold War." In Johns and Statler, *Managing an Earthquake*.

Parmet, Herbert S. *Eisenhower and the American Crusades.* New York: Macmillan, 1972.

Parry-Giles, Shawn J. "'Camouflaged' Propaganda: The Truman and Eisenhower Administrations' Covert Manipulation of News." *Western Journal of Communication* 60 (Spring 1996): 146–67.

————. "The Eisenhower Administration's Conceptualization of the USIA: The Development of Overt and Covert Propaganda Strategies." *Presidential Studies Quarterly* 24 (Spring 1994): 263–76.

————. "Propaganda, Effect, and the Cold War: Gauging the Status of America's 'War of Words.'" *Political Communication* 11 (1994): 203–13.

———. "Rhetorical Experimentation and the Cold War, 1947–1953: The Development of an Internationalist Approach to Propaganda." *Quarterly Journal of Speech* 80 (1994): 448–67.

———. *The Rhetorical Presidency, Propaganda, and the Cold War, 1945–1955.* Westport, Conn.: Praeger, 2002.

———. "The Rhetorical Tension Between Propaganda and Democracy: Blending Competing Conceptions of Ideology and Theory." *Communication Studies* 14 (Summer 1993): 117–31.

Pease, Stephen E. *Psywar: Psychological Warfare in Korea, 1950–1953.* Harrisburg, Penn.: Stackpole Books, 1992.

Pells, Richard. *Not Like U.S.: How Europeans Have Loved, Hated, and Transformed American Culture.* New York: Basic Books, 1997.

Peterson, Tore Tingvold. *The Middle East Between the Great Powers: Anglo-American Conflict and Cooperation, 1952–7.* New York: St. Martin's Press, 2000.

Phillips, David Atlee. *The Night Watch.* New York: Atheneum, 1977.

Pickett, William B. *Eisenhower Decides to Run: Presidential Politics and Cold War Strategy.* Chicago: Ivan R. Dee, 2000.

Pilat, Joseph F., Robert E. Pendley, and Charles K. Ebinger, eds. *Atoms for Peace: An Analysis after Thirty Years.* Boulder, Colo.: Westview Press, 1985.

Pisani, Sallie. *The CIA and the Marshall Plan.* Lawrence: University Press of Kansas, 1991.

Plummer, Brenda Gayle. *Rising Wind: Black Americans and U.S. Foreign Affairs, 1935–1960.* Chapel Hill: University of North Carolina Press, 1996.

Potter, David M. *The American Round Table Discussions on People's Capitalism at Yale University, New Haven, Connecticut, November 16 and 17, 1956.* New York: Advertising Council, 1957.

———. *An Inquiry into the Social and Cultural Trends in America Under Our System of Widely Shared Materials Benefits: [Discussions] at the Yale Club, New York, New York, May 22, 1957.* New York: Advertising Council, 1957.

Prados, John. *President's Secret Wars: CIA and Pentagon Covert Operations from World War II through the Persian Gulf.* Rev. ed. Chicago: Ivan R. Dee, 1996.

Pratkanis, Anthony R., and Elliott Aronson. *Age of Propaganda: The Everyday Use and Abuse of Persuasion.* New York: W. H. Freeman, 1991.

Prevots, Naima. *Dance for Export: Cultural Diplomacy and the Cold War.* Hanover, N.H.: University Press of New England, 1998.

Pruden, Caroline. *Conditional Partners: Eisenhower, the United Nations, and the Search for a Permanent Peace.* Baton Rouge: Louisiana State University Press, 1998.

Pruessen, Ronald W. *John Foster Dulles: The Road to Power.* New York: Free Press, 1982.

Puddington, Arch. *Broadcasting Freedom: The Cold War Triumph of Radio Free Europe and Radio Liberty.* Lexington: University Press of Kentucky, 2000.

Rabe, Stephen G. *Eisenhower and Latin America: The Foreign Policy of Anti-Communism.* Chapel Hill: University of North Carolina Press, 1988.

———. "Eisenhower Revisionism: A Decade of Scholarship." *Diplomatic History* 17 (Winter 1993): 97–115.

Rathmell, Andrew. *Secret War in the Middle East: The Covert Struggle for Syria, 1949–1961.* London: Tauris Academic Studies, 1995.

Rawnsley, Gary D. *Radio Diplomacy and Propaganda: The BBC and VOA in International Politics, 1956–64.* New York: St. Martin's Press, 1996.

Rawnsley, Gary D., ed. *Cold War Propaganda in the 1950s.* New York: St. Martin's Press, 1999.

Reich, Cary. *The Life of Nelson Rockefeller: Worlds to Conquer, 1908–1958.* New York: Doubleday, 1996.

Reuth, Ralf Georg. *Goebbels.* Translated by Krishna Winston. New York: Harcourt Brace, 1993.

Richelson, Jeffrey T. *America's Secret Eyes in Space: The U.S. Keyhole Spy Satellite Program.* New York: Harper & Row, 1990.

Richter, James G. *Khrushchev's Double Bind: International Pressures and Domestic Coalition Politics.* Baltimore: Johns Hopkins University Press, 1994.

———. "Reexamining Soviet Policy Towards Germany During the Beria Interregnum." Cold War International History Project Working Paper No. 3. Washington, D.C.: Woodrow Wilson International Center for Scholars, June 1992.

Roetter, Charles. *Psychological Warfare.* London: Batsford, 1974.

Rogers, Everett M. *A History of Communication Study: A Biographical Approach.* New York: Free Press, 1994.

Roman, Peter J. *Eisenhower and the Missile Gap.* Ithaca, N.Y.: Cornell University Press, 1995.

Rosenberg, Emily S. *Spreading the American Dream: American Economic and Cultural Expansion, 1890–1945.* New York: Hill and Wang, 1982.

Rostow, Walt W. *Europe After Stalin: Eisenhower's Three Decisions of March 11, 1953.* Austin: University of Texas Press, 1982.

———. *Open Skies: Eisenhower's Proposal of July 21, 1955.* Austin: University of Texas Press, 1982.

Roth, Lois. "Public Diplomacy and the Past: The Studies of U.S. Information and Cultural Programs (1952–1975)." United States Department of State: Foreign Service Institute Executive Seminar in National and International Affairs, 23rd Session, 1980–1981.

Rubin, Ronald I. *The Objectives of the U.S. Information Agency: Controversies and Analysis.* New York: Praeger, 1968.

Rubinstein, Alvin Z. *Soviet Foreign Policy Since World War II: Imperial and Global.* 4th ed. New York: HarperCollins, 1992.

Ruffner, Kevin C., ed. *Corona: America's First Satellite Program.* Washington, D.C.. Center for the Study of Intelligence, 1995.

Rydell, Robert W. *All the World's a Fair: Visions of Empire at American International Expositions, 1876–1916.* Chicago: University of Chicago Press, 1984.

————. *World of Fairs: The Century-of-Progress Expositions*. Chicago: University of Chicago Press, 1993.

Salminen, Esko. *The Silenced Media: The Propaganda War Between Russia and the West in Northern Europe*. New York: St. Martin's Press, 1999.

Sangmuah, Egya N. "Eisenhower and Containment in North Africa, 1956–1960." *Middle East Journal* 44 (Winter 1990): 76–91.

Saunders, Frances Stonor. *The Cultural Cold War: The CIA and the World of Arts and Letters*. New York: New Press, 1999.

Schaller, Michael. *Altered States: The United States and Japan Since the Occupation*. New York: Oxford University Press, 1997.

————. *The American Occupation of Japan: The Origins of the Cold War in Asia*. New York: Oxford University Press, 1985.

Schlesinger, Stephen, and Stephen Kizner. *Bitter Fruit: The Untold Story of the American Coup in Guatemala*. Garden City, N.Y.: Doubleday, 1982.

Schmidtz, David. *Thank God They're On Our Side: The United States and Right-Wing Dictatorships, 1921–1965*. Chapel Hill: University of North Carolina Press, 1999.

Schneider, Ronald M. *Communism in Guatemala, 1944–1954*. New York: Praeger, 1958.

Schrecker, Ellen. *Many Are the Crimes: McCarthyism in America*. Boston: Little, Brown, 1998.

Schröder, Hans-Jurgen. "Marshall Plan Propaganda in Austria and Western Germany." In *The Marshall Plan in Austria*, edited by Günter Bischof, Anton Pelinka, and Dieter Stiefel, 212–46. New Brunswick, N.J.: Transaction Publishers, 2000.

Schulzinger, Robert. *A Time for War: The United States and Vietnam, 1941–1975*. New York: Oxford University Press, 1997.

Schumacher, Frank. "Cold War Propaganda and Alliance Management: The United States and West Germany in the 1950s." Paper presented to the Society for Historians of American Foreign Relations, June 2000.

————. "Democratization and Hegemonic Control: American Propaganda and the West German Public's Foreign Policy Orientation, 1949–1955." In *The American Nation, National Identity, Nationalism*, ed. Knud Krakau, 285–316. Münster: Transaction Publishers, 1997.

————. *Kalter Krieg und Propaganda: Die USA, der Kampf um die Weltmeinung und die ideelle Westbindung der Bundesrepublik Deutschland, 1945–1955*. Trier: Wissenschaftlicher Verlag, 2000.

Schwenk, Melinda. "Reforming the Negative Through History: The U.S. Information Agency and the 1957 Little Rock Integration Crisis." *Journal of Communication Inquiry* 23 (July 1999): 288–306.

Schwoch, James. "The Cold War, the Space Race, and the Globalization of Public Opinion Polling." Paper presented to the International Studies Association, March 2002.

Scott-Smith, Giles. *The Politics of Apolitical Culture: The Congress for Cultural Freedom, the CIA, and Post-War American Hegemony*. London: Routledge, 2002.

Scott-Smith, Giles, and Hans Krabbendam, eds. *Boundaries of Freedom: The Cultural Cold War.* London: Frank Cass, 2003.

Shaw, Tony. "Martyrs, Miracles, and Martians: Religion and Cold War Cinematic Propaganda in the 1950s." *Journal of Cold War Studies* 4 (Spring 2002): 3–22.

Sherry, Michael. *In the Shadow of War: The United States Since the 1930s.* New Haven, Conn.: Yale University Press, 1995.

Siekmeier, James F. *Aid, Nationalism, and Inter-American Relations: Guatemala, Bolivia, and the United States, 1945–1961.* Lewiston, N.Y.: Edwin Mellon Press, 1999.

Simpson, Christopher. *Blowback: America's Recruitment of Nazis and Its Effects on the Cold War.* New York: Weidenfeld and Nicolson, 1988.

————. *Science of Coercion: Communication Research and Psychological Warfare, 1945–1960.* New York: Oxford University Press, 1994.

Sisson, Edgar. "Sisson's Account of Wilson's Fourteen Points Speech." In *A Psychological Warfare Casebook*, edited by William E. Daugherty and Morris Janowitz, 89–96. Baltimore: Johns Hopkins University Press, 1958.

Smith-Norris, Martha. "The Eisenhower Administration and the Nuclear Test Ban Talks, 1958–1960: Another Challenge to 'Revisionism.'" *Diplomatic History* 27 (September 2003): 503–41.

Smoke, Richard. *National Security and the Nuclear Dilemma: An Introduction to the American Experience in the Cold War.* 3rd ed. New York: McGraw-Hill, 1993.

Snead, David L. *The Gaither Committee, Eisenhower, and the Cold War.* Columbus: Ohio State University Press, 1999.

Snyder, Alvin A. *Warriors of Disinformation: American Propaganda, Soviet Lies, and the Winning of the Cold War—An Insider's Account.* New York: Arcade, 1995.

Soapes, Thomas F. "A Cold Warrior Seeks Peace: Eisenhower's Strategy for Nuclear Disarmament." *Diplomatic History* 4 (Winter 1980): 57–71.

Soley, Lawrene. *Radio Warfare: OSS and CIA Subversive Propaganda.* New York, Praeger, 1989.

Sorensen, Thomas C. *The Word War: The Story of American Propaganda.* New York: Harper & Row, 1968.

Sproule, J. Michael. *Propaganda and Democracy : The American Experience of Media and Mass Persuasion.* Cambridge: Cambridge University Press, 1997.

Stanke, Jaclyn. "Danger and Opportunity: Eisenhower, Churchill, and the Soviet Union After Stalin, 1953–1955." Ph.D. diss., Emory University, 2002.

Statler, Kathryn. "Building a Colony: The Eisenhower Administration and South Vietnam, 1953–1961." In Johns and Statler, *Managing an Earthquake.*

Steel, Ronald. *Walter Lippmann and the American Century.* New York: Vintage Books, 1981.

Steininger, Rolf. "John Foster Dulles, the European Defense Community, and the German Question." In *John Foster Dulles and the Diplomacy of the Cold War,* edited by Richard H. Immerman, 79–108. Princeton: Princeton University Press, 1990.

Stephens, Oren. *Facts to a Candid World: America's Overseas Information Program.* Stanford: Stanford University Press, 1955.

Stern, John Allen. "Propaganda in the Employ of Democracy: Fighting the Cold War with Words." Ph.D. diss, State University of New York at Stony Brook, 2002.

Steury, Donald P., ed. *Intentions and Capabilities: Estimates on Soviet Strategic Forces, 1950–1983.* Washington, D.C.: Center for the Study of Intelligence, 1996.

Strong, Robert A. "Eisenhower and Arms Control." In *Reevaluating Eisenhower: American Foreign Policy in the 1950s,* edited by Richard A. Melanson and David Mayers, 241–66. Urbana: University of Illinois Press, 1987.

Stueck, William. *The Korean War: An International History.* Princeton: Princeton University Press, 1995.

Sulzberger, C. L. *A Long Row of Candles: Memoirs and Diaries, 1934–1954.* New York: Macmillan, 1969.

Summers, Robert E., ed. *America's Weapons of Psychological Warfare.* New York: H. W. Wilson, 1951.

Sun Tzu. *The Art of Warfare.* Translated by Roger T. Ames. New York: Ballantine, 1993.

Suny, Ronald Grigor. *The Soviet Experiment: Russia, the USSR, and the Successor States.* Oxford: Oxford University Press, 1998.

Suri, Jeremi. "America's Search for a Technological Solution to the Arms Race: The Surprise Attack Conference of 1958 and a Challenge for 'Eisenhower Revisionists.'" *Diplomatic History* 21 (Summer 1997): 417–52.

———. *Power and Protest: Global Revolution and the Rise of Détente.* Cambridge: Harvard University Press, 2003.

Swann, Paul. "The Little State Department: Hollywood and the State Department in the Postwar World." *American Studies International* 29 (April 1991): 2–17.

Takeyh, Ray. *The Origins of the Eisenhower Doctrine: The U.S., Britain and Nasser's Egypt, 1953–57.* London: Macmillan, 2000.

Tarde, Gabriel. *The Laws of Imitation.* New York: Henry Holt, 1903.

Taubman, William. *Khrushchev: The Man and His Era.* New York: W. W. Norton, 2003.

Taylor, Philip M. *Munitions of the Mind: War Propaganda from the Ancient World to the Nuclear Age.* Glasgow: Patrick Stephens, 1990.

Thomas, Damion Lamar. "The Good Negroes: African-American Athletes and the Cultural Cold War, 1945–1968." Ph.D. diss., University of California, Los Angeles, 2002.

Thomas, Martin. "Defending a Lost Cause? France and the United States Vision of French North Africa, 1945–1956." *Diplomatic History* 26 (Spring 2002): 215–47.

Thompson, Kenneth W., ed. *The Eisenhower Presidency.* Lanham, Md.: University Press of America, 1984.

Thomson, Charles Alexander Holmes, and Walter H. C. Laves. *Cultural Relations and U.S. Foreign Relations.* Bloomington: Indiana University Press, 1963.

————. *Overseas Information Service of the United States Government.* New York: Arno Press, 1972.

Thomson, Oliver. *Easily Led: A History of Propaganda.* Stroud, Gloucestershire: Sutton Publishing, 1999.

Tompson, William J. *Khrushchev: A Political Life.* New York: St. Martin's Griffin, 1995.

Trachtenberg, Marc. *A Constructed Peace: The Making of the European Settlement, 1945–1963.* Princeton: Princeton University Press, 1999.

Trotter, Wilfred. *Instincts of the Herd in Peace and War.* New York: Macmillan, 1916.

Tuch, Hans N. USIA: *Communicating with the World in the 1990s.* Washington, D.C.: Public Diplomacy Foundation, 1994.

Tuch, Hans N., and G. Lewis Schmidt, eds. *Ike and the USIA: A Commemorative Symposium.* Washington, D.C.: Public Diplomacy Foundation, 1991.

Tye, Larry. *The Father of Spin: Edward L. Bernays and the Birth of Public Relations.* New York: Henry Holt, 1998.

U.S. Congress. House. Committee on Appropriations. Departments of State and Justice, the Judiciary, and Related Agencies for 1956: Hearings Before the Subcommittee of the Committee on Appropriations. 84 Congress, 1 Session, 1955.

U.S. Congress. House. Committee on Foreign Affairs. *Wining the Cold War: The U.S. Ideological Offensive.* Hearings Before the Subcommittee on International Organizations and Movements. 88 Congress, 2 Session, 1964.

U.S. Congress. House. Committee on Foreign Affairs. Subcommittee on State Department Organization and Foreign Affairs. Review of United States Information Agency Operations. 85 Congress, 2 Session, 1959.

U.S. Congress. House. Permanent Select Committee on Intelligence. Subcommittee on Oversight. *The CIA and the Media.* Hearings Before the Subcommittee on Oversight of the Permanent Select Committee on Intelligence. 95 Congress, 1 and 2 Sessions, 1978.

U.S. Congress. Senate. Committee on Appropriations. *Survey of United States Information Service Operations: Western Europe.* 1957.

U.S. Congress. Senate. Select Committee to Study Governmental Operations with Respect to Intelligence Activities. Hearings. 94 Congress, 1 Session, 1976.

U.S. Congress. Senate. Select Committee to Study Governmental Operations with Respect to Intelligence Activities. *Covert Action in Chile, 1963–1973.* 1975.

U.S. Congress. Senate. Subcommittee of the Committee on Foreign Relations. Subcommittee on Disarmament. Control and Reduction of Armaments: Hearings Before the United States Senate Committee on Foreign Relations, Subcommittee on Disarmament. 84 Congress, 2 Session, 1956.

U.S. Congress. Senate. Subcommittee on National Policy Machinery. *Organizational History of the National Security Council.* 86 Congress, 2 Session, 1960.

U.S. Department of Defense. *Militant Liberty: A Program of Evaluation and Assessment of Freedom.* Washington, D.C.: Government Printing Office, 1955.

U.S. Department of State. *Documents on Disarmament, 1954–1959,* 2 vols. Washington, D.C.: Government Printing Office, 1960.

———. *Foreign Relations of the United States, 1945–1950: Emergence of the Intelligence Establishment.* Washington, D.C.: Government Printing Office, 1996.

———. *Foreign Relations of the United States, 1952–1954: National Security Affairs.* Vol. 2. Washington, D.C.: Government Printing Office, 1984.

———. *Foreign Relations of the United States, 1952–1954: Eastern Europe; Soviet Union; Eastern Mediterranean.* Vol. 8. Washington, D.C.: Government Printing Office, 1988.

———. *Foreign Relations of the United States, 1952–1954: Indochina.* Vol. 13. Washington, D.C.: Government Printing Office, 1982.

———. *Foreign Relations of the United States, 1955–1957: American Republics: Central and South America.* Vol. 6. Washington, D.C.: Government Printing Office, 1987.

———. *Foreign Relations of the United States, 1955–1957: Foreign Economic Policy; Foreign Information Program.* Vol. 9. Washington, D.C.: Government Printing Office, 1987.

———. *Foreign Relations of the United States, 1955–1957: United Nations and General International Matters.* Vol. 11. Washington, D.C.: Government Printing Office, 1988.

———. *Foreign Relations of the United States, 1955–1957: National Security Policy.* Vol. 19. Washington, D.C.: Government Printing Office, 1990.

———. *Foreign Relations of the United States, 1955–1957: Regulation of Armaments; Atomic Energy.* Vol. 20. Washington, D.C.: Government Printing Office, 1990.

———. *Foreign Relations of the United States, 1955–1957: Soviet Union; Eastern Mediterranean.* Vol. 24. Washington, D.C.: Government Printing Office, 1989.

———. *Foreign Relations of the United States, 1958–1960: United Nations and General International Matters.* Vol. 2. Washington, D.C.: Government Printing Office, 1991.

———. *Foreign Relations of the United States, 1958–1960: National Security Policy; Arms Control and Disarmament.* Vol. 3. Washington, D.C.: Government Printing Office, 1996.

———. *Foreign Relations of the United States, 1958–1960: Near East Region; Iraq; Iran; Arabian Peninsula.* Vol. 12. Washington, D.C.: Government Printing Office, 1993.

———. *Foreign Relations of the United States: Guatemala.* Washington, D.C.: Government Printing Office, 2003.

United States Information Agency. *Communist Propaganda: A Fact Book.* 1957–1958.

———. *Facts About the United States.* 1956.

———. *People-to-People: A Program of International Friendship.* 1957.

Ungar, Sheldon. *The Rise and Fall of Nuclearism: Fear and Faith as Determinants of the Arms Race.* University Park: Pennsylvania State University Press, 1992.

Vaughan, James. "The Anglo-American Relationship and Propaganda Strategies in the Middle East, 1953–1957." Ph.D. diss., University of London, 2001.

———. "Propaganda by Proxy? Britain, America, and Arab Radio Broadcasting, 1953–1957." *Historical Journal of Film, Radio and Television* 22 (2002): 157–72.

Von Eschen, Penny M. *Black Americans and Anticolonialism, 1937–1957.* Ithaca, N.Y.: Cornell University Press, 1997.

———. *Satchmo Blows Up the World: Jazz Ambassadors Play the Cold War.* Cambridge: Harvard University Press, 2004.

Wagnleitner, Reinhold. *Coca-Colonization and the Cold War: The Cultural Mission of the United States in Austria After the Second World War.* Chapel Hill: University of North Carolina Press, 1994.

Wala, Michael. "Selling the Marshall Plan at Home: The Committee for the Marshall Plan to Aid European Recovery." *Diplomatic History* 10 (Summer 1986): 247–65.

Wall, Irwin M. *France, the United States, and the Algerian War.* Berkeley: University of California Press, 2001.

Wallas, Graham. *The Great Society: A Psychological Analysis.* New York: Macmillan, 1917.

Wang, Zuoyue. "American Science and the Cold War: The Rise of the U.S. President's Science Advisory Committee." Ph.D. diss., University of California, Santa Barbara, 1994.

Warner, Michael. "Origins of the Congress for Cultural Freedom." *Studies in Intelligence* 38 (Summer 1995). http://odci.gov/csi/studies/95unclas/war.html.

Warshaw, Shirley Anne. *Reexamining the Eisenhower Presidency.* Westport, Conn.: Greenwood Press, 1993.

Washburn, Abbott. Remarks at conference on Public Discourse in Cold War America. Texas A&M University, March 1998.

Weart, Spencer R. *Nuclear Fear: A History of Images.* Cambridge: Harvard University Press, 1988.

Welch, David. "Power of Persuasion." *History Today* 49 (August 1999): 24–26.

Wenger, Andreas. *Living with Peril: Eisenhower, Kennedy, and Nuclear Weapons.* Lanham, Md.: Rowman & Littlefield, 1997.

Westad, Odd Arne, ed. *Reviewing the Cold War: Approaches, Interpretations, Theory.* London: Frank Cass, 2000.

Whitaker, Urban G., Jr. *Propaganda and International Relations.* San Francisco: Chandler Publishing, 1960.

Whitfield, Stephen J. *The Culture of the Cold War.* 2nd ed. Baltimore: Johns Hopkins University Press, 1991.

Wilcox, Dennis L., Phillip H. Ault, and Warren K. Agee. *Public Relations: Strategies and Tactics.* 3rd ed. New York: HarperCollins, 1992.

Wilford, Hugh. *The CIA, the British Left, and the Cold War: Calling the Tune?* London: Frank Cass, 2003.

———. "Playing the CIA's Tune? The *New Leader* and the Cultural Cold War." *Diplomatic History* 27 (Winter 2003): 15–34.

Williams, William Appleman. *The Tragedy of American Diplomacy.* New York: W. W. Norton, 1972.

Winand, Pascaline. *Eisenhower, Kennedy, and the United States of Europe.* New York: St. Martin's Press, 1993.

Winkler, Allan M. *Life Under a Cloud: American Anxiety About the Atom.* New York: Oxford University Press, 1993

———. *The Politics of Propaganda: The Office of War Information, 1942–1945.* New Haven, Conn.: Yale University Press, 1978.

Wittner, Lawrence S. *One World or None: A History of the World Nuclear Disarmament Movement Through 1953.* Stanford: Stanford University Press, 1993.

———. *Resisting the Bomb: A History of the World Nuclear Disarmament Movement, 1954–1970.* Stanford: Stanford University Press, 1997.

Wood, Nelson Ovia. "Strategic Psychological Warfare of the Truman Administration: A Study of National Psychological Warfare Aims, Objectives, and Effectiveness." Ph.D. diss., University of Oklahoma, 1982.

Yaqub, Salim. *Containing Arab Nationalism: The Eisenhower Doctrine and the Middle East.* Chapel Hill: University of North Carolina Press, 2004.

Yergin, Daniel. *The Prize: The Epic Quest for Oil, Money, and Power.* New York: Simon and Schuster, 1991.

Young, John W. *Winston Churchill's Last Campaign: Britain and the Cold War, 1951–5.* Oxford: Clarendon Press, 1996.

Yurechko, John J. "The Day Stalin Died: American Plans for Exploiting the Soviet Succession Crisis of 1953." *Journal of Strategic Studies* 3 (May 1980): 44–73.

———. "From Containment to Counter-Offensive: Soviet Vulnerabilities and American Policy Planning, 1946–1953." Ph.D. diss., University of California, Berkeley, 1980.

Zingg, Paul J. "The Cold War in North Africa: American Foreign Policy and Postwar Muslim Nationalism, 1945–1962." *Historian* 39 (November 1976): 40–61.

Zoubir, Yahia H. "U.S. and Soviet Policies Toward France's Struggle with Anticolonial Nationalism in North Africa." *Canadian Journal of History* 30 (December 1995): 439–66.

Zubok, Vladislav. "Soviet Intelligence and the Cold War: The 'Small' Committee of Information, 1952–53." Cold War International History Project Working Paper No. 4. Washington, D.C.: Woodrow Wilson International Center for Scholars, 1992.

Zubok, Vladislav, and Constantine Pleshakov. *Inside the Kremlin's Cold War: From Stalin to Khrushchev.* Cambridge: Harvard University Press, 1996.

INDEX